Dwyl,

thanks for
your leadership

DESIGN FOR SIX SIGMA

In Technology and Product Development

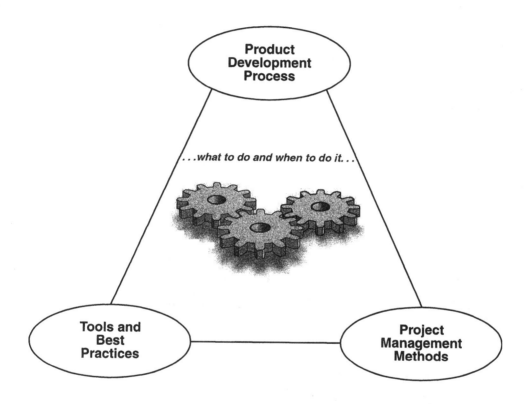

Product Development Process

. . .what to do and when to do it. . .

Tools and Best Practices

Project Management Methods

C. M. Creveling, J. L. Slutsky, and D. Antis, Jr.

ISBN 0-13-009223-1

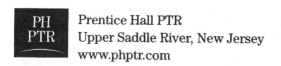

Prentice Hall PTR
Upper Saddle River, New Jersey
www.phptr.com

Library of Congress Cataloging-in-Publication Data

Creveling, Clyde M.
   Design for Six Sigma : in technology & product development / C.M. Creveling, J.L.
   Slutsky & D. Antis.
      p. cm.
   Includes bibliographical references and index.
   ISBN 0-13-009223-1
      1. Quality control--Statistical methods. 2. Experimental design. I. Slutsky, Jeff, 1956-
   II. Antis, D. (Dave) III. Title.

TS156 .C74 2003
658.5'62dc21   2002027435

Publisher: Bernard Goodwin
Editorial Assistant: Michelle Vincenti
Marketing Manager: Dan DePasquale
Manufacturing Manager: Alexis R. Heydt-Long
Cover Director: Jerry Votta
Cover Designer: Nina Scuderi
Art Director: Gail Cocker-Bogusz
Full-Service Production Manager: Anne Garcia
Composition/Production: Tiffany Kuehn, Carlisle Publishers Services

© 2003 Pearson Education, Inc.
Publishing as Prentice Hall PTR
Upper Saddle River, New Jersey 07458

Prentice Hall books are widely used by corporations and government agencies for training, marketing, and resale.

For more information regarding corporate and government bulk discounts please contact: Corporate and Government Sales (800) 382-3419 or corpsales@pearsontechgroup.com

Crystal Ball® is a registered trademark of Decisioneering. Process Improvement Methodology® is a registered trademark of Sigma Breakthrough Technologies, Inc. Post-it® is a registered trademark of 3M Corporation. Other company and product names mentioned herein are the trademarks of registered trademarks of their respective owners.

Printed in the United States of America
10 9

ISBN 0-13-0092231

Pearson Education LTD.
Pearson Education Australia PTY, Limited
Pearson Education Singapore, Pte. Ltd.
Pearson Education North Asia Ltd.
Pearson Education Canada, Ltd.
Pearson Educación de Mexico, S.A. de C.V.
Pearson Education—Japan
Pearson Education Malaysia, Pte. Ltd.

We would like to dedicate this book to our families.

For Skip Creveling
    Kathy and Neil Creveling
For Jeff Slutsky
    Ann, Alison, and Jason
For Dave Antis
    Denise, Aaron, Blake, Jared, and Shayla Antis

Without their patience and support, this book would have been much thinner.

Brief Contents

Contents

PART II Introduction to the Major Processes Used in Design For Six Sigma in Technology and Product Development 45

PART III   Introduction to the Use of Critical Parameter Management in Design For Six Sigma in Technology and Product Development   249

Foreword

Once in a while an idea evolves from merely a good concept to the inspiration for real change. The continued search for the "new, best way" can sometimes divert attention from the development of existing ways to a higher standard. Design For Six Sigma (DFSS) is such a development. It marries the notion of doing things right the first time with the belief in the importance of leadership in making things better.

DFSS is based on the foundation that design is a team sport in which all appropriate organizational functions need to participate. Each key function needs to apply its appropriate science. The key function in product design is typically the core engineering team. In information systems design it will be the core information technology team. In people systems design it will be the human resources team. The "science" any of these teams deploy must revolve around a well-balanced portfolio of tools and best practices that will enable them to develop outstanding data and thus deliver the right results.

In our company's experience, most of the design projects have been product related so I will refer to our learnings in that area—although my guess is that the lessons apply more broadly.

Before DFSS we had a number of initiatives around new product introductions. In addition to the typical milestone monitoring process (project management within our Phase/Gate process) there was a great deal of emphasis placed in co-location and cross-functional teams. The outcomes were fairly predictable, in hindsight. While leadership around co-location shortened the lines of communication, it did little for the content or the quality of those communications. We needed to communicate with better data.

In this team sport of design using DFSS we would expect many functions to have important inputs to the design process—and in the case of the product design, we now expect engineering to apply its science with the added benefit of the right balance of DFSS tools to create optimum solutions for our customers.

DFSS sets forth a clear expectation of leadership roles and responsibilities. This includes clear definition of expectations measures, deliverables, and support from senior management.

Accountability is clarified with DFSS, enabling recognition of proper performance. We focus on all this, while applying the latest design techniques in a disciplined, preplanned sequence.

This may sound like the latest "new way" if not for the fact that it really is more like the development of the old way to a higher standard that feels a lot like the right way.

—Frank McDonald
VP and General Manager
Heavy Duty Engines
Cummins, Inc.

Preface

In its simplest sense, DFSS consists of a set of needs-gathering, engineering and statistical methods to be used during product development. These methods are to be imbedded within the organization's product development process (PDP). Engineering determines the physics and technology to be used to carry out the product's functions. DFSS ensures that those functions meet the customer's needs and that the chosen technology will perform those functions in a robust manner throughout the product's life.

DFSS does not replace current engineering methods, nor does it relieve an organization of the need to pursue excellence in engineering and product development. DFSS adds another dimension to product development, called Critical Parameter Management (CPM). CPM is the disciplined and focused attention to the design's functions, parameters, and responses that are critical to fulfilling the customer's needs. This focus is maintained by the development team throughout the product development process from needs gathering to manufacture. Manufacturing then continues CPM throughout production and support engineering. Like DFSS, CPM is conducted throughout and embedded within the PDP. DFSS provides most of the tools that enable the practice of CPM. In this light, DFSS is seen to coexist with and add to the engineering practices that have been in use all along.

DFSS is all about preventing problems and doing the right things at the right time during product development. From a management perspective, it is about designing the right cycle-time for the proper development of new products. It helps in the process of inventing, developing, optimizing, and transferring new technology into product design programs. It also enables the subsequent conceptual development, design, optimization, and verification of new products prior to their launch into their respective markets.

The DFSS methodology is built upon a balanced portfolio of tools and best practices that enable a product development team to develop the right data to achieve the following goals:

1. *Conceive* new product requirements and system architectures based upon a balance between customer needs and the current state of technology that can be efficiently and economically commercialized.
2. *Design* baseline functional performance that is stable and capable of fulfilling the product requirements under nominal conditions.

3. *Optimize* design performance so that measured performance is robust and tunable in the presence of realistic sources of variation that the product will experience in the delivery, use, and service environments.

4. *Verify* systemwide capability (to any sigma level required, 6σ or otherwise) of the product and its elements against all the product requirements.

DFSS is managed through an integrated set of tools that are deployed within the phases of a product development process. It delivers qualitative and quantitative results that are summarized in scorecards in the context of managing critical parameters against a clear set of product requirements based on the "voice of the customer." In short it develops clear requirements and measures their fulfillment in terms of 6σ standards.

A design with a critical functional response (for example, a desired pressure or an acoustical sound output) that can be measured and compared to upper and lower specification limits relating back to customer needs would look like the following figure if it had 6 sigma performance.

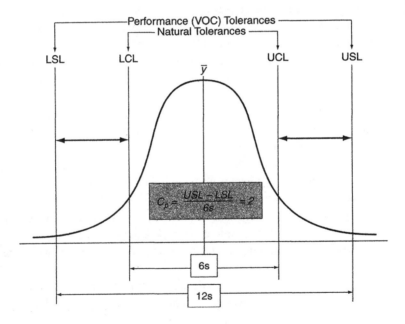

The dark black arrows between the control limits (UCL and LCL, known as *natural tolerances* set at +/− 3 standard deviations of a distribution that is under statistical control) and the specification limits (USL and LSL, known as VOC-based performance tolerances) indicates design latitude that is representative of 6 sigma performance. That is to say, there are 3 standard deviations of latitude on each side of the control limit out to the specification limit to allow for shifts in the mean and broadening of the distribution. The customer will not feel the variability quickly in this sense. If the product or process is adjustable, there is an opportunity to put the mean back on to the VOC-based performance target or to return the distribution to its desired width within its natural tolerances. If the latitude is representative of a function that is not serviceable or adjustable, then the latitude is suggestive of the reliability of the function if the drift off target or distribution broadening is measured over time. In this case, Cp (short-term distribution broadening with no mean shift) and Cpk metrics (both mean shifting and distribution broadening over long periods of time) can be clear indicators of a design's robustness (insensitivity to sources of variation) over time. DFSS uses capability metrics to aid in the development of critical product functions throughout the phases and gates of a product development process.

Much more will be said about the metrics of DFSS in later chapters. Let's move on to discuss the higher level business issues as they relate to deploying DFSS in a company.

At the highest level, any business that wants to excel at product development must have the following three elements in strategic alignment:

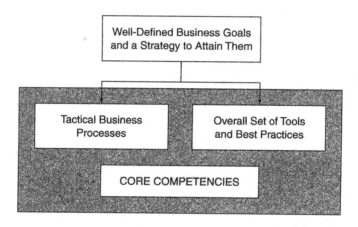

Design For Six Sigma fits within the context of a key business process, namely the product development process. DFSS encompasses many tools and best practices that can be selectively deployed during the phases of a product development process. Specifically, DFSS integrates three major tactical elements to help attain the ubiquitous business goals of low cost, high quality, and rapid cycle-time from product development:

1. A clear and flexible product development process
2. A balanced portfolio of development and design tools and best practices
3. Disciplined use of project management methods

The product development process controls the *macro-timing* of what to do and when to do it using a flexible structure of phases and gates. A balanced portfolio of tools and best practices are what to do within each phase of the product development process. The disciplined application of project management in the form of PERT charts of work breakdown structures defines the *micro-timing* for the critical path of applying tools and best practices within each phase.

DFSS works equally well in technology development organizations and in product design organizations. This book will demonstrate complete approaches to applying DFSS in both a technology development process and a product design process.

The metrics of DFSS break down into three categories:

1. Cycle-time (controlled by the product development process and project management methods)
2. Design and manufacturing process performance capability of critical-to-function parameters (developed by a balanced portfolio of tools and best practices)
3. Cost of the product and the resources to develop it

DFSS is focused on CPM. This is done to identify the few variables that dominate the development of baseline performance ($Y_{avg.}$), the optimization of robust performance (S/N and σ), and the certification of capable performance (Cp and Cpk) of the integrated system of *designed parameters*. DFSS instills a system integration mind-set. It looks at all parameters—within the product and the processes that make it—as being important to the integrated performance of the system elements, but only a few are truly critical.

DFSS starts with a sound business strategy and its set of goals and, on that basis, flows down to the very lowest levels of the design and manufacturing process variables that deliver on those goals. To get any structured product development process headed in the right direction, DFSS must flow in the following manner:

Define business strategy: Profit goals and growth requirements

Identify markets and market segments: Value generation and requirements

Gather long-term voice of customer and voice of technology trends

Develop product line strategy: Family plan requirements

Develop and deploy technology strategy: Technology and subsystem platform requirements

Gather product specific VOC and VOT: New, unique, and difficult needs

Conduct KJ analysis: Structure and rank the VOC

Build system House of Quality: Translate new, unique, and difficult VOC

Document system requirements: New, unique, and difficult, and important requirements

Define the system functions: Functions to be developed to fulfill requirements

Generate candidate system architectures: Form and fit to fulfill requirements

Select the superior system concept: Highest in feasibility, low vulnerability

DFSS tools are then used to create a hierarchy of requirements down from the system level to the subsystems, subassemblies, components, and manufacturing processes. Once a clear and linked set of requirements is defined, DFSS uses CPM to measure and track the capability of the evolving set of Ys and xs that comprise the critical functional parameters governing the performance of the system. At this point DFSS drives a unique synergy between engineering design principles and applied statistical analysis methods. DFSS is not about statistics—it is about product development using statistically enabled engineering methods and metrics.

DFSS does not require product development teams to measure quality and reliability to develop and attain quality and reliability. Product development teams apply DFSS to analytically model and empirically measure fundamental functions as embodied in the units of engineering scalars and vectors. It is used to build math models called *ideal or transfer functions* [Y = f(x)] between fundamental ($Y_{response}$) response variables and fundamental (x_{inputs}) input variables. When we measure fundamental ($Y_{response}$) values as they respond to the settings of input (x_{inputs}) variables, we avoid the problems that come with the discontinuities between continuous engineering input variables and counts of attribute quality response variables.

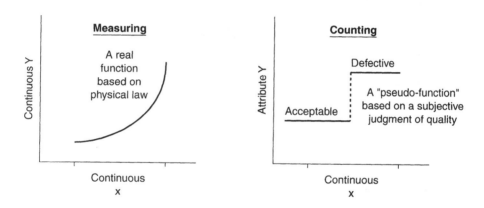

DFSS avoids counting failures and places the engineering team's focus on measuring real functions. The resulting fundamental models can be exercised, analyzed, and verified statistically through Monte Carlo simulations and the sequential design of experiments.

Defects and time-to-failure are not the main metrics of DFSS. DFSS uses continuous variables that are leading indicators of impending defects and failures to measure and optimize critical functional responses against assignable causes of variation in the production, delivery, and use environments. We need to prevent problems—not wait until they occur and then react to them.

If one seeks to reduce defects and improve reliability, avoiding attribute measures of quality can accelerate the time it takes to reach these goals. You must do the hard work of measuring functions. As a result of this mind-set, DFSS has a heavy focus in measurement systems analysis and computer-aided data acquisition methods. The sign of a strong presence of DFSS in a company is its improved capability to measure functional performance responses that its competitors don't know they should be measuring and couldn't measure even if they knew they should! Let your competitors count defects—your future efficiencies in product development reside in measuring functions that let you prevent defective design performance.

DFSS requires significant investment in instrumentation and data acquisition technology. It is not uncommon to see companies that are active in DFSS obtaining significant patents for their inventions and innovations in measurement systems. Counting defects is easy and cheap. Measuring functions is often difficult and expensive. If you want to prevent defects during production and use, you have to take the hard fork in the metrology road back in technology development and product design. Without this kind of data, CPM is extremely difficult.

The technical metrics of Critical Parameter Management in DFSS are as follows:

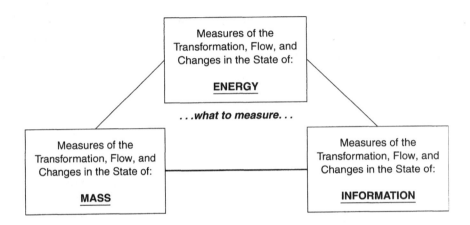

Information is represented by analog and digital logic and control signals.

What to measure is the mass, energy, and controlling signals within and across your systems. *When to measure* is defined by your micro-timing diagram (*critical path*) of tool and best practice applications within the phases of your product development process.

The underpinnings of DFSS deserve a brief review before we turn you loose on the rest of the book. DFSS, like Six Sigma for Production Operations, follows a roadmap. Six Sigma for Production Operations follows a process roadmap outlined by the **MAIC** acronym, which stands for Measure, Analyze, Improve, and Control. This is based, in large part, on the historic work of Walter Shewhart when he devised the underlying principles of statistical process control for production processes. Unfortunately this has little to do with the *process* of product development. Many in the Six Sigma business have tried to tell the R&D community that all they need to do is put a "D" in front of the MAIC process and voilà! you get DFSS. NOT TRUE!!! Define, measure, analyze, improve, and control is not a proper process recipe for product development. We know many have started DFSS within this SPC context, but there is a better, more appropriate process context in which to conduct DFSS.

This book is written by technology development and product design engineers for readers with the same or similar backgrounds. A major part of the book's intent is to establish a proper set of roadmaps that fit the paradigms and process context of technology development and product development. These roadmaps are set up in the format of a Phase/Gate product development process structure.

The I²DOV Technology Development Process Roadmap*:

Invent and Innovate Phase and Gate

  Develop Phase and Gate

    Optimize Phase and Gate

      Verify Phase and Gate

The CDOV Product Design Process Roadmap*:

Concept Development Phase and Gate

  Design Development Phase and Gate

    Optimization Phase and Gate

      Verification of Capability Phase and Gate

As much as we love and respect the MAIC process for production and transactional processes, it simply has no rational application context for DFSS, if you run your company based on a modern product development process. Leaders such as Admiral Raborn of the Polaris program or later proponents such as Cooper, Wheelwright and Clark, or Clausing and Pugh might be reasonable candidates to be the patron saints of modern Phase/gate product development processes, but it surely is not and should not be Walter Shewhart! Shewhart and his process provide great historical underpinnings for production operations; however, we will not lean too heavily on his work, at least as far as running the phases and gates of a product development process, until the final steps in transitioning from product design into production. In that sense, then, the I²DOV technology development process roadmap flows into the CDOV product design process roadmap, which in turn flows into the DMAIC production process roadmap.

This book is organized in seven sections:

1. *Organizational Leadership, Financial Performance, and Value Management Using Design For Six Sigma*
2. *Product Development Processes Using Design For Six Sigma*
3. *Critical Parameter Management in Design For Six Sigma*
4. *Tools and Best Practices for Invention, Innovation, and Concept Development*
5. *Tools and Best Practices for Design Development*
6. *Tools and Best Practices for Optimization*
7. *Tools and Best Practices for Verifying Capability*

These sections will build on this brief introduction to the disciplined and rigorous world of DFSS for technology and product development. We hope you enjoy this overview describing what "hard stuff" your technology and product development teams need to do (and when they need to do it) in order to take your company to the next level of success in our evolving world of product development excellence.

---

*I²DOV and CDOV are derivatives of the I²DOC and CDOC roadmaps used by Sigma Breakthrough Technologies, Inc., and are used by permission.

How to Get the Most From This Book

This text on DFSS was written to serve several types of readers:

1. Executives, R&D directors and business leaders
2. Program managers, project managers and design team leaders
3. Technical practitioners who comprise design teams

If you are an executive, R&D director, or some other form of business leader, we wrote the Introduction and Part I for you.

If you are a program manager, project manager, or a design team leader, we wrote Parts II and III primarily for you.

If you are a technical practitioner who will be applying the tools of DFSS on technology and product design programs and projects we wrote the Tool chapters in Parts IV through VII for you.

An extensive glossary at the end of the book is intended for all readers.

Parts II through VII of this book were designed to serve as a reference to be used over and over as needed to remind and refresh the reader on what to do and when to do it during the phases and gates of technology development and product design. These parts can be used to guide your organization to improve discipline and rigor at gate reviews and to help redesign your product development process to include Six Sigma metrics and deliverables.

If you want to understand any DFSS tool and its deliverables prior to a gate review, we recommend reading the appropriate tool chapter(s) prior to the gate review.

—Skip Creveling
Jeff Slutsky
Dave Antis

Specific Chapter Reading Recommendations

	Introduction to Design For Six Sigma	Part I (Chapters 1–3): Organizational Leadership, Financial Performance and Value Management using DFSS	Part II (Chapters 4–7): Product Development Processes Using DFSS	Part III (Chapters 8–13): Critical Parameter Management Using DFSS
Primary Audience	All readers	Executives, R&D directors, business leaders, design program managers, design project managers, and design team leaders	Program managers, project managers, and design team leaders	Project Managers design team leaders, master black belts, black belts, green belts, and any technical professional interested in how to manage critical requirements and the technical deliverables that fulfill them during the phases and gates of a technology and product design process
Secondary Audience		Master black belts, black belts, green belts as well as any technical professional interested in leadership, financial performance, and the management of value	Executives, R&D directors, business leaders, master black belts, black belts, green belts and any technical professional interested in how Phase/Gate processes can be used to help manage new product development in a Six Sigma context	Executives, R&D directors, and design program managers who want to gain a general understanding of the process and tools associated with Critical Parameter Management

(continued)

Specific Chapter Reading Recommendations

	Part IV (Chapters 14–18): Tools and Best Practices for Invention, Innovation, and Concept Development	Part V (Chapters 19–27): Tools and Best Practices for Design Development	Part VI (Chapters 28–30): Tools and Best Practices for Optimization	Part VII (Chapters 31–34): Tools and Best Practices for Verifying Capability
Primary Audience	Design team leaders, master black belts, black belts, green belts, and any technical professional interested in how the tools of DFSS are applied and linked during the conceptual design phase of technology development and product design	Design team leaders, master black belts, black belts, green belts, and any technical professional interested in how the tools of DFSS are applied and linked during the design development phase of technology development and product design	Design team leaders, master black belts, black belts, green belts, and any technical professional interested in how the tools of DFSS are applied and linked during the optimization phases of technology development and product design	Design team leaders, master black belts, black belts, green belts, and any technical professional interested in how the tools of DFSS are applied and linked during the verification of capability phases of technology development and product design
Secondary Audience	Executives, R&D directors, design program managers, and design project managers who want to gain a general understanding of the tools associated with concept design	Executives, R&D directors, design program managers, and design project managers who want to gain a general understanding of the tools associated with technology or design development	Executives, R&D directors, design program managers, and design project managers who want to gain a general understanding of the tools associated with technology or design optimization	Executives, R&D directors, design program managers, and design project managers who want to gain a general understanding of the tools associated with capability verification

Acknowledgments

The authors would like to acknowledge the role of Sigma Breakthrough Technologies, Inc. for providing the opportunity to test and validate the Design For Six Sigma methodology documented in this book. SBTI supported the use of the methods and tools described here within a wide range of companies during their corporate Six Sigma deployments. The cycles of learning were of true benefit to the final product you see here.

There are a few leaders, colleagues, mentors, students, and friends that we would like to recognize and thank for their influence on our thinking—but more importantly, on our actions.

While the body of knowledge known as Design For Six Sigma is still evolving and forming, some elements are becoming accepted standards. In the area of Customer Interviewing and the KJ method, Joe Kasabula has truly changed the earliest phases of DFSS in new Product Development. Joe, thanks for letting us stand on your shoulders; and in the words of Dave Antis "we love you, man!" In the area of Concept Engineering, the efforts of Dr. Don Clausing and the late Dr. Stuart Pugh in making the tools and best practices for managing the fuzzy front end of Product Development practical and actionable are unsurpassed. Our thinking on Product Development as a process has been greatly enhanced by the works of Dr. Don Clausing and Mr. Robert Cooper. Probably the single most important person in the last two decades when it comes to defining how an engineer can properly develop tolerances in a Six Sigma context is our great friend Mr. Reigle Stewart. We simply have never seen or heard a better teacher on the subject of developing tolerances for Six Sigma performance. We believe a major characteristic of a Six Sigma design is its "robustness" to sources of variation. We thank the world's best mentors on Robust Design: Dr. Genichi Taguchi; his son, Shin Taguchi; and Dr. Madhav Phadke.

We would also like to thank Dr. Stephen Zinkgraf, CEO of Sigma Breakthrough Technologies Inc., for his support. Steve's contributions to the Six Sigma industry are vast and his influence and efforts to provide us with an environment in which we could build on our ability to link Six Sigma to Product Development is gratefully acknowledged. We further acknowledge that the I^2DOV and CDOV roadmaps, used as the backbone of this book, are derivatives of the I^2DOC and CDOC roadmaps which have been successfully deployed by Sigma Breakthrough Technologies, Inc., for technology development and product development.

Thanks to all our peers and the students of DFSS who read our manuscripts and offered advice. Mr. Steve Trabert (from StorageTek Corp.), Mr. David Mabee, and Mr. Joseph Szostek were especially helpful for their detailed review and comments on improvements.

We would also like to recognize and express our heartfelt thanks to our colleagues, Mr. Randy Perry (Chapter 2), Mr. Scott McGregor (Chapter 3), Mr. Ted Hume (Chapters 20 and 33), and Dr. Stephen Zinkgraf (Epilogue), for contributing to this text. Their work helped us round out this comprehensive book on DFSS. Each of them added a unique perspective that we were lacking. Our thanks to each of you for helping us make this a better book.

Finally, we want to say thank you to the world's most patient and long-suffering editor, Mr. Bernard Goodwin, and his staff at Prentice Hall. Thanks also to Tiffany Kuehn and her team at Carlisle Publishers Services. You all comprised an outstanding publishing team.

PART I

Introduction to Organizational Leadership, Financial Performance, and Value Management Using Design For Six Sigma

These three topical areas are the basis on which DFSS, at the executive and managerial level, achieves success. Corporate support is not good enough. DFSS initially requires a driving impulse from the highest levels in a company, well beyond verbal support. It then requires an active, frequent, and sustaining push from the top down. Clear and measurable expectations, required deliverables, and strong consequences for high and low performance must be in place for DFSS to work. If you want to read how this was handled at GE, just read the chapter on Six Sigma from Jack Welch's autobiography *Jack: Straight from the Gut.*

The reason to do DFSS is ultimately financial. It generates shareholder value based on delivering customer value in the marketplace. Products developed under the discipline and rigor of a DFSS-enabled product development process will generate measurable value against quantitative business goals and customer requirements. DFSS helps fulfill the *voice of the business* by fulfilling the *voice of the customer:*

- DFSS satisfies the voice of the business by generating profits through new products.
- It satisfies the voice of the customer by generating value through new products.
- It helps organizations to meet these goals by generating a passion and discipline for product development excellence through active, dynamic leadership.

Let's find out how. . . .

1

The Role of Executive and Management Leadership in Design For Six Sigma

Leadership Focus on Product Development as a Key Business Process

Let's begin with a great example of an executive who successfully led a company into organic growth using Design For Six Sigma (DFSS), Jack Welch of General Electric (GE). Jack Welch took advantage of the cost reduction approach of operational Six Sigma pioneered by Motorola and Allied Signal, which guided GE to over $1.5 billion in savings during its third full year of deployment. However, Jack saw the incredible power of Six Sigma in terms of growth and realized that limiting it to a cost-reduction focus was short-sighted. He saw the numbers and their impact inside of GE and noticed that GE's customers were not yet feeling the results of their Six Sigma initiatives. Never complacent, Jack set a new goal where the customer would feel the impact of Six Sigma as GE developed new products and services. In the end, GE became the most successful company to date to deploy the DFSS methodology, as you'll soon see.

In 1996, GE Power Systems launched an innovation in their gas turbine design that was so reliable that 210 units were produced without a single unplanned reliability-based power outage. This put them into the enviable position of market leader at exactly the time power demand surged in the late 1990s. Even in 2002, Power Systems is one of the most significant contributors to GE's corporate profits, all due to a deliberate focus on product reliability and customer satisfaction. GE Medical Systems launched a CT Scanner called LightSpeed in 1998. The time for a chest scan was reduced from 3 minutes to a mere 17 seconds. GE Medical Systems launched 22 new designs between 1998 and 2001 using Design For Six Sigma. Jack drove the transition from operational Six Sigma to Design for Six Sigma as a matter of strategic intent. GE management did the most effective thing a company can do to move the benefits of Six Sigma upstream into their company's strategic business processes—they aligned the tools and best practices of DFSS with their product development process, and they did it with rigor and discipline.

Welch was relentless about getting the results of Six Sigma out to the customer. This behavior made him open enough to listen to a suggestion from GE Plastics and set the mark for a new vision for his Six Sigma investment. GE launched metrics of performance into their scorecard system which were measured as the customers saw them. Armed with the customers' view, GE has not only applied Six Sigma to their products but also to the way their business processes deliver services to their customers. This brings us to our next major insight about properly leading a company using DFSS: Creation of management scorecards that measure new product development performance data in comparison to things customers directly state they care about.

GE's current CEO, Jeffrey Immelt, has taken this one step further, withholding a portion of executive bonuses unless the executives can prove that they have added value to their customers. At GE, the Six Sigma story is still evolving. Our third major insight for managing in a DFSS context is that executive management will continue to drive Six Sigma into the upstream organizations that do advanced platform and product planning, as well as applied research and technology development. A new form of DFSS called Technology Development For Six Sigma (TDFSS) has emerged in the last year. This book contains the first disciplined approach to its deployment within a structured Phase/Gate process context.

In contrast, consider a different company who launched Six Sigma in 1997. In the late 1980s, Polaroid Corporation was considered one of most enviable companies in America. They had over $600 million in cash and an excellent balance sheet. Their market share was secure and growth seemed to be guaranteed into the next century. Polaroid took on some debt to avoid a hostile takeover by Shamrock Holdings before the end of 1988. They then conducted two very expensive product launches into carbon-based films for medical imaging and small-format SLR instant cameras. Both of these initiatives failed miserably, which resulted in a change to their top leadership. With a new CEO focused exclusively on cost reduction, Polaroid initiated a Six Sigma program in their operations.

Although official savings were never made public, it is believed that Polaroid Corporation (a $2 billion company at the time) saw savings in relative proportion to those posted by Motorola, Honeywell, and GE. In April of 2000, Polaroid's former Executive Vice President of Global Operations told a Six Sigma team of a major appliance manufacturer that their Six Sigma cost reduction initiative had kept them out of bankruptcy for the past 4 years. Unfortunately, in the fall of 2001, Polaroid, with over $900 million in debt, filed for Chapter 11 protection. How could a company that embraced Six Sigma get into such a position? Primarily, they failed to take it upstream into the most important business processes for organic growth—their technology development and product commercialization processes.

Real estate agents, insurance adjusters, and other businesses that used instant photography for the past 50 years had all switched to digital cameras. Polaroid saw this trend coming too late. They finally tried to enter this emerging market but failed to stand up to the marketing and design competition of Sony, HP, and Kodak. There is no doubt that Polaroid would have stayed alive longer and maybe even survived with a focus on linking together key up-front Design For Six Sigma elements such as market forecasting and segmentation, voice of technology evaluations and benchmarking, gathering and processing the voice of the customer, product platform architecting, and focused technology building based on all the previously mentioned activities. Things would have likely been different if they had deployed these as key tools within the earliest phases of their product development process.

One key lesson to be learned from Polaroid is the risk that accrues from focusing on internal metrics and executive opinions rather than external data and customer needs. An internal focus can bring short-term gains if you are lucky, but customers and technology change rapidly, as Polaroid learned too late. Just like the shift away from slide rules and typewriters, we have witnessed the passing of another era in technology and the company which brought it to fruition.

We are reminded of the efforts taken at Motorola to achieve the goal of Six Sigma in their new product introductions. In 1988 and 1989, they achieved remarkable results and saw manufacturing labor costs reduce to almost half. Scrap was reduced by 65 percent per year, and Motorola immediately became more competitive in their market segment. As they moved into the 1990s and attempted to achieve 5 and 6 sigma designs, they often heard the following argument, *"Due to the exponential relationship of sigma scores to defects, there is not nearly as much savings when improving from 5 to 6 sigma as there is when improving from 3 to 4 sigma."* They were convinced that they could no longer achieve the remarkable profit margin improvements of before. What made this scenario worse was that their teams actually believed it.

Internal metrics, such as sigma improvement levels based upon cleaning up designs in defects and waste, are only half correct. Consider a company which produces paper products. Through some manufacturing and production process focused improvements, the company can reduce the scrap and the amount of paper wasted through paper breaks and reel changes. However, as long as the focus is on reducing the amount of trim, the return per Six Sigma project will only get smaller. This is because the paradigm of trim and waste reduction has been established as the primary goal for this product. Contrast this with a focus on product platform application in alignment with market characteristics which cause their customers to choose their products over the competition across differentiated segments. New, breakthrough designs using different substrates, densities, materials, and processing technologies will largely be ignored if the cost of poor quality (COPQ) focus becomes the overriding goal for their business processes. Business processes must account for both cost control as well as investment of capital to increase growth. Six Sigma has evolved to the point where its new frontiers are on the growth side. This means Six Sigma has to be coupled with the business processes that govern technology development and product commercialization.

As consultants, we have been able to work with over 35 different companies in the deployment of Six Sigma. Many of these went on to deploy DFSS as a natural follow on to operational Six Sigma. We have been privileged to see some companies achieve results that match or rival GE. We have seen others fail despite their best efforts. For us, one thing is very clear in the deployment of DFSS: You must focus on acquiring true customer needs and then apply the discipline of Design For Six Sigma within the phases and gates of your product development process to efficiently transform those needs into new products and services.

A company which we have been able to work with since 2000 is Samsung SDI. Samsung SDI produces various displays including ones for color televisions and computer monitors, STN LCD, VFD, and rechargeable batteries. Samsung SDI kicked off their Six Sigma program with a vengeance in January of 2000. Within 24 months, they saw their profits more than triple to $530 million USD and their sales grow from $4 billion USD to $4.4 billion USD. In December of 2000 Samsung SDI was the first recipient of the annual South Korean Six Sigma Award.

SDI has deployed Six Sigma across all areas of management, one of which was their product development organization, with numerous DFSS projects focused on improving design as well as completely new innovations. SDI led Samsung Electronics to become the largest producers of televisions and flat panel display monitors in the world (*Newsweek,* 2002). Their flat panel displays for TVs and computer monitors were rated the number one value independently by two large electronic chains in the United States. Samsung SDI and Samsung Electronics, both as world leading companies, have maintained a mutually supportive relationship. The two launched an amazing string of technologies and products that are setting them up to become a powerhouse multinational company in their own right.

As Samsung SDI launched their DFSS program, they went through a rigorous process of matching the broad set of DFSS tools with their existing Product Development Process (PDP). As we will discuss in Chapters 5 and 6, generic product development TDFSS and DFSS roadmaps have been proven to help align companies' technology development and product design processes with the right mix of DFSS tools and best practices. Samsung SDI fully embraced the Critical Parameter Management (CPM) methodology we discuss in Chapters 8–13 to manage and balance the voice of customer and the voice of physical law throughout their development process. Within 6 months of launching DFSS, SDI had a well designed system of scorecards and tool application checklists to manage risk and cycle-time from the voice of the customer through to the launch of products that meet customer and business process demands. The culture of SDI embraced this disciplined approach and they have realized tangible benefits in a very short period of time. Just visit your local computer and electronics stores and look for Samsung products—they are the direct result of their recent DFSS initiatives. SDI is literally putting DFSS developed products on store shelves today and using product development methods they initiated just 2 years ago. One of them is OLED, which is recently emerging as a new type of display. As a result of a successful DFSS project, Samsung SDI became the first company in the world to produce the 15.1" XGA AMOLED display.

The product development process should be one of the top three strategic business process issues for any executive leadership team. Even if you have excellent technological platforms and world class sales and marketing capabilities, you must always make the management of the PDP (sometimes referred to as the product pipeline) a key part of the way you manage your business. The DFSS roadmap and tool sets presented in this text can, if followed with sincere and active leadership discipline, help your teams design cycle-times and manage risks that consistently deliver products and services that truly delight your customers.

You can launch a DMAIC (Define, Measure, Analyze, Improve, and Control) Six Sigma program with significant initial success without much regard for your business processes.

Chances are good that each function and operation has enough internal waste to yield numerous financial breakthrough projects during the first few years of a traditional Six Sigma initiative. However, achieving your top-line growth goals will be strongly improved by integrating Six Sigma tool applications within the core business processes which align key customer metrics with your flow of product offerings.

The best performance of launching products under the influence of DFSS comes from using the right tools at the right time from start to finish during your Phase/Gate product development cycle. Just like Samsung SDI, you must map the generic DFSS roadmap into your specific Phase/Gate process, develop scorecards for quick and easy measurable go/no-go decisions, and rigorously manage your PDP with data from properly selected and applied tools. This brings us to our next major recommendation: Executive management cannot take a hands-off approach to the design and structure of the product development process and the tools that are required to develop the data they need to manage risk and cycle-time.

We have seen some of the best-designed product development processes in the world. Many of these processes were making almost no impact on the company's cycle-time or commercial success. The usual reason is passive indifference to actually using the Phase/Gate discipline the process was supposed to embody. Few teams or individuals were ever held accountable for using specific tools and delivering specific results. We have also seen well-designed DFSS tools integrated into these Phase/Gate processes with little effect on success. Again, management indifference to the process, its deliverables, and its metrics lay at the root of the anemic performance. While we can't expect senior leadership to know all the intricacies of the DFSS tools and roadmaps, we must expect regular and disciplined gate reviews with real go/kill decisions based on business, customer, and capability metrics. The key issue for senior management is to establish a disciplined product development process, require and personally participate in its use, and structure a system of performance metrics that are routinely summarized and used to manage risk and drive growth at their operational level.

The first three parts of this book will provide an outline for a systematic approach of DFSS integration to product development process management in your company.

Again, we admonish you to take an active role in developing and using your PDP. Hold executives, managers, and teams accountable for the deliverables that come from proper tool utilization. Attend the regular gate reviews—ask yourself "When was the last time I sat in on a gate review and got involved with the teams and shared in the positive and negative implication of their results ?" Demand data, reject unfounded suppositions and guessing, and require statistical validation to back up the recommendations proposed at gate reviews. Make the hard decisions to kill the low probability of success projects early and manage short falls in project performance as proactively as you can. If you get involved with your PDP correctly, you will achieve PDP cycle-time reductions that are predictable and reproducible.

How is it possible to attain meaningful cycle-time reductions in your PDP if we add more customer needs and requirements definition tasks up front such as DFSS requires? Most commercialization cycle-time problems are attributed to correcting errors in judgment and problems created by undisciplined use of a limited set of tools that underdeveloped critical data. Few programs are late because teams focused on preventing technical and inbound marketing problems. DFSS is about preventing problems and providing breakthrough solutions to well-defined requirements—not about fixing problems your people created in the earlier phases. In fact, as much as two-thirds of the cycle-time in new product development is spent correcting such problems.

There are two distinct areas that the DFSS roadmap will address, two "voices" which must be understood to minimize the redesign and design rework loop (build-test-fix engineering): The voice of the customer (VOC) and the voice of the product/process (VOP). Conducting Critical Parameter Management will remove the conditions that promote design rework.

Often a team will complete the design of an exciting new product only to hear the customer say, "That is not what I asked for." We will spend considerable time addressing this issue in Parts III and IV (Critical Parameter Management and Concept Development) and provide a roadmap to minimize this type of expensive commercialization error. We have seen countless new designs and concepts that no customer really wanted. The greatest opportunity to increase the efficient utilization of product development team resources is to reduce the number of projects that have no true "voice of the customer." In these cases, little or no data exists to back up the voice of the corporate specialists.

The other issue is that the product itself is incompatible with its manufacturing processes and the end use operating environment. How many times have we seen a product that is patentable but not manufacturable or fails to function as expected when the customer puts it to use? Parts V through VII cover the tools within the DFSS roadmap that are designed to prevent and minimize these kinds of problems. As an executive, you need to be in touch with these tools and how they provide the data you and your management teams need to manage risk.

The Strategic View of Top-Line Growth

The following list of project types illustrates the transition a business must undertake to shift from bottom-line Six Sigma operations projects to top-line DFSS projects. Type A projects give you the most control over the market and your profit margins (assuming the customer values the new product), while Type D and E projects yield the lowest market control. Most DFSS programs start with Type C through E projects and evolve to Type A and B. If your DFSS program remains focused on Type D and E, you may still suffer the same fate as Polaroid and others. You can only cut costs so much—then you have to spend money to make money.

Top-Line Growth Using Long-Term DFSS-Based Commercialization Projects

Type A Project: Initiate market share by opening new markets with new designs (*establish the price*)

Type B Project: Obtain market share in new markets with an existing design (*at the current market price*)

Type C Project: Improve margins and revenue growth in existing markets with a new design (*premium price*)

Bottom-Line Growth Using Short-Term Hybrid DFSS/Operations Six Sigma Projects

Type D Project: Decrease cost of an existing design

(*hold the current price*)

Type E Project: Decrease product rework and warranty costs

(*reduce cost of poor quality*)

GE's former CEO, Jack Welch, stated, "You can't manage (grow) long term if you can't eat short term . . . any fool can do one or the other. The test of a leader is balancing the two" (Welch, 2001). While hundreds of companies will speak to their cost-reduction success using Six Sigma, only a fraction will state that it helped them achieve remarkable growth. Most of those actually attain growth through the additional capacity from production and transactional process improvement projects. Very few companies have been able to use Six Sigma to develop new products and services which "so delight the customer" that they impact the annual reports through increased growth and revenues. This book is about changing organizational behavior to grow the top line.

The model for strategic business growth can be illustrated as a matrix:

	Market	
	Existing	New
Product Existing	Type D/E (lower costs)	Type B (deliver value)
Product New	Type C (deliver value)	Type A (deliver value)

A business can eat short term with cost-reducing Type D/E projects and then grow over the long term with value-driving Type A, B, and C DFSS-based projects and programs.

Operations (DMAIC) Six Sigma will help you solve your current problems and "eat short term." DFSS will prevent many of those same problems from ever occurring again and will leverage your resources to drive value into your future designs. If you do not develop new products that possess strong value in your customer's eyes, you will indeed starve. If much of your new product development resources are distracted from their mission because they are constantly fixing poor designs that have managed to get out into the plant and into your customer's hands, how can you ever find the time to fulfill your business strategy?

In some of the following chapters in this text, you will begin to see the strategic layout illustrated in the following list, defining the context in which a business deploys DFSS to generate new product development initiatives. It is a recipe for getting product development started right. These are the key elements that help drive value out to customers and return the required profits to the business and its shareholders.

Define business strategy: profit goals and requirements

Identify markets and market segments: value generation and requirements

Gather long-term voice of customer and voice of technology trends

Develop product line strategy: family plan requirements

Develop/deploy technology strategy: technology and subsystem platform requirements

Gather product-specific VOC and VOT: new, unique, and difficult needs

Conduct KJ analysis: structure and rank the NUD* VOC needs

Build system House of Quality: translate NUD VOC needs

Document system requirements: create system requirements document

Define the system functions: modeling performance

Generate candidate system architectures: concept generation

Select the superior system concept: Pugh concept selection process

You can see that the strategy transitions quickly into tactical applications of specific tools and best practices. Financial performance will flow from your core competencies found within this list. Fail to fund, support, and execute on any one of these line items and your company could be in big trouble over the long term. It may look good to take short cuts in the short term but in the long term you will underperform. DFSS tools and best practices will greatly aid your efforts to gain significant improvements in financial performance. The gains come from your leadership and personal, frequent, visible support of these elements of success.

Enabling Your Product Development Process to Have the Ability to Produce the Right Data, Deliverables, and Measures of Risk within the Context of Your Phase/Gate Structure

There is a fine balance that must be maintained between the following elements of a product development strategy:

1. A clear, flexible, and disciplined product development process
   - Most companies have a formal, documented product development process.
   - Most companies do not follow their own process with discipline and rigor.
   - The process allows for the insufficient development and documentation of the critical parameter data required to properly assess risk at a gate review.
2. A broad portfolio of tools and best practices
   - Most product development teams use an anemic, limited set of tools mainly due to time constraints imposed by management deadlines for technology transfer and product launch.

---

*NUD stands for New, Unique, and Difficult

- Product development teams have widely varying views regarding what tools should be deployed during product development. This causes the development of insufficient data required to manage phase-by-phase risk accrual.
- Management has avoided being prescriptive in requiring specific tool use, thereby requiring a predominantly reactive approach to risk management because they don't get the right data at the right time.

3. A rigorous approach to project management
   - Projects are funded without a clear, up-front, integrated project plan that illustrates a balanced use of tools and best practices on a phase-by-phase basis.
   - Project managers make up their plan as they react to problems—project plans are architected by a reactive strategy rather than a proactive strategy.
   - Project tasks are not underwritten by a broad portfolio of tools and best practices, assuring a series of self-perpetuating build-test-fix cycles.
   - Project task timelines are estimated using best case deterministic projections rather than using reasonable statistical estimates of the shortest, likeliest, and longest estimated task cycle-times. Critical path cycle-times need to be analyzed using Monte Carlo simulations.

Chapter 4 provides a detailed discussion of how these three items integrate to estimate, design, and manage cycle-time within the context of DFSS.

Executive Commitment to Driving Culture Change

Being a proactive leader requires disciplined follow-through on product development process execution. DFSS can and will become a major source of success in your business if you spend the time to plan a disciplined implementation within the context of a well-designed and managed product development process. The goal is to transition your business from short-term cost efficiency to sustained economic growth over the long term. To do so, you must lead your organization to deliver value to your customers by giving them what they want. Then they will happily provide you with what you want.

Lead boldly!

Summary

1. Six Sigma has to be coupled with the business processes that govern technology development and product commercialization.
2. Align the tools and best practices of DFSS with your product development process phases and gates.
3. Create management scorecards that measure new product development performance data in comparison to things customers directly state they care about.

4. Executive management must drive Six Sigma into the upstream organizations that do advanced platform and product planning and applied research and technology development.

5. Executive management cannot take a hands-off approach to the design and structure of the product development process and the tools that are required to develop the data they need to manage risk and cycle-time.

6. Establish a disciplined product development process, require and personally participate in its use, and structure a system of performance metrics that are routinely summarized and used to manage risk and drive growth at the operational and strategic levels.

References

Welch, J. (2001). *Jack: Straight from the gut.* New York: Warner Business Books.
Samsung in bloom. (2002, July 15). *Newsweek,* 35.

Measuring Financial Results from DFSS Programs and Projects

This chapter was written by Randy Perry, an expert in Six Sigma financial metrics and a former Design For Six Sigma specialist at Allied Signal Corporation.

The purpose of Six Sigma in design organizations is to allow businesses to accomplish well-defined business objectives effectively and efficiently as measured by financial results. Specifically, Design For Six Sigma implementation can be characterized by three major goals, all of which result in improved financial performance. First, DFSS implementation allows companies to better meet the needs of customers by deploying "the voice of the customer" throughout the phases and gates of a product design and commercialization process. Secondly, DFSS methodology enables businesses to more efficiently execute the development and scale-up of new products, thus improving time-to-market and reducing development rework. Finally, DFSS allows companies to successfully transfer new technologies from the R&D organization to the design organization. It then enables the transfer of new products from the design organization to the manufacturing organization with minimal production waste once these new products are produced in a company's manufacturing facilities. The use of Design For Six Sigma in each of these areas can result in significant and measurable financial benefits.

A Case Study

Throughout this chapter we will consider the example of a new chemical product, which is being commercialized, with a projected sales price of $1.60 per pound. Since this is a new chemical product without current commercial sales, the marketing group has, with 95 percent confidence, estimated that the actual sales price could be anywhere from $1.45 to $1.75 per pound. Marketing also indicates that the customer will not accept the product with a strength of less than 88 pounds per square inch (psi) or greater than 100 psi. The total operating expense for production of the product is projected to be $1.00 per pound of product produced with an operating yield of

90 percent first grade product per pound of production. It is projected that 10 million pounds of the product will be sold next year and each succeeding year for the next 10 years. After 10 years it is believed that the product will no longer be viable in the marketplace.

The total investment, including technical development resources and capital equipment, is estimated at $10 million prior to any sales revenue being generated. The company's cost of capital is currently 10 percent and its tax rate 40 percent. Sales of the new product are scheduled to begin on January 1 of next year. Today is July 1, and the R&D and marketing groups have estimated that the product can be ready for commercial sales from the manufacturing facility six months from today. As mentioned earlier, we will refer to this case study example as we examine various financial aspects of the DFSS implementation process. Our intent is to build linkage between DFSS tools and meaningful financial returns to the business.

Deploying the Voice of the Customer

Many new product commercialization efforts fail due to lack of early customer involvement in defining product requirements and their resulting specifications. Requirements are based on customer needs while specifications are developed during the phase of the product development process to fulfill the requirements as closely as possible through the use of design tools.

Clearly, customers buy products that can be purchased at pricing that allows them to make their required profits while meeting a specific functional need. Also, suppliers develop products that allow them to make desired profits and grow their strategic businesses. A product development process that integrates Design For Six Sigma methods addresses customer need and product value requirements in part through the execution of the Quality Function Deployment (QFD) process as shown in Figure 2–1.

Execution of **Quality Function Deployment,** as part of the concept engineering process, enables customers to define specific needs. These needs are then rigorously translated into functional requirements that the product must fulfill. Through the application of numerous DFSS tools the product design specifications are developed. The hierarchy of specifications defines the measurable design characteristics that are developed in an attempt to fulfill the product requirements. The final use of QFD is in the definition of the requirements by which the new product will be manufactured. Failure to execute the concept engineering process—and more specifically the QFD process—effectively can result in a failed commercialization effort and significant financial losses. Other serious consequences to the business can also occur, such as damaged customer relationships, loss of marketplace credibility, and the possible loss of current market share. A conservative estimate of the cost of failure at this early stage of product development can be the loss of technology and design staff time, raw materials, operating expenses, and perhaps even wasted major capital investment for new production processing equipment. Should a newly developed product not meet customers' needs or pricing requirements, a major investment of time and money may result in the generation of minimal new sales revenue.

As shown in Figure 2–2, commercialization expenses typically follow a traditional investment S-curve with a small percentage of total commercialization spending occurring early and late in the project execution process.

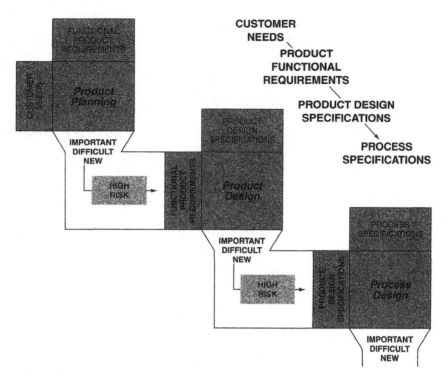

Figure 2–1 Quality Function Deployment Process

Source: Used by permission of Sigma Breakthrough Technologies, Inc.

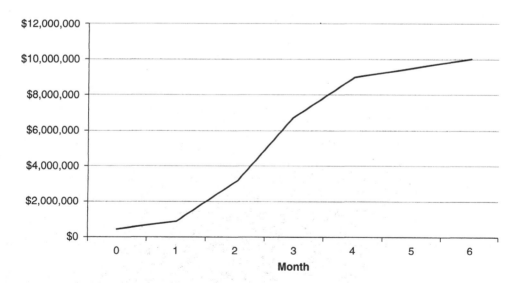

Figure 2–2 Product Development Cost

15

Early in the commercialization effort, concepts for the new product are being formed and reviewed with customers. During this **concept engineering** phase, technical and marketing personnel typically conduct relatively inexpensive customer interviews, benchmarking, and laboratory bench scale trials to assess new product concept viability. Much larger expenditures occur in the middle of the project Phase/Gate process. Soon after a concept has been selected and product and process definitions completed, capital project scoping begins, followed by costly major equipment purchases. Capital equipment installation and checkout then begin. Once major equipment checkout is completed, relatively inexpensive start-up costs are then incurred. For the reasons described previously, determination that a new product will not be successful in the marketplace *early in the development process* can save a company a great deal of money. In the case study example presented earlier, determination within the first month of the development process that our new product will not successfully meet the needs of the customer or supplier will allow our costs to be contained to less than $1 million of the total development budget of $10 million. As can be seen in Figure 2–2, the cost of killing our project increases rapidly over the next 3 months to approximately $9 million! The total cost of a failed commercialization effort in our example could range up to $10 million with additional undefined costs for lost opportunity to develop other products, loss of marketplace credibility, loss of current market share, and damage to key customer relationships. Early customer involvement and properly executed DFSS methods can help minimize these costs and allow key development resources to be reallocated to more strategic projects as needed.

DFSS Project Execution Efficiency

The second area of financial benefit that comes from the implementation of Design For Six Sigma principles is the efficient execution of the product development process and the transitional scale-up to steady-state production of new products. Once a "system" (product, manufacturing, and support process) concept has been defined and agreed upon through the QFD process, numerous DFSS tools are used to identify and quantify key relationships and control variables. We call this **Critical Parameter Management (CPM).** Through use of these tools, significant cost savings can be realized by improving the product time-to-market and reducing product rework. To illustrate these points we will return to our case study example. For example, let's suppose that through the use of DFSS tools we can reduce the time-to-market for our new product by 3 months, reducing the commercialization time from 9 months to our projected 6 months. Under this scenario we can use traditional financial measures of net present value (NPV) and internal rate of return (IRR) to assess the value of this improved time-to-market.

Net present value is defined as the total value, in current dollars, of a series of future cash flows. In our specific case, the series of cash flows will consist of an initial expenditure of $10 million in year zero (today's dollars after investment tax credits) and the collection of net after-tax profits in years 1 through 10. The future cash flows in years 1 through 10 will be discounted for the time value of money using the company's cost of capital, which has been given at 10 percent. In order to calculate the financial benefit associated with our expedited time-to-market we must first estimate the after-tax profits associated with commercialization of our new product. We will use the following equation to calculate annual after-tax profits.

$$\text{After-Tax Profit} = \text{Volume} \times (\text{Sales Price} - \text{Cost per Pound/Yield}) \times (1 - \text{Tax Rate})$$

For our new product we know that the marketing group has estimated that the product will sell 10 million pounds per year for the next 10 years at a sales price of $1.60 per pound. We also have estimates from our development team that the production cost is estimated at $1.00 per pound of material produced and that the process should run with a 90 percent yield of first quality product per pound of material produced. More realistic cases will involve a slower volume ramp-up with pricing changes being forecast throughout the product life cycle. This type of analysis would require a spreadsheet model using the best estimates of volume, price, yield, and production cost by period. We will continue with our simplified example for illustrative purposes. The after-tax profit for our new product can now be obtained as:

After-Tax Profit =

$$10 \text{ M lbs/year} \times \left(\$1.60/\text{lb product} - \frac{\$1.00/\text{lb production}}{.9 \text{ lbs product/lb production}} \right) \times (1 - .4)$$

$$= \$2.933 \text{ M/year}$$

With knowledge of the product's life cycle (10 years), after-tax cash flow generated from sales of our new product, initial required investment, and the cost of capital, we can now calculate the net present value for our new product as:

$$NPV_1 = -10M + \frac{(2.9333)}{(1 + .10)^1} + \frac{(2.9333)}{(1 + .10)^2} + \frac{(2.9333)}{(1 + .10)^3} + \frac{(2.9333)}{(1 + .10)^4} + \frac{(2.9333)}{(1 + .10)^5}$$

$$+ \frac{(2.9333)}{(1 + .10)^6} + \frac{(2.9333)}{(1 + .10)^7} + \frac{(2.9333)}{(1 + .10)^8} + \frac{(2.9333)}{(1 + .10)^9} + \frac{(2.9333)}{(1 + .10)^{10}}$$

$$NPV_1 = \$8.02 \text{ M}$$

In other words, the expected value of our new product in current dollars is expected to be approximately $8 million if the product is commercialized on time and with all current assumptions. A commercialization delay of 3 months, given a finite product life cycle of 10 years, will reduce the after-tax revenues generated from the product in the first year to

$$9/12 \times \$2.933 \text{ M} = \$2.2 \text{ M}$$

Calculating the new NPV for our product with the delayed commercialization timing yields the following results:

$$NPV = -10M + \frac{(2.2)}{(1 + .10)^1} + \frac{(2.9333)}{(1 + .10)^2} + \frac{(2.9333)}{(1 + .10)^3} + \frac{(2.9333)}{(1 + .10)^4} + \frac{(2.9333)}{(1 + .10)^5}$$

$$+ \frac{(2.9333)}{(1 + .10)^6} + \frac{(2.9333)}{(1 + .10)^7} + \frac{(2.9333)}{(1 + .10)^8} + \frac{(2.9333)}{(1 + .10)^9} + \frac{(2.9333)}{(1 + .10)^{10}}$$

$$NPV_2 = \$7.36 \text{ M}$$

$$NPV \text{ Difference} = NPV_1 - NPV_2 = \$8.02 \text{ M} - \$7.36 \text{ M} = \$.66 \text{ M}$$

Direct comparison of the NPVs demonstrates that the cost of commercialization delays can be substantial, in this case approximately $660,000 for a delay of only 3 months. Another financial measure often used in assessing the value of new products is the internal rate of return (IRR). The IRR is defined as the cost of capital that will yield a net present value of zero. In the case of our delayed product commercialization, the 3-month delay reduces the IRR of our project from 26.5 percent to 24.6 percent. Again, a relatively small delay of 3 months in bringing this product to market produces a significant reduction in IRR of 1.9 percent. The analysis also assumes that no damage has been done to the customer's desire to purchase the product. In some cases a delay in development can have substantially higher costs as customers seek alternative solutions or competitors enter the marketplace first, thus reducing available market share and pricing power.

Reduction of design rework is another major financial benefit of the DFSS process. Rework not only slows the cycle time of bringing a product to market, but also increases the cost of development by requiring work to be done more than once. DFSS greatly reduces the seemingly endless build-test-fix cycles that can sap months out of already tight development schedules. Rework costs could be relatively minor, such as repetition of experiments in a laboratory or pilot unit facility, or very costly, such as the redesign and modification of major portions of the design or capital production equipment. These costs depend upon the circumstances of the rework required and could range from thousands of dollars to millions of dollars. Needless to say, rework is a costly form of waste in project execution with costs arising from both delayed commercialization timing and additional development costs. Proper execution of the DFSS process will help minimize the cost of waste associated with rework.

Production Waste Minimization

The third type of financial benefit typically associated with implementation of Design For Six Sigma is the minimization of production waste once a new product is transferred to manufacturing for full-scale production. The DFSS process is designed (especially through the deployment of the lower level Houses of Quality in QFD) to identify and control key manufacturing process variables that result in the product consistently meeting customer expectations as defined by the production process specification limits. As discussed earlier, production process specification limits

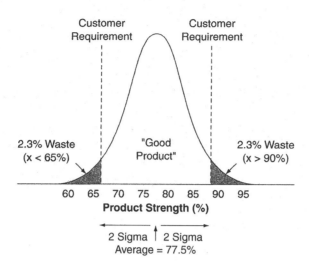

Figure 2–3   Product Strength Waste Analysis

originate from the customer. These limits must be established early in the product commercialization process. DFSS strongly supports—in fact, it demands—the concurrent engineering of the product and its supporting network of production processes. Failure to establish production requirements early can result in delays in the commercialization schedule, costly rework, or translation of a design into manufacturing that is not capable of meeting customer requirements.

Let's suppose for a moment that a product and its associated manufacturing processes are developed with only minimal input from the customer regarding product requirement limits. As shown in Figure 2–3, we will assume that our best knowledge of the customer requirement is that the product should never have product strength of less than 65 percent or above 90 percent. Using this information, the company's R&D staff has designed a product with strength of 77.5 percent, halfway between what are believed to be the customer requirements. The process is determined to have a product strength standard deviation of 6.25 percent and so is positioned 2 standard deviations from the customer specifications. Under the assumption of a normal distribution of product strength around the mean, we can estimate the total waste in this process as approximately 4.6 percent. But what happens to this process in the long term once the process is transferred to manufacturing? Process variation is likely to be substantially higher over a period of years than that observed in the short-term development process. Also, what happens if the customer decides to tighten the product specification limits prior to finally accepting the product? These questions represent a few of the risks associated with this product since customer specifications were not specifically agreed to early in the development process and the process was not developed with a reasonable view of long-term variation. We will now examine these concepts in more detail with our case study example.

The process for production of our new product is projected to produce a first quality yield of 90 percent with an average strength of 94 psi based on short-term development data. Additional data from the R&D effort indicates that the product's strength standard deviation in the short term is approximately 3.64 psi. Based on this information we can perform some important financial assessments for our new process. First, using the distance of the average strength from the specification limits, we can calculate the sigma level of our new process. The **sigma level** will be defined as the number of standard deviations that the mean is located from the specification limits. In our case, the mean of 94 psi is located 6 psi from both the lower and upper specification limits. Given our estimation of the standard deviation at 3.64 psi we can see that we currently have a 1.65 sigma process. Using the standard normal curves, we can estimate the waste associated with this process at approximately 10 percent, 5 percent both below the lower specification and above the upper specification limits (see Figure 2–4).

The cost associated with this short-term waste can be estimated by calculating the entitlement (or best possible result) of the process at 100 percent yield.

Entitlement After-Tax Profit $=$

$$10 \text{ M lbs/year} \times \left(\$1.60/\text{lb product} - \frac{\$1.00/\text{lb production}}{1 \text{ lbs product/lb production}} \right) \times (1 - .4)$$

$$= \$3.6 \text{ M/year}$$

We can also now calculate the entitlement net present value of our new product as:

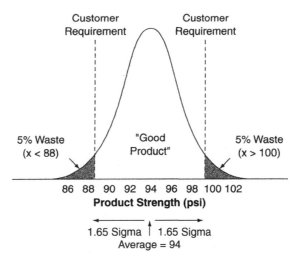

Figure 2–4 Case Study Example–Initial Specifications

$$\text{Entitlement NPV}_1 = -10\text{M} + \frac{(3.6)}{(1 + .10)^1} + \frac{(3.6)}{(1 + .10)^2} + \frac{(3.6)}{(1 + .10)^3} + \frac{(3.6)}{(1 + .10)^4}$$

$$+ \frac{(3.6)}{(1 + .10)^5} + \frac{(3.6)}{(1 + .10)^6} + \frac{(3.6)}{(1 + .10)^7} + \frac{(3.6)}{(1 + .10)^8} + \frac{(3.6)}{(1 + .10)^9} + \frac{(3.6)}{(1 + .10)^{10}}$$

$$\text{NPV}_{\text{Ent}} = \$12.1 \text{ M}$$

From this analysis we can see that the short-term loss of 10 percent yield in our new process with after-tax annual profit of $2,933,000 and NPV of $8,020,000 is costing us approximately $667,000 per year with a projected net present value penalty of approximately $4 million. Calculation of the entitlement internal rate of return indicates an IRR_{Ent} of 34 percent versus an IRR of 26.5 percent with 10 percent yield loss. But what if the customer changes the specifications once the product is in commercial production in our manufacturing facility? Suppose for a moment that the customer tightens the specification limits just slightly to a lower specification of 90 psi and an upper specification of 98. With our average of 94 psi, we can now calculate the sigma level of our new process as:

Sigma Level = 4.0 psi/3.64 psi = 1.1

Using the standard normal distribution we can calculate the expected waste of this new process to be 27.1 percent, as shown in Figure 2–5.

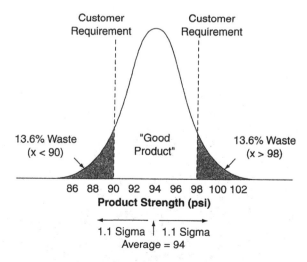

Figure 2–5   Case Study Example–Tightened Specifications

Also, we can calculate the after-tax profit for our product as:

$$10 \text{ M lbs/year} \times \left(\$1.60/\text{lb product} - \frac{\$1.00/\text{lb production}}{.729 \text{ lbs product/lb production}} \right) \times (1 - .4)$$

$$= \$1.37 \text{ M/year}$$

We can now calculate the net present value of our new product with the tightened process specifications as:

$$\text{NPV} = -10\text{M} + \frac{(1.37)}{(1 + .10)^1} + \frac{(1.37)}{(1 + .10)^2} + \frac{(1.37)}{(1 + .10)^3} + \frac{(1.37)}{(1 + .10)^4} + \frac{(1.37)}{(1 + .10)^5}$$

$$+ \frac{(1.37)}{(1 + .10)^6} + \frac{(1.37)}{(1 + .10)^7} + \frac{(1.37)}{(1 + .10)^8} + \frac{(1.37)}{(1 + .10)^9} + \frac{(1.37)}{(1 + .10)^{10}}$$

$$\text{NPV} = -\$1.58 \text{ M}$$

The IRR of our new product now becomes 6.2 percent, which is below our cost of capital. The results of the customer tightening the specifications results in a new product that is no longer financially viable. In fact, we see that the value of the product over a 10-year life cycle is now negative, indicating that the company will now lose money by investing in this product under these conditions.

This analysis is the "best case" estimate of the penalty associated with production yield loss for our new product. As mentioned earlier, long-term variation of a process in a manufacturing environment is generally substantially larger than the short-term variation defined in R&D. While each process and product is different, conventional estimates indicate that a process can shift by as much as 1.5 standard deviations in either direction from the mean in the long term.

Let's consider this scenario as the "worst case" financial penalty associated with translation of a process from R&D to manufacturing without testing to assess the long-term sigma of the process. As shown in Figure 2–6, if the process to manufacture our new product shifts down by 1.5 sigma (5.46 psi), the new operating average for the process will be 88.5 psi.

This operating region is just 0.5 psi above our lower specification limit and will result in an estimated 44.5 percent waste in the long term. An estimate of the financial impact of such a shift in the process can be calculated as follows:

After-Tax Profit =

$$10 \text{ M lbs/year} \times \left(\$1.60/\text{lb product} - \frac{\$1.00/\text{lb production}}{.555 \text{ lbs product/lb production}} \right) \times (1 - .4)$$

After-Tax Loss = $-\$1.21$ M/year

(Notice that we take credit for a tax loss to help soothe the pain!)

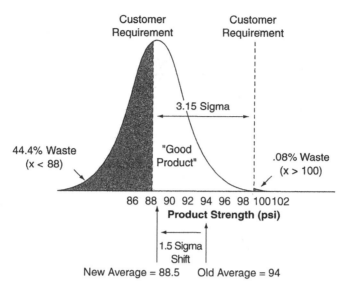

Figure 2–6   Case Study Example–Long-Term Process Shift

We can also now calculate the net present value of our new product as:

$$\text{NPV} = -10\text{M} + \frac{(-1.21)}{(1 + .10)^1} + \frac{(-1.21)}{(1 + .10)^2} + \frac{(-1.21)}{(1 + .10)^3} + \frac{(-1.21)}{(1 + .10)^4} + \frac{(-1.21)}{(1 + .10)^5}$$

$$+ \frac{(-1.21)}{(1 + .10)^6} + \frac{(-1.21)}{(1 + .10)^7} + \frac{(-1.21)}{(1 + .10)^8} + \frac{(-1.21)}{(1 + .10)^9} + \frac{(-1.21)}{(1 + .10)^{10}}$$

$$\text{NPV} = -\$17.4\,\text{M}$$

This is indeed a financial disaster that must be avoided! First, we must firm up the customer specifications prior to committing to full-scale commercialization. The customer must realize the importance of committing to specifications early for the financial well-being of the new product in the long term. Secondly, we must not continue with full-scale production scale-up unless we can tighten the short-term standard deviation of our process. How small should the standard deviation of our new process be? We can answer this question by calculating the waste, after-tax profit, NPV, and IRR for our new product in the long term at varying levels of sigma. If we assume that the process target is initially 94 psi and that it will shift by 1.5 standard deviations in the long term we can estimate the financial results at different values of process standard deviation as shown in Table 2–1.

As can be seen from the results presented in Table 2–1, the full financial benefit of $12.12 million from our new product will only be realized if we can reduce the process standard

Table 2–1   Financial Results versus Process Sigma

Sigma Level	ST Sigma	LT Mean Shift	LT Waste	After-Tax Profit ($M)	LT NPV	LT IRR
0.5	3.00	89.5	30.9%	$0.920	–$4.348	–1.5%
1.0	2.40	90.4	15.9%	$2.468	$5.167	21.0%
1.5	2.00	91.0	6.7%	$3.170	$9.481	29.3%
2.0	1.71	91.4	2.3%	$3.460	$11.262	32.5%
2.5	1.50	91.8	.62%	$3.563	$11.890	33.7%
3.0	1.33	92.0	.135%	$3.592	$12.071	34.0%
3.5	1.20	92.2	.0233%	$3.599	$12.112	34.07%
4.0	1.09	92.4	.00317%	$3.600	$12.119	34.08%
4.5	1.00	92.5	.00034%	$3.600	$12.120	34.08%

deviation from the current estimate of 3.64 psi to approximately 1.0 psi. In fact, we see from the analysis that a standard deviation of 1.0 psi corresponds to a long-term process sigma level of 4.5. Given the assumed 1.5 sigma shift in the long term, we can now see that the new process will be a short-term Six Sigma process with a standard deviation of 1.0 psi. Should the new process fall below the 5 sigma level, significant financial penalties can quickly result.

Pricing and Business Risk

Having seen the importance of minimizing production waste in maximizing financial returns, we will next analyze the sensitivity of these returns to sales price. As mentioned earlier in describing our new product commercialization case study, the marketing group has determined that the sales price for our new product will be between $1.45 and $1.75 per pound with 95 percent certainty. Using the standard normal curve at 95 percent probability, we can estimate the projected standard deviation for the pricing of our product as follows:

One Standard Deviation = ($1.60 − $1.45)/1.96 = $.0765

Calculating the price at values ranging from −2 to +2 standard deviations from the mean, we can assess the sensitivity of the net present value of our new product to changes in product pricing. The results of this analysis are presented in Figure 2–7.

As can be seen by the analysis presented, for each change of 7.65 cents in sales price the net present value changes by $2.8 million. Suppose for a moment that our R&D team has completed work on our new product and we now have a 1.8 sigma long-term process, which is projected to sell for $1.60 per pound. We can see from the analysis in Figure 2–7 that the NPV of this project is expected to be $9.4 million. If the actual sales price of the product once commercial sales begin is $1.53, we can see that the NPV gap caused by the decreased sales price can be recovered by moving the process to a long-term 4.5 sigma process. In this situation process im-

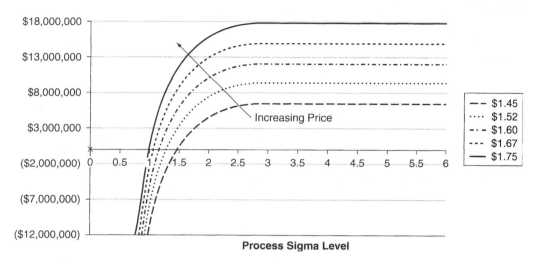

Figure 2–7 NPV Price Sensitivity

provement can offset an initially overly optimistic estimate of the sales price and the project will still bring the $9.4 million expected value. Unfortunately, process improvement cannot always be used as a means of recovering lost market value due to inaccurate estimates of selling price. As shown in Figure 2–7, once a process reaches the 4.5 long-term sigma level, pricing impacts will directly influence product value with few options to mitigate these effects. For this reason, the concept of *Six Sigma Pricing* is introduced. Simply put, the risk of pricing changes should be estimated through direct dialogue with the customer. After these conversations, an assessment of the pricing sigma level should be made. Once the target price and sigma level are established, a project risk analysis can be run to assess the risk of pricing changes on the potential NPV of the project. As shown in Figures 2–8 and 2–9, production cost and sales volume also have major impact on the value of newly commercialized products.

In order to assess the total business risk for a newly commercialized product, estimates of target value and standard deviation for yield, sales price, production cost, and sales volume should be established and analyzed using Monte Carlo simulation techniques. The result of such an analysis for our case study product is shown in Figure 2–10 using the following assumptions.

	Target Value	**Standard Deviation**
Sales Price	$1.60	$0.0765
Sales Volume	10 M lbs/yr	1 M lbs/yr

(*Note: Capped at maximum of 10M lbs/yr by capacity constraints*)

Production Cost	$1.00	$0.051
Yield	90%	2.6%

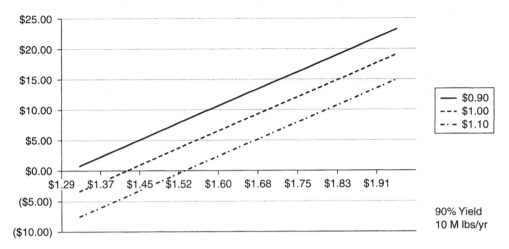

Figure 2–8　Sensitivity of NPV To Cost and Price

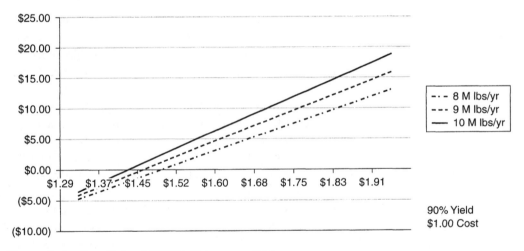

Figure 2–9　Sensitivity of NPV To Volume and Price

The currently expected NPV for our new product given our best estimate of total risk is approximately $6.5 million. We can also see that there is a 3.5 percent risk that our product could actually result in a negative NPV. In order to move closer to our previously calculated entitlement NPV of $12.1 million we need to examine the expected value of the key financial drivers (yield, price, cost, and volume) and reduce the variation in the estimate of these values.

In summary, Design For Six Sigma techniques can help us define and manage the financial risks associated with newly commercialized products. Using DFSS methods, the voice of the cus-

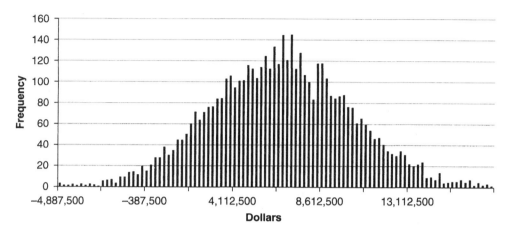

Figure 2–10 NPV Forecast

tomer is captured early in the process, and risks of producing a product that will not meet customer expectations are minimized. Use of DFSS within a product development process also results in improved time-to-market, minimization of design rework, and reduction of production
process waste. Finally, Design For Six Sigma provides a framework through which companies
can manage general business risks associated with new product commercialization efforts.

Managing Value with Design
For Six Sigma

Written by Scott A. McGregor, Marketing Consultant for Design For Six Sigma and
Vice President of Marketing at Uniworld Consulting, Inc.

Value is a driving function of design. What an interesting concept. It seems that this would be
a simple rule in the design of new features or products. In fact, it is one of the hardest rules
for a company to follow. It should be a heuristic of every organization that has product and service designs at the heart of its long-term business. Why don't companies embrace this as an internal rule? It is not because extracting value from the voice of the customer is hard or too vague.
In fact, many different processes make the process efficient and productive. Good marketing personnel and design personnel can extract that voice and turn it into an excellent detailed design
that speaks to customer needs with weight and priority. Companies fail to extract value as a driving function for a number of reasons: time, pet projects, politics/internal culture, poor process,
and budget. There are others, I'm sure, but those are the ones I see most frequently.

Extracting Value

Extracting value is not an engineering science. Extracting value is not solely a marketing science.
Many marketing departments don't even do the kind of value analysis that would position their
products in the market. The corporate marketing machines that do are often the winners in the market game. They dominate the market with products that win share and influence their position in
the market. At the very least, using an analytic value model will help an organization position itself in the marketplace with metrics that go beyond share and volumes. Indeed, share is the measure of market acceptance but often that measure comes late or not at all in the data we collect.

Real winners extract value measures as a means of discovering market acceptance, find
gaps in the market mind-set for the offering, and lead a company to making the right decisions in
market performance and product changes. Sometimes, measuring market perceptions of value
leads a company to engage in innovative design.

In recent years I have seen a number of products enter into the market where it is apparent that value is being managed as a function of marketing and R&D departments. PRILOSEC®/ LOSEC® (omeprazole), a pharmaceutical for the treatment of a number of gastrointestinal ailments, is a product with which I had extensive experience in managing customer value. Through diligent management of the value proposition, this medication went from last place in the anti-ulcer/GERD (gastroesophageal reflux disease) market to first place. Originally marketed in the United States by Merck & Co., Inc., and Astra Merck, Inc. (currently it is owned by AstraZeneca L.P.), it is the premier product in its class and is currently the world's most prescribed pharmaceutical product. Value became a driver for the product from the prelaunch phase of new product indications to postlaunch market activities. On another level, value became a driver in determining business case decisions for budget and resource allocation at a portfolio level. I am convinced that understanding and managing value internally and externally is the key driver for a marketing department in the success of a product.

PRILOSEC® is an example of a product being managed at all levels. Some would argue that value management as a function of design is a waste of time, especially those who work in non-consumer products and services. Too often I have heard business managers say, "Value management would be great for consumer products like toothpaste or soap, but we sell our products to other manufacturers or to other corporate clients. Why should we do this kind of research for a business-to-business buyer?" When I first started hearing this argument, I thought it had some validity. Honestly, why would you commit to research and data measures on the value of products that are often specified and constrained by other businesses? It seems it would be a wasted effort. A beer can manufacturer will tell you, "A beer can is a beer can is a beer can!" Is there really much room to extract value and design improvements to a beer can? How many changes have there been since the pop-top opener? Some would argue that the beer and soft drink can market is a great example of only designing to customer specifications and making sure we have enough margin to make a profit when it is manufactured. Having recently worked with a major manufacturer of this kind of product, I became aware that this mentality has led the company to a far worse position than focusing on margins. It has let the customer (soft drink and beer bottlers) determine what their profit margins will be. They search for margin in the "down-gauging" of can thickness, thus saving them money in aluminum costs. All other improvements are in incremental improvements in preventing error. They have lost control not only of the value proposition but of the internal value proposition for the product itself. The only management left to them is the management of margins. The profit is no longer latent in the market; now it is just extracted from internal capability and margins. If there were a word for a product that would describe it as something less than a commodity, I would use it here. However, the picture is not all bleak for the can industry, and I will return to it later to illustrate that there is hope for the old beer can after all.

Cummins, Inc., is a premier example of a company that strives to prevent the overall commoditization of its core product lines. You find marketing intimately tied to design and manufacturing throughout the organization. Cummins has customers from end consumers to corporate buyers that it strives to satisfy by meeting specifications and to delight by exceeding expectations with novel and robust designs. One example of this is through the continued management of customer perceptions of value in light duty truck engines (Dodge Ram 2500-Cummins diesel trucks). Daimler-Chrysler could, with relative ease, start migrating trucks to its standard Mercedes diesels that have satisfied global markets for years. By all appearances, it would be a wise move to an internal capability that would result in savings to the bottom line, but Cummins has created a stronger and more valued perception, in my opinion, for its product in the market and with the manufacturer. The company not only manages value with the customer base but constantly looks for new market positions against which to build its portfolio. It is an example of great product management at all levels for an industrial-based product. I can only praise Cummins further for its skillful handling of regulatory issues while managing a portfolio of value-added industrial products. Cummins is my greatest argument against disregarding value as an industrial manufacturer for business buyers.

From PRILOSEC® to beer cans to engines, from consumer to industrial buyers, from innovative to commodity products, value is the function that determines market acceptance and success. It must therefore be the key driver in designing products and services and their subsequent improvements.

Value as a Formula

Yes, it's possible to realize the concept of value as a formula. How could we express perceived value to the customer as a measurable calculation? To begin, let's start with the simplest expression of value.

$$\text{Value} = \Sigma(\text{Quality components}) - \text{Price}$$

Value is the sum of the worth of the quality relative to the price. As a rule customers always buy on value and quality. Buyers may say they buy on price, but that is a failure on the seller's and the buyer's part to realize the true value proposition before them. Well, now we have a formula for value, so how do we define quality into the formula?

$$\text{Quality} = \Sigma(\text{Product dimensions}) + \Sigma(\text{Service dimensions})$$

In short, quality is the sum of the features and benefits of the product and of the service that comes with the product. Each product can be broken down into its fundamental dimensions of quality. Garvin (1987) best outlined dimensions of quality from the customer's point of view:

1. Performance—primary operating characteristic of a product or service
2. Features—secondary characteristics of a product or service
3. Reliability—frequency with which a product or service fails
4. Conformance or consistency—match with specifications or standards
5. Durability—product life
6. Serviceability—speed, courtesy, and competence of repair
7. Aesthetics—fits and finishes
8. Perceived quality—reputation (pp. 49–50)

Dimensions of quality may change for each product line, but these eight clearly and concisely capture the vision of dimensional quality. Pharmaceutical companies traditionally have gone another route, using safety, tolerability, efficacy, ease of use, and price as their key five dimensions. They have been forced to rethink that with the advent of stronger buying positions of customers like health maintenance organizations and group purchasing organizations, but the value proposition still focuses on dimensions of quality that are important to the buyer.

Service also has dimensions of quality. Consider these 10 dimensions of service listed by Berry et al. (1985):

1. Reliability—consistency of performance
2. Responsiveness—timeliness of service
3. Competence—possession of required skills and knowledge
4. Access—approachability and ease of contact
5. Courtesy—politeness, respect, consideration, and friendliness
6. Communication—listening to customers and keeping them informed in a language they can understand
7. Credibility—trustworthiness, believability, honesty
8. Security—freedom from danger, risk, or doubt
9. Understanding of the customer—efforts to understand the customer's needs
10. Tangibles—physical evidence of service (pp. 45–46)

Some companies use specific dimensions of quality or service to find their place in the market. For instance, Steinway has domineered the market for high-end pianos with the elegant design and substantive quality of its grand pianos. People buy them because they want the same sound and look of a piano that would be in front of a symphony. Yamaha, on the other hand, looks to provide pianos to people who want a good-sounding, affordable piano in their home. Yamaha serves more of the common buyer market. Each provides quality that more than satisfies its customer base, but each is specific to that base at the same time.

Being first in all dimensions of quality is seldom necessary. Quality Function Deployment helps one see this. A well-managed, well-designed product can find its place in the market by focusing on some of the key dimensions and executing to them without fail. Dell™ Computers and Gateway® both exemplify this. For Dell™ the dimensions have been building machines with an eye to business requirements and the service level that is required by those businesses (performance, reliability, serviceability, responsiveness, and competency of repair). Gateway® delivers a quality product for the average consumer who wants to be involved in the design of their end product. Gateway® delivers on similar dimensions but with different measurable components for each of those dimensions (performance, features, reliability, responsiveness, communication, and courtesy). Each company actively manages these dimensions and ensures the overall quality of the product for the consumers they understand best. The dimensions can be subtle but are nonetheless effective.

Competing on quality can often come with some tricky mistakes. Let's take a moment and address a few. First is the fundamental mistake of directly attacking the industry's quality leader. Why would that be a mistake? Well, there are a number of reasons. The quality leader may not be the market leader. We can't lose sight of the fact that we are in market competition, not quality competition. We are merely using quality as a weapon to change the market perceptions. The easiest way to change market perception is by changing price, but that is a gross error that turns a market of quality leaders into a commodity market overnight. Another key mistake is competing on quality dimensions that are unimportant to consumers. I laughed heartily at Apple when it took a normal Macintosh computer and put it in a colored, stylized shell. As a business user of machines I found its offering inadequate to my needs and the needs of the corporation. My error was not seeing the quality dimensions for the market they were pursuing. For college-age students with computing needs, a need for style, and a need for space savings, the iMac spoke volumes to them. That effort alone brought Apple back once again from the edges of extinction.

Other obvious mistakes include assuming reliability and conformance are less important than customers think they are; an ancillary error is relying on the wrong quality metric. In the 1970s GM relied on its dedication to single platform quality with moderate modification to style to sell the same general car through multiple brands. GM focused on these platforms while virtually ignoring dimensions such as reliability, performance, consistency, and durability. Buyers taught Detroit a lesson by buying on the quality dimensions described in Japanese automotive offerings. Certainly, the cars were smaller and offered less features, but they did offer consistent quality in engine reliability, fit and finish, interiors that did not have loose screws under the carpet, and excellent service records. Fulfilling the designs of the value formula, Japanese auto makers offered their products at a price that the consumer saw as superior value.

Measuring Value in the Marketplace

We have established value, quality, price, and dimensions of quality in the value formula. Where do we go next? We have to operate against competitors at a level that we can quantify and at which we can operate. In fact, what we seek to do is to understand the value proposition in the marketplace to a quantifiable level that allows us to design or innovate around it. That is the goal and the place where the statistical metrics of Six Sigma become quite important. Next, we will treat these concepts abstractly and begin to move toward some measurable view of them. We will abstract each of them to see them as relative components of the great value formula. We will now use product and service value and quality measures to guide our business and R&D decisions.

Two new questions now emerge. What do we measure and how do we use it meaningfully? I won't attempt to answer how we get the value data—there are a number of means of obtaining market data, not the least of which is customer surveys. I will discuss extracting voice for a design project, but time and space constraints do not allow a discussion of extracting quantifiable market perception data. What do we measure? We measure comparative performance against our competitors. We begin with a simple comparison of three or four simple data points. Believe it or not, a few data points tell an incredible story. A wise product manager for the world's largest

Table 3–1 Comparison of Light Bulb Data Points

	Britelite	**Competitor A**	**Competitor B**	**Competitor C**
Market Share	37%	34%	24%	5%
	Stable	Growing	Shrinking	Stable
Price—Relative to ours	100	104%	91%	85%
Direct Cost—Relative to ours	100	101%	100%	90%
Technology (Equal to ours, ahead of ours, behind ours)	Equal	Ahead	Equal	Behind

antihypertensive medicine once told me, "Always look for the story in the data." He was right. Reliable data points can paint a meaningful picture of our position and opportunity in the marketplace. The crucial data points are share, relative price, relative direct cost, and relative technology. These four fundamental points paint a telling picture of where you are. Look at the example in Table 3–1 I have constructed with a comparison of light bulbs. This data is from a prominent U.S. manufacturer we'll refer to as Britelite in this example.

These four data points, when interpreted together, give an enlightening quick picture of our situation. This picture is usually validated by further perceived value analysis, but in a crunch these are the four basics. Let's take a look at the data for this example. While Britelite owns the market for these particular light bulbs, it is in a competitive market. You'll notice that Britelite has a close competitor (A) that has a technology ahead of Britelite's. Not only is it ahead but the competitor is selling it for a premium. It also costs competitor A a little more to manufacture. In general, A is a stable competitor, but it has a competitive advantage. Competitor B is selling technology equal to Britelite but at a discount. This is telling because B's costs are equal to Britelite's. Competitor B has less market share and one can assume that it is discounting to maintain or gain share. This gives rise to a fifth data point for our matrix—trend. If we could put an arrow beside each share to reflect the direction share has been moving in a recent time frame—lets say 6 months—it could enrich our situational analysis story. Assuming B's share has been shrinking, we can conclude that the price discount is a tactical move to preserve share.

Finally, we have competitor C, which makes an inferior product, has poor share, and offers a deep discount compared to other companies in the market. Clearly, C is selling on price only. It is just trying to get a piece of the pie by selling a weaker product at a discount. Its small share says it is having a rough time. Competitors in this position bring up some important issues. First, a company with small share and low price faces only a few options if the situation persists. It may scavenge the market with price to get share. It can also destroy the market with large volume at a very low price to extract as much dollar value as possible out of the situation. Or it can get out of the market. I am always afraid of these kinds of competitors. They can do a lot of damage to everyone in a short time. In fact, any company competing on price and not quality issues places all other competitors in a potentially bad situation.

We need to understand, with the data we have, the kind of market in which we are competing. Are we competing on price or quality? The data points in Table 3–1 should reveal most of the answer to us. Based on the data we have two quality competitors (Britelite and A), and we have two price competitors (B and C). Generally, competitors that offer a price discount of at least 10 percent or more are price competitors. I consider B a price competitor because of price, shrinking share, and equal technology. In general, it is in a position of only playing with price. Britelite is sustaining price on existing technology while A is selling at a premium for premium technology. We could interpret this abstracted data, once we have put it in the form of a usable ratio, to outline our market. To do this we sum the shares of those competing on quality, and then we sum the shares of those competing on price. These two sums will give us the ratio we seek.

Price ratio component: B + C = 24% + 5% = 29%

Quality ratio component: Britelite + A = 37% + 34% = 71%

Quality-to-Price Ratio = 71:29

This may seem simple, but it is quite revealing about a market situation. In this case the data says that the market is in a pretty good position. Most of the market is responding to quality competition issues and is satisfied dominantly by quality competitors. Yet, we can't ignore that almost a full third of the market is responding to price. Markets become unstable when they compete on price. Buyers love unstable markets because they win on the price game. As designers, manufacturers, and marketers we should strive to maintain stable, quality, competition-based markets. When the dominant portion of the market migrates to price, the market is operating with no perceived difference in the products. You are playing in a commodity market.

The simple fact about commodity behavior markets is that they never deal with premium margins in their price. Commodity market manufacturers only design to maintain or incrementally improve margin: A light bulb is a light bulb is a light bulb. Commodity markets never stabilize unless there is a leap in the quality of the product and design—and the consumer recognizes that leap. This explains why we see new light bulb designs like "soft white," "natural light," and even "true blue" lights. New designs and new and better quality offerings help to stabilize the market by sustaining price and margin. Companies that innovate and improve the quality of their product in a stable market and then sell that quality at a discount without a strategy need to fire their marketing managers for not seeing the internal quality of their own designs.

The Japanese for years have discounted novel and improved designs for only one reason—share. One aspect of Japanese marketing philosophy is "win the war, then dominate the conquered." Crude and highly accurate in its description, it captures a divide-and-conquer mentality. Japanese companies only discount price to gain as much share as possible, destabilize the market, and drive away competition. Then, when the market is "conquered," they slowly begin to raise the price and bring the margins in to a more profitable situation. Fortunately, the world does not always subscribe to this philosophy.

We want to strive for quality-based competition in the marketplace. This engages and entitles our design teams to the right balance of effort. In this context DFSS has the most to offer.

DFSS builds value into our design so we can play the quality game. We can use DFSS to make existing products more price competitive but that is not its best use or highest purpose.

Identifying the Purposes of Design

To enhance our vision of this concept, let's describe the types of designs toward which we could work. I will describe three types of intrinsic purposes behind our designs in a quality-based market. The first is *fitness to standard* quality. It is synonymous with the kind of quality people expect to be provided when they buy a product. If you buy a candy bar, you expect it to contain sugar. If you buy a car, you expect it to come with tires. When customers are surprised by a lapse in a product's fitness to standard, the sale is usually lost.

The next level of quality purpose is *fitness to use* quality. Fitness to use describes a level of quality that is clearly voiced by the customer. Customers usually see or envision something they like and ask for it in the version of the product they are seeking. Clearly, customers don't expect power seats in a vehicle, but they often request the feature because they have seen it elsewhere. It is the kind of quality for which all companies maintain design efforts to stay competitive. When Ford introduced cruise control in the United States a number of years ago, the other automakers rushed to provide a similar solution to stay competitive.

The final level of quality purpose is the level of purpose to which every designer aspires. Every design engineer hopes to work on these kinds of projects. Every customer searches for this level of quality as well. It comprises what is often referred to as "surprisers" and "delighters," and it is called *fitness to latent expectation.*

Most companies have told me they spend at least 85 to 95 percent of their design effort time on fitness to standard and fitness to use. When the market is predominantly commodity based, they spend even more time in fitness to standard efforts alone. Unfortunately, the longer a company spends focused on fitness to standard, the more likely it will stay there and forever be focused on margin. You might as well call those design efforts "fitness to margin." In this scenario the morale of the engineering teams devolves, then the morale of marketing devolves, and then the company begins slow erosion toward a lifetime business in managing margins. The can manufacturer I mentioned earlier is a classic example of this. Many companies thrive for years in this persona, but rarely do they reemerge from that situation without a commitment to improved quality in their existing products and improved quality in the form of new and innovative designs. These kinds of designs can be marketed on the strength of their value proposition. Palmtop computers are a recent and fine example of this truth. A modem manufacturer, 3Com®, saw that the growth for the high-speed phone line modem business was going to stabilize and eventually destabilize toward commodity behavior. 3Com® subsequently commissioned design innovation in a whole new area. The result of its effort was the first really usable and robust palmtop computers. It continues to thrive while the modem business generally devolves into generic competitive behavior. Palm computing continues to enhance the value proposition to its market by enhancing these designs and communicating that value to the customer.

Cummins continues to maintain a strong relationship with its industrial buyers by not only improving margin on cost issues but also enhancing value by designing in quality that constantly delights customers: more power, greater fuel economy, better emissions, and reduced repairs. Cummins manages its design efforts with a highly trained design team, strong R&D philosophy and testing facilities, and a constant eye to the inherent market situation.

Frank Purdue of Purdue Chicken fame created value in the market for his chicken many years ago by improving the quality of the chicken, meat-on-the-bone, color of the meat, and the company's public image. The result was a stronger perception of quality in the customer's mind. Purdue sold more chicken at a higher price. I always emphasize that point: More chicken at a higher price! What is the moral of the chicken story? If Frank can do that with chicken, what can you do with your product?

Design Based on the Voice of the Customer

The place to start is to begin a discussion on extracting the voice of the customer (VOC) and turning those needs into meaningful requirements for design. This has to be done in such a way that we always keep the customer at the heart of the effort. If we were to ask the different departments of a company what was important to a new product design project, we would find the answers are quite varied. R&D is concerned with functionality and performance; finance focuses on cost; sales is concerned with features, service, and support; and manufacturing might be overly concerned with the assembly of a new product. If a design team is divided on the critical issues for a product, how can it resolve these different opinions? Well, let's first consider the source of these opinions. R&D staff consider functionality important because that is what they do— design functionality. Finance, by the same token, is concerned with the cost of scrap, rework, and warranty claims, and sales might be making more commission on service contracts than on the product itself. This is just an example, but it reflects that although we like to think the company is unified in design, each department is compelled by different drivers. We need a way to manage and balance all the critical parameters that are at play. With that said, where is the customer voice?

If we heed the voice of the customer, it will serve the requirements of all the internal voices if it is properly mined from the customer base. Those leading a design project, whether they realize or not, always want the best result for their project, and they want a list of design issues (requirements) to be weighted and prioritized. Why weighted and prioritized? Because in the business of fulfilling these requirements, budget and resource constraints usually arise that force trade-offs and hard decisions. Constraints like time, budget, human resources, manufacturing capability, and competitive pressures force the design team to focus on what is critical. We can't do everything, so having the design requirements placed into a weighted and prioritized list is imperative. Customer voice should serve as a primary source of the design requirements. The various other organizations and constituencies that contribute to the design team serve as supportive incoming voices. True, we can't ignore the internal voice of the business, but the bigger issue is whether we pay attention to the external customer voice in the first place.

On a tangential note, the enthusiasm for fulfilling internal departmental priorities still needs to be managed from a value and development standpoint. We can't ignore that these internal departments have relevant input and responsibility to the product whether it is in the design

> A who/what matrix links key quality attributes to the business processes that drive performance on those attributes and shows who is the process owner.

Quality Attribute	Development	Components	Release	Assembly	Sales	Service
Image Quality	⊠	X	X			
Blemishes		⊠	X	X		
Fading	⊠					
Turnaround Time	⊠					
Service				X	X	⊠
Availability			X	X	⊠	
Reprints	⊠					

Figure 3–1 Sample What/Who Matrix

phase or in the postlaunch phase of its life cycle. A simple, effective tool for managing these responsibilities is the What/Who Matrix. This elegant tool helps take the value criteria set forth in the value/quality analysis and internalize it by departmental responsibility. We could take Garvin's eight dimensions and use them as the criteria or use whatever criteria we specify. See the matrix in Figure 3–1 for a film manufacturer as an example.

You will note that the left-hand column of Figure 3–1 lists the quality attributes while the top row contains the names of the internal departments responsible for the quality attribute. An X represents some accountability to the issue. A box around an X shows lead responsibility for the quality attribute.

Putting Concept Engineering to Work

Once we have set value, quality, and customer voice as guideposts for our efforts, the key issue is selecting a methodology for championing these issues while pulling out the most achievable, market-attentive, and innovative designs we can. In fact, we need to ensure that we have a process that is documented and reflects a thorough and considered approach to the market, the customer, and the goals of the company concurrently.

The overarching process that controls this is the product development process. Within the product development process resides one of the greatest best practices I have ever seen in using the voice of the customer—concept engineering. Over the last year alone I have probably seen several "revolutionary" methodologies that "release innovation" and bring "excitement" to new designs. I hate these kinds of adjectives for design methodologies. People purporting this kind of excitement in training others in their "methods" are often worried about their method working in the first place. They overhype it because they really have not seen their own process through to completion on anything significant. Many methodologies are empty to me for one reason only: They are centrally focused on market testing after the design is completed as evidence of considering the customer. They use build-test-fix methods. In other words, presenting a model or concept to customers for input *after* initial design is the key means of extracting voice. *That is not extracting voice—that is trying to get customers to buy in to an internally focused design.*

I was recently presented a case study on satellite/cellular hybrid phones. As you probably are aware, the largest company in this business went bankrupt in 1999. Its internal voice told the company that many business people would be thrilled to have a phone that would work anywhere. There was almost no consideration of the customer and the market in this costly project. The old adage of a 1930s car manufacturing executive still holds true, "It is not that we build such bad cars: it's just that *they* are such lousy customers." The company designed this system of satellites, phones, and beepers with the target of selling them to business users. Along the way it failed to discover that cellular business users don't care *how* their call is connected, just that it *is* connected. In most developed countries, cellular towers went up much easier and faster than a satellite. The satellite/cellular company was much more interested in putting satellites up over the United States than let's say Siberia. In fact, Siberia was where the market was—a place where phone lines don't exist and without tower-based cellular systems.

Oil fields, ocean-based oil rigs and ships, and remote locations really needed these phones. You could ask for and get $7.00 a minute for satellite calls there, but not in New York where you had to be outside in direct sight of a satellite. Tower-based cellular airtime is less than 50 cents a minute on an expensive day. In its last days, this manufacturer starting facing comments such as, "Analysts wonder if, in its rush to get satellites in orbit, [Company X] failed to flesh out its marketing strategy." In the end, the most telling comment is the one all design-based companies have faced in their failures: "What they didn't do was to focus on, analyze and pay attention to the market they could serve. [Company X] had a series of major (technical) accomplishments along with some extraordinary (voice of the customer) failures" (Clarissa Bryce Christensen, space industry consultant). Market and customer voice would have made all the difference.

This story exemplifies the need for concept engineering. What I like about this methodology is its diligence in assuring customer voice is respected at all phases and gates of the product development process. It emerged out of MIT and a consortium of Polaroid, Bose, and others in the 1980s. Since then it has seen an evolution of changes and improvements (by the virtue of its mission, it is ever self-improving) that have made it a robust and sound method for uncovering design requirements by placing customer requirements at the heart of the effort. It incorporates methods that are discussed in this book, which I will only outline here.

Concept Engineering Methods

Preparing for customer interviews

Customer interviewing in teams

Image KJ

Requirements KJ

Requirements translation

Product-level quality function deployment (QFD)

Concept ideation

Pugh Concept Selection

Lower-level quality function deployment

Many companies use one or two of these methods, most frequently QFD and Pugh Concept Selection. Concept engineering (CE) puts these tools into larger context in a process that is thorough and well thought out. In fact, it is so diligent many engineers have told me they feel it is overkill because they already knew "most" of the information that comes from the process. The point and value of CE is that it disregards nothing and assures a complete set of critical requirements for the design project. It also assures that you use the rational, market-perceived quality data you've worked on to assess value in the marketplace. I don't know exactly why, but design engineers, manufacturing engineers, and other R&D personnel generally resist talking to the customer. I theorize two general reasons. First is a general fear and dislike of consuming valuable "design" time in extracting this information from customers. Second is a perceived environment in every company where someone, a leader or a dominant design team person, says, "I know what the customer wants. We don't need to do all this customer-focused work because we've done this so long we know what they need. Besides, I am in charge here because I already know the customer needs, and I have some new ideas that I know the customer will love." You may find yourself nodding in agreement that this exists to some level in your company.

You can't ignore the quality component of being diligent with your customers. This is why CE starts with customer interviews that are not rigidly structured and tied to Likert scales. Instead, they are focused on extracting information, images of use, and emotion from the customer. After a complete set of customer interviews, this data can be compiled into a meaningful set of weighted customer issues using KJ analysis. *KJ* stands for Jiro Kawakita, a Japanese anthropologist who said that if mathematicians can treat variables data in a grouped and prioritized manner then we should be able to do the same with attributes/language data. Some people look at the results of a KJ diagram and see an affinity diagram. In fact, it is much more than that. It is the process that gets to the "output" that is unique and different. Space doesn't allow a full description of the process here, but I will say that it is incredibly logical and intuitive, and there is never disagreement by the time you reach the end of a KJ. Chapter 14 will illustrate what I am talking about here. This structuring of VOC data is important to a team in a design project because the KJ links the critical customer priorities to team understanding and consensus. The beauty is that 3 months into the design phase if someone wants to change direction on the project, the KJ serves as the customer-focused guidepost that keeps the team on track.

The Image and Requirements KJ bring together images and voices of user needs that were offered by customers and documented by designers in their interviews. The images (context of use and need) are brought into the Requirements KJ as a supporting element that validates customer need. When the KJs are done, the design team then translates the customer "voiced requirements," which now have weight and priority, into language and descriptions that are understandable to the product development organization. We generally use translation worksheets that integrate the imaged use and need context with the stated need context. The final translation of critical (ranked by weight and priority) customer needs into technical requirements is done using Quality Function Deployment.

I can't emphasize enough the importance of keeping CE self-documenting and the data set complete. If a CE project is documented well, no one can approach the team and gut the effort in lieu of their "expert" opinion of what is needed without a customer-championed fight. Enough of my plea to keep good documentation for a complete CE project. Data, real customer data, is the key to getting it right and keeping the team focused on what is real as opposed to what is opinion.

Once customer requirements have been translated, we then can start the process of focusing on the critical requirements, aligning them with the value analysis that we have done earlier, and looking for comparisons against competitive quality winners in the class. To achieve this goal we will continue our use of Quality Function Deployment (QFD). QFD is a thorough means of aligning the measures mentioned previously and translating the requirements from "general and weighted" to "*detailed* and weighted" with an added analysis of the internal capability to deliver.

A good QFD application should give us a view of what we are currently capable of delivering and what we need to do to fill the detailed technical requirements. See Figure 3–2 for a graphic depiction of this application.

QFD effectively brings together market data, customer input data, and internal capability data to develop a workable set of detailed, critical requirements. We often limit the focus of QFD to those requirements that are new, unique, and difficult. With a workable set of detailed, critical requirements, we can use concept ideation as the creative means of coming up with many different kinds of designs to fulfill the requirements—both those that are routine and those that are new, unique, and difficult.

The goal of Concept Ideation is to create a rich set of creative design bases, compare them against the detailed requirements list, and ultimately hybridize the best elements that lie within them using the **Pugh Process** of concept selection. The graph in Figure 3–3 illustrates what we are trying to achieve in creating many useful concepts and then selecting down to the most feasible one or two to take back out to our customers to see if we are on the right track.

Simply put, concept ideation entails sitting down with the requirements and creatively coming up with many ways of fulfilling each requirement. When you have several options for each requirement, you start putting them together into a concept. You do this until you have several concepts to benchmark against an existing best-in-class product or a previous design. With Pugh

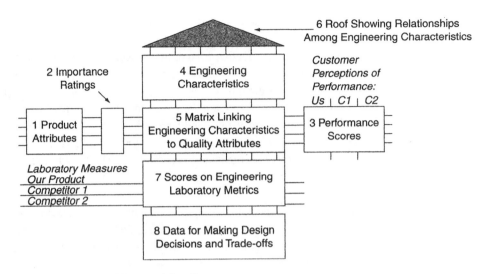

Figure 3–2   QFD: The House of Quality
Source: Used with permission from Sigma Breakthrough Technologies, Inc.

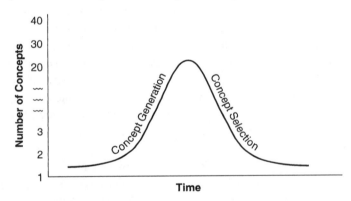

Figure 3–3   Concept Engineering Cycle
Source: Used with permission from Sigma Breakthrough Technologies, Inc.

Process concept selection, you whittle down (hybridize) all these innovative design concepts into the best, and most achievable, choice. The work plan for doing this is shown with the Pugh matrix in Figure 3–4. You rate the ability of each concept to fulfill the detailed, critical requirements that came from your QFD exercise along with other important business and external requirements. Those that fulfill those prioritized requirements best can be selected, or you could hybridize the concepts into a best option.

From the Pugh Process we can opt for a second round of QFD to assure that the detailed, critical customer requirements that were converted to design requirements are thoroughly translated into specific design requirements embedded in the selected, superior design. The focus is on workable requirements that the design team can understand.

RATING CODE S= Same + = Advantage - = Disadvantage	Current Can	Competitor	Best in class	Concepts						
NAME OF CONCEPT										
Code Combination										
• Colorful										
• Unique										
• Novel										
• Easy to manufacture										
• Round top										
• Easy to assemble										
• Easy to distribute										
• Certifiable/Repeatable										
• Texas appeal										
• California appeal										
• Florida appeal										
• U.S. appeal										
• Europe appeal										
• Germany appeal										
• UK appeal										
• Good for promotions										
• Good for year round sale										
• Good for seasonal sale										
• 18-25 appeal										
• All other ages appeal										
• Adaptable to various brands										
• Good for soft drinks										
• Good for beer										
• Shape is novel										
• Volume is affected negatively										
• Volume is affected positively										
Sigma +'s										
Sigma -'s										
Sigma S's										

Figure 3–4 Pugh Matrix

I am only briefly outlining what can be done with value analysis and CE, but I hope you can appreciate that this is a thorough and diligent process. It really is not victim to random and creative ideas. It is a way of capturing those ideas and finding out if they are meaningful to the voice of the customer. In a few brief paragraphs I have outlined a way to travel down a road to product design requirements that are completely customer focused and driven by market perceived value. Recapping the highlights:

- Understanding and defining market value levers
- Competitive market analysis
- Measuring market-perceived quality for current situation
- Defining internal quality goals

- Designing project goals
- Random customer input with focus on new design goals
- Weighted customer value and product requirements
- Detailed customer requirements (in customer language)
- Detailed customer requirements (in design language)
- General product/technical requirements
- Concepts to fulfill product/technical requirements
- Best concept selected
- Translate specific concept product/technical requirements to detailed design requirements

Our goal in marketing and product/service design is to bring a rational, disciplined, and thorough process to identifying market value, working toward that value as an internal quality goal. We seek to thoroughly extract the voice of the customer and design to that voice with weight, priority, and a commitment to both the organization and the customer in achieving a set of world-class design requirements in our quest to manage value to Six Sigma standards.

References

Berry, L. L., Zeithaml, V. A., & Parasuraman, A. (1985, May–June). Quality counts in services, too. *Business Horizons,* 45–46.

Garvin, D. A. (1987). *Managing quality: The strategic and competitive edge.* New York: Simon & Schuster.

PART II

Introduction to the Major Processes Used in Design For Six Sigma in Technology and Product Development

Design For Six Sigma integrates three major elements to attain the goals of product development (*low cost, high quality, and rapid cycle-time*):

1. A clear and flexible product development process
2. A balanced portfolio of tools and best practices
3. A disciplined use of project management methods

The product development process controls the macro-timing of what to do and when to do it using a flexible structure of phases and gates. The use of project management methods controls the micro-timing of the application of specific tools and best practices within any given phase's critical path.

In this section we look at how DFSS influences the process of product development in four differing contexts:

1. Program and project management
2. Technology development

3. Product design
4. System architecting, engineering, and integration (within product design)

Chapter 4 explores how to develop, design, model, and simulate product development processes and project management plans that comprise the macro- and micro-timing options to develop a product to Six Sigma cycle-time capability standards.

Chapter 5 illustrates how the phases of a technology development process are structured relative to deploying DFSS tools and best practices to fulfill gate deliverables.

Chapter 6 illustrates how the phases of a product design or commercialization process are structured relative to deploying DFSS tools and best practices to fulfill gate deliverables.

Chapter 7 illustrates how to conduct system architecting, engineering, and integration using the tools and best practices of DFSS in the CDOV process. Heavy emphasis is placed on the gate deliverables from deploying DFSS within a system engineering organization.

A balanced portfolio of tools and best practices is aligned to each phase of the technology development, product design, and system engineering process. The program management team uses project management in the form of PERT charts of work breakdown structures to custom design the micro-timing for the critical path of tools and best practices used to generate the deliverables at each gate.

These four chapters will greatly aid any company that wants to improve its current product development processes to be compatible with Six Sigma standards of excellence.

Management of Product Development Cycle-Time

Before we get into the mechanics and details of product development processes in the next few chapters, we need to set the stage for how to act as program and project managers in a Design For Six Sigma context.

This book is structured around three components that are critical to successful product development (see Figure 4–1):

1. Disciplined use of a well-structured product development process
2. Rigorous application of a balanced portfolio of tools and best practices
3. Detailed use of project management methods within and across the phases of a product development process

As we proceed through this chapter, first we will consider *processes,* second we will discuss the role of *tools,* and third we will develop detailed *project management* methods to help us design the "right" cycle-time for our product development initiatives.

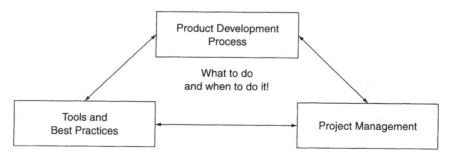

Figure 4–1   Critical Components in Successful Product Development

The Product Development Process Capability Index

The metrics of Six Sigma can be used to improve one's ability to manage product development projects. The metric that is most associated with Six Sigma performance is the **capability index:**

$$Cp = (USL - LSL)/6\sigma$$

The capability index is quite flexible in its ability to represent the ratio of *required* performance to *measured* performance. We can apply this metric to the capability of a management team to meet its product development cycle-time goals. The Cp index can be restated for measuring product development process capability:

$$Cp = (\text{Cycle-Time Limits})/6\sigma \text{ of Measured Cycle-Time}$$

The **upper** and **lower specification limits** for product development cycle-time are defined as:

USL = The time it takes to complete either one phase or a complete set of phases within a product development process when using a full and complete set of tools and best practices to fulfill the system requirements (i.e., the longest allowable cycle-time)

LSL = The time it takes to complete either one phase or a complete set of phases within a product development process when using a minimum set of tools and best practices to fulfill the system requirements (i.e., the shortest allowable cycle-time)

The definition of time limits to conduct the phases of product development is considered the *voice of the business.* In that context, the USL and LSL are just like any other set of requirements established to satisfy the voice of some constituency. The time allotment for single-phase cycle-time or complete product development process cycle-time (all eight phases in the I^2DOV and CDOV processes, outlined later in this chapter) is defined by the advanced product planning team. This team is usually made up of business leaders from inbound marketing, sales and support (outbound marketing), R&D, product design, supply chain management, and manufacturing management.

The cycle-time to complete a phase or the complete product development process can be both estimated and literally measured as the product is developed. The standard deviation of cycle-time is easily calculated by performing Monte Carlo simulations on the phase-by-phase cycle-times within a program. Actually quantifying the real standard deviation of a business's product development cycle-time can take years. The key is to realistically estimate task cycle-times within each phase. Every critical task can be underwritten by a tool or best practice that delivers measurable results that fulfill specific product requirements.

The worst thing a business can do is to arbitrarily establish, or guess without data, a product launch date and then mandate that the product development community meet it. This is a recipe for the development of schedules that are simply unattainable. The amazing thing is everybody knows these schedules are unrealistic, but year in and year out this game is played. Up the chain of command no one is held accountable for this kind of anemic and undisciplined management behavior. The application of a Six Sigma approach to cycle-time capability measurement is a reasonable way to break this cycle of dysfunctional behavior (see Figure 4–2).

What we suggest is a disciplined, realistic, detailed estimation of task cycle-times on a phase-by-phase basis. These time estimates form a phase-by-phase project management plan, sometimes referred to as an *integrated program plan*. Within each phase, a *critical path* emerges. This critical path is a serial set of tasks that define the time it will take to complete the phase. The application of Monte Carlo simulations to the critical path cycle-times allows the program management team to "rehearse" the tasks within each phase under many changing time estimates. It is easy to run thousands of practice runs through any given phase and see the resulting best, median, and worst-case cycle-time projections.

When a management team says, "We don't have time to do these estimates, we have to get this product out the door," you can pretty much guarantee that they will be late in delivering the product. They have a weak plan to get the product developed on a realistic time scale. For large projects, it is routine for the lateness to encompass the better part of a year! Maybe part of "getting the product out the door" should begin to include taking the time to properly plan the detailed tasks to do the right things at the right time in the first place. The following 9 steps illustrate a

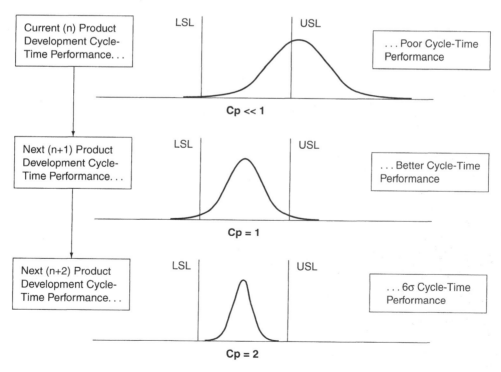

Figure 4–2   Designing Rational Product Development Cycle-Times

proven approach for designing rational product development cycle-times that can be measured and held accountable to Six Sigma performance standards. The goal is to gain control of product development cycle-time and systematically improve it from one product to the next.

A general, 9-step process model for the disciplined management of product development cycle-time can be deployed as follows:

1. Build the phases and gates structure that defines your product development process (what to do at the *macro* level).
2. Define the full set of tools and best practices you will use to deliver the required results during product development (specific tasks and deliverable results).
3. Align an appropriate (somewhere between a full set and minimum set) set of tools and best practices within each phase of your product development process (what to do at the *micro* level).
4. Estimate start and end dates for each task that is enabled by a tool or best practice within each phase.
5. Define who is responsible for using specific tools to deliver specific results that certify the completion of each task within each phase.
6. Create a PERT chart of the tool-based tasks for each phase (using Microsoft® Project).
7. Identify the critical path of tasks and the USL-LSL that control the cycle-time for each phase's PERT chart.
8. Construct a Monte Carlo simulation model to calculate the earliest, median, and latest phase completion times and Cp index created by the Critical Path tasks within each phase (using Crystal Ball® in conjunction with Excel).
9. Create an integrated program plan by linking all the phase cycle-time estimates together. Construct statistical forecasts and Cp index for the aggregated cycle-time for the whole program (using Crystal Ball® in conjunction with Excel).

Product Development Process

1. Build the phases and gates structure that defines your product development process (what to do at the **macro** *level).*

A well-structured **product development process** (PDP) enables executives, program managers, project managers, and engineering team leaders to control what to do and when to do it from a macro-timing perspective. A good PDP will in large measure establish the "poise" of the organization. This is particularly true when all the troubles come flowing in as a normal matter of course in product development. Products tend to mirror the processes that develop and deliver them. Not all companies have intentionally designed product development phases. But most companies have at least two major "phases" that they must manage:

- WAHBL phase (When All Hell Breaks Loose)
- SHTF phase (S%#! Hits the Fan)

We hope to give you some better suggestions for phase titles as we proceed.

A Phase/Gate structure for conducting and completing development and design tasks enables the management team to manage risk carefully in reasonable "packets" of time. Each packet of time is called a phase. A **phase** is made up of a rational subgroup of tasks or actions that are enabled by a specific set of tools and best practices. A reasonable number of tasks are aggregated and conducted to deliver a predetermined set of results that move the goals of the product development program to a rational point of completion and risk assessment.

A **gate** covers a very short period of time, usually less than a day. The program management team assesses progress against goals, understands the current state of accumulated risk, and plans for the successful deployment of the next phase tasks and actions. The gate should be roughly broken down into 20 percent of the time used to look back at risks that have accrued and what is going to be done about them. Eighty percent of the time should be devoted to planning, preventing problems, and enabling the success of the next phase.

We recommend architecting a lean and flexible product development process that acts a lot like a demand flow manufacturing process. Any size product development project or program (PDP) should be able to be taken through your phases and gates — big ones and little ones. The phases and gates should be no more complex than your product and manufacturing process requirements demand — but no simpler (Einstein's approach!). Your product development process should be carefully architected to be lean and flexible. If it's too lean, your customers are quite likely to get mean so be careful. One way to know if it's too lean is if nobody pays serious attention to the gate reviews.

Another characteristic of your PDP is that it should be the key influence on how you manage your program's **macro-timing cycle.** Macro-timing is cycle-time constraint and management *across* the phases. In contrast, **micro-timing** is cycle-time constraint and management of the specific *within-phase* tasks that are enabled by specific tools and best practices. Micro-timing is managed using the methods of project management. We will discuss those shortly.

A reasonable product development process contains two distinct processes:

1. A technology development process
2. A product commercialization (design) process

The reason for two separate Phase/Gate processes is to reduce the risk associated with new technology development. Many companies develop new, unproven technology right in the middle of a product design program. Often, the new technology is difficult to tame, and the whole commercialization program suffers in terms of poor quality, budget overruns, and uncontrollable cycle-time. These are symptoms of poor discipline within an organization. Fortunately these symptoms can be mitigated, but not by measuring cost, quality, and cycle-time as primary metrics. The chapters on Critical Parameter Management will explain what to measure to help assure these macro-measures come out reasonably well.

The approach to technology development we have used to great benefit is to first develop and certify new technologies before they are allowed into a product design program. The timing between technology development and product design processes can either be purely serial or set up in staggered form of parallel timing. The more you conduct Technology Development For Six Sigma (TDFSS) in parallel without a fall-back design in the wings, the more risk you heap upon your commercialization program. The phases and gates for technology development are intended to produce essentially equivalent designs to those current designs that have experienced actual use in the application environment by actual customers. Your technology development process has to stand in as a reasonable surrogate of the misuse, abuse, and stresses encountered in the real world. What can happen "out there" must be induced "in here." Your technology development teams must act like customers as they stress test their new designs prior to transfer into the design organization.

We will thoroughly introduce you to the four phases and gates of a TDFSS process in the next chapter. If you are familiar with Design For Six Sigma, this is a similar process refined for the up-stream R&D activities that are used to *I*nvent, *I*nnovate, *D*evelop, *O*ptimize, and *V*erify new platforms and subsystems. We call it the **I²DOV process.**

The DFSS process in this text is focused on the design of new products and production processes. It includes four generic phases: *C*oncept design, *D*esign development, *O*ptimization, and *V*erify certification. We call it the **CDOV process.** For reference, the production operations process for Six Sigma is often called **MAIC,** for *M*easure, *A*nalyze, *I*mprove, and *C*ontrol, which is largely based on Walter Shewhart's famous contributions to statistical process control methods.

We are not optometrists, but we do believe that when the I²DOV and CDOV process elements are properly integrated into a product development process, they will help you see clearly how to develop great products with the right cycle-time and performance. The result is complete and balanced products that look and perform like the processes that created them.

Tools and Best Practices

2. Define the full set of tools and best practices you will use to deliver the required results during product development (specific tasks and deliverable results).

A good portion of this text is dedicated to laying out the flow of tools and best practices used in the TDFSS (I²DOV) and DFSS (CDOV) processes. These tools and best practices are the focus of "what to do" within each phase. The gate reviews look closely at the deliverables from each application of tools and best practices. If you have an anemic set of tools and best practices that you are always short-circuiting because you have no time to use them properly, the following suggested tools and best practices will help you.

The chapters in this text offer specific suggestions for defining a full suite of tools and best practices. These suggested tools and best practices will go a long way in helping you determine where, within your design functions, you need Six Sigma performance and then how to attain it and prove it with data at a gate review. Please note that not all design functions or parameters need to be at Six Sigma levels of performance; some can be less and a few will need to be much greater.

Here is a summary list of the numerous tools and best practices you should seriously be considering for application across the phases of your product development programs:

Phase/Gate processes
Customer value and marketing segmentation analysis
Concept engineering
Voice of the customer gathering
KJ (affinity diagramming) methodology
Quality Function Deployment (QFD)
Houses of Quality (EQFD)
Concept generation methods (TRIZ, etc.)
System architecting
Practical design principles
Pugh concept selection
Critical Parameter Management
Project management techniques
Basic statistical techniques
    Descriptive statistics
    Normal distribution parameters
    Graphical techniques
        box plots
        histograms
        scatter plots
        time series plots
        run charts
        pareto charts
        inferential statistics
        central limit theorem
        hypothesis testing
        confidence intervals
        sample size determination
Measurement systems analysis
    Gage R&R and P/T studies
Design process capability analysis
Design for manufacturability and assembly
Minitab data analysis methods
Failure modes and effects analysis (FMEA)
    Design failure analysis
    Manufacturing process failure analysis
Design of experiments for empirical modeling of "ideal functions and transfer functions"
Screening studies and sequential experimentation
Full and fractional factorial experiments
    Degrees of freedom
    Main effects and interaction studies
    Center points and blocking methods

Analysis of variance (ANOVA)
Simple and multiple regression
Response surface methods
Taguchi robust design techniques
    QLF and S/N metrics
    Design for additivity
    Static methods
    Dynamic methods
Tolerance design
    Analytical (WC, RSS, 6s, Monte Carlo)
    Empirical (DOE, ANOVA, QLF, and Cp)
Reliability analysis (Prediction, HALT, HAST)
Multi-vari studies (ANOVA)
Statistical process control charts
Manufacturing process capability analysis
Data transformations
    Box-Cox transformation for nonnormal data

You can see that we have many tools and best practices to consider for application as we conduct our product development tasks. Each of these items is capable of delivering certain, specific results. If your design needs these results to satisfy customer needs or to fulfill critical requirements, you must seriously consider deploying the tool or best practice. Now, we have to face the practical fact that we have a finite amount of time to conduct product development. Only a specific number of these tools and best practices can actually be conducted before everybody gets old and dies! You need to select a rational set of these tools and best practices and commit to applying them rigorously. The key thing you will have to manage to keep your risks in balance is what tools and best practices you are *not* going to use! Choose wisely.

Aligning Tools within Phases

3. Align an appropriate (somewhere between a full set and a minimum set) set of tools and best practices within each phase of your product development process (what to do at the micro *level).*

The next thing we need to do is align the appropriate tools and best practices to the appropriate phases of a product development process. We have established the I^{2}DOV and CDOV product development processes as general models that are easy to understand and relate to your specific product development process vocabulary. Your words may be a bit different but everybody does some form of conceptual design, design development, optimization, and verify certification. Most companies have more than four general phases. We have laid out a sample set of tools and best practices in the following pages to fit within the CDOV process as an example. A full set of tools and best practices are developed in Chapters 5 and 6. We have broken Phases 3 and 4 into two segments because that is typical of the way most successful companies we have seen actually structure their product development processes. You will see a bit of tool overlap as we progress through the phases.

Concept Phase (Phase 1)

Market segmentation analysis

Economic and market trend forecasting

Business case development methods

Competitive benchmarking

VOC gathering methods

KJ analysis

QFD/House of Quality – Level 1 (system level)

Concept generation techniques (TRIZ, brainstorming, brain writing)

Modular and platform design

System architecting methods

Knowledge-based engineering methods

Pugh concept evaluation and selection process

Math, graphical, and prototype modeling (for business cases and scientific and engineering analysis)

Design (Phase 2)

Competitive benchmarking

VOC gathering methods

KJ analysis

QFD/Houses of Quality – Levels 2-5 (subsystem, subassembly, component, and manufacturing process levels)

Concept generation techniques (TRIZ, brainstorming, brain writing)

Pugh concept evaluation and selection process

Design for manufacture and assembly

Value engineering and analysis

Design failure modes and effects analysis

Measurement systems analysis

Critical parameter management

Knowledge-based engineering methods

Math modeling (business case and engineering analysis to define ideal functions and transfer functions)

Design of experiments (full and fractional factorial designs and sequential experimentation)

Descriptive and inferential statistical analysis

ANOVA data analysis

Regression and empirical modeling methods

Reliability modeling methods

Normal and accelerated life testing methods

Design capability studies

Multi-vari studies

Statistical process control (for Design Stability and Capability Studies)

Optimization (Phase 3a: Subsystem and Subassembly Robustness Optimization)

Subsystem and subassembly noise diagramming

System noise mapping

Measurement system analysis

Use of screening experiments for noise vector characterization

Analysis of means (ANOM) data analysis techniques

Analysis of variance (ANOVA) data analysis techniques

Baseline subsystem and subassembly CFR Signal-to-Noise Characterization

Taguchi's methods for robust design

Design for additivity

Full or fractional factorial experiments for robustness characterization

Generation of the additive S/N model

Response surface methods

Design capability studies

Critical parameter management

Subsystem/subassembly/component reliability analysis

Life testing (Normal, HALT, and HAST)

Optimization (Phase 3b: System Integration and Stress Testing)

Measurement system analysis

System noise mapping

Nominal system CFR design Cp studies

Stress case system CFR design Cp studies

System-subsystem-subassembly transfer function development

System level sensitivity analysis

Design of experiments (screening and modeling)

ANOVA

Regression

Analytical tolerance analysis

Empirical tolerance analysis

Statistical process control of design CFR performance

Reliability Assessment (normal, HALT, and HAST)

Critical Parameter Management

Verify (Phase 4a: Product Design Verification)

Measurement system analysis

System noise mapping

Analytical tolerance design

    Worst case analysis

    Root sum of squares analysis

    Six Sigma analysis

    Monte Carlo simulation

Empirical Tolerance Design

    System Level sensitivity analysis

Design of experiments

ANOVA

Regression

Multi-vari studies

Design capability studies

Nominal system CFR design Cp studies

Stress case system CFR design Cp studies

Reliability assessment (Normal, HALT, and HAST)

Competitive benchmarking studies

Statistical process control of design CFR performance

Critical parameter management (with capability growth index)

Verify (Phase 4b: Production Process and Launch Verification)

Measurement system analysis

Multi-vari studies

Design and manufacturing process capability studies

Nominal system CFR design Cp studies

Reliability assessment (Normal, HALT, and HAST)

Statistical process control of design and process CFR performance

Critical parameter management (with capability growth index)

We now have aligned numerous tools and best practices to the phases of the CDOV product development process. The phases of CDOV define the macro-timing segments that can be used to control the program's overall cycle-time. The tools and best practices can be carefully matched to "within-phase" actions or tasks that help fulfill specific gate requirements. How long will it all take? Let's begin to answer that question with the next section.

Project Management

We have established the need for the macro-timing of the phases and gates of product development. We have also illustrated the need for bringing a carefully selected set of tools and best practices into that structure. It is time to constrain and manage the detailed serial and parallel flow of specific tasks that are enabled by these tools and best practices.

We will briefly introduce a few key principles of project management at this point. This will establish a common vocabulary to work with as we "design" the cycle-time of our programs and projects in a Six Sigma context. Phase timing can be structured somewhat like the various segments or stages of a manufacturing process. The tools are different because the deliverables are different, but using tools and best practices at the right time in the process is the same. Statistical process control techniques help keep the measured output of a process, over time, on target, or at least within reasonable boundaries (upper and lower control limits). In a similar fashion, statistical analysis can be applied to the product development tasks that are tracked using project management methods. If you get the sense that we are taking you into the world of capability analysis for product development cycle-time management, you are quite right.

Too many companies manage programs to "all-success" schedules. They tend to use deterministic (single-point-in-time estimates) modeling of program cycle-time. This is often sterile and unrealistic. An emerging best practice is the use of an integrated, phase-by-phase project management plan. It involves the disciplined tracking of a critical path that is built up from carefully selected, tool-based tasks that are timed using statistically based Monte Carlo simulation models. This is the basis for Six Sigma program management.

The statistical and probability-based methods of Six Sigma start to become important in project and program management when we fit statistical distributions to start and end dates for within-phase tasks. We can use these stochastic methods to estimate and manage the start and end dates for the phases. Ultimately, we can link all the phases as probabilistic, as opposed to deterministic, events that can be analyzed as an integrated program plan. Integrated program plans are made up from the project plans from each phase that are usually documented in the form of a PERT chart.

4. Estimate start and end dates for each task that is enabled by a tool or best practice within each phase.

We are ready to assign cycle-time estimates to our within-phase tasks. The program or project manager can now ask his or her design teams to estimate the start and end dates for the specific tool and best practice tasks that will fulfill the program's gate deliverables. It is imperative that the team members within the product development organization understand that they are expected to use these tools and best practices and will be held personally accountable for the results gained from their proper and rigorous deployment.

We are now getting into the micro-mechanics of cycle-time design and management. Most people don't think about it this way, but a good project manager is really an expert in designing cycle-times. In reality, the project manager is in the profession of architecting "designer cycle-times," crafting molding, and shaping the architecture of the cycle-time structures!

The following illustrations are examples of a template we use throughout this text to show how specific tasks, enabled by one or more tools or best practices, can be linked in serial or parallel networks that are referred to as a *work breakdown structure*. In fact, this text is loaded with them! These form the structural basis we will use as we proceed into the design of phase-by-phase cycle-times.

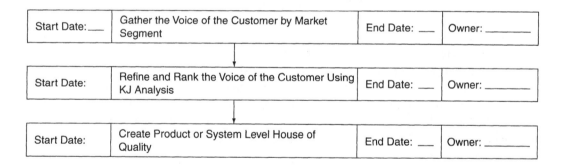

The specific tools deployed in these three tasks are:

1. Customer interviewing
2. KJ analysis (affinity diagramming)
3. Quality function deployment (QFD)

Specific deliverables come from using these tools:

1. Raw voice of the customer database
2. Grouped and ranked VOC data in the form of an KJ (affinity) diagram
3. System level House of Quality (a matrix structure diagram)

The next two chapters demonstrate, through a reasonable amount of detail, how to account for the major tasks in technology development and product design. Laying out the structure and flow of the major tasks that your teams will conduct during product development will leave little to the imagination as we attempt to be good cycle-time designers and time management accountants. So now a project manager is not only a designer, but an architect and accountant of time as well!

5. Define who is responsible for using specific tools to deliver specific results that certify the completion of each task within each phase.

Each task has a start date and end date, and now we add who specifically is responsible—personally accountable—for the delivered results from using a tool or best practice. This approach to project management during product development holds specific people accountable for specific deliverables. If you are a manager and are uncomfortable holding people accountable for delivering specific results using carefully selected tools and best practices within the agreed time allocation, then this approach to product development is definitely not for you! If what we have read about Jack Welch, former CEO and the Six Sigma czar of General Electric, is to be believed, he likely would find this level of accountability to be quite appropriate for weeding out the bottom 10 percent of low performers from the product development community every year! Of course you need to replace these people with new and capable talent so you have enough resources to apply the right number of tools and best practices rather than heaping the extra responsibility on the remaining resources.

Many managers like to leave the decision of how results are attained up to their people. This is fine if the people are properly trained and enabled and actually choose to apply a balanced set of tools and best practices. It has been our experience that this almost never happens. Usually cycle-time is forced to be unrealistically tight and the teams simply revert to build-test-fix methods and quickly get behind in fulfilling their desire to do things right. Their design releases get many iterations ahead of their actual data analysis! It is rarely their fault, though. They were not governed by the right or reasonable expectations, so they guess and bad things start to happen. Quite frankly, it is mainly attributable to poor project management technique. Although we've never asked him, it is a good bet that this kind of behavior would get you on Jack Welch's "sorry but you've got to go" list.

The right thing to do as a manager is to reach consensus about reasonable expectations with your teams on the right balance of tools and best practices that absolutely must be deployed. Then give your people the time and resources they need to use them properly. As a result, you will not need to watch them do things over a third, fourth, or fifth time. You won't be able to deploy all these tools and best practices, but you can carefully choose a rational subset and then see to it that they get used properly.

The Power of PERT Charts

6. Create a PERT chart of the tool-based tasks for each phase (using Microsoft® Project).

A program or project manager can easily have each of his or her team leaders assist in the construction of PERT charts that act as flow diagrams to illustrate exactly what tools and best practices are going to be used within each phase of technology development and product design.

You can use three major software tools to gain proficiency at project management in a Six Sigma context:

1. Microsoft® Project
2. Microsoft® Excel
3. Decisioneering's Crystal Ball® Monte Carlo Simulator

Tool #1 for Six Sigma Project Management: Microsoft® Project

Project management is a common topic that has been taught and practiced in industry for many years. The approach presented in this text is enabled by Microsoft® Project software, which is easy to use and readily available. Microsoft® Project allows one to easily build a project time line using both PERT and Gantt charting methods. We recommend doing your project management plans in PERT chart format. PERT charts are essentially serial and parallel flow charts comprised of block diagrams such as the sample in Figure 4–3.

PERT charts are good at showing how numerous tasks are linked to one another. They are also good at illustrating who is "on the hook" for delivering the results. When an entire phase of tasks is linked, the software can illustrate the critical path of tasks and the time required to complete the phase deliverables, as in Figure 4–4.

With such a diagram we can easily identify the tasks that add up to define the Critical Path and the total cycle-time for the phase we are evaluating. You should create a PERT chart for each phase of your technology development and product design processes. It is not uncommon for engineers to have personal PERT charts that essentially act as an agreement or "contract" between them and their manager or team leader. These personal PERT charts get aggregated to comprise the overall phase PERT charts. All the phase PERT charts are linked to form what is called the *integrated program plan (IPP)*.

Task based on the application of a tool or best practice that has a specific deliverable.	
Task Owner:	Duration:
Start Date:	End Date:

Figure 4–3 Sample PERT Chart Task Box

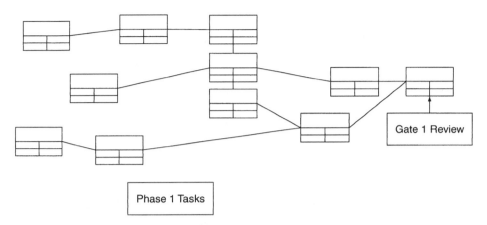

Figure 4–4   PERT Charts Linked in Phase

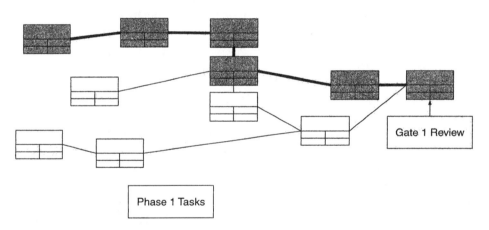

Figure 4–5   Identifying the Critical Path

7. Identify the critical path of tasks and the USL-LSL that control the cycle-time for each phase's PERT chart.

When an entire phase of tasks are linked, the software can illustrate the critical path of tasks and the time required to complete the phase deliverables, as in Figure 4–5.

When we arrange the tasks into a network, we see serial and parallel tasks flowing across the phase. A certain "string" of tasks will constrain how long it will take to complete the phase. If we focus on this Critical Path, we can help assure cycle-time is under "nominal" control. After this is done, we can use the other two software tools to assure the critical path is under a form of "statistical" control.

Each phase can have specification limits defined for its allowable range of Critical Path cycle-time. The Upper and Lower Specification Limits for product development cycle-time are defined as:

USL = The time it takes to complete either one phase or a complete set of phases within a product development process when using a full and complete set of tools and best practices to fulfill the system requirements (i.e., the longest allowable cycle-time)

LSL = The time it takes to complete either one phase or a complete set of phases within a product development process when using a minimum set of tools and best practices to fulfill the system requirements (i.e., the shortest allowable cycle-time)

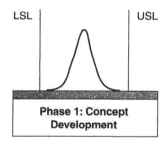

8. Construct a Monte Carlo simulation model to calculate the earliest, median, and latest phase completion times and Cp index created by the Critical Path tasks within each phase (using Crystal Ball® in conjunction with Excel).

Tool #2 for Six Sigma Project Management: Microsoft® Excel

We now take the tasks that are on the critical path and lay them out in an Excel spreadsheet that is also linked to Crystal Ball® Monte Carlo Simulation software (www.decisioneering.com). These tasks will be evaluated mathematically by summing their individual cycle-times. The cycle-times are estimates from engineering team judgment. We will illustrate the popular *triangular distribution* approach to estimating cycle-time variation. The time estimates for each task are assigned three possible or likely durations. One way to use this approach is to list each task and then enter three time estimates in the Excel columns:

1. Median or Most Likely Time Duration for the task
2. Minimum or Least Time Duration for the task
3. Maximum or Longest Time Duration for the task

The final column contains the type of distribution we choose to represent the probabilistic, statistical, or "likely" variation that characterizes the nature of how the task duration time might vary. See Table 4–1.

Table 4–1 Estimating Cycle-Time Using Triangular Distributions

Phase 1 Concept Development Tasks:	Median Task Time (days)	Minimum Time (days)	Maximum Time (days)	Type of Distribution
Gather the voice of the customer	10	7	13	Triangular
Refine & rank the voice of the customer using KJ analysis	2.83	1.5	4	Triangular
Create product or system level House of Quality	5	4	6	Triangular
Conduct competitive product benchmarking	8.33	7	10	Triangular
Generate product or system level requirements document	14	12	16	Triangular
Define the functions that fulfill the system requirements	1.83	1	2.5	Triangular
Generate system concept evaluation criteria	1.67	1	2	Triangular
Generate system concepts that fulfill the functions	25	20	30	Triangular
Evaluate system concepts	5.67	4	7	Triangular
Select superior system concept	3	2	4	Triangular
Analyze, characterize, model, & predict nominal performance	60	54	66	Triangular
Develop system reliability reqts, initial reliability model, & FMEA	20	18	22	Triangular
Total Phase 1 Development Time	**157.33333**			

The total Phase 1 development time is the dependent (Y) variable. The independent (x) variables are the 12 Phase 1 task time estimates.

Tool #3 for Six Sigma Project Management: Crystal Ball®

A software program called Crystal Ball®, from Decisioneering, which runs as an add-on to Excel, can now be used to "activate" the Excel cells that contain the task duration estimates (x variables are called *assumptions*) and the predicted Critical Path cycle-time (the Y variable is called the *forecast*).

Crystal Ball® is a richly featured Monte Carlo and Latin Hypercube simulator. It allows one to build mathematical models (Y = f(x)) in Excel and conduct what-if studies using samples from prescribed probability distributions (in this case triangular distributions) as the assumption inputs to the model. In project management cycle-time forecasting, the models are simple summations of the task durations that lie on the critical path.

Thus, the model for Phase 1 cycle-time is: $Y = Task_1 + Task_2 + Task_3 + \ldots + Task_{12}$

Each task is not treated as a deterministic event! Each is represented by the triangular distribution, and its value is randomly determined by Crystal Ball® for each run through the equation.

We could choose a number of distributions from inside Crystal Ball's® Menu to represent duration time variation for each task. The normal, uniform, and triangular are the most common options.

The **normal distribution** requires one to know the mean and standard deviation of the task duration time. We find most teams do not usually know these parameters.

The **uniform distribution** basically assumes that there is an equally likely chance that the task duration will lie somewhere between two points in time:

The shortest likely duration

The longest likely duration

The **triangular distribution,** as in Figure 4–6, assumes you can estimate three separate likelihood values for the task duration:

Median or most likely time duration for the task

Minimum or shortest likely time duration for the task

Maximum or longest likely time duration for the task

We recommend the triangular distribution because it is a conservative and reasonable estimator of likelihood without trying to overstretch statistical parameter assumptions. Crystal Ball® allows you to skew your minimum and maximum duration inputs—your duration estimates do not need to be symmetrically spaced. Figure 4–6 is an example of a Crystal Ball® assumption window for inputting the Phase 1 task duration estimates (symmetrically spaced) for "Gather the Voice of the Customer."

Crystal Ball® can be set up to run through your phase tasks anywhere from 500 to many tens of thousands of practice runs. You can literally run your Phase 1 tasks through their estimated durations thousands of times. Recall that your forecast Y will be the total Phase 1 cycle-time. So, 10,000 trips through our math model for Phase 1 looks like this as far as Crystal Ball® is concerned:

Math Version of Phase 1: $Y = Task_1 + Task_2 + Task_3 + \ldots + Task_{12}$

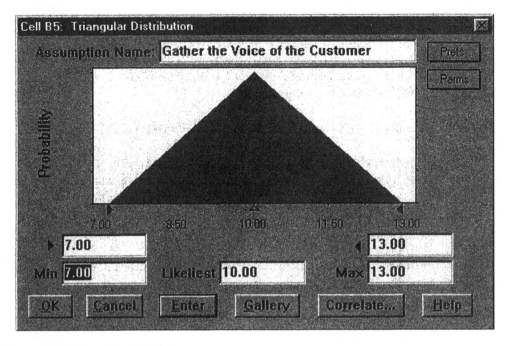

Figure 4–6 Triangular Distribution

Crystal Ball® Version of Phase 1:

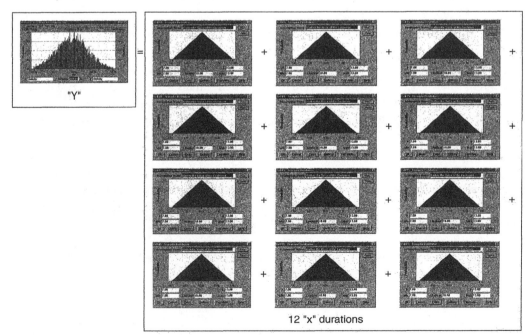

12 "x" durations

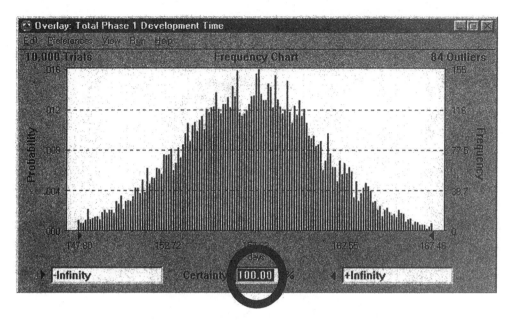

Figure 4–7   Frequency Chart from Crystal Ball®

Here we see that 10,000 random samples from 12 independent, triangular task duration distributions produce a *normal* distribution of 10,000 Phase 1 cycle-times.

This happens because of a statistical truth called the *central limit theorem*. What is important is the *statistical output parameters* associated with the Y forecast distribution. These output statistics can be evaluated for several interesting characteristics:

Forecast range minimum, maximum, and width

Forecast mean, median, and standard deviation

Figure 4–7 and Figure 4–8 show an example of Crystal Ball's® output in the form of a frequency chart and table of descriptive statistics for the Phase 1 cycle-time forecast.

Phase 1 cycle-time lies somewhere between the range of 143 and 170 days duration. There appear to be 27 days between the worst and best case cycle-time for Phase 1. The most likely cycle-time for Phase 1 is estimated to be 157 days. This is helpful information. We can rerun this model with various task duration estimates in any symmetrical or asymmetrical geometry we choose. When we assign upper and lower control limits to the forecasted Phase 1 cycle-time and estimate its standard deviation, we can calculate Phase 1's Cp index. Now, we can start talking about Six Sigma standards of performance in project management terms!

We can control the width of the Phase 1 cycle-time distribution and begin to minimize variability in our product development process. We have just built a method to predict product development process capability.

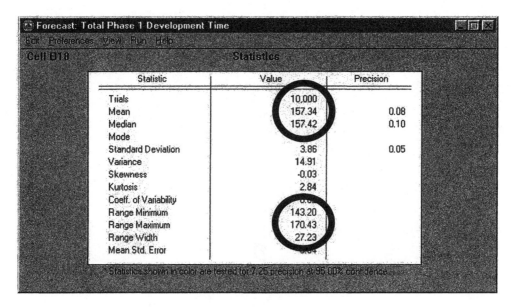

Statistic	Value	Precision
Trials	10,000	
Mean	157.34	0.08
Median	157.42	0.10
Mode		
Standard Deviation	3.86	0.05
Variance	14.91	
Skewness	-0.03	
Kurtosis	2.84	
Coeff. of Variability		
Range Minimum	143.20	
Range Maximum	170.43	
Range Width	27.23	
Mean Std. Error		

Figure 4–8   Table of Descriptive Statistics from Crystal Ball®

The Cp index for each phase can be calculated from the Specification Limits (USL-LSL) and the estimated standard deviation from the Monte Carlo simulation (see Figure 4–9).

9. Create an integrated program plan by linking all the phase cycle-time estimates together. Construct statistical forecasts and Cp index for the aggregated cycle-time for the whole program (using Crystal Ball® in conjunction with Excel).

We conclude the process for Six Sigma product development cycle-time management by rolling up the various phase cycle-times into the integrated program cycle-time. Crystal Ball® provides the phase-by-phase forecasts needed to estimate the IPP cycle-time:

 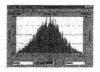  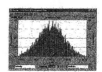

Using the descriptive statistics tables from Crystal Ball® once again, we can look at the rolled-up worst, best and, most likely (mean) cycle-time. The beauty of this approach is that we can, in a few minutes, emulate our phase timing designs in any way we like. We can add or re-

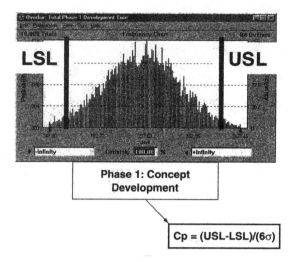

Figure 4–9   Calculating the Cp Index

move tasks, change duration estimates, and assess the outcome. A program management team can easily sit down together, look into the future, and come up with a phase-by-phase, task-by-task, tool-by-tool plan for meeting its cost, performance, and cycle-time objectives. Choose to deploy these tools and you get this performance. Eliminate those tasks and you get these risks. Around and around you can go until you emerge with a balanced plan that lets you sleep at night, knowing you cut the right corners for what you know at that time in your product development process. Later when you know more, you can add new tasks to react to an unforeseen problem. The common theme in your planning for the right cycle-time should be "it's cheaper to prevent a problem than it is to solve one."

The illustration in Figure 4–10 shows how the integrated program plan can be measured against Six Sigma capability standards.

Product development capability performance can be tracked and, more importantly, improved with the Six Sigma metric in Figure 4–10 in place.

We hope this overview of using Microsoft® Project, Excel, Crystal Ball®, and product development capability growth metrics sets the stage for success as you design the right cycle-times for your programs and projects. These methods are easy to learn and apply. We close this chapter by reminding you all that the shortest cycle-time is not always the right cycle-time. Balance and proportion are essential to proper risk management, which is really what this is all about. Sleep well.·

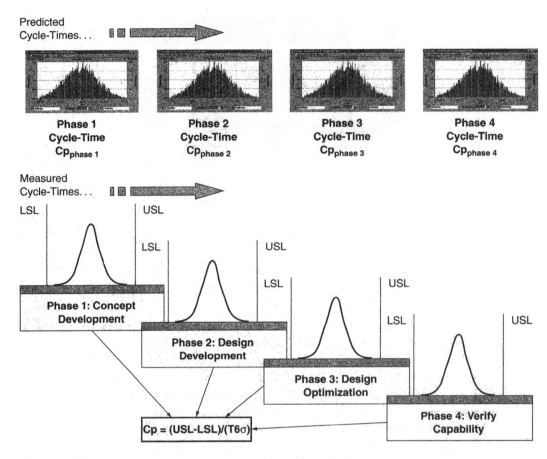

Figure 4–10 Measuring the IPP Against Six Sigma Standards

References

Lowery, G., & Stover, T. S. (2001). *Managing Projects with Microsoft Project 2000*. New York: Wiley & Sons.

Shtub, A., Bard, A. F., & Globerson, S. (1994). *Project Management: Engineering, technology, and implementation.* Englewood Cliffs, NJ: Prentice Hall.

Technology Development Using Design For Six Sigma

The development of new technology is the fuel that powers the timely delivery of new products and new production processes. Within this portion of the total product development process, advances in performance and quality are actually born. Any company that does not view its R&D team as its strategic starting point for the development of quality and for laying the foundation for efficient cycle-time in product development is very much off base.

Product development encompasses two distinct categories in this text: the first is technology development, and the second is product design, sometimes referred to as **commercialization.** This text takes the position that the teams residing within research & development organizations have the best and earliest opportunity to develop the foundations for Six Sigma performance within the detailed processes used to generate new and leveraged technologies. If you were to ask, "Are there Six Sigma methods that can be deployed in the day-to-day activities within an R&D context?" the answer is a definitive yes! This chapter constructs an architecture for the deployment of such methods in the development and transfer of new and leveraged technologies.

Industry is now beginning to recognize that technology development projects should be largely independent from product commercialization projects. This is never an easy thing to do but if you want to minimize technical difficulties as you approach design certification and launch, this is precisely what must be done. It is risky to develop new technology within a product design project. There are big differences in what needs to be done and known during technology development and product commercialization. Technologies that are under development are immature and overly sensitive to sources of manufacturing, distribution, and customer use variability ("noise factors"). When they are forced into the high-pressure time line of a commercialization project, corners get cut and sensitivities go undetected. Immature, underdeveloped technologies can greatly slow the progress of the product due to their inability to integrate smoothly and function correctly in the presence of **noise factors,** or any source of variation from external, unit-to-unit, or deterioration sources. In a word, they are not "certified" for safe commercial or design application. Deploying new technologies prior to their certification is analogous to launching a new product design that has not been made "capable" from a manufacturing and assembly perspective. To the statistician, it is similar to running a process before one has proven it to be under a state of statistical control. We must quantitatively prove that a new technology is both robust and tunable before it can be used in a new product. By *robust,* we mean capable short-term performance as measured by Cp indices, even in the presence of noise factors. By tunable we mean capable long-term performance as measured by Cpk indices (Cpk is the measure of capability after the mean has shifted off target by "k" standard deviations). This chapter lays a foundation for the proper development and transfer of new technologies into a product commercialization project.

Technology development and transfer must be properly controlled and managed by facts generated using a system integration process we call *Critical Parameter Management* (CPM).

CPM is fully explained and developed in a later section. The proper place to initiate CPM in a business is during advanced product planning and technology development. If CPM is practiced at this earliest stage of product development, then a certified portfolio of critical functional parameters and responses can be rapidly transferred as a "platform" into the design of families of products during numerous product commercialization programs.

Critical Parameter Management is a system engineering and integration process that is used within an overarching technology development and product commercialization roadmap. We will first look at a roadmap (I^2DOV, "eye-dov") that defines a generic technology development process, and then move on to define a similar roadmap (**CDOV,** pronounced "see-dov") that defines a rational set of phases and gates for a product commercialization process. These roadmaps can be stated in terms of manageable phases and gates, which are useful elements that help management and engineering teams to conduct product development with order, discipline, and structure.

The I^2DOV Roadmap: Applying a Phase/Gate Approach to Technology Development

The structured and disciplined use of phases and gates to manage product commercialization projects has a proven track record in a large number of companies around the globe. A small but growing number of companies are using phases and gates within their R&D organizations to manage the development of new and leveraged technologies *prior to commercial design application.* In the next two chapters we will show the linkage between the use of an overarching product development process, deployment of a full portfolio of modern tools and best practices, and a well-structured approach to project management from R&D all the way to product launch.

With this triangle of structural stability depicted in Figure 5–1, one can drive for breakthroughs that deliver Six Sigma results because management, technology, and commercialization teams all know what to do and when to do it.

An Introduction to Phases and Gates

A phase is a period of time during which specific tools and best practices are used in a disciplined manner to deliver tangible results that fulfill a predetermined set of technology development requirements. We often use the term *breakthrough* when discussing results. What we mean by this term are results that are hard to get without using the kinds of tools and processes we are suggesting in this text. Six Sigma methods have a legacy of producing results that were elusive prior to its application.

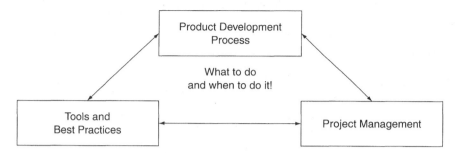

Figure 5–1 Critical Components in Successful Product Development

A gate is a stopping point within the flow of the phases within a technology development project. Two things occur at this stopping point:

1. A thorough assessment of deliverables from the tools and best practices that were conducted during the previous phase (specifically the high-risk areas that did not completely fulfill the gate requirements)
2. A thorough review of the project management plans (typically in the form of a set of PERT charts) for the next phase

Gates often focus on preventative measures to keep design flaws and mistakes from occurring in the next phase. Gates are also used to manage work-around plans to keep things moving forward after difficulties and problems have been identified. A well-designed gate review takes into account that some things will go wrong and enables the team to deal with these inevitabilities.

It is to be expected that some technology subsystems will go through a gate on time while other subsystems may experience unexpected difficulties in meeting the gate criteria. It is common for engineering teams to use a three-tiered system of colors to state the readiness of their subsystems to pass through a gate.

- *Green* means the subsystem has passed the gate criteria with no major problems.
- *Yellow* means the subsystem has passed some of the gate criteria but has numerous (two or more) moderate to minor problems that must be corrected for complete passage through the gate.
- *Red* means the subsystem has passed some of the gate criteria but has one or more major problems that preclude the design from passing through the gate.

Gate passage is granted to subsystems that are green. Gate passage is provisionally granted to subsystems that are yellow. Gate passage is denied to subsystems that are red. Time and resources are planned to resolve the problems; it is not unusual to see the time ranging between 2 and 6 weeks and in some cases longer. Work-around strategies are put in place to keep the whole project moving forward while the subsystems that have failed gate passage are corrected. This situation requires engineering managers to develop phase transition strategies that allow for early and late gate passage. The ideal state is when all subsystems pass all the gate criteria at the same time for every gate. This, of course, almost never happens. For complex projects it is foolish to expect anything other than a series of unexpected problems. For technology managers who prefer a preventative approach, the second order of business in developing a Phase/Gate process is the inclusion of a phase transition strategy that effectively deals with managing the "unexpected." The first order of business is structuring each phase with a broad portfolio of tools and best practices that help prevent unexpected design problems. This is done by focusing on the tools and best practices that use forward-looking metrics that measure functions (scalar and vector engineering variables) instead of counting quality attributes (defects and failures). The key is to prevent the need to react to quality problems later. Quality must be designed in to the new technologies by careful, disciplined use of the right tools and best practices that measure the right things to fulfill this "prevent strategy" from an engineering perspective. Parts IV through VII of this text discusses such tools and metrics.

Quality is certainly one of the primary requirements of a technology development project, but it is clear that the most efficient path to quality does not flow through a series of "quality metrics." They tend to drive "react" strategies as opposed to "prevent" strategies. We will discuss the antidote to the inefficiencies of quality metrics in the upcoming chapters on Critical Parameter Management.

There is broad acceptance that using an overarching product development process to guide technology development and product commercialization is the right thing to do. What is often overlooked is the value of a broad, balanced portfolio of tools and best practices that are scheduled and conducted within each phase using the methods of Project Management.

The best way to think about establishing the architecture of any product development process is to structure it as a macro-timing diagram to constrain "what to do and when to do it" across the entire product development life cycle. With such a macro-timing diagram established, the product development management team can then conduct the phases and gates using checklists and scorecards based on a broad portfolio of tools and best practices. The techniques of project management can then be used to construct balanced, micro-timing diagrams (often in the form of a PERT chart) that establish discipline and structure within the day-to-day use of the tools and best practices for each person on the product development team. We recommend PERT charts because they help define personal accountability for using the right tools and best practices on a day-to-day basis within the phases. They also greatly enhance the management of parallel and serial activities, work-around strategies, and the critical path tasks that control the cycle-time of each phase. Managing a technology development project in this manner helps assure that all the right things get done to balance the ever-present cost, quality, and cycle-time requirements.

Checklists

Checklists are brief summary statements of tools and best practices that are required to fulfill a gate deliverable within a phase. These actions are based on work everyone must do based on a broad portfolio of tools and best practices. You know you have a good checklist when it is easy to directly relate one or more tools or best practices to each gate deliverable. Another way of recognizing a good checklist is that it can be used to build project management PERT and Gantt charts for day-to-day management of tool use. Tools and best practices produce deliverables that can be tracked on a daily, weekly, or monthly basis and can be easily reviewed using a summary scorecard when a gate is reached. Checklists are used during project planning of tool use prior to entering the next phase. This is done to prevent problems in the next phase.

An example of a Six Sigma Tool and Best Practice Checklist follows.

Checklist of Phase 1 Tools and Best Practices:

Technology roadmapping

VOC gathering methods

KJ analysis

QFD/House of Quality—Level 1 (system level)

Patent analysis and exposure assessment

Market segmentation analysis

Economic and market trend forecasting

Business case development methods

Competitive benchmarking

Scientific exploration methods

Knowledge-based engineering methods

Math modeling (business cases, scientific, and engineering analysis)

Concept generation techniques (TRIZ, brainstorming, brain-writing)

Modular and platform design

System architecting methods

Upcoming chapters on tools have an extensive, more detailed set of checklists of step-by-step action to be done within each tool or best practice.

Scorecards

Scorecards are brief summary statements of deliverables from specific applications of tools and best practices. They are useful to help react to problems encountered within the previous phase.

Scorecards state what deliverable is to be provided, as seen in the example in Figure 5–2. They show who is specifically responsible for the deliverable so they can be held accountable for their behavior. The original due date for the deliverable is listed along with the actual date the deliverable is completed. The status of the deliverable in terms of completeness of fulfilling its requirements is documented in the Completion Status column. The actual, measured performance attained against the required criteria is documented in the Performance vs. Criteria column. A column for the recommendation for exiting the gate is also provided in the scorecard. A final column is used to attach notes to provide details concerning any positive or negative issues for the deliverables.

Phase 1 Invention & Innovation

Deliverable	Owner	Original Due Date vs. Actual Deliv. Date	Completion Status Red Green Yellow	Performance vs. Criteria Red Green Yellow	Exit Recommendation Red Green Yellow	Notes
Market Trend Impact Analysis			O O O	O O O	O O O	
Market Assessment			O O O	O O O	O O O	
Market Segmentation Analysis			O O O	O O O	O O O	
Market Competitive Analysis			O O O	O O O	O O O	
Market Opportunity Analysis			O O O	O O O	O O O	
Gap Analysis			O O O	O O O	O O O	
Long Range VOC Document			O O O	O O O	O O O	
Product Family Plan			O O O	O O O	O O O	
Technology Response Deliverables			O O O	O O O	O O O	
Technology Roadmap Document			O O O	O O O	O O O	
Patent Trend Impact Analysis			O O O	O O O	O O O	
Patent Position & Risk Assessment Document			O O O	O O O	O O O	
System & Subsystem Technology Houses of Quality			O O O	O O O	O O O	
Technology Requirements Documents			O O O	O O O	O O O	
Documented List of Functions That Fulfill the Technology Requirements			O O O	O O O	O O O	
Documented Inventions & Patent Strategy			O O O	O O O	O O O	
Portfolio of Candidate Technology Concepts That Fulfill the Functions & Requirements			O O O	O O O	O O O	
Documented Critical Parameter Database & Scorecards			O O O	O O O	O O O	
Phase 2 Project Mgt. Plans			O O O	O O O	O O O	

Figure 5–2 Example of a General Gate Review Scorecard

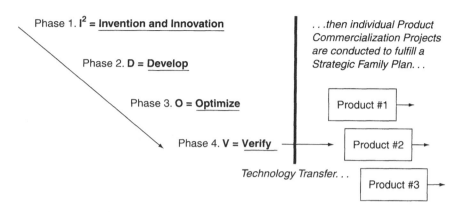

Figure 5–3 I²DOV Phases

The scorecard helps the gatekeepers manage risk as they attempt to keep the project on schedule. They can see the critical path items and deploy the resources at their disposal to solve key issues. The scorecard is a useful element in the process of Critical Parameter Management. The tools help identify critical parameters, and the scorecards highlight the progress in the development of the critical parameters.

No one technology development process architecture will meet the needs and priorities of all companies. Every company must define a product development process that fits its culture, business process, and technological paradigm.

Let's move on to review the details of the phases and gates of the **I²DOV** process. As we do this, we will begin to develop the Critical Parameter Management context that is helpful in development of Six Sigma performance.

The I²DOV phases and gates structure for technology development outlined in this chapter follows four discrete project management steps, as shown in Figure 5–3.

Beginning in Figure 5–4, we will define each of these phases and describe how they flow as part of the overarching product development process that leads to successful product launches.

I²DOV and Critical Parameter Management During the Phases and Gates of Technology Development

Figure 5–4 illustrates the major elements for **I²DOV** Phase 1: **Invention and Innovation** based on Market Segmentation and a Product Line Strategy.

Here we see a flow of actions that can be conducted with the help of a set of tools and best practices under the temporal constraint of the principles of project management. Typically a technology manager constructs a PERT chart to plan the timing of these activities as they relate to use of a balanced portfolio of tools and best practices.

Start Date: ___	Define business goals, business strategies, and mission	End Date: ___	Owner: _____
Start Date: ___	Define markets and market segments	End Date: ___	Owner: _____
Start Date: ___	Construct technology roadmaps and document technology trends	End Date: ___	Owner: _____
Start Date: ___	Gather and translate "over-the-horizon" voice of the customer	End Date: ___	Owner: _____
Start Date: ___	Define product line strategies (PLS) and family plans	End Date: ___	Owner: _____
Start Date: ___	Create technology Houses of Quality	End Date: ___	Owner: _____
Start Date: ___	Conduct technical benchmarking	End Date: ___	Owner: _____
Start Date: ___	Generate technology system requirements document	End Date: ___	Owner: _____
Start Date: ___	Define functions that fulfill the technology requirements	End Date: ___	Owner: _____
Start Date: ___	Enable, support, and conduct invention (generate new knowledge)	End Date: ___	Owner: _____
Start Date: ___	Enable, support, and conduct innovation (enhanced/leveraged reuse of existing knowledge)	End Date: ___	Owner: _____
Start Date: ___	Refine, document, and link discoveries, learning, and basic models to product line strategy	End Date: ___	Owner: _____
Start Date: ___	Transfer refined, linked, and documented knowledge into technology concept design	End Date: ___	Owner: _____
Start Date: ___	Generate technology concepts that fulfill the functions	End Date: ___	Owner: _____

Figure 5–4   Flow of Phase 1 of Technology Development

I²DOV Phase 1: Invention and Innovation

The specific issues related to each action item (start date for each action item, the action item, end date for each action) will be reviewed in this section. Each of these action items can be conducted by using a tool or best practice. The flow of actions can be laid out in the form of a PERT chart for project management purposes.

Define Business Goals, Business Strategies, and Mission

We generally view the invention and innovation phase as the beginning of technology development. It is also the beginning of the Critical Parameter Management process. Before any technology development work can begin, the foundations of the business must be well defined. The businesses goals, strategy, and mission need to be clearly established. Corporate leaders must document short-term and long-term business and financial targets that constitute their goals. They must have a documented plan that will achieve the goals. This constitutes their strategy. The leaders of the business must work with the board of directors to document a mission statement that defines their core values that provide boundaries for the businesses strategic framework and goal setting process. Everyone must understand why the business exists.

Define Markets and Market Segments

At this step the market and technology teams discover where to gather the longer-term, "over-the-horizon" voice of the customer. Defined markets and market segments are essential to guiding the advanced product planning team as it goes out to discover what customers need so team members can apply their core competencies properly. Once it is clear what market segments exist, the advanced product planning team can conduct more specific market-focused voice of the customer gathering activities. Technologies may be developed to apply across or within market segments. It is important to define when technology needs are synergistic across market segments because this will clearly signal to the R&D leaders that a major platform opportunity exists. If a technology platform can be developed to serve "across market" commercialization requirements, great economies of scale can be attained. This scenario leads to the single investment of capital that can deliver profit from numerous family lines of products. To a lesser extent, profit amplification can occur within a market segment by developing a technology platform that can serve within market product family plan requirements. Market segmentation is a great help in product family planning. Without it companies don't know when or how to divide R&D resources to help attain market share and when to integrate resources to attain economies of scale in the use of their R&D resources.

Construct Technology Roadmaps and Document Technology Trends

Technology roadmaps are concurrently developed with the market segmentation documents. They help the business and technical leaders structure what technology development projects should be undertaken during the first phase of technology development.

Technology roadmaps document past and current technology patents, cost structures, performance, and development cycle-times. They also develop future technology forecasts within and outside the company. A good technology roadmap will contain the following three items:

1. **Historical technologies** that have completed their life cycles. They should contain documented development cycle-times, costs vs. generated revenue streams, and the details of their functional performance during their life cycles.

2. **Current technologies** that are yet to complete their life cycles. Their performance, cost, and development cycle-times should all be documented.

3. **Future technologies** should be scoped for estimated cost, development cycle-time, and functional performance capabilities. These projections should be based both on technology trend forecasting and the long-range voice of the customer assessment. Latent customer needs are estimated within this form of documentation.

Gather and Translate "Over-the-Horizon" Voice of the Customer

The use of voice of the customer data is essential for circumspect advanced product planning. Pure invention and innovation that is not in some way linked to customer needs can produce technology that is protected by a deep portfolio of uncommercializable patents. This is referred to as *design for patentability.* If your company is not careful, it can produce what is called a *6-foot-deep patent portfolio.* This occurs when unbridled R&D organizations develop technology independent from the inbound marketing group and the advanced manufacturing organization. The company becomes rich in patents that are incapable of being successfully transferred into revenue-bearing commercialization projects. A good question to ask about your company's patent portfolio is, "What is the revenue bearing product hit rate based on the patent portfolio?"

It is essential that the technologists participate in gathering, ranking, and translating customer needs as part of the advanced product/technology planning process. This helps them be circumspect as they seek balance between their internal inventive activities and the development of technologies that fulfill explicit, external market needs. However, it is entirely possible to follow the voice of the customer into a business failure (Christenson, 2000). Invention of new technologies that surpass known customer needs, often called *latent needs,* will often play a key role in new product development.

Part of the technology development process includes the invention and discovery of unique phenomena that may be able to be refined into a safe and mature technology independent of any known customer need. The focus in this case is on "the observation, identification, description, experimental investigation and theoretical explanation of natural phenomena." This is the definition of science from the American Heritage Dictionary and is largely what embodies the process of invention. The voice of the customer data may or may not play a strong role in the "discovery of science" process. Scientists and engineers who participate in the invention process may use VOC data to guide their thinking but many do not like any constraints on their thought processes. There are really only two choices to make during technology development:

1. Recognize market needs and then drive the inventive processes so that the discovery of physical law is directly targeted to be commercially useful according to an advanced product planning strategy.

2. Just invent in an unrestrained fashion and if what you come up with somehow is commercially useful, *Bravo! You have good luck.*

Define Product Line Strategies (PLS) and Family Plans

Long-range customer needs are initially defined in broad terms and are then refined into product line technology requirements based on market segmentation. The R&D organization should help gather and process this customer information. Once the long-range VOC data is used to define market segment needs, additional VOC data can be acquired to develop specific product line or family requirements. R&D teams use documented product line requirements as the basis for developing platform technologies, system and subsystem technologies that can be used as a basis for a family of products. The platforms are intended to fulfill the voice of the customer requirements over a family of commercial products. Requirements for individual technologies that may or may not support a platform are also identified and documented. Unfortunately, family planning that is tactically fulfilled by a technology development process is largely nonexistent in many modern companies.

Create Technology Houses of Quality

The VOC data for a specific market segment will be used to document a product line strategy and a series of embryonic family plans. The specific needs projected by the long-range VOC data and the technology roadmaps will provide the balanced input needed to construct a series of technology Houses of Quality. The balance comes from the integration of the voice of the customer and the voice of technology.

A technology House of Quality is created using a matrix approach while translating technology *needs* into technology *requirements*. Once a group of technology requirements is defined for a family of products, the R&D team can begin the highly technical task of defining and developing physical functions that fulfill the requirements. These physical functions lie at the heart of conceptual development of technology architectures that can ultimately become a platform. A technology House of Quality is the foundation upon which a platform is conceived and architected.

Conduct Technology Benchmarking

A major portion of a technology House of Quality is dependent upon three forms of comparative information:

1. Projected customer opinions and perceptions of existing and new trends in technology that will drive fulfillment of their needs in product performance.
2. Projected technology progressions based on current and foreseeable technological innovations. Frequently these are forecasted in the latest patents and invention disclosures.
3. Current technology benchmarking that assesses the state of the art in technology being applied in new products that are currently on the commercialization scene.

The technology House of Quality will use the projections and benchmark data to help assign priorities and rank criticality during the translation of technology needs (family plan needs) into technology requirements ("hard" technical metrics associated with need fulfillment).

Generate Technology System Requirements Document

Once the technology House of Quality is built and filled out, it forms the basis for the development of a formal document that contains the technology requirements, which the appropriate teams of technologists within the R&D organization will fulfill. Not all requirements have to come from the technology House of Quality. The HOQ typically focuses attention on those things that are new, unique, or difficult. Some technology requirements may fall into less-demanding categories and obviously still need fulfillment even though they were not a point of focus from the perspective of the HOQ. For example, a technology may need to fulfill some readily known and easily fulfilled regulatory or safety requirements. These requirements need to be integrated along with the new, unique, and difficult requirements that flow from the HOQ.

A new system of technology made up of integrated subsystem technologies is called a *platform*. The platform is considered a system. Platform development is well served by approaching it as one would any engineered system. Here we see strong similarities between a system or product requirements document and the platform requirements document. Later in text, an entire chapter is dedicated to the detailed process of system engineering in the context of DFSS.

If one is focusing the technology development activity to just one or more subsystems, the requirements document is still developed but just for the subsystem technology. In both cases a balanced portfolio of tools and best practices is used to fulfill these requirements.

Define Functions That Fulfill the Technology Requirements

The real technical work of the R&D scientists and engineers begins at this step. Here actual physical and engineering principles are identified to fulfill the requirements. Specific functions are described to clearly identify the underwriting physics that define how the requirements will be fulfilled. This is the time for technology development teams to define what must be "made to happen" to enable the function behind the technology. The actual form and fit concepts come later during the final stages of concept engineering. What we are talking about here are functions independent of how they are attained. A force may be required, an acceleration, a displacement, a reaction—whatever dynamic transformation, flow, or state changes of mass, energy, and information (analog or digital logic and control signals) necessary to satisfy a requirement. Once the team understands the nature of the fulfilling functions, it can enter into conceptual development with real clarity of purpose. The team can innovate freely to develop multiple architectures, or "forms and fits," to fulfill the functions that in turn fulfill the requirements.

Enable, Support, and Conduct Invention (Generate New Knowledge)

If a requirement demands a function that is not found in the current physics or engineering knowledge base, then a new invention may be required to meet the need. It is far more common that innovation is used to integrate and extend applications of known physical principles and engineering science. It is extremely important to note that the risk associated with this set of activities forces us to recommend that one stay clear of conducting technology development within product commercialization processes. When you get to the point of a required invention, it is likely that developing the right function may be extremely difficult or impossible. Invention can burn up a lot of time—time you simply do not have in a product commercialization project that is tied to a launch date. When this happens to a "design" that is really just an uncertified technology that has been prematurely inserted into a commercialization project, the project schedule is usually compromised and profound implications arise for meeting the market window of the product. This form of "untamed technology" holds the impending revenues hostage! Our recommendation is not to mix uncertified technologies into your product commercialization projects.

Enable, Support, and Conduct Innovation (Enhanced/Leveraged Reuse of Existing Knowledge)

Innovation is by far the most popular form of developmental activity that characterizes R&D processes as we enter the 21st century. Some call it applied research and development. The operative word is *applied*. What that word means is that there is a preexisting technology knowledge base available for reuse. This documented knowledge is available to be applied by either using it in a novel way or by extending or advancing the state of the art. Most baccalaureate and masters level engineers are trained to extend the state of the art while most doctoral level engineers and scientists are prepared to create and discover new knowledge. To some, this may not be a clear point of distinction, but those of us who have been in a situation where time is of the essence know that invention of new knowledge is simply not a timely option. We must take what is known and use it to get the technology developed to meet the needs of our colleagues in product design.

Refine, Link, and Document Discoveries, Learning, and Basic Models to PLS

Once new knowledge and reused, but newly extended, knowledge have been developed, it is time to complete the construction of an important bridge. That bridge links what the team has documented as a knowledge base to a set of well-defined application requirements related to the family of products that will flow from the product line strategy. This is how we integrate the practical results being born out of the technology strategy with the product line strategy, ultimately fulfilling the business strategy.

The long-range voice of the customer and our benchmarked voice of technology lead us to construct a technology House of Quality that leads us to a technology requirements document that leads us to define technology functions that enabled the fulfillment of these diverse needs. The knowledge of "functionality fulfillment" must be held up to the product line strategy in a technology review and answer the questions about whether and how we can get there from here. This is a test of feasibility prior to investing a lot of time and resources in concept generation. The business leaders should see a rational flow of fulfillment at this point. If it is not clear, then further work needs to be done to bring all the fulfillment issues into alignment with the product line strategy.

Transfer Refined, Linked, and Documented Knowledge into Technology Concept Design

It is time to move all the knowledge forward into a format that enables the next stage of technology development known as technology concept development. The functions and requirements must be documented so that a rational set of concept generation, evaluation, and selection criteria can be rapidly developed. If there is ambiguity about what the concepts must do and fulfill, then the concepts will inevitably suffer from competitive vulnerability and will be functionally inferior. Concept selection is yet another test of feasibility, but this time it has to do with both functional and architectural potential to fulfill the technology requirements.

Generate Technology Concepts That Fulfill the Functions

The technology team now focuses the full weight of its individual and corporate development skill to generate candidate architectures to fulfill the functions and requirements that have been tied back to the product line strategy.

For integrated subsystems of technologies, the goal of the team is to first define a number of candidate system architectures, or platforms. Platforms are used to produce a number of products from one investment of capital within the R&D organization. Platforms are necessary to make efficient use of business capital to fulfill product line strategies. It is hard to fulfill long-range business strategies through R&D projects that deliver one subsystem at a time for refitting onto aging products. It is fine if a company uses a mixed technology development strategy of delivering some new subsystems to extend existing product families while new platforms are on the way. Eventually a new platform must emerge to sustain the business as markets and technologies evolve. In some business models technology turnover is so rapid that the notion of a reusable platform does not apply.

Numerous candidate subsystem or platform concepts need to be generated. If a single concept dominates from one strong individual or "clique" within the R&D organization, the likelihood of the identifying best possible concept is not going to be attained. The entire team should break up and generate several concepts as individuals. After many concepts are generated, then the team convenes to evaluate and integrate the best ideas into a superior "super-concept" that is low in competitive vulnerability and high in functional feasibility. This state of superiority is achieved through consensus by the technology development team using concept selection criteria from the previously developed functions and requirements.

Gate 1 Readiness

At this early point in technology development a business should have a proprietary, documented body of science, documented customer needs for future product streams, and the embryonic technical building blocks that can be developed into candidate technologies. The term *fuzzy front end* is appropriate to characterize the nature of this phase of technology development. We don't have mature technologies yet—just the science, customer needs, and candidate concepts from which we hope to derive them.

With a viable cache of candidate technical knowledge and a well-planned commercial context in which to apply that knowledge, we approach a rational checkpoint to assess readiness to move on into formal technology development.

Whether technology is built on completely new knowledge or integrated from existing knowledge (something we call *leveraged technology*), critical parameter relationships will need to be identified, stabilized, modeled, documented, desensitized to sources of variation, and integrated across functional and architectural boundaries. Many of the delays that are incurred during product commercialization are due to technologies that have incomplete understanding and a lack of documentation of critical parameter relationships. As far as this text is concerned, no difference is made between newly invented technologies or innovations that reuse existing technologies. Both situations require structured, disciplined R&D teamwork to safely deliver "certified" technologies to the commercialization teams. Once a certified technology is in the hands of a strong commercialization team, the team can rapidly transform it into capable designs. These issues must be foremost in the minds of the decision makers as the technology development process flows through its phases and gates.

At Gate 1 we stop and assess the candidate technologies and all the data used to arrive at them. The tools and best practices of Phase 1 have been used, and their key results can be summarized in a Gate 1 scorecard (Figure 5–5). A checklist of Phase 1 tools and their deliverables can be reviewed for completeness of results and corrective action in areas of risk. Figure 5–6 defines the processes of I^2DOV Phase 2.

General Checklist of Phase 1 Tools and Best Practices

Technology roadmapping

VOC gathering methods

KJ analysis

QFD/House of Quality—Level 1 (system level)

Patent analysis and exposure assessment

Market segmentation analysis

Economic and market trend forecasting

Business case development methods

Competitive benchmarking

Scientific exploration methods

Knowledge-based engineering methods

Math modeling (business cases, scientific and engineering analysis)

Concept generation techniques (TRIZ, brainstorming, brain-writing, etc.)

Modular and platform design

System architecting methods

Phase 1 Invention and Innovation

Deliverable	Owner	Original Due Date vs. Actual Deliv. Date	Completion Status Red Green Yellow	Performance vs. Criteria Red Green Yellow	Exit Recommendation Red Green Yellow	Notes
Market Trend Impact Analysis			O O O	O O O	O O O	
Market Assessment			O O O	O O O	O O O	
Market Segmentation Analysis			O O O	O O O	O O O	
Market Competitive Analysis			O O O	O O O	O O O	
Market Opportunity Analysis			O O O	O O O	O O O	
Gap Analysis			O O O	O O O	O O O	
Long Range VOC Document			O O O	O O O	O O O	
Product Family Plan			O O O	O O O	O O O	
Technology Response Deliverables			O O O	O O O	O O O	
Technology Roadmap Document			O O O	O O O	O O O	
Patent Trend Impact Analysis			O O O	O O O	O O O	
Patent Position & Risk Assessment Document			O O O	O O O	O O O	
System & Subsystem Technology Houses of Quality			O O O	O O O	O O O	
Technology Requirements Documents			O O O	O O O	O O O	
Documented List of Functions That Fulfill the Technology Requirements			O O O	O O O	O O O	
Documented Inventions & Patent Strategy			O O O	O O O	O O O	
Portfolio of Candidate Technology Concepts That Fulfill the Functions & Requirements			O O O	O O O	O O O	
Documented Critical Parameter Database & Scorecards			O O O	O O O	O O O	
Phase 2 Project Mgt. Plans			O O O	O O O	O O O	

Figure 5–5   General Scorecard for I²DOV Phase 1 Deliverables

Start Date: ___	Generate and refine technology concept evaluation criteria	End Date: ___	Owner: _____
Start Date: ___	Evaluate and select superior technology concepts	End Date: ___	Owner: _____
Start Date: ___	Analyze, characterize, model, and stabilize nominal and tunable performance of superior technology	End Date: ___	Owner: _____
Start Date: ___	Verify and prepare superior technology concepts to take into optimization	End Date: ___	Owner: _____

Figure 5–6   Flow of Phase 2 of Technology Development

I²DOV Phase 2: Develop Technology Concept Definition, Stabilization, and Functional Modeling

Figure 5–6 illustrates the major elements for I²DOV Phase 2: **Develop** Technology Concepts. This phase of technology development is designed to produce three distinct deliverables:

1. **Superior technology and measurement system concepts** that are worth taking on into the technology optimization phase
2. **Underlying math models** (deterministic and statistical) that characterize the nominal and tunable performance (baseline) of the new technology concepts and their measurement systems
3. **Prototype hardware, firmware, and software** that provides a physical model of the new technology subsystem concepts and their measurement systems

The nature of this second phase of technology development is one of characterization of nominal and tunable performance relationships between input or independent (x) variables and output or dependent (Y) variables. This phase require all three major forms of modeling to be conducted:

1. Analytical or mathematical models
2. Graphical and spatial models (2D and 3D sketches, layouts, and CAD files)
3. Physical models (hardware, firmware, and software prototypes)

In standard Six Sigma terminology this is the point in the process where the technology development team quantifies the (Y) variables as a function of the (x) variables. In the vocabulary of Critical Parameter Management, the (Y) variables are called **critical functional responses** and the (x) variables are candidate **critical functional parameters.** Sometimes the (x) variables are not totally independent. At times (x) variables have codependencies and/or correlated effects on a (Y) variable existing between them. The former are referred to in statistical jargon as *interactions*. Additionally, the (x) variables may not have a linear relationship with the (Y) variable(s). It is important to be able to quantitatively state whether the relationship is linear, nonlinear, and the extent of the nonlinearity if proven to exist. Lastly, we need to know which (x) variables have an ability to adjust or tune the mean value of (Y) without having a large effect on the standard deviation of the (Y) variable. The statistical variable used to quantify this relationship is the coefficient of variation (COV): (σ/y_{bar}) $\times$ 100. The COV is also the basis for an important measure of robustness, the signal-to-noise (S/N) ratio, during the optimization phase.

The (x) variables are often broken down into two distinct categories: control factors and noise factors. Control factors are (x) variables that can readily be controlled and specified by the technology development team. Noise factors are (x) variables that are not easily controlled or specified by the engineering team. Sometimes the (x) variables are impossible to control. This phase of technology development provides the first cut at distinguishing control factors from noise factors. The (x) factors that are found to be marginally controllable or extremely difficult or expensive to control can be isolated for special attention. If the situation is bad enough, the technology concept may need to be refined or abandoned. If the concept is found to be nominally stable and tunable, the team can treat these poorly behaved control factors as noise factors during the optimization phase of technology development.

The multifunctional team that initiates technology concept development is typically made up of some or all of the following personnel:

- Scientists from various technical domains
- Development engineers from various technical domains
- Advanced manufacturing engineers
- Service specialists
- Product planning and inbound marketing specialists
- Supply chain development specialists
- Health, safety, environmental, & regulatory specialists
- Technology development partners from industry, academia, and government
- Selected customers

Generate and Refine Technology Concept Evaluation Criteria

Concepts must be evaluated against a set of criteria that is directly related and traceable to the technology requirements. In fact, the concept evaluation criteria can literally be the technology requirements with little or no refinement from the statements found in the technology requirements document. It is perfectly acceptable to develop the evaluation criteria prior to generating the concepts. Whether one uses the technology requirements document or the typically broader and more detailed evaluation criteria to develop technology concepts is up to the team.

Concept evaluation criteria consist of all key requirements that must be satisfied by the technology concept. These criteria must comprehensively represent VOC, VOT, and product line strategy requirements. One of the reasons KJ analysis (affinity diagramming) is so important is that it helps state VOC needs and VOT issues in a hierarchy of relationships and in a rank order of importance. This information will help the technology development team form a comprehensive list of critical evaluation criteria. All concepts will be evaluated against a "Best-in-

Class" benchmark datum concept. It is not uncommon for one or more of the new candidate concepts to bring new criteria to the forefront. This new criteria springs from innovation and newness embodied in the candidate concepts. Care should be taken not to miss these new criteria that help discriminate and refine conceptual feasibility during the Pugh concept evaluation process.

Evaluate and Select Superior Technology Concepts

At this point the **Pugh concept selection process** is used to evaluate the set of candidate concepts against the datum concept. The Pugh process uses an iterative converging–diverging process to evaluate concepts. The ranking of each candidate concept vs. the datum is nonnumeric and uses simple $(+)$ better than the datum, $(-)$ worse than the datum, or (S) same as the datum, ranking designations against the evaluation criteria. The iterative converging–diverging process is used to weed out weak concept elements, enabling the integration of hybrid, super-concepts from the remnants of candidates that have strong elements against the datum. Iterative "pulses" of concept screening and synthesis build up a small, powerful set of newly integrated hybrid concepts that typically have the capacity to equal or surpass the datum relative to the superior fulfillment of the evaluation criteria (see Chapter 17 on the Pugh process).

Analyze, Characterize, Model, and Stabilize Nominal and Tunable Performance of Superior Technology

Once a superior concept or two have been defined from iterations through the Pugh concept selection process, the team is ready to conduct an extensive technical characterization. It is not a bad idea to have a backup concept as a safety valve in case the superior concept possesses a fatal flaw that was missed during the Pugh process. The superior concepts are only proven to be low in competitive vulnerability and high in technological superiority relative to the datum. They are worth the investment required to certify the new technology's nominal and tunable performance prior to optimization. Once in a while a superior concept passes the concept evaluation process only to be found unusable under the deeper analytical scrutiny of this next step in the development of the technology. As you can see, Phase 2 is crucial for weeding out technologies that are often "time bombs" when allowed to creep into a product commercialization process without proper development and scrutiny. Conducting failure modes and effects analysis on the superior concepts at this point is extremely valuable in exposing risks.

The analysis and characterization must include the development of all key Y as a function of x relationships. As previously mentioned each relationship must be quantified in terms of nominal and tunable performance. Linear and nonlinear relationships must be defined. Interactions between control factors must be identified. If these relationships are a problem, the interactive relationships must be reengineered so that the codependency between control factors will cause severe variation control problems later in the product commercialization process. This reengineering to survive interactivity between control factors, called *design for additivity,* can first be approached in Phase 2 of technology development and finalized in the Phase 3 optimization activities. In Phase 3 we add the additional characterization of interactions between control factors and noise factors, the main focus of robust design.

The most important knowledge that must come from this phase of technology development is the quantification of the **ideal/transfer functions** that underwrite the basic performance of the technology. In the next phase we will evaluate the controllable (x) variables that govern the performance and adjustability of the ideal/transfer function for their robust set points in the presence of the uncontrollable (x) variables known as noise factors. If we have incomplete or immature forms of the "Y is a function of x" relationships and fail in our attempt to attain robustness, we can pass along yet another "time bomb" for our peers in product commercialization. By "time bomb" we mean a technology that gets integrated into a product system and "blows up" by producing unacceptable or uncontrollable performance with little or no time left to correct the problem. Technology that is poorly developed and characterized is a major reason why product delivery schedules slip.

Certify and Prepare Superior Technology Concepts to Take into Optimization

It is challenging to know when a team has enough knowledge developed to state that a new technology is truly ready to go forward into the optimization phase of technology development. Let's look at a list of evidence that could convict us in a court of law of having conducted a thorough characterization of a technology's nominal and tunable performance.

1. Math models that express the "ideal function" of the technology:
    a. First principles math models
        i. Linear approximation of $Y=f(x)$
        ii. Nonlinear approximation of $Y=f(x)$
    b. Empirical math models
        i. Linear approximation of $Y=f(x)$
        ii. Interactions between control factors
        iii. Nonlinear approximation of $Y=f(x)$
2. Graphical models
    a. Geometric/spatial layout of the subsystem and platform technologies
    b. Component drawings in enough detail to build prototypes
3. Physical models
    a. Prototype subsystems
    b. Prototype components
    c. Measurement systems
        i. Transducers
        ii. Meters
        iii. Computer-aided data acquisition boards and wiring
        iv. Calibration hardware, firmware, and software
4. Critical Parameter Management data
    a. Mean values of all critical functional responses
    b. Baseline standard deviations of all critical functional responses (without the effects of noise factors)

    c. Cp of all critical functional responses (without the effects of noise factors)

    d. Coefficient of variation (COV) of all critical functional responses (without the effects of noise factors)

    e. List of control factors that have a statistically significant effect on the mean of all critical functional responses, including the magnitude and directionality of this effect

    f. List of control factors that have a statistically significant effect on the standard deviation of all critical functional responses, including the magnitude and directionality of this effect (without the effects of noise factors)

    g. List of control factors that have a statistically significant effect on the coefficient of variation (COV) of all critical functional responses (without the effects of noise factors)

    h. List of control factors that have statistically significant interactions with one another, including the magnitude and directionality of this effect

    i. List of CFRs that have statistically significant interactions (correlations) with one another at the subsystem and system level (including the magnitude and directionality of this effect)

We must be able to answer a simple question: If we change this variable, what effect does that have on all the critical functional responses across the subsystem? We have to get this nominal performance knowledge first, in the absence of statistically significant noise factors. We will refine this knowledge in Phase 3 where the major emphasis is on intentional stress testing of the baseline subsystem technology with statistically significant noise factors. But before we can optimize robustness of the subsystem technology, we must characterize its control factors for their nominal contributions to "Y is a function of x" relationships.

We must be able to account for critical correlations, sensitivities, and relationships within each new subsystem technology under nominal conditions. If we lack knowledge of these fundamental, critical relationships, the technology is risky due to our partial characterization of its basic functional performance. The development team has no business going forward until these things are in hand in a summary format for key business leaders and managers to assess. If they don't assess this evidence, they are putting the future cycle-time of the product commercialization projects at risk.

Gate 2 Readiness

If technology is allowed to escape Phase 2 with anemic characterization, it is almost a sure bet that someone down the line will have to pay for the sins of omission committed here. In a very real sense the capital being invested in the new technology is being suboptimized. A great deal more capital will have to be spent to make up for the incomplete investment that should have occurred right here. You can either pay the right amount now (yes, it always seems a high price to pay) and get the right information to lower your risks or be "penny wise and pound foolish" and pay really big later when you can least afford it in both time and money! Phase 2 of technology development is truly the place to rigorously practice an old adage: Do it right the first time.

The following checklist outlines Phase 2 tools and best practices, while Figure 5–7 offers a Phase 2 scorecard. Then we move on to Phase 3 of I^2DOV in Figure 5–8.

General Checklist of Phase 2 Tools and Best Practices

Pugh concept selection process

Knowledge-based engineering methods

Math modeling (functional engineering analysis)

Concept generation techniques (TRIZ, brainstorming, brain-writing)

Measurement system development and analysis

Hypothesis formation and testing

Descriptive and inferential statistics

Design of experiments

Regression analysis

Response surface methods

FMEA, fault tree analysis

Modular and platform design

System architecting methods

Prototyping methods (e.g., soft tooling, stereo lithography, early supplier involvement)

Cost estimating methods

Risk analysis

Critical Parameter Management

Project management methods

I^2DOV Phase 3: Optimization of the Robustness of the Subsystem Technologies

Figure 5–8 illustrates the major elements for I^2DOV Phase 3: **Optimize** subsystems.

With the approval to enter the third phase of technology development, the new technologies have been proven to be worth the further investment required to optimize their performance. Optimization of a new technology focuses on two distinct activities:

1. **Robustness optimization** of the critical functional responses (CFRs) against a generic set of *noise factors* that the design will likely see in a downstream product commercialization application (per the deployment schedule based on the family plan strategy)
2. **Certification of adjustment factors** that have the capability to place the mean of the technology's CFRs on to a desired target within the dynamic range specified by the application requirements from the family plan

Phase 2 Develop Technology

Deliverable	Owner	Original Due Date vs. Actual Deliv. Date	Completion Status Red Green Yellow			Performance vs. Criteria Red Green Yellow			Exit Recommendation Red Green Yellow			Notes
Project Cost Magnitude Estimate			○	○	○	○	○	○	○	○	○	
Project Type (Platform or Subsystem Technology Development)			○	○	○	○	○	○	○	○	○	
Documented & Verified Analytical & Empirical Models (Ideal Functions)			○	○	○	○	○	○	○	○	○	
Documented Empirical Data to Support the Verification of the Analytical Models			○	○	○	○	○	○	○	○	○	
Documented Critical Functional Responses & Data Acquisition Means with %R&Rs & P/T Ratios Certified			○	○	○	○	○	○	○	○	○	
Documented Ideal Functions of the Superior Technology Concepts			○	○	○	○	○	○	○	○	○	
Documented FMEA Analysis			○	○	○	○	○	○	○	○	○	
Documented Dynamic Adjustment Factors for Platforms & Singular Technologies			○	○	○	○	○	○	○	○	○	
Documented Critical Parameter Database & Scorecards												
Documented Critical Parameters for Manufacturing Process Technologies			○	○	○	○	○	○	○	○	○	
Risk Assessments for Functions and Parameters that are unstable or overly sensitive			○	○	○	○	○	○	○	○	○	
Prototype Hardware, Firmware, & Software			○	○	○	○	○	○	○	○	○	
Phase 3 Project Mgt. Plans			○	○	○	○	○	○	○	○	○	

Figure 5-7 A General Scorecard for I²DOV Phase 2 Deliverables: Develop Technology Concepts

Start Date: ___	Develop subsystem noise diagrams and platform noise maps	End Date: ___	Owner: _____
Start Date: ___	Conduct noise factor experiments	End Date: ___	Owner: _____
Start Date: ___	Define compounded noises for robustness experiments	End Date: ___	Owner: _____
Start Date: ___	Define engineering control factors for robustness dev. experiments	End Date: ___	Owner: _____
Start Date: ___	Design for additivity and run designed experiments	End Date: ___	Owner: _____
Start Date: ___	Analyze data, build predictive additive model	End Date: ___	Owner: _____
Start Date: ___	Run verification experiments to certify robustness	End Date: ___	Owner: _____
Start Date: ___	Document critical functional parameter nominal set points and CFR relationships	End Date: ___	Owner: _____

Figure 5–8   Flow of Phase 3 of Technology Development

The superior technology concepts have been converted into prototype form and are proven to possess stable, tunable, functional behavior under nominal conditions within the comparative context of both analytical and empirical models.

The nominally characterized, conceptually superior technology prototypes are now refined to have the capacity to be evaluated as **robustness test fixtures.** Robustness test fixtures are made up of adjustable or easily changeable engineering parameters, commonly called *control factors.* These prototype fixtures are used to evaluate critical functional responses at various levels of control and noise factor set points. Their purpose is to investigate and measure interactions between control and noise factors. They are heavily instrumented to measure both control and noise factor set points (inputs) and critical functional responses (outputs). Noise factors can be changed during this phase of technology development. The goal in evaluating such fixtures is to identify control factor nominal set points that leave critical functional responses minimally sensitive to the debilitating effects of compounded noise factors.

The superior concepts are certified as "safe" technologies into which additional resources can be invested to make them insensitive to commercial noise factors. Many R&D organizations fall short of thorough noise testing by just evaluating new technologies in the presence of "lab noise." Lab noise is not anything like use noise. Lab noise is usually random and not stressful. Use noise is many times stronger than lab noise and is not random. This is yet another source of risk that promotes lengthy schedule slips when partially robust and untunable technology is delivered into a commercialization project.

Attempting to make poorly behaved and marginally characterized technology robust later in product design can be a waste of time and money. This is why Genichi Taguchi and others recommend the development of dynamic technology based on well-characterized ideal functions, as we discussed in Phase 2. The certified math models represent the ideal functions to which Taguchi so frequently refers. As a rule, robustness optimization activities should not come before the development of some form of credible analytical model. How rigorous and detailed the model should be is a matter of healthy debate that is based on economics as well as the application context. The rule is don't waste your time trying to make a design that is unstable under nominal conditions and robust under stressful conditions. Only stable, well-characterized designs are worth taking into the robustness optimization process.

Review and Finalize CFRs for the New Subsystem Technologies

Before starting any optimization activities, the development team must have a well-defined set of critical functional responses. In addition, each CFR must have a measurement system defined with its capability documented in terms of repeatability and reproducibility, known as an *R&R study,* and precision-to-tolerance ratio, known as a *P/T ratio.* To produce viable results, continuous variables that are scalar or vector measures of functions that directly fulfill requirements must be used. CFRs are what we measure and make "robust" when quantifying the effects of control factor and noise factor interactions. We prefer not to count defects but rather to measure functions within and across the integrated subsystems being developed as new platform technologies. This assures we measure fundamental Y outputs as we change fundamental x inputs.

During optimization, we assess the efficiency of the transformation, flow, or change of state of mass, energy, and information (logic and control signals that provide information about the flow, transformation, or state of the mass and energy of the subsystems). The efficiency is measured under the conditions of intentionally induced changes to both control and noise factors. To measure the flow, transformation, and state of these physical parameters we must move beyond a counting mentality to a measurement context. Counting defects and assessing time between failures, while being measures of "quality," are reactive metrics that are only useful after the function has deteriorated. *We need to measure impending failure.* This requires the measurement of functions in the presence of the noise factors that cause them to deteriorate over a continuum of time. We simply cannot wait until a failure occurs. This preventative strategy leads us to our next topic—the description of the activities that focus the development team's attention on the "physics of noise" that disrupt the flow, transformation, or state of mass, energy, and information. This is the science of reliability development. We value reliability prediction and evaluation, but they have little to do with the actual development of reliability.

Develop Subsystem Noise Diagrams and Platform Noise Maps

Noise has been described as any source of variation that disrupts the flow, transformation, or state of mass, energy, and information. More specifically noise factors break down into three distinct categories:

1. External sources of variation
2. Unit-to-unit sources of variation
3. Deterioration sources of variation

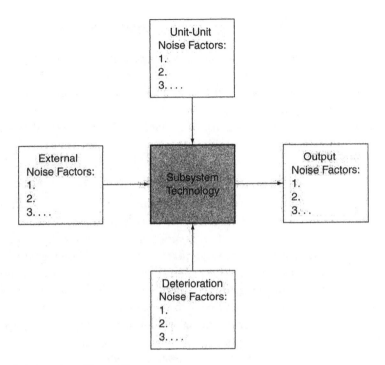

Figure 5–9 A Subsystem Noise Diagram

It is extremely useful to create a noise diagram that documents specific noise factors that fall under each of the categories shown in Figure 5–9.

The cross-functional development team, and downstream advisors from design, production, service/support organizations, and selected customers must identify all the noise factors that potentially have a statistically significant effect on each of the subsystem critical functional responses. Once a subsystem has a noise diagram constructed for each CFR, linkage can be established between each of the subsystem noise diagrams. It is important to map noise transmission paths within and between subsystem technologies. When two or more subsystem technologies are integrated to form a platform or some sort of higher-level system, then we must characterize how noise from one subsystem affects the others. We also need to characterize the flow of noise from the platform out to other subsystems that may connect or interact with it. You can think of an engine as a platform made up of various subsystems, with the platform itself assembling into a car or truck or whatever type of product is being developed. If noise transmits beyond the platform to reach a customer, service engineer, or other external constituency, we must document that potential negative relationship. When we account for these paths of noise transmission, we begin to understand how to alter the control factors to help guide the CFRs into a state of robustness against the noise. Insensitivity to noise is another way of communicating the meaning of robustness. If you don't have a comprehensive, systemwide noise map, composed of subsystem noise diagrams, how can you possibly begin the task of making a new subsystem technology or platform robust and know when you have finished?

Conduct Noise Factor Experiments

We just mentioned that subsystem noise diagramming and system noise mapping documents noise factors that might be statistically significant. We can't tell whether any one of the noise factors is statistically significant (*if it has an effect on the CFR that is statistically distinguishable from random events within a given confidence interval in the functional environment of the subsystem technology or platform*) by sitting around a table talking about them in a conference room. It is imperative that the development team design an experiment to quantify the magnitude and directional effect of those noise factors selected for experimental evaluation. Failure modes and effects analysis can be used to help identify and screen noise factors for inclusion in the noise experiment.

A **noise experiment** is usually a two-level, fractional factorial designed experiment. It can be set up as a full or fractional factorial design. If strong interactivity is suspected then a minimum of a Resolution V Fractional Factorial Design is recommended. If moderate-to-low interactivity is anticipated between the noise factors, then a Resolution III Plackett-Burman Screening Experiment (an L12 array) is capable of providing a good set of data upon which to conduct **ANOVA,** an acronym for *an*alysis *of var*iance results. This analysis will provide the information required to tell if any noise factors are statistically significant.

The subsystem prototype is set up at its baseline set point conditions and held there while the noise factors are exercised at their high and low set point conditions. The mean values of the CFRs are calculated. Analysis of means will provide the data to explain the magnitude and directionality of each noise factor on the CFR.

Define Compounded Noises for Robustness Experiments

The results from a noise experiment provide the evidence required to go forward in defining how to properly and efficiently stress test any given subsystem technology for robustness optimization. The statistically significant noise factors can now be isolated from the rest of the noise factors—they are the only noises worth using during robustness optimization experiments. The others are simply not producing a significant effect on the mean value of the CFRs. Using the analysis of means data and their main effects plots, the development team can literally see the directional effect the noise factor set points have on the mean values of the CFRs. The team can quantitatively demonstrate the effect of the significant noise factors. One more piece of guesswork is removed from the technology development process!

Noise factor set points that cause the mean of a subsystem CFR to rise significantly are grouped or compounded together to form a **compound noise factor,** referred to as **N1.** Noise factor set points that cause the mean of a subsystem CFR to fall significantly are grouped or compounded together to form a compound noise factor referred to as **N2.** Both compounded noise factors have the same set of statistically significant noise factors within them. The difference is the specific set points (high or low) that are associated with noise factor alignment with **N1** or **N2** (see Chapter 28 on Robust Design for examples).

This noise compounding strategy facilitates the unique and efficient gathering of two data points (data taken for each experimental treatment combination of control factors at N1 and N2 compounding levels) during the exploration of interactivity between control factors and noise factors. The reason we need to be sure we have real noise factors identified is that we want to induce interactivity between control factors and noise factors to locate robust control factor set points. If the noises are not truly significant, we are not going to be exposing the control factors to anything that will promote the isolation of a true robust set point. Much time can be wasted testing a control factor against noises that are little more than random sources of variation that are not representative of what is going on in the actual use environment. Many R&D organizations only evaluate "nominal optimal performance" in the sterile and unrealistic environment of random and insignificant levels of noise. Their technology is then transferred to design in a state of relative immaturity and sensitivity. The design teams are left to deal with the residual problems of incapability of performance, which is certainly a design defect! The usual response is, "That's funny—it worked fine in our labs." For those on the design team it is anything but funny.

Define Engineering Control Factors for Robustness Development Experiment

We now turn our attention to the control factors that will be candidates for changes to their baseline set points to new levels that leave the CFRs in a state of reduced sensitivity to the significant, compounded noise factors (N1 & N2). The selection of these control factors is not a trivial event. Not all control factors will help in inducing robustness. In fact only those control factors that possess interactions with the noise factors will have an impact in reducing sensitivity to noise (see Figure 5–10).

We begin control factor selection using the information about the ideal functions $[Y = f(x)]$ gained during Phase 2. Any control factor that had a significant effect in the underlying functional model is a candidate control factor for robustness improvement. Actually we will be looking for three kinds of performance from the candidate control factors:

1. Control factors that have moderate to strong interactions with the compounded noise factors are good robustizing factors (good for minimizing the standard deviation even with the noises freely acting in their uncontrolled fashion in the use environment).

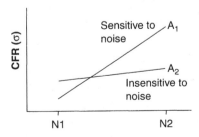

Figure 5–10 A Control Factor to Noise Factor Interaction Plot

2. Control factors that have relatively weak interactions with the compounded noise factors but have strong effects on shifting the mean value of a CFR are **critical adjustment parameters** (CAPs are good for recentering the function onto a desired target, thus forcing Cpk = Cp).

3. Control factors that have low interactivity with the compounded noise factors and low shifting effect on the mean of the CFR are good economic factors (allowing one to set them at their lowest cost set points).

Design for Additivity and Run Designed Experiment

Using the knowledge from Phase 2 function modeling, assess the nature of the interactions between the control factors (CF × CF interactions). Apply the rules for additivity grouping or sliding levels based on how energy and mass are used in the ideal function to minimize or eliminate the disruptive effects of antisynergistic interactions (see Chapter 28 for example). The experiment must be set up to focus on producing data that exhibits and highlights the interactivity between the control factors and the compounded noise factors. Within these CF × NF interactions the robustness optimization potential is found.

An appropriately sized array is selected to evaluate the control factor main effects and any control factor × control factor interactions (usually a few two-way interactions that were statistically significant in the Phase 2 analysis). The typical array will be a three-level Fractional Factorial format. If additivity grouping or sliding levels are used to suppress or remove the effects of CF × CF interactions that are not useful, it is common to use a Resolution III design (typically an L18 Plackett-Burman experimental array). If useful two-way interactions between control factors are to be evaluated, then a Resolution V or Full Factorial design must be used (*remember, a Resolution IV design will likely confound the two-way interactions with one another,* as covered in more detail in Chapter 22). Since the compound noises are exposed equally to each row from the Fractional Factorial designed experiment, the control factor treatment combinations are aligned in a Full Factorial exposure to the noise. A full noise factor exposure to the Fractional Factorial control factor combinations assures that the noise was faithfully and fully delivered to (interact with) the fraction of the possible design set points you selected. Using a Fractional Factorial noise exposure strategy would understress the control factors and leave the design at risk of missing a CF × NF interaction that could be exploited on behalf of your customer's satisfaction! For those who think Taguchi methods are optional or ineffective, you may want to rethink that position. You are probably ignoring a whole class of interactions (CF × NF) that can make a big difference for your customers.

Analyze Data, Build Predictive Additive Model

The robust design experiment will produce a rich set of data that is capable of being analyzed by two common statistical methods:

1. Analysis of means (ANOM)
   a. Provides insight on which level of each CF is best for robustness (largest S/N) or variance minimization (smallest σ or σ^2) of the CFR

b. Provides insight on which control factors have the largest effect on the mean value of the CFR (critical adjustment parameters)

2. Analysis of variance (ANOVA)

a. Provides insight on which control factors have the largest effect relative to one another and to the error variance in the entire data set (*optional for S/N metric but quite useful for variance and mean analysis*).

The development team will use the control factor set points that produced maximum improvements in CFR robustness to construct a unique math model referred to as Taguchi's **additive model** (see Chapter 28). It is used to predict overall robustness gain in units of the decibel. The additive model is a prediction of the expected robustness for the CFR under evaluation.

Run Verification Experiments to Certify Robustness

The optimized control factor set points must be verified in the final step of robust design. The same set of compounded noise factors are reused during the verification experiments. It is common to run a minimum of five replicate verification tests to prove that the predicted S/N, mean, and variance values are reproducible at these new set points. Baseline set point verification tests are also commonly conducted to see if the design acts the same way that it did back in Phase 2 testing prior to robustness optimization. This helps assure the development team is not missing some additional control or noise factors that are important to the safe transfer of the technology for commercial application.

It is also common to verify the dynamic range of the critical adjustment parameters that affect the adjustment of the mean value of the CFRs. The best approach is to conduct further optimization of these CAPs by using regression or response surface methods.

Document Critical Functional Parameter Nominal Set Points and CFR Relationships

Once the additive model and dynamic tuning capability of the subsystem technology has been verified, it is time to document the new set points and tuning range limits. At this point the Critical Parameter Management database is updated with the results of the Phase 3 quantitative deliverables. The traceability of critical parameter relationships will be greatly enhanced with the data from Phase 3.

Gate 3 Readiness

Phase 3 in the technology development process is used to take technology that is safe, stable, and tunable into a state of maturity that is insensitive to noise. Gate 3 is the place for assessing the proof that the new technology is robust to a general set of noise factors that will likely be encountered in commercial application environments. It is also the place to assess the tunability of the CFR to global optimum targets using one or more critical adjustment parameters (see the Phase 3 scorecard in Figure 5–11).

General Checklist of Phase 3 Tools and Best Practices

Subsystem noise diagramming

System noise mapping

Measurement system analysis

Noise experiments

Taguchi methods for robust design

Design of experiments

Analysis of means

Analysis of variance

Regression

Response surface methods

Critical parameter management

Project management methods

I²DOV Phase 4: Verification of the Platform or Subsystem Technologies

Figure 5–12 illustrates the major elements for I²DOV Phase 4: **Verify** Integrated System (Platform) or Subsystem

The final phase of technology development (Figure 5–12) is primarily focused on integrating and verifying the newly robust and tunable subsystem technologies into an existing product architecture or into a new platform architecture. The development of subsystems and platforms is verified for commercial readiness at the Phase 4 gate. The transfer of the new technologies to product design is dependent on transferability criteria (technology requirements) that will be fulfilled in this final phase of technology development.

The evaluations that are conducted during this phase are *stressful* by nature to help assure the technologies are mature enough to transfer. Some technologists are averse to stress testing their "pampered" designs because they fear that they will be rejected and will not be used in the next new product. If these individuals succeed in protecting their designs, they place every other employee in the company at risk and do great disservice to the customers of the business. Most technology problems show up in product design when technologies have been prematurely deployed in a commercialization project. They were not thoroughly stress tested for robustness (S/N and σ) and capability (mean, Cp, and Cpk). They look and work fine until they are used in a context where real noise factors are applied to them, intentionally or otherwise. At this point their true character emerges, and all hell breaks loose! They were not completely characterized, usually to save face, time, and money, so they were transferred at great risk. When we reactively use the checklists and scorecards at the end of each of the four gates of technology development, we can expect to routinely uncover the risk that is building due to anemic and incomplete characterization. Risk can be managed much more efficiently if one proactively uses the checklists and scorecards to preplan for the successful application of a balanced set of tools and best practices. If one

Phase 3 Optimize Technology

Deliverable	Owner	Original Due Date vs. Actual Deliv. Date	Completion Status Red Green Yellow	Performance vs. Criteria Red Green Yellow	Exit Recommendation Red Green Yellow	Notes
Quantification & Identification of Statistically Significant Noise Factors			○ ○ ○	○ ○ ○	○ ○ ○	
Directionality & Magnitude of the Noise Vectors			○ ○ ○	○ ○ ○	○ ○ ○	
Quantification & Identification of the Baseline (before robustness optimization) and New (after robustness optimization) S/N Values for the Critical Functional Responses			○ ○	○ ○ ○	○ ○ ○	
Identification of Control Factors That Can be Used to Improve Robustness (includes the Additive Model Parameters That Are Critical to Robustness			○ ○ ○	○ ○ ○	○ ○ ○	
Identification & Optimization of the Control Factors That Can be used to Adjust the Mean onto a Given Target Without Overly Compromising Robustness			○ ○ ○	○ ○ ○	○ ○ ○	
Verification Results from Additive Model Confirmation Experiments			○ ○ ○	○ ○ ○	○ ○ ○	
Cp & Cpk Values for the Pre- & Post-Robustness Optimization Critical Functional Responses						
Updated Critical Parameter Management Database & Scorecards						
Project Mgt. Plans for Phase 4						

Figure 5–11 A General Scorecard for I²DOV Phase 3 Deliverables

Start Date: ___	Integrate subsystems into platform test fixtures	End Date: ___	Owner: _____
Start Date: ___	Refine and confirm "between and within" subsystem CFRs	End Date: ___	Owner: _____
Start Date: ___	Develop, integrate, and verify platform data acquisition systems	End Date: ___	Owner: _____
Start Date: ___	Design platform or integrated subsystem nominal performance tests	End Date: ___	Owner: _____
Start Date: ___	Conduct nominal performance tests	End Date: ___	Owner: _____
Start Date: ___	Evaluate data against technology requirements	End Date: ___	Owner: _____
Start Date: ___	Design platform or integrated subsystem stress tests	End Date: ___	Owner: _____
Start Date: ___	Conduct stress tests	End Date: ___	Owner: _____
Start Date: ___	Perform ANOVA on data to identify sensitivities	End Date: ___	Owner: _____
Start Date: ___	Refine and improve platform or integrated subsystem performance	End Date: ___	Owner: _____
Start Date: ___	Rerun stress tests and perform ANOVA on data to verify reduced sensitivities	End Date: ___	Owner: _____
Start Date: ___	Conduct reliability evaluations for new subsystem technologies and platforms	End Date: ___	Owner: _____
Start Date: ___	Document CFPs and CFR relationships and verify that the technology is transferable	End Date: ___	Owner: _____

Figure 5–12 Flow of Phase 4 of Technology Development

chooses to bypass certain tools and best practices, the technology will accrue certain risks. The gates are designed to assess the accrual of risk and to stop or alter development when the risk gets too high.

The fourth, "integrative" phase is extremely important and the one that most companies bypass with reckless abandon. When development teams fail to integrate their new subsystem technologies into an existing product or the new platform and stress test them with statistically significant noise factors, they fore-go obtaining the knowledge that assures the risk of transferring the new technology is acceptable. Many say, "We can't afford to develop technology in this rigorous fashion." They go on to claim, "We like to integrate and test (usually with very low levels of statistically insignificant noise) early during product design and find out where our problems are and then fix (react to) them." This is called **build-test-fix** engineering and is antithetical to preventing functional performance problems. This approach assures there *are* problems and seeks comfort in isolating and "fixing" them—typically over and over again. It is a recipe for high warranty costs; late product launches; many rounds of costly post-shipping modification kits; aggravated, overworked service engineers; and the loss of once-loyal customers. Even worse, from a strategic perspective, is the fact that this approach ties up R&D resources and design resources on lengthy "reliability emergency assignments" that keep them away from working on the next design project that is critical to future revenue generation. Build-test-fix engineering robs the corporation of its future generation of value, productivity, and economic growth. All the excuses for build-test-fix engineering end up being the most costly and inefficient way of doing business.

Integrate Subsystems into Platform or Product Test Fixtures

The optimized subsystems are now ready to be integrated into one of two types of systems for stress testing and performance certification:

1. An existing product
2. A new platform that will form the basis of many new products

When the new subsystem or subsystems have been developed for integration into an existing product that is currently in the marketplace, the R&D organization must obtain a sufficient number of these systems for stress testing and verification of integrated performance capability.

When the new subsystems are developed for integration into a new platform, then a significantly different effort must be made to integrate the subsystems into a functioning system we recognize as a platform. This is harder and requires more effort because there is no existing product to be reworked to accept the new subsystems. The platform is a whole new system. It is entirely possible that numerous existing subsystem designs will be reused and integrated into the new platform architecture during product design.

Refine and Confirm "Between and Within" Subsystem CFRs

Prior to conducting certification tests, the development team needs to identify exactly how the critical functional responses are going to be measured as they transmit energy, mass, and information (analog and digital logic and control signals) across subsystem boundaries. Every subsystem has an "internal or within" set of CFRs defined for it already. These were developed and measured in the first three phases of technology development. Now that the team integrates all the new subsystem technologies together, it is highly likely that there will be some need to measure how the CFR from one subsystem affects the mating subsystem. If one subsystem receives mass from another subsystem, then the mass flow rate will need to be measured as a "between" subsystem CFR. If the mechanism for mass flow or transfer is through the generation of a magnetic flux, then that would be classified as a "within" subsystem CFR. The important issue here is for the team to be comprehensive in accounting for CFRs that measure functional interface relationships. One helpful way to approach this accounting of CFRs at integration is to write energy balancing equations to account for the flow, transfer, or state of change between energy and mass across the interface boundary. These "between" subsystem CFRs may have been defined previously during Phase 2 and Phase 3 but not measured in the real context of an integration test. The task now becomes one of measurement system development and analysis—our next topic.

Develop, Integrate, and Verify Platform Data Acquisition Systems

A data acquisition system capable of measuring all within and between subsystem, platform, or product level CFRs needs to be developed, integrated, and verified at this point. These systems include numerous transducers that physically measure the CFRs, wiring, interface circuit boards, computers, and software that facilitate the collection, processing, and storage of the CFR data.

The data acquisition system will need to have its repeatability, reproducibility, and precision-to-tolerance ratio evaluated to assure that the measurements are providing the level precision and accuracy required to verify the performance of all the CFRs.

Design Platform or Integrated Subsystem Nominal Performance Tests

The team will need to structure a series of performance tests to evaluate the Cp and Cpk of the CFRs under nominal conditions. This test is designed to look at the performance of all of the CFRs under nonstressful conditions. It represents the best you can expect from the subsystems and the integrated platform or product at this point in time. These tests usually include normal use conditions and scenarios. All critical-to-function parameters within the subsystem and critical functional specifications at the component level are set at their nominal values. Environmental conditions are also set at nominal or average conditions.

Conduct Nominal Performance Tests

With a comprehensive, nominal condition test plan in hand, the team can move forward and conduct the nominal performance tests. The data acquisition system is used to collect the majority of the performance data from the CFRs. It is likely that some independent, special forms of metrology may be needed to measure some of the CFRs, typically at the platform or product level. These are often called *off-line* evaluations. In some cases data must be gathered on equipment in another lab or site equipped with special analysis equipment.

Evaluate Data Against Technology Requirements

The data is now summarized and evaluated against the technology requirements. Here we discover just how well the CFRs are performing against their targets that were established back in Phase 1. This is likely to be the best data the team is going to see because it was collected under the best or nominal conditions. With the nominal capabilities of the CFRs in hand, the team is ready to get an additional set of capability data run under the real and stressful conditions of noise.

Design Platform or Integrated Subsystem Stress Tests

The exposure of new technologies to a statistically significant set of noise factors has been discussed in the Phase 3 robustness optimization process. We return to that topic within the new context of the integration of the robust subsystems. To be circumspect with regard to realistically stressing the new platform technologies or existing products that are receiving new subsystem technologies, the team has to expose these systems to a reasonable set of noise factors.

The typical noise factor list includes numerous sources of variation that are known to provoke or transmit variation within the system integration context. A designed experiment is the preferred inspection mechanism that is capable of intentionally provoking specific levels of stress within the integration tests. Designed experiments help assure that thorough and properly controlled amounts of realistic sources of noise are intentionally introduced to the integrated subsystems. Without this form of designed evaluation of the various CFRs associated with the integrated subsystems, the technology is not fully developed and cannot be safely transferred into product commercialization.

Conduct Stress Tests

Conducting a stress test under designed experimental conditions requires a good deal of planning, careful set-up procedures, and rigorous execution of the experimental data acquisition process. As is always the case when measuring CFRs, the data acquisition equipment must be in excellent condition with all instruments within calibration and measurement system analysis standards.

Perform ANOVA on Data to Identify Sensitivities

The data set from the designed experiment can be analyzed using Analysis of Variance methods. ANOVA highlights the statistical significance of the effects of the noise factors on the various CFRs that have been measured. It is extremely important that the team be able to document the sensitivities of each CFR to specific integration noise factors. This information is extremely valuable to the product design engineers that will be in receipt of the transferred technologies. If the integrated system of subsystems is too sensitive to the induced noise factors, then it may be necessary to return the technologies to pass through Phase 2 and 3 where the subsystems can be changed and recharacterized for added capability and robustness.

Refine and Improve Platform or Integrated Subsystem Performance

If the integrated subsystems are shown to have sensitivities that can be de-tuned without returning the subsystems to Phase 2 and Phase 3, then this should be done at this point. It is not unusual to find that subsystems need to be adjusted to compensate for interface sensitivities and noise transmission. Sometimes a noise factor can be removed or its effect minimized by some form of noise suppression or compensation. Typically it is possible but expensive to control some noise factors. A common form of noise compensation is the development of feedback or feed-forward control systems. These add complexity and cost to the technology but there are times when this is unavoidable.

Rerun Stress Tests and Perform ANOVA on Data to Verify Reduced Sensitivities

Once changes have been made to the subsystems and noise compensation strategies have been deployed, the subsystems can be integrated again and a new round of stress testing can be conducted. The ANOVA process is conducted again and new sensitivity values can be evaluated and documented.

Conduct Reliability Evaluations for New Subsystems Technologies and Platforms

Each subsystem technology and integrated platform has its reliability performance evaluated at this point (see Chapter 33).

Document CFPs and CFR Relationships and Verify That the Technology Is Transferable

The ANOVA data also holds capability (Cp) information for each CFR. This data is in the form of variances that can be derived from the mean square values within the ANOVA table by dividing the total mean square by the total degrees of freedom for the experiment to obtain the total variance for the data set. If certain noises are applied to carefully selected critical functional parameters, usually as either tolerance extremes or deteriorated functional or geometric characteristics, then the sensitivities of the changes at the critical functional parameter level (x inputs) can be directly traced to variation in the CFRs (Y outputs). This is an important final step in critical parameter characterization within the final phase of technology development.

Gate 4 Readiness

This last phase of technology development requires additional summary analysis and information related to the following topics:

- Patent stance on all new technologies
- Cost estimates for the new technologies
- Reliability performance estimates and forecasts (see Chapter 20)
- Risk summary in terms of functional performance, competitive position, manufacturability, serviceability, and regulatory, safety, ergonomic, and environmental issues

Every company is different, so your specific areas of summary for technology development information transfer may contain additional items beyond this list. The scorecard in Figure 5–13 covers the specific deliverables of Phase 4.

Phase 4 Verify Technology

Deliverable	Owner	Original Due Date vs. Actual Deliv. Date	Completion Status Red Green Yellow	Performance vs. Criteria Red Green Yellow	Exit Recommendation Red Green Yellow	Notes
Summary of Final Technology Requirements vs. Actual Performance Specifications Attained			○ ○ ○	○ ○ ○	○ ○ ○	
Mean & Standard Deviations for each CFR			○ ○ ○	○ ○ ○	○ ○ ○	
F Ratios associated with Control Noise Factor contributions for each Subsystem and System Level CFR			○ ○ ○	○ ○ ○	○ ○ ○	
Cp Values for each Subsystem and System Level CFR			○ ○ ○	○ ○ ○	○ ○ ○	
Percent Contribution to Overall Variation from Critical Noise Factors						
Summary Sensitivity Analysis for All Critical Functional Parameters Relative to their Critical Functional Responses			○ ○ ○	○ ○ ○	○ ○ ○	
Summary Engineering Analysis Report containing all analytical & empirical math models (including any Monte Carlo Simulations & CAE results), FMEAs, Noise Diagrams, System Noise Maps, Results from All Designed Experiments including Robustness Optimization Activities. This report is to contain the summary of all knowledge to be transferred to the design engineering teams—it is their "design guide."			○ ○ ○	○ ○ ○	○ ○ ○	

Documented & Transferable Data Acquistion Systems				
Documented Critical Parameter Database for the Integrated Technologies				
Critical Functional Responses				
Critical Adjustment Parameters				
Critical Functional Parameters				
Critical Noise Factors				
Patent Applications for New Technologies				
Reliability Estimates for New Technologies				
Hardware, Firmware, & Software that Embodies the new Subsystem or Platform Technologies				

Figure 5–13 General Scorecard for I^{2}DOV Phase 4 Deliverables

General Checklist of Phase 4 Tools and Best Practices

Nominal performance testing methods

Measurement system analysis

Stress testing methods using designed experiments

ANOVA data analysis

Empirical tolerance design methods

Sensitivity analysis

Reliability estimation methods

Cost estimation methods

Patent search and analysis

Cp characterization

References

Christenson, C. M. (2000). *The innovator's dilemma.* New York: Harper Collins.

Clausing, D. (1993). *Total quality development.* New York: ASME.

Cooper, R. G. (2001). *Winning at new products* (3rd ed.). New York: Perseus.

Meyer, M. H., & Lehnerd, A. P. (1997). *The power of product platforms.* New York: The Free Press.

Utterback, J. M. (1994). *Mastering the dynamics of innovation.* Boston: Harvard Press.

Wheelwright, S. C. & Clark, K. B. (1992). *Revolutionizing product development.* New York: The Free Press.

Product Design Using Design For Six Sigma

The use of phases and gates to manage the timing and flow of product commercialization projects has shown strong results in many companies. Some skeptics contend that such approaches limit creativity, constrain the early discovery of problems, and hold back the speedy launch of new products. We have found that companies with such spokespeople possess some common and repeatable traits:

1. A single personality or a small clique of insiders claims to represent the voice of the customer. This driving influence is used to define new product concepts and requirements in relatively ambiguous terms that slowly enable the development team to "back themselves into a corner" to a default system architecture and "final" product specifications relatively late in the product design time line.

2. A focus on the early, "speedy" integration of immature and underdeveloped subsystems which require seemingly endless cycles of corrective action.

3. A loosely defined set of skills enabled by the undisciplined practice of a limited portfolio of tools and best practices is centered on reacting to problems discovered (. . .actually created) by trait 2.

4. A legacy of residual postlaunch problems ties up a significant portion of the company's development and design resources, precluding it from working on the future technologies and designs that will be needed to create the future revenue stream for the business.

5. A "modification kit" stream contains the true optimized and final designs that send a message to the customer that "you have been our guinea pig while we finish designing this product—after you have bought and paid for it."

Most companies have embraced a Phase/Gate process but deploy it in a passive, undisciplined manner. They have a reasonable process—they just don't use it very well. They have made a small step forward but are still clinging to many of their past bad habits. Their legacy is similar to the one we just described: poor system reliability, numerous subsystem redesign cycles, a modification kit strategy to finish the design while it is in the customers' hands, cost overruns with residual high cost of ownership, and missed market windows.

This chapter will develop an active and disciplined approach to the use of phases and gates during product commercialization. The goal is simple: Prevent the scenario we just described.

An Introduction to Phases and Gates

A phase is a period of time during which specific tools and best practices are used in a disciplined manner to deliver tangible results that fulfill a predetermined set of product commercialization deliverables. A gate is a stopping point within the flow of the phases within a product commercialization project. Two things occur at this stopping point: (1) a thorough assessment of deliverables from the tools and best practices that were conducted during the previous phase, specifically the high-risk areas that did not fully fulfill the gate requirements, and (2) a thorough review of the project management plans, typically in the form of a set of PERT charts, for the next phase. Gate reviews mostly focus on preventive measures to keep design flaws and mistakes from occurring in the next phase. Gates are also used to manage work-around plans to keep things moving forward after difficulties and problems have been identified. A well-designed gate review takes into account that things will go wrong and enables the team to deal with these inevitabilities.

It is to be expected that some subsystems will go through a gate on time while other subsystems may experience unexpected difficulties in meeting the gate criteria. It is common for engineering teams to use a three-tiered system of colors to state the readiness of their subsystem to pass through a gate.

- *Green* means the subsystem has passed the gate criteria with no major problems.
- *Yellow* means the subsystem has passed some of the gate criteria but has numerous (two or more) moderate to minor problems that must be corrected for complete passage through the gate.
- *Red* means the subsystem has passed some of the gate criteria but has one or more major problems that preclude the design from passing through the gate.

Gate passage is granted to subsystems that are green. Gate passage is provisionally granted to subsystems that are yellow. Gate passage is denied to subsystems that are red. Time and resources are planned to resolve the problems, which may take 2 to 6 weeks or even longer. Work-around strategies are put in place to keep the whole project moving forward while the subsystems that have failed gate passage are corrected. This situation requires engineering managers to develop phase transition strategies that allow for early and late gate passage. The ideal state is when all subsystems pass all the gate criteria at the same time for every gate. This, of course, almost never happens. For complex projects it is foolish to expect anything other than a series of "unexpected" problems. For engineering managers who are preventative in their thinking, the second order of business in developing a Phase/Gate commercialization process is the inclusion of a phase transition strategy that effectively deals with managing the unexpected.

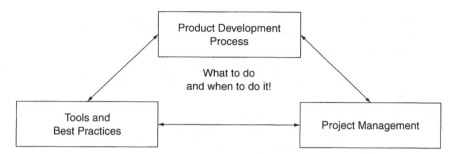

Figure 6–1 Components of Successful Product Development

The first order of business is structuring each phase with a broad portfolio of tools and best practices that prevent unexpected design problems. This is done by focusing on the tools and best practices that use forward-looking metrics that measure functions (scalar and vector engineering variables) instead of counting quality attributes (defects). The key is to prevent the need to react to quality problems.

Quality must be designed in by careful, disciplined use of the right tools and best practices that measure the right things to enable a preventive strategy from an engineering perspective. The latter chapters of this text discuss such tools and metrics.

Quality is certainly one of the primary requirements of a product commercialization project, but it is now clear that the most efficient path to quality does not flow through a series of quality metrics. They tend to enable "react" strategies as opposed to "prevent" strategies. We will discuss the antidote to the inefficiencies of quality metrics in Chapter 12.

There is near-universal acceptance that using an overarching product development process to guide technology development and product commercialization is the right thing to do. What is often overlooked is the value of a broad portfolio of tools and best practices that are scheduled and conducted within each phase using the methods of project management (Figure 6–1).

When commercialization teams structure "what to do and when to do it" using PERT charts under the guiding principles of project management, engineering teams have their best chance of meeting cycle-time, quality, and cost goals. If the commercialization process is open to personal whims and preferences, chaos will soon evolve and the only predictable things will become schedule slips, the need for design scrap and rework, and perhaps "career realignment" for the culpable engineering managers (if they are actually held accountable for their mismanagement of the company's resources!).

Preparing for Product Commercialization

A few key components must be in place, in this order, before proper product commercialization can take place:

1. Business goals defined and a strategy to meet them
2. Markets and segments that will be used to fulfill the business goals

3. Voice of the customer and voice of technology defined based on specific market segments
4. Product line strategy and family plans defined to fulfill specific market segment needs
5. Technology strategy defined to fulfill product line strategy
6. New platform and subsystem technologies developed and verified within their own independent technology development process (I^2DOV—see Chapter 5)

With these things in place, it is much easier to see what product flow to commercialize and when each individual product from the family plan should be delivered to the market. This is how value is created for the business and its shareholders. Design For Six Sigma quantifies "top-line growth" financial results through its ability to enable and deliver results that fulfill the specific product requirements that underwrite the validity of the product's business case.

How do we value the use of Six Sigma in product commercialization? Through the fulfillment of the product's business case and the avoidance of downstream cost of poor quality. In this context the Taguchi Loss Function can be helpful in quantifying the cost avoidance induced by preventing design problems. Design For Six Sigma can also be valued through the direct cost savings realized by the redesign of products and production processes that are causing warranty, scrap, and rework costs.

In terms of delivering a complete, well-integrated system, the methods of system engineering and integration must be in place as a governing center for the product commercialization project (see Chapter 7 on System Engineering and Integration). The whole project must be managed from the technical perspective of system engineering, not from the parochial perspectives of a loosely networked group of subsystem managers. If you are sensing a form of dictatorship that runs the project with a firm, disciplined hand from a project management perspective, that is much preferred over the "bohemian-existential" approach of poorly linked subsystem teams. A "governing legal authority" should be in place to keep personal agendas and political behavior from running the show. If this is allowed to happen, it's back to those five traits we mentioned at the beginning of this chapter.

Defining a Generic Product Commercialization Process Using the CDOV Roadmap

We will use a generic and simple model to illustrate a Phase/Gate structure to conduct product commercialization. The **CDOV** process phases we are suggesting in this text serve as guideposts to help you align the tools and best practices of Design For Six Sigma to the actual phases in your company's commercialization process. **CDOV** stands for the first letters in the four phases featured in Figure 6–2.

It is not uncommon to add additional subgates within each of the four phases. This is particularly true of Phases 3 and 4. Phase 3 often requires subgates due to the need to first develop subsystems and then to integrate them into a system. Phase 4 requires subgates to verify the product design and then moves on to verify production ramp-up, launch, steady-state operations, and support processes. We will illustrate how they flow as part of the overarching product commercialization process that leads to successful product launches.

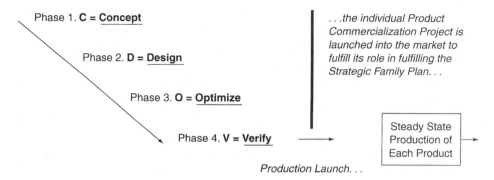

Figure 6–2 CDOV Phases

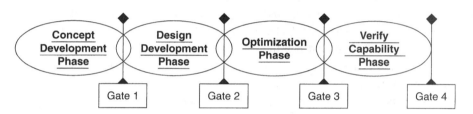

Figure 6–3 Overlaps in Phase/Gate Transmissions

The use of gates requires that the date for a gate be firmly established. The criteria for gate passage are flexible. Thus we illustrate the nature of Phase/Gate transition as an overlapping interface. Some deliverables will be late while others will come in early.

The program management team can, at their discretion, allow some activities to start early or end late as the diagram in Figure 6–3 indicates.

We will define the general nature of each of these phases and describe the detailed steps within each phase. Because this book is focused on the technical development of products in the context of Six Sigma tools, we will not spend a lot of time on business case development and some other details that must be included in a comprehensive product development process. For the sake of being circumspect in making the reader aware of these additional issues, we will briefly illustrate what a general phase and gate should include by topical area.

Phase Topics

- Business/financial case development
- Technical/design development
- Manufacturing/assembly/materials management/supply chain development
- Regulatory, health, safety, environmental, and legal development
- Shipping/service/sales/support development

We will focus most of our discussions on the technical/design development and manufacturing/assembly/materials management/supply chain development areas. It is important to recognize though, that a broad range of Six Sigma methods can be applied across all the phase topics listed here. That broader range includes Six Sigma methods in transactional business processes, manufacturing operations processes, and service processes, all of which deploy the MAIC (*M*easure, *A*nalyze, *I*mprove and *C*ontrol) roadmap.

The best way to think about establishing the architecture of any product development process is to structure it as a macro-timing diagram to constrain "what to do and when to do it" across the entire product development life cycle. With such a macro-timing diagram established, the product development management team can then conduct the phases and gates with the help of checklists and scorecards based on a broad portfolio of tools and best practices and their deliverables. The techniques of project management can then be used to construct micro-timing diagrams, preferably in the form of PERT charts, that establish discipline and structure within the day-to-day use of the tools and best practices for each person on the product development team. We recommend PERT charts because they help define personal accountability for using the right tools and best practices within the phases. They also greatly enhance the management of parallel and serial activities, work-around strategies, and the critical path tasks that control the cycle-time of each phase. Managing a commercialization project in this manner helps assure that all the right things get done to balance the ever present cost, quality, and cycle-time requirements.

Checklists

Checklists are brief summary statements of tools and best practices that are required to fulfill a gate deliverable within a phase. These are actions based on work everyone must do based on a broad portfolio of tools and best practices. You know you have a good checklist when it is easy to directly relate one or more tools or best practices to each gate deliverable. Another way of recognizing a good checklist is that it can be used to build project management PERT and Gantt charts for day-to-day management of tool use. Tools and best practices produce deliverables that can be tracked on a daily, weekly, or monthly basis and can be easily reviewed using a summary scorecard when a gate is reached. Checklists are used during project planning of tool use in the current phase to prevent problems in the next phase. See Figure 6–4 for an example of a tools and best practices checklist.

Scorecards

Scorecards are summary results derived from data that was produced from the use of a full portfolio of tools and best practices within a phase. Scorecards can be developed for any level of summary detail required to pass a gate. Typically senior management will see high-level scorecards that are simple "traffic lights" of red, yellow, or green scores. These high-level scorecards also contain summary statements of risks and issues. Scorecards align with checklists and contain whatever level of detail is required to prove the risks that constrain the project are being managed properly. Scorecards are used after a phase to react to problems that occurred in the current phase or past phases. Scorecards are used before the next phase to assure the team has the resources it

---

Checklist of Phase 1 Tools and Best Practices
Market segmentation analysis
Customer value management
Product portfolio development
Economic and market trend forecasting
Strength, weakness, opportunity, and threat (SWOT) analysis
Business case development methods
Competitive benchmarking
VOC gathering methods
KJ analysis
QFD/House of Quality—Level 1 (system level)
Concept generation techniques (TRIZ, brainstorming, brain-writing)
Modular and platform design
System architecting methods
Knowledge-based engineering methods
Pugh concept evaluation and selection process
Math modeling (business cases, scientific and engineering
    analysis)

Figure 6–4   Example of a Tool and Best Practice Checklist

needs to measure the appropriate data to fulfill the next gate review criteria. See the example in Figure 6–5.

No one product development process architecture will meet the needs and priorities of all companies. Every company must define a product development process that fits its culture, business process, and technological paradigm.

Let's move on and review the details of the phases and gates of the **CDOV** process. As we do this, we will begin to develop the Critical Parameter Management paradigm that is helpful in development of Six Sigma performance.

The CDOV Process and Critical Parameter Management During the Phases and Gates of Product Commercialization

The diagram in Figure 6–6 illustrates the major elements for CDOV Phase 1: Develop a system concept based on market segmentation, the product line, and technology strategies.

In this section we will review the specific details related to each action item: start date for each action item, the action item, and end date for each action. Each of these action items can be conducted by using a tool or best practice. The flow of actions can be laid out in the form of a PERT chart for project management purposes (see Chapter 4).

Gather the Voice of the Customer by Market Segment

The best way to start any commercialization project is to know two key things well:

1. The voice of the customer: what your customers need
2. The voice of technology: what product and manufacturing technology is currently capable of providing

Phase 1 Concept Development

Deliverable	Owner	Original Due Date vs. Actual Deliv. Date	Completion Status Red Green Yellow	Performance vs. Criteria Red Green Yellow	Exit Recommendation Red Green Yellow	Notes
Ranked & Documented VOC Database			O O O	O O O	O O O	
System Level House of Quality			O O O	O O O	O O O	
Competitive System Benchmarking Data			O O O	O O O	O O O	
System Level Requirements Documents			O O O	O O O	O O O	
Documented List of System Level Functions That Fulfill the System Requirements			O O O	O O O	O O O	
Documented Superior System Concept			O O O	O O O	O O O	
Portfolio of Candidate Technologies That Help Fulfill the System Functions & Requirements			O O O	O O O	O O O	
Documented Critical Parameter Database with estimated System CFR Interaction Transfer Functions			O O O	O O O	O O O	
Documented Reliability Model and Predicted System Reliability Performance with Gap Analysis			O O O	O O O	O O O	
Estimated System Cost Projection			O O O	O O O	O O O	
Phase 2 (Design Development) Project Management Plans			O O O	O O O	O O O	

Figure 6–5   Example of a General Gate Review Scorecard

Start Date:___	Gather the voice of the customer by market segment	End Date: ___	Owner: _____
Start Date:___	Refine and rank the voice of the customer using KJ analysis	End Date: ___	Owner: _____
Start Date:___	Create product or system level House of Quality	End Date: ___	Owner: _____
Start Date:___	Conduct competitive product benchmarking	End Date: ___	Owner: _____
Start Date:___	Generate product or system level requirements document	End Date: ___	Owner: _____
Start Date:___	Define the functions that fulfill the system requirements	End Date: ___	Owner: _____
Start Date:___	Generate system concept evaluation criteria	End Date: ___	Owner: _____
Start Date:___	Generate system concepts that fulfill the functions	End Date: ___	Owner: _____
Start Date:___	Evaluate system concepts	End Date: ___	Owner: _____
Start Date:___	Select superior system concept	End Date: ___	Owner: _____
Start Date:___	Analyze, characterize, model, and predict nominal performance of the superior system	End Date: ___	Owner: _____
Start Date:___	Develop reliability requirements, initial reliability, model, and FMEA for the system	End Date: ___	Owner: _____

Figure 6–6   Flow of Phase 1 Product Commercialization

We will lay out the importance of gathering detailed information regarding both of these foundational areas in the next few paragraphs. These two sources of information will form the basis of what the entire product development community will "design to fulfill or surpass." Critical Parameter Management is initiated by generating a database of ranked, prioritized, stated, and observed customer needs.

It is helpful to begin product design activities by asking questions like:

- What do customers specifically say they need?
- What do their actions suggest they need?
- What technology is being used today to meet their current needs?
- What new technology (manufacturing and product design technology) is capable of extending or improving their satisfaction with a new product?

Notice all the actions implied in these questions are focused on fulfilling stated or observed customer needs.

One of the biggest problems in starting the process of product development is balancing the voice of the corporate marketing specialist with the voice of the customer. Most companies underinvest in gathering an up-to-date and thorough database of actual customer needs. It is much cheaper and easier to get a little data once in a while and then use intuition, experience, personal opinion, and well-intentioned "best guesses" to define a surrogate body of information that is submitted as representative of the voice of the customer. It has been our experience that many marketing groups have weak market research processes, a poorly developed set of tools and best practices, limited budgets, and little cross-functional linkage to the engineering community. All of this adds up to the most common form of VOC data being characterized by the following traits:

1. The "data," such as it is, is frequently old—often by 1 or more years.
2. The marketing team is incapable of showing the actual words taken down from a customer interview. "Data" is usually documented in the memory of the market specialist or written down in the "voice" of the market specialist.
3. The "data" is not structured and stored in a format that is easy to translate and convert into product requirements.
4. The information in hand is from a limited set of "key" customers that is not truly (or statistically) representative of the true market segment being targeted.
5. The "data" is the "considered opinion" of a strong-willed and powerful executive (the CEO, R&D or engineering vice president, or marketing or sales executive).
6. Engineers who are developing the product and manufacturing service and support processes were not involved at all in gathering the voice of the customer data (nor have they been heavily involved in competitive performance benchmarking).

The real problem is that developing products and their supporting manufacturing processes in this anemic context places the business at great risk. The risk lies in the under- or overdevelopment of a product based on a poor foundation of incoming customer need information and competitive product design and manufacturing process technology. The business is essentially guessing or relying on intuition. Of course, if you have unlimited financial resources, this is fine. Developing hunches into products can be a tremendous waste of time and money.

A major antidote to the improper development of product requirements is to go out into the customer use environment and carefully gather the real voice of the customer. Product development teams can accomplish this by planning the use of a proven set of VOC gathering tools and best practices during Phase 1 of product development.

Numerous methods are available to facilitate the gathering of customer needs data. That data can be grouped in two general categories:

1. Customer needs defined by *what they say*
2. Customers needs defined by *what they do*

One of the most effective methods for gathering data about stated customer needs is to form a small team of two or three people and hold a face-to-face meeting with a reasonable cross-section of customers. One person leads the discussion, one takes detailed notes or records the conversations, and one listens and adds to the discussion to help keep a diverse and dynamic dialog in motion. This approach is characterized by an active and dynamic exchange between the customers and the interview team whom we will call the inbound marketing team. The inbound marketing team is typically a cross-functional group made up from a reasonable mix of the following professions:

1. Market research and advanced product planning
2. Technical (R&D, system or subsystem design engineers, production operations, etc.)
3. Industrial design
4. Service and support
5. Sales
6. Program management

The inbound marketing team first prepares a customer interview guide that will enable team members to set a specific flow of discussion in motion during the customer meeting. The interview guide is not a questionnaire but rather a logical "flow diagram" containing specific topics that are designed to stimulate free-flowing discussions about customer needs and issues that help align the company's advanced product planning activities with the customers' future needs. If a questionnaire is all that your business needs to define future product development requirements, then phone calls, handout questionnaires, and mailing methods are much more economical. Questionnaires are quite limited in the kind of detailed data that can be obtained, because you get little exchange and clarification of information to verify that what was stated was correctly received. The power of face-to-face discussions with a reasonable number of customers (current, former, and potential) is unsurpassed in assuring that you obtain many key VOC attributes based on the **KANO** model of customer satisfaction (see Figure 6–7):

1. Basic needs that must be in the future product
2. Linear satisfiers that produce proportional satisfaction with their increased presence in the future product
3. Delighters that are often generally unknown by the customer until you stimulate some sense of latent need that was "hidden" in the mind of the customer

The other key data that is available to the inbound marketing team is the observed behavior of customers in the product acquisition and use environment. These are relatively passive events for the inbound marketing team but obviously dynamic events on the customer's side. It is here that the team watches and documents customer behavior as opposed to listening to what they say they need. This approach has been called *contextual inquiry* (Clausing, 1994). This

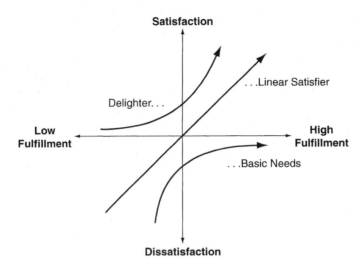

Figure 6–7   The KANO Customer Satisfaction Model

method also requires careful preplanning so that meaningful vantage points and contexts for observation can be arranged. The team must find a way to participate in use experiences. Obviously not all products are amenable to this approach, but when the product is used in an observable environment, it can be quite useful in assisting in the determination of unspoken needs. In the world of KJ analysis (our next topic), this data is profoundly useful in developing "images" of customer problems and needs. A well-spread story tells how Honda engineers positioned themselves in parking lots and watched how people used their trunks and hatchbacks to help gather data for improving the next hatchback design. Eastman Kodak hands out one-time-use cameras to their employees and assesses their use habits. There are many creative ways to align your teams with use experiences. Never underestimate the knowledge you can gain by getting inside the situational events that are associated with the actual use of your products and processes, especially in cases where disruptive sources of variation, or causes of loss, are active within the use environment.

Refine and Rank the Voice of the Customer Using KJ Analysis

Once customer data is gathered in a sufficient quantity to adequately represent the scope of the market segment, the team refines and documents the raw data by structuring it into a more useful form. The KJ method, named after the Japanese anthropologist Jiro Kawakita, is used to process the VOC data into rational groups based on common themes. It employs higher levels of categorizing the many details of expressed or observed/imaged need. The name *affinity diagramming* has been used to characterize this process.

KJ analysis is quite useful when you are working with nonnumerical data in verbal or worded format. The mass of VOC input can, at first glance, seem to be a jumble of multidimensional and multidirectional needs that is quite hard to analyze. In practical terms, KJ analysis helps make sense out of all these words, observations, and images by aligning like statements in a hierarchical format. Once the raw VOC is segmented, aligned underneath higher-level topical descriptors (higher levels of categorizing the many details of expressed or observed/imaged need), and ranked under these topical descriptors, it can be loaded into the system or product level House of Quality using quality function deployment techniques.

Create Product or System Level House of Quality

A process known as **quality function deployment** (QFD) is commonly used to help further refine the affinitized (or grouped by similar characteristics) VOC data. QFD uses the data, mapped using the KJ method, to develop a clear, ranked set of product development requirements that guide the engineering team as they seek to determine what measures of performance need to be at six Sigma levels of quality.

The QFD process produces a two-dimensional matrix called a House of Quality. The **House of Quality,** at the system level, is used to transform VOC data into technical and business performance requirements for the new product or process. Many teams overdo the structuring of the House of Quality by including too many minor details in it. The House of Quality should only be in receipt of VOC data that falls under one or more of the following categories:

1. **New needs:** features or performance your product has not historically delivered
2. **Unique needs:** features or performance that are distinctive or highly desired beyond the numerous other less-demanding needs that must be provided
3. **Difficult to fulfill needs:** features or performance that is highly desired but is quite difficult for your business to develop and will require special efforts/investments of resources on your part

Cohen (1995) provides a graphical depiction of QFD, featured in Figure 6–8.

Everything that is to be included in the product design requirements document is important, but not everything is critical. The House of Quality is a tool for critical parameter identification and management. It helps the development team see beyond the many important requirements to isolate and track the few critical requirements that matter the most to successful commercialization.

Conduct Competitive Product Benchmarking

Following the voice of the customer blindly can lead to a failure in the commercialization of a new product. Let's use an analogy to illustrate this point. When a band solicits musical requests from the audience, it had better be able to play the tune! Taking requests needs to be a constrained event unless there is nothing you cannot do. One way to establish market segmentation or VOC constraint is by establishing limits to whom you choose to develop and supply products. The limitation is usually set by your company's capability in a given area of technology, manufacturing, outbound marketing, distribution channels, and service capability.

Within your established boundaries of limitation, which some people on a more positive note label a *participation strategy,* you need to be well informed as to what your competition is

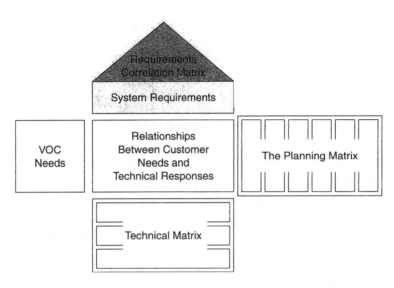

Figure 6–8 House of Quality

developing and delivering. This requires that some form of competitive benchmarking be conducted. Building a balanced approach to what customers say they want, what you can provide, and what your competition can provide is a reasonable way to conduct product development.

From a Six Sigma performance perspective, we need to decide what critical functional responses need to be at Six Sigma quality. We must use competitive performance data to help define critical functional requirements and then develop designs that can be measured with CFRs. This chain of critical requirements and their fulfilling critical responses is the essence of Critical Parameter Management and can be helpful in defining competitive options. Competitive benchmarking adds one more dimension to what candidate critical requirements and responses need to be considered as we set up the system House of Quality. It can also be one more form of what we call the "voice of physics" or the "voice of the process." If a competitor is doing something new, unique, or difficult relative to what you are currently able to do, then you must seriously consider what you are going to do and what you are *not* going to do.

The House of Quality has a specific section where these trade-offs can be ranked and documented. This area of competitive vs. your performance trade-offs is often where the initial identification of Six Sigma requirements are born. The competition is found to be good at a measurable function, and customers say they want more performance in this area, so we have an opportunity to surpass the competition by taking that specific, candidate-critical function to a whole new level—perhaps to Six Sigma functional performance or beyond. It is important to note that not everything needs to be at a Six Sigma level of performance. If you find a strategic function that is worth taking to the level of Six Sigma in terms of value to the business case and customer, you had better be able to measure it against the requirements and against your competition!

Many of the tools of Six Sigma are useful for comparative benchmarking. The following are examples:

1. Design and process failure modes and effects analysis
2. Taguchi noise diagramming and system noise mapping
3. Pugh's concept evaluation process
4. Functional flow diagramming and critical parameter mapping
5. Taguchi stress testing
6. Highly accelerated life tests and highly accelerated stress testing
7. Multi-vari studies
8. Statistical process control and capability studies (for both design functions and process outputs)

Generate Product or System Level Requirements Document

The system House of Quality contains the information that relates and ranks the *new, unique, and difficult* (NUD) VOC needs to the system level technical requirements that are to be fulfilled by the design engineering teams. These NUD items, along with all the many other system level requirements, must be organized and clearly worded in the **system requirements document** (SRD). The SRD contains all of the product level design requirements. It is important to recognize that three distinct types of requirements must be developed and documented in the SRD:

1. Candidate-critical system level requirements from the system House of Quality
2. Candidate-critical system level requirements that were not in the system House of Quality because they were not new, unique, or difficult
3. Noncritical but nonetheless important system requirements that must be included to adequately complete the product level design

The system requirements document is the design guide and "bible" that the system engineering team, the subsystem design teams, and all other cross-functional participants in product design and development must seek to fulfill. If a product requirement is not covered in the SRD, it is highly likely that the omitted requirement will not be fulfilled. On the other hand, the SRD is the governing document for what does get delivered. When it is well managed, it prevents design feature and function "creep." SRDs are frequently reused and added to as a family of products is developed as the serial flow of product line deployment is conducted.

Design Functions That Fulfill the System Requirements

Once the system requirements are clearly worded and documented, it is now time for the engineering teams to integrate their talents to identify what functions must be developed at the system level to fulfill the system requirements. Defining these functions is the job of a cross-functional team of subsystem engineering team leaders, service engineers, industrial designers, system engineers, and other appropriate product development resources all led by the system engineering and integration team leader. For organizations that do not have a system engineering team leader, the chief engineer or project manager often leads this effort.

System level functions are defined as physical performance variables that can be measured to assess the fulfillment of the system requirements. The functions are ideally defined

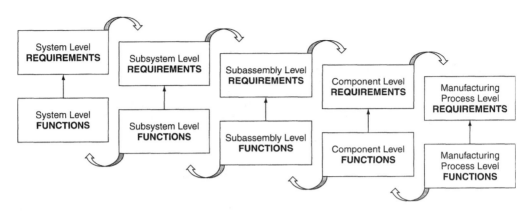

Figure 6–9   System Connectivity

independent of how they are accomplished. The focus is on a measurable response that can be specified as either an engineering scalar or vector. It is important to define continuous variables that will be directly and fundamentally affected by the soon-to-be-defined subsystem functions. The system functions have a strong influence on the definition of the subsystem functions. The subsystem functions are strongly dependent on subassembly and component functions, which in turn are strongly affected by manufacturing process functions. These interconnections are depicted in Figure 6–9.

Critical Parameter Management derives from a carefully defined architectural flow down of requirements that can be directly linked to functions that are engineered to flow up to fulfill the requirements.

Later we will discuss a method of analytically modeling these functional hand-offs across the architectural boundaries within the system. We call these analytical models *ideal/transfer functions*.

The system level functions will be used to help define the many possible system architectures that can be configured to fulfill the functions. Here we see how customer needs drive system level technical requirements. The system level technical requirements drive the system level engineering functions. The final linkage is how the system level engineering functions drive the definition of the system level architectural concepts (Figure 6–10). Once a system architecture is estimated from this perspective, the inevitable trade-offs due to subsystem, subassembly, and component architectures will begin.

As you can see in Figure 6–10, the system drives the design of everything within its architecture.

Generate System Concept Evaluation Criteria

The criteria used to assess all of the candidate system concepts, against a best-in-class benchmark datum, must come from the system requirements and the system functions. These criteria must

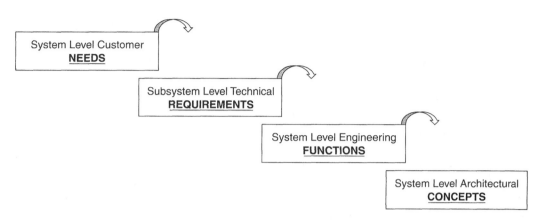

Figure 6–10   System Level Linkages

be detailed in nature. Stating low cost as criteria is too general. One must decompose the elements that drive cost and state them as criteria. This rule applies to most requirements. It is OK to state a system reliability goal, but it may be better to state the more detailed criteria that lead to overall reliability. In this context, low part count would be reasonable criteria.

The goal of concept development is to define a superior architecture that is high in feasibility and low in competitive vulnerability. A major mistake that many engineering teams make is confusing feasibility and optimization. A major focus in the concept phase of the product commercialization process is developing the attribute known as feasibility. Feasibility is characterized by the following traits:

1. Projected or predicted ability to fulfill a requirement
2. Predicted capacity to perform, as opposed to *measured* capability (Cp) to perform
3. Inherent robustness to customer use variation and environments
4. Logical in an application context
5. Likelihood or possibility of balancing requirements

Engineers are typically undertrained in the tools and best practices for developing feasibility. They tend to have extensive analytical and mathematical training in developing optimality. Feasibility is measured in more generic and general terms against a broad set of system level requirements. Optimality is typically measured in specific engineering units against specific performance targets, usually at the system and subsystem level. We often find engineering teams overly focused on subsystem optimization much too early in the commercialization process. This leads to subsystems that work well as stand-alone units under nominal laboratory conditions. When these "optimized" subsystems are integrated into the system configuration, they tend to suffer significant functional performance problems and must undergo significant redesign efforts. Once in a while, they actually must be scrapped altogether and a new subsystem concept must be developed. In cost of poor quality terms, design, scrap, and rework costs are often found to have their root cause back in this concept development phase of commercialization.

Feasibility metrics are focused on requirements that are embryonic and developmental in nature. Requirements will undergo some evolution and refinement as the product commercialization process moves forward. Optimality metrics become clear later in the CDOV process, specifically in the optimization phase (Phase 3). Many engineers become impatient, jump the gun on feasibility development, prematurely define just one vague system concept, and immediately begin analytical optimization across the various subsystems. The claim is, "We are fast, we get right at the job of engineering the product." The results of taking shortcuts like this are underdeveloped system architectures that are underdefined with regard to their balanced capacity to fulfill all the system requirements and functions. This being the case, you can see how a number of subsystems can end up being poorly developed simply due to weakly defined system concept criteria.

Generate System Concepts That Fulfill the Functions

System architecting is a body of knowledge comprised of both science and art. System concepts are "architected," or structured, through a set of steps that synthesize the following items:

1. System requirements that fulfill the VOC needs
2. System functions that fulfill the system requirements
3. Engineering knowledge of system design, interface development, and integration
4. Prior experience with system performance issues, successes, and failures

A superb reference for conducting this unique discipline is *Systems Architecting,* by Eberhardt Rechtin (1990). It contains many heuristics, or "rules of thumb," that are extremely useful in guiding the process of defining system architectural concepts.

One of the most important aspects of developing candidate system concepts is to develop more than one. If just one system concept is developed, it is highly likely it will be lacking in its ability to possess balanced fulfillment of the system requirements. If numerous system concepts are developed and compared to a best-in-class benchmark datum concept, using a comprehensive set of criteria based on the system requirements and system functions, then it is much more likely that a superior conceptual architecture can be identified. The power of this approach, known as the Pugh concept selection and evaluation process, is in its ability to integrate the best elements from numerous concepts into a "super-synthesized" superior concept.

This approach drives strong consensus around a team-developed system architecture. It is fine to have a chief system architect, but no one personality should dominate the definition of the actual system concept. It is desirable to have one person lead a disciplined process of system architectural development.

The results of the system architecting process must come from the team, but the focus on the disciplined use of system conceptual development tools frequently comes from a strong leader of the team. The leader should not dictate a concept but should insist on using the right tools and process so the team can develop the right concept.

Evaluate System Concepts

System concepts, in the plural, are essential to the proper convergence on a single, superior system concept that will serve as the basis for all subsystem, subassembly, component, and manufacturing process requirements. Too many products have their system defined and driven by the subsystems. This obviously works but is a recipe for system integration nightmares. Thus, we need to evaluate numerous system concepts.

The system concepts we evaluate are developed by individuals or small groups. If one person generates all the concepts, a personal bias or other form of limitation can greatly inhibit the potential to fulfill and balance all the system requirements. You need many concepts with many approaches from diverse perspectives.

Each system concept is compared to the best-in-class benchmark datum using the criteria. It is typical to have between 20 and 30 criteria, depending on the nature and complexity of the system requirements and functions. All team members should have one or two candidate concepts that they have personally brought to the evaluation process. The concepts are evaluated by using nonnumeric standards of comparison:

(+) is assigned to concepts that are better than the datum for a given criteria

(−) is assigned to concepts that are worse than the datum for a given criteria

(S) is assigned to concepts that are the same as the datum for a given criteria

These evaluation standards are efficient for the purpose of assigning value in terms of feasibility. If numbers are used, the evaluation can easily drift away from feasibility and become a typical engineering argument over optimal concepts.

Select Superior System Concept

The evaluation is meant to be iterative. Divergent concepts are now integrated. It is good to step back after a first pass through the evaluation process and integrate the best features and attributes of the various "winning" concepts. This will generate a smaller, more potent set of synthesized concepts that can then go through a second round of evaluation. If one of the concepts clearly beats the datum in the first round, it becomes the new datum. The synthesized concepts are then compared to the new datum. Once a second round of comparisons is done, a superior concept tends to emerge.

A superior concept is defined as the one that scores highest in feasibility against the criteria—in comparison to the datum. The team should be able to clearly articulate exactly why the selected concept is truly superior to the others and why it is low in competitive vulnerability.

Analyze, Characterize, Model, and Predict Nominal Performance of the Superior System

With a superior concept in hand, the detailed engineering analysis at the system level can begin. It is now reasonable to invest time and resources in system modeling in depth. This is the time when a system level "black box network model" is formally laid out. It is where the System Level Critical Parameter Management model is initially developed.

Up to this point, the Critical Parameter Management database just has the structured VOC data (from KJ Analysis) residing at the highest level with direct links established to the appropriate system requirements that were developed, in part, by using the system level House of Quality from quality function deployment.

The CPM database can now be structured to include the system level network model of "ideal/transfer functions" that help represent the sensitivity relationships between system level requirements and system level critical functional responses. These are typically estimated first order equations that enable system level performance trade-off studies to be quantitatively modeled. These are equations that express correlations and the magnitude and directionality of sensitivity between system level CFRs. They are commonly referred to as **transfer functions** in the jargon of Six Sigma users because they quantify the hand-off or transfer of functional sensitivity. It is important to invest in developing predictions of potentially interactive relationships between the system level critical functional responses and their requirements. Remember that the "roof" of the system House of Quality should provide our first clue that certain requirements may indeed possess interactive or codependent relationships. The system level transfer functions should account for the actual magnitude and directionality of these predicted codependencies. A little later in the design phase, the subsystem level sensitivity relationships will be identified as they relate up to the system level critical functional responses and the system level requirements they attempt to fulfill.

These system level transfer functions will have to eventually be confirmed by empirical methods. This usually occurs during the design, optimization, and verify phases as part of the system engineering and integration process. Designed experiments, analysis of variance, regression, robust design, empirical tolerancing, and response surface methods are the leading approaches for confirming these sensitivity models between the various system level CFRs.

Develop Reliability Requirements and Initial Reliability Model for the System

Now that we have a System Concept defined, a first cut at an integrated model of system requirements and system functions in hand, we can initiate the development of a system reliability model. This model will contain the required system reliability targets and the predicted reliability output at the system level. With system reliability targets and projected reliability performance the team will have an estimate of the reliability gap between the required and predicted reliability of the system.

This process inevitably requires the superior subsystem concepts to be developed somewhat concurrently with the system so that their effect on the system performance can be estimated. In this context, subsystem definition must be considered during system definition. The hard part is to refrain from letting predetermined subsystem concepts define the system requirements. The system must drive the subsystems, which in turn must drive subassembly, component, and manufacturing process requirements. Refinements, trade-offs, and compromises must be made in a concurrent engineering context to keep everything in balance. It is never easy to say exactly what the system requirements are until the subsystems are defined because of inevitable dependencies. The system requirements and functions change as the subsystem architectures are firmed up. It is crucial to systems architecting and engineering goals to have a strong focus on the system requirements as a give-and-take process takes place between system and subsystem architecture and performance trade-offs. The key thing to remember is system requirements should take precedence over and drive subsystem requirements. This rule can obviously be broken successfully, but in a DFSS context it should be done only under the constraints and insights from the Critical Parameter Management process.

Gate 1 Readiness

At Gate 1 we stop and assess the full set of concept development deliverables and the summary data used to arrive at them. The tools and best practices of Phase 1 have been used, and their key results can be summarized in a Gate 1 scorecard (Figure 6–11). A checklist of Phase 1 tools and their deliverables can be reviewed for completeness of results and corrective action in areas of risk.

Prerequisite Information to Conduct Phase 1 Activities

Market segmentation analysis

Market competitive analysis

Market opportunity analysis

Market gap analysis

Product family plan

General Phase 1 Gate Review Topics

- Business/financial case development
- Technical/design development
- Manufacturing/assembly/materials management/supply chain development
- Regulatory, health, safety, environmental, and legal development
- Shipping/service/sales/support development

Checklist of Phase 1 Tools and Best Practices

Market segmentation analysis

Economic and market trend forecasting

Business case development methods

Competitive benchmarking

VOC gathering methods

KJ analysis

QFD/House of Quality—Level 1 (system level)

Concept generation techniques (TRIZ, brainstorming, brain-writing)

Modular and platform design

Phase 1 Concept Development

Deliverable	Owner	Original Due Date vs. Actual Deliv. Date	Completion Status			Performance vs. Criteria			Exit Recommendation			Notes
			Red	Green	Yellow	Red	Green	Yellow	Red	Green	Yellow	
Ranked & Documented VOC Database			○	○	○	○	○	○	○	○	○	
System Level House of Quality			○	○	○	○	○	○	○	○	○	
Competitive System Benchmarking Data			○	○	○	○	○	○	○	○	○	
System Level Requirements Documents			○	○	○	○	○	○	○	○	○	
Documented List of System Level Functions That Fulfill the System Requirements			○	○	○	○	○	○	○	○	○	
Documented Superior System Concept			○	○	○	○	○	○	○	○	○	
Portfolio of Candidate Technologies That Help Fulfill the System Functions & Requirements			○	○	○	○	○	○	○	○	○	
Documented Critical Parameter Database with estimated System CFR Interaction Transfer Functions			○	○	○	○	○	○	○	○	○	
Documented Reliability Model and Predicted System Reliability Performance with Gap Analysis			○	○	○	○	○	○	○	○	○	
Estimated System Cost Projection			○	○	○	○	○	○	○	○	○	
Phase 2 (Design Development) Project Management Plans			○	○	○	○	○	○	○	○	○	

Figure 6–11 A General Scorecard for CDOV Phase 1 Deliverables

> System architecting methods
>
> Knowledge-based engineering methods
>
> Pugh concept evaluation and selection process
>
> Math modeling (business cases, scientific and engineering analysis)

CDOV Phase 2: Subsystem Concept and Design Development

The diagram in Figure 6–12 illustrates the major elements for CDOV Phase 2: Develop subsystem concepts and designs.

Transitioning into Phase 2 of the CDOV product commercialization process requires that the major deliverables from the tools and best practices from Phase 1 are adequately met. It is also predicated on proper, advance planning for use of the design tools and best practices. This is all part of the preventative strategy of Design For Six Sigma.

Create Subsystem Houses of Quality

It is becoming relatively common to see companies that are deploying DFSS develop a system level House of Quality. Unfortunately the use of the lower level Houses of Quality tends to drop

Start Date: ___	Create subsystem Houses of Quality	End Date: ___	Owner: _____
Start Date: ___	Conduct subsystem benchmarking	End Date: ___	Owner: _____
Start Date: ___	Generate subsystem requirements documents	End Date: ___	Owner: _____
Start Date: ___	Develop functions that fulfill the subsystem requirements	End Date: ___	Owner: _____
Start Date: ___	Generate subsystem concept evaluation criteria	End Date: ___	Owner: _____
Start Date: ___	Generate subsystem concepts that fulfill the functions	End Date: ___	Owner: _____
Start Date: ___	Evaluate subsystem concepts	End Date: ___	Owner: _____
Start Date: ___	Select superior subsystem concepts	End Date: ___	Owner: _____
Start Date: ___	Analyze, characterize, model, predict, and measure nominal performance of superior subsystem (including DFMA, initial tolerances, and cost analysis)	End Date: ___	Owner: _____
Start Date: ___	Develop reliability model and DFMEA for each subsystem	End Date: ___	Owner: _____

(continued)

Figure 6–12   Flow of Phase 2 of Product Commercialization

Start Date: ___	Create subassembly, component, and manufacturing process Houses of Quality	End Date: ___	Owner: _____
Start Date: ___	Conduct subassembly, component, and manufacturing process benchmarking	End Date: ___	Owner: _____
Start Date: ___	Generate subassembly, component, and manufacturing process requirements documents	End Date: ___	Owner: _____
Start Date: ___	Develop functions that fulfill subassembly, component, and manufacturing process requirements	End Date: ___	Owner: _____
Start Date: ___	Develop functions that fulfill subassembly, component, and manufacturing process requirements	End Date: ___	Owner: _____
Start Date: ___	Generate subassembly, component, and manufacturing process evaluation criteria	End Date: ___	Owner: _____
Start Date: ___	Generate subassembly, component, and manufacturing concepts that fulfill the functions	End Date: ___	Owner: _____
Start Date: ___	Evaluate subassembly, component, and manufacturing process concepts	End Date: ___	Owner: _____
Start Date: ___	Select superior subassembly, component, and manufacturing process concepts	End Date: ___	Owner: _____
Start Date: ___	Analyze, characterize, model, predict, and measure nominal performance of superior subassemblies, components, and manufacturing processes (including DFMA, initial tolerances, and cost analysis)	End Date: ___	Owner: _____
Start Date: ___	Develop reliability models and DFMEAs for reliability critical subassemblies and components	End Date: ___	Owner: _____

Figure 6–12 *Continued*

off precipitously after the completion of the system House of Quality. The system House of Quality can and should be used to develop subsystem Houses of Quality for each subsystem within the product architecture. The subsystem House of Quality is extremely useful in the development of clear, well-defined subsystem requirements documents. The subsystem requirements must align with and support the fulfillment of the system requirements. In the same context, the subsystem functions must align with and fulfill the system functions. It is our opinion that one of the weakest links in the chain of product commercialization best practices is the cavalier approach design teams take relative to developing clear subsystem, subassembly, component, and manufacturing process requirements. We know from years of experience that it is quite rare to find design teams with clear, well-documented requirements much below the system level. Things tend to get pretty undisciplined and sloppy down in the design details. This is yet another reason why most companies suffer from moderate to severe performance problems that require costly redesign at the

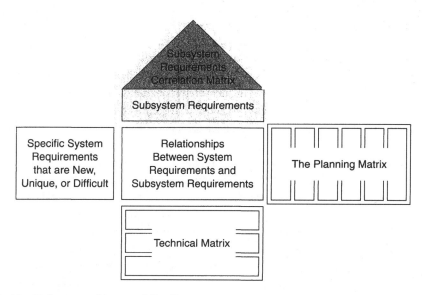

Figure 6–13 Subsystem House of Quality

subsystem level. The design phase of **CDOV** is specifically arranged to instill much needed discipline in this early but crucial area of requirements definition. It is yet another key element in Critical Parameter Management.

At a general level, the subsystem Houses of Quality (Figure 6–13) are designed to illustrate and document specific system requirements as inputs, specific fulfilling subsystem requirements as outputs, and a variety of system-to-subsystem criticality quantification and correlation relationships. They also have "rooms" that quantify competitive benchmarking relationships, both from the customer's and the engineering team's perspective, as they align with the system and subsystem requirements.

It bears repeating that only that which is new, unique, or difficult from the system requirements should be brought down into the subsystem Houses of Quality.

If the subsystem Houses of Quality are not generated, it is highly likely that there will be some form of system integration problem stemming from this lack of assessment and documentation that must be reacted to later on in the commercialization process. It just makes good sense to apply the discipline of QFD to the system-to-subsystem requirements that are new, unique, and difficult. Without this focus, these things may transform into that which is *expensive, embarrassing, and treacherous.* We view the construction of the subsystem Houses of Quality and the subsystem requirements documents as key steps in the DFSS strategy of design problem prevention.

Conduct Subsystem Benchmarking

When appropriate, subsystems should be benchmarked from the customer's perspective. If the product design is such that the customer has no possible access to or interest in one or more of the subsystems, then the benchmarking activity will be solely conducted by the engineering team from a technical perspective. If the product is serviceable or repairable, then the benchmarking perspective of the "customer in the form of the repair technician" would be required.

A clear case where subsystem benchmarking from a customer's perspective would be necessary is found in the replacement process of inkjet cartridges in a personal desktop printer. A case where the customer would have little role in subsystem benchmarking would be the efficiency of a sealed cooling subsystem within a light projector used for business presentations.

Subsystem benchmarking is important because it adds information that can be used to help establish which subsystems possess more critical, value-adding functions from those that are of less consequence. The customer benchmarking input helps the team identify advantages in competitive features. The technical benchmark data can uncover performance or architectural design information that can be targeted, surpassed, or exploited if a weakness is found. Benchmarking essentially is all about exposing vulnerabilities and opportunities. The ensuing concept generation, evaluation, and selection processes are all about averting, minimizing, or at least recognizing competitive vulnerabilities. Sometimes it is wise to allow yourself to be equaled or surpassed by a competitor in certain areas that are not strategic to your business case. In the copier industry, Six Sigma quality is strategic in the area of image quality but not in cabinetry fit and finish (at least not yet). The auto industry is another matter—fit and finish of interior and exterior body elements is a matter for Six Sigma performance metrics.

Generate Subsystem Requirements Documents

We cannot overemphasize the importance of creating clear, well-defined subsystem requirements documents. This is done by combining the subsystem House of Quality data with all the other requirements associated with each of the subsystems that comprise the integrated system. The subsystem Houses of Quality should not contain all the subsystem requirements. They cover just those requirements that are new, unique, or difficult. The subsystem teams are responsible for developing designs that fulfill all of the requirements—those that are "critical" to the new, unique, and difficult requirements that have the largest impact on fulfilling the VOC needs as well as the rest of the requirements that are of general importance. If a subsystem is being reused from a previous product, then many of the requirements are simply repeated from the last design. A small set of additional or "upgraded" requirements is added to the list of "reused" requirements. All requirements are important, but a few are absolutely critical in identifying the subsystem's functional performance. The team may not know exactly which ones are truly critical, but it's a good bet that many will come from the subsystem Houses of Quality.

Develop Functions That Fulfill the Subsystem Requirements

The subsystem teams need to assign a means of measuring the functions that prove that a requirement has been fulfilled. These functions are identified independent of the concepts that will soon be generated to fulfill them. Once a subsystem concept is defined, the final critical functional responses can be identified along with their specific data acquisition methodology.

Defining a generic set of subsystem functions initiates another key element of the Critical Parameter Management process. The generic set of functions will imply a system of metrics within each subsystem that will eventually evolve into the hardware, firmware, and software used to measure the nominal performance, robustness, tunability, and capability of the subsystem.

Generate Subsystem Evaluation Criteria

Concept evaluation criteria, which will be used in the Pugh concept selection process, must be based on the subsystem requirements. It is typical for the concept criteria to be detailed in nature. The team frequently breaks general or broadly stated requirements down into two or more focused or descriptive criteria statements that provide enough detail to discriminate superiority and low vulnerability between the datum concept and the numerous candidate subsystem concepts. As was the case in system criteria generation, feasibility is what is being evaluated, not optimality. The scope and depth of the criteria need to reflect this goal. It is OK to develop the evaluation criteria before, during, and after the concepts have been generated. In fact, it is quite common for criteria to be added, removed, or modified during the concept evaluation process.

Generate Subsystem Concepts That Fulfill the Functions

This step defines the "form and fit" options that can potentially fulfill the requirements and functions. These are commonly known as *subsystem concepts.* They are also referred to as candidate subsystem architectures. As is always the case in concept development, the team needs to generate numerous subsystem concepts to be compared to a best-in-class benchmark datum.

Concept generation tools range from group-based brainstorming techniques to individual methods that help stimulate creativity in developing concepts. TRIZ is a computer-aided approach to stimulating inventions and innovations. A lone innovator often uses TRIZ as he or she develops numerous concepts. Individuals commonly do concept generation, while concept evaluation is done in small groups (7–10 people). It is important to set some standards for describing and illustrating the concepts. It is undesirable to have some concepts that are underdefined or poorly illustrated while others are highly defined and clearly illustrated. It is therefore necessary for the team to agree to a common standard of thoroughness and completeness of concept documentation. Areas to consider are as follows:

1. **Textual** description of the concepts
2. **Graphical** representation of the concepts
3. **Mathematical equations** that underwrite the basic principles of functional performance
4. **Physical models** of the concepts

However the team chooses to document each concept, everyone must adhere to reasonably similar representations for the sake of fair comparisons.

Evaluate Subsystem Concepts

Each subsystem concept is compared to the best-in-class benchmark datum using the criteria. It is typical to have between 20 and 30 criteria, depending on the nature and complexity of the subsystem requirements and functions. Each team member should have one or two concepts that he or she has personally brought to the evaluation process. The concepts are evaluated by using nonnumeric standards of comparison:

(+) is assigned to concepts that are better than the datum for a given criteria

(−) is assigned to concepts that are worse than the datum for a given criteria

(S) is assigned to concepts that are the same as the datum for a given criteria

These evaluation standards are efficient for the purpose of assigning value in terms of feasibility. If numbers are used, the evaluation can easily drift away from feasibility and become a typical engineering argument over optimal concepts.

Select Superior Subsystem Concepts

The numerous, divergent subsystem concepts are compared to the datum during the first round of evaluation. Once this is done, those subsystem concepts that score high in the comparison can have their best attributes integrated with similar concepts. This concept synthesis process causes a convergence of criteria-fulfilling features, functions, and attributes that often enable the team to find one or more concepts that surpass the datum. One of the superior candidate concepts is used to replace the original datum. A second round of evaluation is then conducted; comparing a smaller, more powerful group of "super-synthesized" subsystem concepts to the new datum. Once a clearly superior subsystem concept emerges, it now is worth taking forward in the design process. If the concept evaluation process is bypassed, it is highly likely that the subsystem that is taken into design will suffer from an inability to provide balanced fulfillment of the subsystem requirements. It is easy to go with one "pet" concept to the exclusion of generating any others, but it is hard to fix its conceptual vulnerabilities later!

Analyze, Characterize, Model, Predict, and Measure Nominal Performance of Superior Subsystems

With a superior set of subsystem concepts defined, it is now time to add a great deal of value to them. They are worth the investment that will be made in them through the application of a number of tools and best practices. In traditional Six Sigma methods, the Measure, Analyze, Improve, and Control (MAIC) process is applied to previously designed and stabilized production processes. In new or redesigned subsystems, the approach is different. The subsystem designs are not yet statistically stable "processes" in the context of Walter Shewhart's approach to statistical process control. In fact, they are just subsystems within a system design, not production processes. The MAIC approach has its historical basis in manufacturing process control charting based on the fact that a production process is in a state of "statistical control." This assumes all assignable causes of variation are known and removed. It also assumes that the variation in the function of the process is due to random, natural causes that are small in comparison to the assignable, non-natural causes (the kind induced in a Taguchi robustness experiment) of variation.

We simply don't know enough about the subsystem parameters and measured responses to state whether the subsystem is or is not under statistical control from a steady-state production point of view. We have design and optimization work to do on the new subsystem concepts before they are ready for the type of long-term capability characterization that comes from the traditional MAIC process. It is possible during the design phase to generate and track short-term design capability performance. It is possible during the optimize and verify phases to emulate long-term shifts by intentionally inducing special cause sources of variation using Taguchi methods. In this sense, the design team can simulate Cpk data. One must recognize that this is a good thing to do, but it is only a simulation of the actual variation to come when real production and customer use begins.

The MAIC process fits nicely within the verify phase of the **CDOV** commercialization process. By then the subsystem designs are mature enough to be:

1. **Measured** against final design requirements
2. **Analyzed** for statistical performance in a production context
3. **Improved** by adjusting the mean of the design critical functional responses onto the desired target using critical adjustment parameters that were developed during the design and optimization phases of CDOV
4. **Controlled** during the steady-state production and customer use process

Analysis, Characterization, and Modeling in the Design Phase

Each subsystem is ready to be formally analyzed, characterized, and modeled. This is done using a blend of knowledge-based engineering tools and best practices along with the Design For Six Sigma tools and best practices. The chief goals within the design phase are to analytically predict and then empirically measure the nominal or baseline performance of the superior subsystems. It is here that the subsystem critical parameters are identified and developed. The modeling focuses on developing basic relationships called *ideal/transfer functions*. These linear and nonlinear systems of equations are comprised of dependent design parameters, (Y) variables, and independent variables, the subsystem (x) variables. Actually, the models also must account for the fact that some of the (x) variables may not be totally independent from one another in their effect on the (Y) variables. We call these codependent x-to-x variable relationships *interactions*. The most popular method of developing these ideal/transfer functions is through the use of a blend of analytical modeling methods and experimental modeling methods.

Measurement Systems in the Design Phase

Probably the most value-adding feature that Six Sigma methods bring to the design, optimization, and verify phases of product commercialization is its rigorous approach to measurement. Within DFSS, the Critical Parameter Management approach leads one to focus on the measurement of critical functional responses (usually in engineering units of **vectors** of magnitude and direction) and critical-to-function specifications (usually in engineering units of **scalars** of magnitude only). We will go into great depth on these measures in the chapters on Critical Parameter Management (particularly in Chapter 12 on the metrics for CPM).

Before the evolving subsystems can be experimentally evaluated, they must undergo detailed development for data acquisition. Design For Six Sigma builds value and prevents problems by measuring things that avoid the cost of poor quality (COPQ). The root causes of COPQ reside in the ability of the mean and standard deviations of the critical functions to possess low variability while maintaining on-target performance. We would have a hard time attaining that kind of performance by measuring defects. *What is the meaning of the mean and standard deviation of a defect?* Design teams must invest in the development and certification of measurement systems that provide direct measures of functional performance. We cannot afford to wait until a failure occurs to assess design quality. We must instill discipline in our teams to avoid measures of attributes, yields, go-no events, defects, and any other measure of quality that is not fundamental to the flow, transformation, or state of energy and mass within the function of the design.

Measurement system development and analysis is a key tool in DFSS. It certifies the capability of instruments that are designed to measure continuous variables that represent the functions of the design as opposed to the more technically ambiguous measures of quality after the function is complete. We caution against waiting to measure design performance after all the value-adding functions are over. When we link this function-based metrology approach to designed experimentation, powerful results based on functional data are delivered at the CDOV gate reviews.

Statistically designed experiments, called **design of experiments** (DOE), are some of the most heavily used tools from the methods of Six Sigma. An approach known as *sequential experimentation* can be used to efficiently and effectively develop critical parameter knowledge and data:

1. **Screening experiments** define basic, linear, main effects between a *Y* variable and any number of *x* values. This approach usually just looks at the effects of independent *x* variables.

2. **Main effects and interaction identification experiments** define how strong the main effects and the interactions between certain main effects (*x*s) are on the *Y* variables.

3. **Nonlinear effects experiments** use a variety of experimental data gathering and analysis structures to identify and quantify the strength of nonlinear effects of certain *x* variables (second order effects and beyond as necessary).

4. **Response surface experiments** study relatively small numbers of *x*s for their optimum set points (including the effects due to interactions and nonlinearities) for placing a *Y* or multiple *Y*s onto a specific target. They can also be used to optimally reduce variability of a critical functional response in the presence of noise factors but only on a limited number of *x* parameters and noise factors.

These four general classes of designed experimentation somewhat understate the power resident in the vast array of designed experimental methods that are available to characterize, model, and enable many forms of analysis for the basic and advanced development of any subsystem, both in terms of mean and variance performance.

Produceability in the Design Phase

Included in the design characterization and analysis work is the process of designing for manufacturability and assembleability, the development of the initial functional and assembly tolerances that affect the output of the subsystem critical functional responses, and, of course, design cost analysis.

Even though some consideration for manufacturing and assembly is a necessity during the subsystem concept generation process, there is always more to do as the design phase adds depth and maturity to the subsystems' form, fit, and function. Formal application of design for manufacturability and assembleability methods is an essential element at this point in the progressive development of each subsystem design. It is also an excellent point in the commercialization process to establish initial tolerances along with a detailed cost estimate for each subsystem.

Design and manufacturing tolerances are embryonic at this point in the CDOV process. They are mainly estimates. Relationships within and between subsystem functions are being quantified and their transfer functions documented. Relationships with subsystem, subassembly, and component suppliers are also in their early stages of development. Two classes of tolerances need to begin to be balanced to have a chance at achieving affordable Six Sigma performance: (1) functional performance tolerances on critical functional responses within and between subsystems and subassemblies and (2) assembly/component tolerances. One class focuses on functional sensitivities and the tolerances required to constrain functional variation. The other class focuses on form and fit sensitivities and the tolerance limits that balance components and subassemblies as they are assembled into subsystems. Subsystem-to-system integration tolerances also must be projected and analyzed for "form, fit and function" at this point. Initial tolerances for both classes can be established using the following analytical tolerance design tools and best practices:

1. Worst case tolerance studies
2. Root sum of squares tolerance studies
3. Six Sigma tolerance studies
4. Monte Carlo simulations

It is also an opportune time to begin to build a detailed database of supplier manufacturing and assembly process capabilities (Cp and Cpk) to minimize the risk of a process-to-design capability mismatch. Here again we see the preventative, proactive nature of the Design For Six Sigma process.

Develop Reliability Model and FMEA for Each Subsystem

The system reliability model from Phase 1 should contain an overall reliability growth factor based on the projected reliability gap between required system reliability and the current system reliability. It is common practice to derive a reliability allocation model from which a reliability budget flows down to each major subsystem. Each subsystem needs to be analyzed for its projected ability to meet the system budget. It is then possible to define a subsystem reliability gap that must be closed for the system to meet its goals.

After the subsystem reliability requirements are documented, the team must embark on defining where the reliability-critical areas reside within the subsystems. A strong method for identifying critical reliability areas in any design is called design failure modes and effects analysis (DFMEA). A more rigorous form includes a probability of failure value known as *criticality*. Thus we see the DFMECA method includes a "probability of occurrence" term. The DFMEA data helps the team predict the reliability of the developing subsystem. It highlights what the likely failure modes are and what specific effect they will have on the reliability of the subsystem and ultimately on the system. DFMEA data is not only useful in defining preventative action in the development of reliability, it is extremely useful in the process of developing robustness, which will be discussed in the optimization phase.

It is worth stating that DFSS, in the context of the CDOV process, has an extremely aggressive approach to developing reliability. The CDOV process contains a heavy focus on reliability requirement definition, modeling, and prediction at the front end of the process and a strong focus on assessing attained reliability at the tail end of the process. The tools and best practices in the design and optimize phases also do a great deal to develop and measure surrogate, forward-looking measures of performance that directly affect reliability (means, standard deviations, signal-to-noise metrics, and CP/Cpk indices). Waiting to measure mean time to failure and other time-based measures of failure is not an acceptable Six Sigma strategy. DFSS is not fundamentally reactive. We do everything possible at every phase in the CDOV process to get the most out of the DFSS tools and best practices for the sake of preventing reliability problems.

Now, the design process steps down to the subassembly, component, and manufacturing process levels.

Create Subassembly, Component, and Manufacturing Process Houses of Quality

With subsystem requirements defined, the team can proceed into a "concurrent engineering" mode of design. After structuring the flow-down of the Houses of Quality, the development of subassemblies, components, and manufacturing processes can proceed in any mix of serial and parallel activities, based on tools and best practices, that make sense to the subsystem design teams. The design teams must resist short-circuiting the development of the subordinate Houses of Quality. Otherwise, these teams will be highly likely to miss something relative to what is new, unique, or difficult down in the details of the subsystem designs. The overarching rule for conceptual development is to let the hierarchical flow-down of system-to-subsystem-to-subassembly-to-component-to-manufacturing process requirements lead to the sequencing of concept generation, evaluation, and selection. The design development tasks will then become a series of trade-offs being made on a daily basis between the form, fit and functions within and between the hierarchy of design elements. The Critical Parameter Management process, under the leadership of the system engineering team and its process, will be your guide to balancing sensitivities and solving the "many-to-many" parametric problems you will encounter within and between the subsystems and their elements.

Conduct Subassembly, Component, and Manufacturing Process Benchmarking

Now that we are getting down into the details of the subsystems and their subordinate requirements, architectures, and functions, not every subassembly, component, or manufacturing process will need to go through competitive or comparative benchmarking. This is a judgment call within the context of what is deemed critical to function and reliability for a given cost.

As part of Critical Parameter Management, the Houses of Quality and the resulting requirements will help make it apparent when a subassembly, component, or manufacturing process is critical enough to require the benchmarking of alternatives for your consideration. When the subassembly or component can be effectively outsourced, then benchmarking should certainly be considered. Outsourcing should be done when a supplier can make designs or components better and cheaper than you can. Many companies are reducing their vertical integration—some for good reasons, but others are walking away from things that should remain as core competencies. Making certain critical subassemblies and components in-house is a strategic decision, which DFSS and Critical Parameter Management can and should greatly influence. Developing specific capability in critical manufacturing and assembly processes falls into the same scenario. Benchmarking in the context of Design For Six Sigma and Production Operations For Six Sigma will help focus what you choose to keep inside your company and what to safely outsource. Great care should be exercised in the sourcing of critical-to-function or critical-to-reliability subassemblies and components.

Benchmarking is often used to help identify good ideas for how competitors develop functions as well as how they develop form and fit architectures that generate functions. Benchmarking architectures is common but we also highly recommend separating the form from the function so your team gets a good sense for both. We have seen successful, revenue-bearing patents come from copying competitive functions without copying their form and fit.

Generate Subassembly, Component, and Manufacturing Process Requirements Documents

It is just as important to develop and fulfill a comprehensive set of requirements at this detailed level of design and manufacturing as it is at the system and subsystem levels. The network of critical parameter relationships can get quite complex down at this level. It has been our experience that some design teams relax the rigor of requirement definition after the system level requirements document is produced. Some of the biggest problems to result from this lack of follow-through on requirements definition are anemic metrics and measurement systems. When subassemblies, components, and manufacturing processes are undermeasured, they tend to underperform later in the system integration, production certification, and customer use phases of the product's life cycle. The system requirements often are difficult to fulfill because the design elements that must support them are underdefined and consequently incapable of delivering their portion of required performance. Each of these three design elements should have their requirements documents defined from a blend of the new, unique, or difficult things that translate out of their respective Houses of Quality along with all other important requirements.

Develop Functions That Fulfill the Subassembly, Component, and Manufacturing Process Requirements

The flow of functions that are essential to the fulfillment of the hierarchy of the product's requirements must be supported by these three elements of the design. Manufacturing processes provide functions to convert mass and energy into components. The components, in turn, provide an integrated set of contributions in support of subassembly and subsystem functionality.

Subassembly functions are developed based on the subassembly requirements. They, too, must indicate how the requirements will be fulfilled without specifically stating how the function will be accomplished. Subassemblies are typically required to create a mechanical, electrical, or chemical function based on an integrated set of components.

Components are a little different because they do not, by themselves, transform or change the state of energy and mass. They must be integrated into a subassembly or subsystem to add up to the accomplishment of the subassembly or subsystem functions. When we discuss additivity in the robust design material, it directly relates to this notion of design elements "adding up" their contribution to facilitate functional performance. Components tend to possess parametric characteristics that are best quantified under some form of specification nomenclature. This is why we call critical functional forms, fits, and features of a component "critical-to-function *specifications.*" These typically are documented as dimensions and surface or bulk characteristics. A structural component may have a deflection requirement that is fulfilled through its ability to provide stiffness. This does not suggest how the component will provide stiffness, just that whatever component architecture is selected must be able to be measured in the units of stiffness to fulfill a deflection requirement [area moment of inertia (in.^{4}) controlling a deflection (in.)].

Manufacturing processes create functions to make components. They control mass and energy states, flows, and transformations that ultimately become components. If the functions within manufacturing processes are directly linked to the manufacturing process requirements, which came from component, subassembly, and subsystem requirements, we have a much better chance of flowing functional integrity up through the system. How often do you link and relate the functions inside your production process to the functions within your designs?

Generate Subassembly, Component, and Manufacturing Process Evaluation Criteria

Criteria to evaluate the subassembly, component, and manufacturing process concepts must be developed at this point. It is not uncommon to develop the criteria during concept generation. The subassembly, component, and manufacturing process requirements documents are used to help define the concept evaluation criteria.

Generate Subassembly, Component, and Manufacturing Process Concepts That Fulfill the Functions

With the detailed functions outlined, we can effectively generate the concepts for the subassemblies, components, and manufacturing processes. This process tends to go faster than system and subsystem concept generation, because there are fewer, less complex requirements down at this level. The tools and best practices to develop these concepts are linked to the physical and structural nature implied in the requirements and functions. Creativity, brainstorming, knowledge-based engineering design methods, and TRIZ are a few examples of the tools used to generate concepts at this level. The need for numerous concepts is just as important as it was with the higher level subsystem and system concepts.

Evaluate Subassembly, Component, and Manufacturing Process Concepts

The Pugh concept evaluation and selection process is a common and popular tool used to arrive at a set of superior concepts. This approach helps prevent designs and production processes from being developed or selected with incomplete forethought. Many times it is tempting to just go with one person's "gut feel" or experience. This has proven over time to be an incomplete approach to concept selection. When a cross-functional team brings its accumulated knowledge, experience, and judgment to bear on a number of concepts, it is much more likely that a superior concept will emerge from the concept evaluation process. Integrating the best elements of numerous concepts into a design that is low in competitive vulnerability and high in its ability to fulfill the criteria is by far the best way to help drive the design to fulfill its requirements.

Select Superior Subassembly, Component, and Manufacturing Process Concepts

The final selection of the superior subassembly, component, and manufacturing process concepts can often require many compromises as design functionality vs. produceability trade-offs are made in a "concurrent engineering" context. As long as the criteria are sufficiently able to account for the range of requirements for these three elements, the likelihood of the most feasible designs and processes being selected is high.

Just as it is important to evaluate numerous concepts, it is equally important to be decisive and go forward with confidence in the selected design or process. When two relatively equal concepts come out of the selection process, it may be wise to codevelop both concepts until a clear winner emerges during the more formal development of the models of the concepts. Data is the tiebreaker!

Analyze, Characterize, Model, Predict, and Measure Nominal Performance of Superior Subassemblies, Components, and Manufacturing Processes

The analysis, characterization, modeling, predicting, and measurement of the various concepts is usually developed in the following general order:

1. **Math models** (functional performance, tolerance stack-ups, and cost estimates)
2. **Geometric models** (spatial form and fit and tolerance stack-ups)
3. **Physical models** (prototypes to evaluate function and fit)

Math and geometric modeling is often performed in parallel as these models support and enable one another.

The most common tools from DFSS in support of this activity are:

1. Design for manufacture and design for assembly
2. Measurement system analysis (design for testability)
3. Designed experiments, ANOVA data analysis, and regression [deriving $Y = f(x)$ equations]
   a. Ideal/transfer functions within and between subsystems and subassemblies
4. Critical Parameter Management
   a. Functional flow diagramming
5. Design capability studies (baseline Cp characterization of CFRs)
6. Design applications of statistical process control (nominal performance stability)
7. Multi-vari studies (screening for "unknown" design influence variables)
8. Value engineering and analysis (cost-to-benefit analysis and FAST diagramming)
9. Analytical tolerance development and simulation (*ideal/transfer functions*)

A great deal of trade-off work is done in this step. Refinement of the initial concepts is done continuously as costs, function, and produceability requirements are initially balanced. Usually the designs and processes are set at the most economical level possible at this point. If basic (nominal) functionality, manufacturability, and assembleability (the latter two terms we often refer to jointly as *produceability*) are attained, we no longer refer to these elements as concepts—they are now full-fledged designs and processes. The optimization phase of CDOV attempts to develop robustness (insensitivity to "assignable cause" sources of variation) in the designs and processes without significant increases in cost. The verify phase of the CDOV process may require significant increases in cost to attain the proper performance and reliability requirements within the scheduled delivery time of the product to the market.

Develop Reliability Models and DFMEAs for Reliability Critical Subassemblies and Components

Concurrent to the development of the designs and processes, the engineering team should be constructing reliability performance predictions. These predictions are used to roll up estimated reliability performance to fulfill overall system reliability requirements. Reliability allocations are made down to the subsystem, subassembly, and component levels within the system architecture.

As the prediction process moves forward, there are always going to be shortfalls or gaps in predicted vs. required reliability. Design failure modes and effects analysis is a major tool used in the earliest stages of reliability prediction and development. It identifies areas of weakness in the designs. It also provides insight as to exactly what is likely to cause a reliability problem. If DFMEA is done early in the design process, the team gains a strategic ability to plan and schedule the deployment of numerous reliability development tools and best practices to prevent reliability problems before they get embedded in the final design. A list of DFSS reliability development tools and best practices follows:

1. Employ DFMEA.
2. Develop a network of ideal and transfer functions using math modeling and DOE methods.
3. Conduct sensitivity analysis on the ideal and transfer functions using Monte Carlo simulations.
4. Conduct Taguchi's robustness optimization process on subassemblies and subsystems prior to system integration (optimized Cp).
5. Use response surface methods to identify critical adjustment parameters to adjust the design functions onto their performance targets to maximize Cpk.
6. Complete normal life testing.
7. Use highly accelerated life testing (HALT).
8. Use highly accelerated stress testing (HAST).

At this point in time, conducting reliability assessments on reliability-critical subassemblies and components is appropriate if they have already been through the supplier's robustness optimization process. If the supplier refuses to put them through robustness optimization, it is highly advisable to begin life testing—*yesterday.* If the supplier offers no proof of confirmed robustness but says it has been done, start life testing immediately! If the supplier agrees to conduct robustness optimization on the subassembly or component, help the supplier do it; then after you are sure the design has been optimized, you can conduct life testing procedures as necessary. It is generally a waste of time to conduct life tests on designs that will soon go through robustness optimization. Wait until that is done and then go forward with life testing.

Gate 2 Readiness

At Gate 2 we stop and assess the full set of deliverables and the summary data used to arrive at them. The tools and best practices of Phase 2 have been used and their key results can be summarized in a Gate 2 scorecard (Figure 6–14). A checklist of Phase 2 tools and their deliverables can be reviewed for completeness of results and corrective action in areas of risk.

Phase 2 Gate Review Topics

- Business/financial case development
- Technical/design development
- Manufacturing/assembly/materials management/supply chain development
- Regulatory, health, safety, environmental, and legal development
- Shipping/service/sales/support development

Phase 2 Design Development

Deliverable	Owner	Original Due Date vs. Actual Deliv. Date	Completion Status Red Green Yellow			Performance vs. Criteria Red Green Yellow			Exit Recommendation Red Green Yellow			Notes
VOC Database Linked to Subsystem, Subassembly, Component, and Manufacturing Process Requirements			○	○	○	○	○	○	○	○	○	
Subsystem, Subassembly, Component, and Manufacturing Process Level Houses of Quality			○	○	○	○	○	○	○	○	○	
Competitive Subsystem, Subassembly, Component, and Manufacturing Process Benchmarking Data			○	○	○	○	○	○	○	○	○	
Subsystem, Subassembly, Component, andManufacturing Process Level Requirements Documents			○	○	○	○	○	○	○	○	○	
Documented List of Subsystem, Subassembly, Component, and Manufacturing Process Level Functions That Fulfill Their Respective Requirements			○	○	○	○	○	○	○	○	○	
Documented Superior Subsystem, Subassembly, Component, and Manufacturing Process Concepts			○	○	○	○	○	○	○	○	○	
Portfolio of Candidate Technologies That Help Fulfill the System Functions and Requirements			○	○	○	○	○	○	○	○	○	
Documented Critical Parameter Database with Estimated Subsystem and Subassembly CFR Ideal Functions			○	○	○	○	○	○	○	○	○	
Documented Critical Parameter Database with Estimated Subsystem and Subassembly CFR Interaction Transfer Functions (between Subsystem and Subassembly Linkage up to the System CFRs)			○	○		○	○		○	○		

Deliverables					
Documented Reliability Model and Predicted System Reliability Performance with Gap Analysis	○	○	○	○	○
Estimated System, Subsystem, Subassembly,Component Cost Projections	○	○	○	○	○
Documentation Supporting Design for Manufacturability and Assembleability	○	○	○	○	○
Documented Design Failure Modes and Effects Analysis for Subsystems, Subassemblies, Reliability Critical Components, and Manufacturing Processes	○	○	○	○	○
Descriptive and Inferential Statistics: HypothesisTesting and Confidence Interval Analysis in Support of the Ideal Functions and Transfer Functions	○	○	○	○	○
Sequential Design of Experiments (Initial Identification of Subsystem, Subassembly, Component, and Manufacturing Process Critical Parameters)	○	○	○	○	○
Analysis of Variance Methods (ANOVA) That Underwrite Statistical Significance of Critical Parameters	○	○	○	○	○
Documented Cp & Cpk Values for all Subsystem Subassembly Design CFRs	○	○	○	○	○
Initial Cp/Cpk Assessment for Candidate Critical-to-Function Specifications from Manufacturing Processes from Supply Chain	○	○	○	○	○
Phase 3A (Optimization) Project (Management) Plans	○	○	○	○	○

Figure 6–14   A General Scorecard for CDOV Phase 2 Deliverables

Checklist of Phase 2 Tools and Best Practices:

Competitive benchmarking

VOC gathering methods

KJ analysis

QFD/Houses of Quality—Levels 2–5 (subsystem, subassembly, component, and manufacturing process levels)

Concept generation techniques (TRIZ, brainstorming, brain-writing)

Pugh concept evaluation and selection process

Design for manufacture and assembly

Value engineering and analysis

Design failure modes and effects analysis

Measurement systems analysis

Critical parameter management

Knowledge-based engineering methods

Math modeling (business case and engineering analysis to define ideal functions and transfer functions)

Design of experiments (Full and Fractional Factorial designs, sequential experimentation)

Descriptive and inferential statistical analysis

ANOVA data analysis

Regression and empirical modeling methods

Reliability modeling methods

Normal and accelerated life testing methods

Design capability studies

Multi-vari studies

Statistical process control (for design stability and capability studies)

CDOV Phase 3A: Optimizing Subsystems

Phase 3 of the CDOV process covers two major parts:

- *Phase 3A* focuses on the robustness optimization of the individual subsystems and subassemblies within the system.
- *Phase 3B* focuses on the integration of the robust subsystems, the nominal performance evaluation, robustness optimization, and initial reliability assessment of the complete system.

The diagram in Figure 6–15 illustrates the major elements of Phase 3A.

Start Date: ___	Review and finalize the critical functional responses for the subsystems and subassemblies	End Date: ___	Owner: _____
Start Date: ___	Develop subsystem noise diagrams and system noise map	End Date: ___	Owner: _____
Start Date: ___	Conduct noise factor experiments	End Date: ___	Owner: _____
Start Date: ___	Define compounded noises for robustness experiments	End Date: ___	Owner: _____
Start Date: ___	Define engineering control factors for robustness optimization experiments	End Date: ___	Owner: _____
Start Date: ___	Design for additivity and run designed experiments or simulations	End Date: ___	Owner: _____
Start Date: ___	Analyze data, build predictive additive model	End Date: ___	Owner: _____
Start Date: ___	Run verification experiments to certify robustness	End Date: ___	Owner: _____
Start Date: ___	Conduct response surface experiments on critical adjustment parameters	End Date: ___	Owner: _____
Start Date: ___	Run verification experiments to certify tunability and robustness parameters for subsystems and subassemblies	End Date: ___	Owner: _____
Start Date: ___	Document critical functional parameter nominal set points, CAP and CFR relationships	End Date: ___	Owner: _____
Start Date: ___	Develop, conduct, and analyze reliability/capability evaluations for each subsystem and subassembly	End Date: ___	Owner: _____

Figure 6–15   Flow of Phase 3A of Product Commercialization

Review and Finalize the Critical Functional Responses for the Subsystems and Subassemblies

Prior to investing a lot of time and resources in conducting the steps of robust design, it is important to step back and review all of the critical functional responses for the subsystems and subassemblies. These direct measures of functional performance inform the teams whether their designs are insensitive to sources of variation or not. Each CFR must have a capable measurement system that can be calibrated and certified for use during the optimization phase of the commercialization process. Again, we emphasize that the CFRs being measured should be continuous engineering variables. They should be scalars (magnitude of the measured variable) and vectors (magnitude and direction of the measured variable) that express the direct flow, transformation, or state of mass, energy, and the controlling signals that reflect the fundamental performance of the designs being optimized.

Robust design is all about how the CFRs react to sources of "noise" (variation) for a given set of control factor set points. These control factor set points will become part of the final design specifications after optimization is complete. Their relationships to the noise factors must be evaluated and exploited. Noise factors are expensive, difficult, or even impossible to control. There is a special, often ignored, class of interactions between control factors and noise factors that show up in the measured CFRs at the system, subsystem, and subassembly level in all products. *The purpose of robust design is to investigate, quantify, and exploit these unique "control factor to noise factor" interactions.* Measuring defects, yields, and go–no/go quality metrics is inappropriate for this physical, engineering context. Recall that it is quite rare to hear a lecture in physics, chemistry, or just about any engineering topic centered on "quality" metrics—they are all focused on basic or derived units that we refer to as *engineering* metrics (scalars and vectors). This is how one must approach DFSS metrics for Critical Parameter Management and robust design in particular.

Develop Subsystem Noise Diagrams and System Noise Map

The identification of noise factors is a key step in the optimization of any system. A noise factor is any thing that can induce variation in the functional performance of a design. The critical functional response is the variable used to measure the effects of noise factors. Noise factors are typically broken down into three general categories:

1. *External Noises*—variation coming into the design from an external source
2. *Unit-to-unit noises*—variation stemming from part-to-part, batch-to-batch, or person-to-person sources
3. *Deterioration noises*—variation due to wear or some other form of degraded state or condition within the design

A noise diagram can be constructed to illustrate the noise factors that are thought to have a large impact on the CFRs for any given subsystem or subassembly. They are most often used to aid in the structuring of two types of designed experiments:

1. **Noise experiments:** used to identify the magnitude and directional effect of statistically significant noise factors that will later be used to induce stress in a robustness optimization experiment

2. **Robustness optimization experiments:** used to investigate and quantify the interactions between statistically significant control factors and statistically significant noise factors.

Statistically significant control factors, along with the interactions between other control factors, are identified during the design phase of the CDOV process. Statistically significant noise factors are identified during this, the optimization phase of the CDOV process. The noise diagram (Figure 6–16) is always associated with subsystems and subassemblies, not the system.

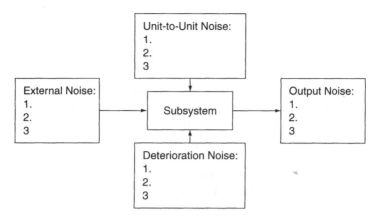

Figure 6–16 Example of a Noise Diagram Template

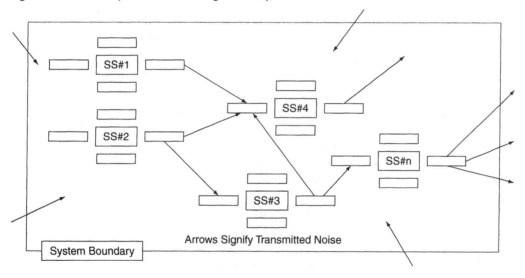

Figure 6–17 Transmitted Noise Within a System Noise Map Template

Noises are accounted for at the system level using a system noise map. A system noise map illustrates the nature of how subsystems and subassemblies transfer noises across interface boundaries as well as how noises enter and leave the system from an external perspective. System noise maps show all the noise diagrams as an interconnected network, so they can get quite large for big systems. System noise maps are used to help construct the designed experiments for the system level stress tests, which will be discussed a little later (Phase 3B) after the subsystems and subassemblies have been proven to be robust. System noise maps are often rather frightening things to see. They show just how many noises are running around inside and outside the system—potentially playing havoc on the ability of the integrated system to fulfill its various requirements (Figure 6–17). Failure to construct and look at noise diagrams and system noise maps is failure to consider the design realistically.

Conduct Noise Factor Experiments

The subsystem and subassembly noise diagrams are now used to construct a designed experiment to test the statistical significance of the noise factors. The team simply cannot afford to be testing the robustness of CFRs with noise factors that are indistinguishable from random events taking place in the evaluation environment. Noise factors may seem significant in a conference room — this experiment generates the data to find out if they are actually significant. Opinions are often wrong about noise factors. They need to be tested to see if they are truly significant using statistically sound methods of data gathering and analysis.

The noise experiment is often done using a special screening experiment called a *Plackett-Burman array*. The data is analyzed using the analysis of variance technique. The result is knowing which noise factors are statistically significant and which are not. These tools can also provide data to indicate the magnitude and directional effect of the noise factors, called a ***noise vector***, on the mean and standard deviation of the CFRs being measured.

Define Compounded Noises for Robustness Experiments

Because robustness optimization experiments cost money, it is common to try to design them to be as short and small as possible. One way to keep a robustness stress test economical is to compound or group the noise factors into logical groups of two. One grouping sets all the noise factors at the set points that cause the mean of the CFR to get smaller. The other grouping takes the same statistically significant group of noise factors and sets them up so that they drive the mean of the CRF high. This approach uniformly stresses the subsystem or subassembly by allowing every combination of control factors to be interacted with the noise factors. This is accomplished by exposing the treatment combinations of the control factors to the full, aggregated effects of the compounded noise factors.

With the scenario in Figure 6–18 we have three control factors at two levels and three noise factors at two levels. Studying every combination of the control factors and noise factors would require 64 experiments. If we compound the noise factors and stress by noise vector effect, as we have done here, we only need 16 experiments. The key is to only use the compounded effect of statistically significant noise vectors.

Once the control factors have interacted with the compounded noise factors, the data is "loaded" with the information needed to isolate control factors that leave the design minimally sensitive to the noise factors.

Define Engineering Control Factors for Robustness Optimization Experiments

Deciding which control factors to place into interaction with the statistically significant noise factors is a matter of engineering science and design principles that are underwritten by the ideal function equations developed in the design phase of the CDOV process. The logical choices are those control factors that play a dominant role in the flow, transformation, or state of energy, mass, and controlling signals within the design.

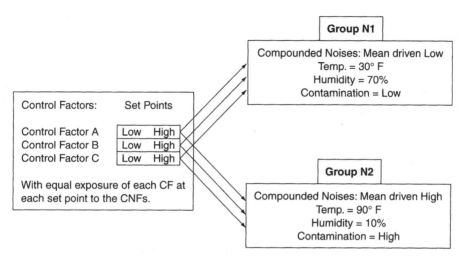

Figure 6–18   Grouping for Robustness Optimization Experiments

Design for Additivity and Run Designed Experiments

Since the Laws of Conservation of Mass and Energy are mathematically defined with additive equations (a form of ideal function), we tend to try and match control factors and critical functional responses that obey these additive relationships. Setting up a robust design experiment according to this logic is called *design for additivity*. It is done to bypass unwanted control factor-to-control factor interactions that tend to arise from measuring quality metrics that possess discontinuous functions with respect to the control factors (see Chapter 28 for more details on additivity).

This is an attempt to further our effort to measure fundamental, functional relationships between specifiable engineering parameters and the critical functional responses they control. Spending the time to set up a robustness experiment this way also tends to require simple experimental arrays because very few control factor-to-control factor interactions end up being evaluated in the robustness evaluations. Recall that the main focus in robust design is to study control factor-to-noise factor interactions.

Analyze Data, Build Predictive Additive Model

Once the designed experiment has been conducted and the CFR data collected using the certified measurement system, the signal-to-noise metrics can be calculated. Signal-to-noise transformations are calculated based on the physics and design principles that underlie the design's ideal function. Signal-to-noise metrics are one way of looking at a summary statistic that quantifies, in numeric form, the strength of the interaction between the control and noise factors. Large differences between S/N values for individual control factor set points indicate an interaction with the noise factors. This is a good thing. If a control factor has no interaction with noise factors, then robustness is impossible to improve. Gains in robustness come directly from interactions between control and noise factors. The larger the S/N gain, the more insensitive the CFR becomes to variation—providing you set the control factor nominal value at its most robust set point and then define the appropriate range of tolerances that help constrain the design's robustness even in the presence of unit-to-unit variation.

The units of the S/N metric are calculated and reported in the logarithmic unit called a decibel (dB). Decibel units are additive, so as we add all the individual control factor contributions to robustness, we build an equation known as the *predictive additive model:*

$$S/N_{predicted} = S/N_{avg} + (S/N_A - S/N_{avg}) + (S/N_B - S/N_{avg}) + \ldots + (S/N_n - S/N_{avg})$$

for *n* number of control factors that actually increase robustness.

Some control factors simply do not interact significantly with noise factors, so they have no "term" to include in the predicted additive model.

Run Verification Experiments to Certify Robustness

It is good design practice to conduct an experiment to quantify the subsystem and subassembly baseline S/N values prior to robustness optimization. That way the new, optimized S/N values can be compared to the baseline S/N values. This helps the team recognize the amount of improvement gained in the optimization process. The gain in the compared S/N values is literal because we have transformed the CFRs from their engineering units into decibels. For every 3 dB of gain in robustness, we see the variance of the CFR drop by a factor of 2 (every 6 dB of gain is equal to lowering the standard deviation by a factor of 4). The standard deviation is lower when we exploit the interactivity between certain control factors and noise factor combinations. The verification experiment provides the data to assure the team that it has correctly found an additive set of control factors that leave the CFR minimally sensitive to the noise factors.

The final "proof" of successful robustness optimization resides in the data gathered from confirmation tests. Verification experiments enable the team to confirm the proper control factors have been identified, with repeatable results in the presence of the statistically significant noise factors. The predicted S/N value must be "matched," or repeated within a range of tolerance. The tolerance range is calculated using one of several options and is commonly set at approximately +/− 2 to 3 dB of the predicted S/N value (see Chapter 28 on robust design for the calculation of S/N confirmation tolerance). If the predicted S/N and the verification test S/N values are repeatable (usually three to five replicates are tested) within the tolerance zone (particularly within the low side), the additive model is confirmed.

Conduct Response Surface Experiments on Critical Adjustment Parameters

Taguchi's approach to robustness optimization includes a method for identifying which control factors have a strong effect on the mean performance of the CFR. At least two kinds of control factors are identified during the optimization phase:

1. Control factors that have a strong effect on robustness of the CFRs
2. Control factors that have a strong effect on adjusting the mean of the CFRs

Two or more of the control factors that have a significant ability to adjust the mean usually undergo additional testing (see Chapters 29 and 30). These parameters are candidates for designation as critical adjustment parameters. These parameters will be used by the assembly and service teams or the customer to adjust the mean of specific CFRs onto a desired target. This, of course, has significance in the long-term capability of the design's performance. Recall that Cpk is the measure of long-term performance capability. When we use statistical process control to track the CFR's performance and we find that the mean has shifted off the target, we use the critical adjustment parameters (CAPs) to put the mean back onto the desired target.

Response surface methods (RSM) are good at identifying and optimizing CAPs. They are a special class of designed experimental methods that provide detailed information about the input-output relationships between the mean of the CFR and a small number of critical adjustment parameters. Response surfaces are actually plots of experimental data that resemble the contour maps often used by hikers. If two CAPs are being studied, the response surface will be in three dimensions and can plot the slope of adjustment sensitivity so team members can clearly see how to precisely adjust the CAPs to put the CFR exactly where they want it. The RSM approach can be used to adjust the standard deviation in a similar manner but is most often used for precision mean adjustment using two, three, or more (rarely more than six) CAPs.

Run Verification Experiments to Certify Tunability and Robustness Parameters for Subsystems and Subassemblies

Now that the control factors have been broken down into two classes, "robustizers" and "mean adjusters," they all can be tested one last time together to prove that the subsystem or subassembly design is both robust and tunable. Taguchi methods have an approach to do this in one step, the *dynamic method*. In many cases it is enough to reach a reasonably optimum, or local optimum, design. For cases where high tuning precision is required (a global optimum is sought) then response surface methods are required. If the design loses its robustness as it is adjusted to a different mean, the team needs to know this as early as possible. In fact, the place to assure that this does not happen to be a characteristic of the design is in the technology development process. This is where Taguchi's dynamic methods are most powerful and preventative of overly sensitive designs. As you can see, we prefer to have relatively high independence between control factors that adjust robustness and those that adjust the mean. This is why we must understand control factor-to-control factor interactions long before we start the optimization phase of the CDOV process. A team has no business conducting Taguchi methods for robust design until it has documented the appropriate deliverables from the tools and best practices within the design phase. With detailed knowledge of the two classes of interactions (CFxCF interactions identified during design phase and CFxNF interactions identified during optimization phase), independent optimization of robustness and mean performance can be efficiently completed. We believe both Taguchi dynamic robustness evaluations and classical RSM techniques are required in their proper sequence to do the job right. Neither is optional or a replacement for the other! If you don't see the wisdom in this, you need to revisit the logic of modeling and robustness in flow of the sequential design of experiments.

Document Critical Functional Parameter Nominal Set Points, CAP, and CFR Relationships

With the optimization data gathered and analyzed, the team can document the results in the Critical Parameter Management database. The ideal/transfer functions, now in their additive S/N model form, can be updated. Additional knowledge of control factor designations can be placed in the hierarchical structure of the database for each subsystem and subassembly. The data for which control factors affect CFR robustness and CFR mean performance must be thoroughly documented for the production and sustaining engineering community. When these communities need to make decisions about cost reduction or design changes, detailed information will be available. Production engineering organizations are almost never given enough design information to enable the ongoing support of the design. Critical Parameter Management corrects this problem.

Develop, Conduct, and Analyze Reliability/Capability Evaluations for Each Subsystem and Subassembly

It is often a waste of time to conduct reliability tests on subsystems and subassemblies that have not yet been optimized for robustness and mean performance. Prior to the gate review for the optimization phase, the subsystems and subassemblies should be placed into appropriately designed life testing. The life tests need not be completed prior to the Gate 3A review but they should be underway. The data from these tests will be heavily reviewed at the Gate 3B review.

Each subsystem and subassembly should also be measured for its CFR design capability performance after robustness optimization is completed. Short-term capability (Cp) evaluation and analysis should be conducted for each critical functional response. This data is to be stored in the Critical Parameter Management database for a capability growth index (see Chapter 12) review at the Gate 3A review.

Gate 3A Readiness

At Gate 3A we stop and assess the full set of deliverables and the summary data used to arrive at them. The tools and best practices of Phase 3A have been used and their key results can be summarized in a Gate 3A scorecard (Figure 6–19). A checklist of Phase 3A tools and their deliverables can be reviewed for completeness of results and corrective action in areas of risk.

Phase 3A Gate Review Topics

- Business/financial case development
- Technical/design development
- Manufacturing/assembly/materials management/supply chain development
- Regulatory, health, safety, environmental, and legal development
- Shipping/service/sales/support development

Phase 3A Optimization

Deliverable	Owner	Original Due Date vs. Actual Deliv. Date	Completion Status			Performance vs. Criteria			Exit Recommendation			Notes
			Red	Green	Yellow	Red	Green	Yellow	Red	Green	Yellow	
Verification of Robustness and Tunability of the Subassemblies and Subsystems and Their Current Design Capability (Cp values) to Fulfill Their Respective Requirements			○	○	○	○	○	○	○	○	○	
Subsystem and Subassembly Noise Diagrams			○	○	○	○	○	○	○	○	○	
System Noise Map			○	○	○	○	○	○	○	○	○	
Complete Set of Statistically Significant Noise Vectors			○	○	○	○	○	○	○	○	○	
Complete Set of Pre- and Post-Robustness Optimization Subsystem and Subassembly CFR S/N values			○	○	○	○	○	○	○	○	○	
Confirmation of the Additive S/N Model			○	○	○	○	○	○	○	○	○	
Complete Set of Critical Robustness Parameters and Their Nominal Set Points			○	○	○	○	○	○	○	○	○	
Complete Set of Critical Adjustment Parameters (incl. Their Ideal "Tuning" Functions)			○	○	○	○	○	○	○	○	○	
Updates to Critical Parameter Database and Scorecards			○	○	○	○	○	○	○	○	○	
Subsystem and Subassembly Reliability Test Plans and Current Data/Results Summary			○	○	○	○	○	○	○	○	○	
Phase 3B Project Management Plans			○	○	○	○	○	○	○	○	○	

Figure 6–19 General Scorecard for CDOV Phase 3A Deliverables

General Checklist of Phase 3A Tools and Best Practices

Subsystem and subassembly noise diagramming

System noise mapping

Measurement system analysis

Use of screening experiments for noise vector characterization

Analysis of means (ANOM) data analysis techniques

Analysis of variance (ANOVA) data analysis techniques

Baseline subsystem and subassembly CFR signal-to-noise characterization

Taguchi's methods for robust design

Design for additivity

Use of Full or Fractional Factorial experiments for robustness characterization

Generation of the additive S/N model

Response surface methods

Design capability studies

Critical Parameter Management

Reliability analysis

Life testing (normal, HALT, & HAST)

CDOV Phase 3B: System Integration

The diagram in Figure 6–20 illustrates the major elements in Phase 3 of the CDOV process: system integration, nominal performance evaluation, robustness optimization, and initial system reliability assessment.

Integrate Subsystems into System Test Units

Once the subsystems and subassemblies have fulfilled their Phase 3A gate requirements, they are safe to proceed into the Phase 3B integration phase of the CDOV process. It is extremely risky to allow nonrobust subsystems with poor adjustment capability to be integrated into the system. Many names are used to define the product as it is integrated into physical hardware, firmware, and software. Some refer to it as the system *breadboard,* while others call the embryonic product the *system integration fixture, test rig,* or *unit.* Whatever you call it, it represents the first completely assembled set of design elements that are attempting to fulfill the system requirements.

It is wise to conduct a fairly rigorous program of inspection on all the components that make up the subsystems and subassemblies that are being integrated into the system. If this is not done, the system integration team has no idea what it has built. Only after testing begins will the anomalies begin to surface. The team must prevent the whole set of errors, mistakes, and false starts that come with "bad parts" getting built into the integration units. It takes a great deal of time to sort these problems out if they are buried in the system from the onset of testing. The famous quote of

Start Date: ___	Integrate subsystems into system test units	End Date: ___	Owner: _____
Start Date: ___	Certify capability of the systemwide data acquisition system	End Date: ___	Owner: _____
Start Date: ___	Conduct nominal performance evaluation on the system	End Date: ___	Owner: _____
Start Date: ___	Conduct system robustness stress tests	End Date: ___	Owner: _____
Start Date: ___	Refine subsystem set points to balance system performance	End Date: ___	Owner: _____
Start Date: ___	Conduct initial system reliability assessment	End Date: ___	Owner: _____
Start Date: ___	Certify system readiness for final product design capability development and assessment	End Date: ___	Owner: _____

Figure 6–20 Flow of Phase 3B of Product Commercialization

"measure twice, cut once" certainly applies here. Design cycle-time will be reduced if the team doesn't have to go through a whole cycle of "mistake finding" due to these kinds of issues.

Certify Capability of the Systemwide Data Acquisition System

As the subsystems and subassemblies are integrated into the system test units, the transducers and data acquisition wiring should be installed, tested, and certified. The system integration team is measuring a select set of subsystem, subassembly, and all the system level critical functional responses.

The critical adjustment parameters must be carefully measured for their effect on their associated CFR variables. This is a major focus of the data acquisition system. The critical adjustment parameters and the critical functional responses they "tune" must be thoroughly measured and their sensitivity relationships documented in the CPM database. Measurement system analysis must be thoroughly applied prior to system testing.

Many subassembly-to-subsystem-to-system transfer functions will be derived from the data taken during the system integration tests. Corrupt data will destroy the integrity of these critical paths of transferred functional performance and variability.

Conduct Nominal Performance Evaluation on the System

After the system is integrated and the data acquisition system is certified, the system integration test team is ready to conduct nominal system performance testing. If the system level CFRs cannot be measured because the system fails to function or the system simply won't run well enough to gather a nominal baseline of performance, there is no sense going on with system stress testing. The primary goal of this first evaluation of integrated system performance is to characterize the initial Cp values for all the system CFRs and the selected subsystem and subassembly CFRs. Additional goals are to identify and document unforeseen problems, mistakes, errors of omission, and their corrective actions across the newly integrated system. The work done in the previous phases was conducted to prevent as many problems, mistakes, errors, and omissions as possible, but some will still be encountered.

Conduct System Robustness Stress Tests

Depending on the nature and severity of the problems stemming from the nominal performance tests, the system integration team can proceed with a series of stress tests. The system stress tests are typically designed experiments that evaluate all of the system CRFs as a well-defined set of critical functional parameters (subsystem and subassembly CFRs) and critical-to-function specifications (component CTF set points) that are intentionally set at all or part of their tolerance limits. The system stress tests also include the inducement of various levels of compounded external and deterioration noise factors.

This form of testing is expected to reveal sensitivities and even outright failures within and across the integrated subsystems and subassemblies. The magnitude and directional effect of the noise factors on the system CFRs is of great importance relative to how the subsystem, subassembly, component, and manufacturing process nominal and tolerance set points will be changed and ultimately balanced to provide the best system performance possible.

Refine Subsystem Set Points to Balance System Performance

Calculating the Cp values for each system level CFR is recommended during this phase. These Cp values will be relatively low as one might expect. It is quite helpful to know, at this early phase in the commercialization process, what the nominal and stress case Cp values are across the system. It is not unusual to take a number of corrective actions and to move a number of subsystem, subassembly, and component set points to help balance and improve system CFR performance. This is one reason why we need robust and tunable designs. Any number of post-corrective action system stress tests can be run as needed to verify performance improvements. The usual number is two, one initial stress test and one just prior to exiting Phase 3B. A minimum of two rounds of stress testing is recommended for your planning and scheduling consideration. Often the first round of stress testing is conducted at the tolerance limits of the critical parameters selected for the test. The second round of stress testing, after design balancing analysis and corrective action is completed, is less aggressive. It is quite reasonable to set the critical parameters at one third of their rebalanced tolerance limits. It is not unusual to go through an initial round of strategic tolerance tightening at the subsystem level during this phase of performance testing and balancing.

Conduct Initial System Reliability Assessment

With the nominal and stress testing completed and corrective actions taken, it is now worth the effort and expense to conduct a formal system reliability evaluation. The system is set to nominal conditions with all components, subassemblies, and subsystems inspected and adjusted to produce the best performance possible at this phase of product commercialization.

Subsystem, subassembly, and component life tests should be ongoing. Environmental stress screening (ESS), highly accelerated life testing (HALT), and highly accelerated stress testing (HAST) should be producing data. Failure modes should be emerging and corrective action should be underway to increase reliability. DFMEAs should be updated.

Verify System Readiness for Final Product Design Capability Development and Assessment

After conducting all the nominal and stress tests and getting a first look at reliability test data, the system has been evaluated to a point where a recommendation can be made relative to the system's "worthiness" of further investment in the next phase (4A and 4B). If the system is too sensitive to noise factors, then corrective action to remove the sensitivities must be done or the program will have to be cancelled or postponed.

The Critical Parameter Management database is updated at this point. All the new sensitivity relationships and changes to the ideal/transfer functions need to be included. The latest system, subsystem, and subassembly CFR Cp performance data also should be added for the Gate 3B review.

Gate 3B Readiness

At Gate 3B we stop and assess the full set of deliverables and the summary data used to arrive at them. The tools and best practices of Phase 3B have been used and their key results can be summarized in a Gate 3B Scorecard (Figure 6–21). A Checklist of Phase 3B tools and their deliverables can be reviewed for completeness of results and corrective action in areas of risk.

Phase 3B Gate Review Topics

- Business/financial case development
- Technical/design development
- Manufacturing/assembly/materials management/supply chain development
- Regulatory, health, safety, environmental, and legal development
- Shipping/service/sales/support development

General Checklist of Phase 3B Tools and Best Practices

Measurement system analysis

System noise mapping

Nominal system CFR design Cp studies

Stress case system CFR design Cp studies

System-subsystem-subassembly transfer function development

Phase 3B Optimization

Deliverable	Owner	Original Due Date vs. Actual Deliv. Date	Completion Status			Performance vs. Criteria			Exit Recommendation			Notes
			Red	Green	Yellow	Red	Green	Yellow	Red	Green	Yellow	
System CFR Nominal Performance Capability Indices			○	○	○	○	○	○	○	○	○	
System CFR Stressed Performance Capability Indices			○	○	○	○	○	○	○	○	○	
System Sensitivity Analysis Results			○	○	○	○	○	○	○	○	○	
System-to-Subsystem-to-Subassembly Transfer Functions			○	○	○	○	○	○	○	○	○	
Refined Subsystem and Subassembly Ideal Functions			○	○	○	○	○	○	○	○	○	
System Reliability Performance			○	○	○	○	○	○	○	○	○	
Subsystem, Subassembly, and Component Reliability and Life Test Performance			○	○	○	○	○	○	○	○	○	
Updated Critical Parameter Database and Scorecards			○	○	○	○	○	○	○	○	○	
Phase 4A Project Management Plans			○	○	○	○	○	○	○	○	○	

Figure 6–21 General Scorecard for CDOV Phase 3B Deliverables

System level sensitivity analysis

Design of experiments (screening and modeling)

ANOVA

Regression

Analytical tolerance analysis

Empirical tolerance analysis

Statistical process control of design CFR performance

Reliability assessment (normal, HALT, & HAST)

Critical Parameter Management

CDOV Phase 4A: Verification of Product Design Functionality

CDOV Phase 4 covers verification of final product design, production processes, and service capability. CDOV Phase 4 contains two major parts:

- *Phase 4A* focuses on the capability of the product design functional performance.
- *Phase 4B* focuses on the capability of production assembly and manufacturing processes, within the business as well as the extended supply chain and service organization.

The diagram in Figure 6–22 presents the major elements in Phase 4A.

Start Date: ___	Conduct final tolerance design on components, subassemblies, and subsystems	End Date: ___	Owner: _____
Start Date: ___	Place all CTF components and CFRs under SPC in supply chain and assembly operations	End Date: ___	Owner: _____
Start Date: ___	Build product design verification units using production parts	End Date: ___	Owner: _____
Start Date: ___	Evaluate system performance under nominal conditions	End Date: ___	Owner: _____
Start Date: ___	Evaluate system performance under stress conditions	End Date: ___	Owner: _____
Start Date: ___	Complete corrective actions on problems	End Date: ___	Owner: _____
Start Date: ___	Evaluate system performance and reliability	End Date: ___	Owner: _____
Start Date: ___	Verify product design meets all requirements	End Date: ___	Owner: _____

Figure 6–22   Flow of Phase 4A of Product Commercialization

Conduct Final Tolerance Design on Components and Subsystems

In Phase 3B, the subsystems and subassemblies passed through their initial system integration evaluations. Numerous adjustments and refinements at the subsystem, subassembly, and component level were made. These were mainly shifts in the nominal set points of the CFRs and CTF specifications to balance overall system CFR performance, both under nominal and stressed conditions. Initial tolerances were defined during the design phase of the CDOV process. Numerous critical parameter (subsystem and subassembly CFRs) and specification (component) tolerances were exercised during the system stress tests. Some of these critical tolerances were adjusted to help mature and improve the system performance to pass Gate 3B requirements.

Now we are ready to develop the final tolerances on all manufacturing process set points, components, subassemblies, and subsystems. This is where Critical Parameter Management comes into full use. The entire network of nominal and tolerance set points must be finalized and verified that they fulfill all requirements. All other nominal and tolerance set points that are of general importance also must be finalized.

The methods of analytical and empirical tolerance design along with design and manufacturing capability studies can be used extensively to verify the tolerances that will constrain the right amount of unit-to-unit noise. Unit-to-unit noise must be held to a level that enables all requirements throughout the integrated system to be fulfilled to the best extent possible. This becomes an economic issue because the tightening of tolerances and improvement of bulk or surface material properties always increases component costs.

Analytical tolerancing is typically conducted on a computer using math (ideal/transfer functions applied within and across design boundaries) and geometric models. Empirical tolerancing is conducted in a design laboratory or production facility using designed experiments to evaluate candidate tolerances on real components, subassemblies, and subsystems. In both cases, numerous tolerances are set at various high and low levels and then the model or prototype exercises the effect of the tolerance swings within the ideal/transfer function being evaluated. The resulting variability exhibits its effects in the measured or calculated CFRs.

Place All CTF Components and CFRs Under SPC in Supply Chain and Assembly

Once all the final tolerances for the components, subassembly adjustments, and subsystem adjustments are defined, those that are critical-to-function are documented and placed under statistical process control (SPC). The critical parameters can be measured when components are assembled and adjusted during the assembly process. Critical specifications are measured during component production as mass, energy, and information are transformed within a production process. SPC methods are used to determine when variation and mean performance have left a state of random variation and have entered a state of assignable cause variation. Engineering analysis, multi-vari studies, capability studies, and designed experimentation can help identify and remove assignable causes of variation.

This approach to variability reduction in the mean and standard deviation of critical parameters and specifications is not cheap. That is why we use CPM. It helps the team spend money only on those design and production process elements that are critical to the functional requirements of the product.

Build Product Design Certification Units Using Production Parts

Once all the elements of the system design are toleranced and the assembly and production process are under statistical control, the first sets of production parts can be ordered, inspected, and assembled. The construction of the first production systems are done by the production teams with the design teams present to document any problems.

The list of critical functional responses that will be measured during the assembly process at the subassembly, subsystem, and system level is prepared at this point. Not all CFRs within the system that were measured during the CDOV process will continue to be measured during steady-state production. Only a small subset of the CFRs need to go forward as part of the manufacturing Critical Parameter Management program for Six Sigma. Many are measured as CTF specifications on components out in the supply chain; a few are measured as CFRs as subassemblies and subsystems are assembled and tested prior to integration into the final product assembly. The remaining CFRs are measured at the system or completed product level, usually just prior to shipping or during service processes.

Evaluate System Performance Under Nominal Conditions

Once the system is built and all CFR measurement transducers are in place, a repeat of the nominal system performance test can be conducted. The Cp of the system CFRs and the production subsystem and subassembly CFRs are measured and calculated. The aggregated effect of these Cp values is reported in a metric called the **capability growth index** or CGI (see Chapter 12).

Evaluate System Performance Under Stress Conditions

The new, "final" tolerances for critical-to-function components and the critical assembly and service adjustments are typically set at one third (for 3 sigma designs) to one sixth (for 6 sigma designs) of their limits for testing purposes. In some cases, this roughly emulates what is known in statistical tolerancing methods as a *root sum of squares* evaluation. It is more statistically likely that the variability that will actually accumulate within and across the system will not be at worst case limits but rather some lesser amount of consumption of the tolerance limits. From experience and from statistical principles we have found it reasonable, over the years, to stress the performance of the critical response variables with components and adjustments at about one third to one sixth of their actual tolerances. The production CFRs are measured during these stress tests with the added stress due to applications of external and deterioration noise factors from the system noise map.

Once again, the Cp of the system CFRs and the production subsystem and subassembly CFRs are measured and calculated. The aggregated effect of these stressed Cp values is reported in the same CGI format. The measured performance becomes the data used in documenting final product design specifications that represents the teams' best attempt to fulfill the system, subsystem, subassembly, and component requirements.

Complete Corrective Actions on Problems

As is always the case after nominal and stress testing, problems surface and must be corrected. The key is to prevent as many of them as possible by using the right tools and best practices at the right time within each phase of the CDOV process. Problems at this point in the process should be relatively minor.

Evaluate System Performance and Reliability

As you can see, this approach postpones traditional reliability assessments until corrective actions are largely completed after nominal and stress testing is done within each major phase. There is no sense assessing reliability when you know full well you don't have the design elements mature enough to realistically assume you are anywhere near the goal you seek to confirm. Developing and evaluating ideal functions, transfer functions, noise vectors, additive S/N models, tolerance sensitivity stack-up models, and capability growth models are all steady and progressive actions that will develop reliability. Reliability tests never grow reliability—they just tell you how much you do or don't have. They are important when conducted at the right time and for the right reasons. They confirm you are attaining your goals but should not be done too early. The rule to follow is: don't consume any resources on predicting or testing reliability that really could be better used in the actions of developing reliability. Many of the tools and metrics of DFSS are there because they are good at creating reliability long before you can actually measure it.

System reliability evaluations typically demonstrate the performance of the functions of the system in terms of *mean time to failure* or *mean time to service*. The underlying matters of importance in reliability testing are:

1. Time
2. Failures

These are common things that customers can easily sense or measure. The product development teams must not ignore these metrics. In DFSS we must have two sets of books that are kept to track the fulfillment of all the product requirements. The *book of requirements* that customers worry about is typically at the system level and is rarely stated in engineering units. To fully satisfy these quality requirements against which customers measure, we must have a set of books that dig deeper, down through the hierarchy of the product and production systems architectures. This "book of translated requirements" is loaded with fundamental engineering metrics that can provide the preventative, forward-looking metrics that assure us that the functions can indeed be replicated, over the required lengths of time that will fulfill the VOC requirements. Our metrics have to be good enough to measure and make us aware of impending failures!

Verify Product Design Meets All Requirements

The ability to verify the system design requires:

1. A *clear set of design requirements* at the system, subsystem, subassembly, component, and manufacturing process level
2. A *capable set of design performance metrics* for the critical functional responses at the system, subsystem, and subassembly levels and a capable set of metrics for the critical-to-function specifications at the component and manufacturing process level.

The construction and use of this integrated system of requirements and metrics is called Critical Parameter Management. It will be much easier to rapidly determine the status of capability to fulfill form, fit, and function requirements if a disciplined approach is taken from the first day of the CDOV process to track these critical relationships. Critical Parameter Management possesses a scorecard that accumulates Cp performance growth across the subsystems and at the system level. This scorecard is populated with Cp values from all the subsystem critical functional responses and a rolled up summary capability growth value, CGI. It is somewhat analogous to "rolled throughput yield (RTY)" that is used in production operations and transactional forms of Six Sigma. The CGI is based on CFRs while RTY is typically based on defects per million opportunities or defects per unit. The CGI is the management team's eyes into the "soul" of the product from a functional perspective. When the CGI scorecard is at its required level of Cp growth, the quality is inherent in the certified product design.

Gate 4A Readiness

At Gate 4A we stop and assess the full set of deliverables and the summary data used to arrive at them. The tools and best practices of Phase 4A have been used and their key results can be summarized in a Gate 4A scorecard (Figure 6–23). A checklist of Phase 4A tools and their deliverables can be reviewed for completeness of results and corrective action in areas of risk.

Phase 4A Gate Review Topics

- Business/financial case development
- Technical/design development
- Manufacturing/assembly/materials management/supply chain development
- Regulatory, health, safety, environmental, and legal development
- Shipping/service/sales/support development

Phase 4A Verify

Deliverable	Owner	Original Due Date vs. Actual Deliv. Date	Completion Status Red Green Yellow	Performance vs. Criteria Red Green Yellow	Exit Recommendation Red Green Yellow	Notes
System CFR Nominal Performance Capability Indices			○ ○ ○	○ ○ ○	○ ○ ○	
System CFR Stressed Performance Capability Indices			○ ○ ○	○ ○ ○	○ ○ ○	
System Sensitivity Analysis Results			○ ○ ○	○ ○ ○	○ ○ ○	
System Reliability Performance			○ ○ ○	○ ○ ○	○ ○ ○	
Subsystem, Subassembly, and Component Reliability and Life Test Performance			○ ○ ○	○ ○ ○	○ ○ ○	
Updated Critical Parameter Database and Scorecards			○ ○ ○	○ ○ ○	○ ○ ○	
Capability Growth Index Data			○ ○ ○	○ ○ ○	○ ○ ○	
Phase 4B Project Management Plans			○ ○ ○	○ ○ ○	○ ○ ○	

Figure 6–23 General Scorecard for CDOV Phase 4A Deliverables

General Checklist of Phase 4A Tools and Best Practices

Measurement system analysis

System noise mapping

Analytical tolerance design

    Worst case analysis

    Root sum of squares analysis

    Six Sigma analysis

    Monte Carlo simulation

Empirical tolerance design

    System level sensitivity analysis

Design of experiments

    ANOVA

    Regression

Multi-vari studies

Design capability studies

    Nominal system CFR design Cp studies

    Stress case system CFR design Cp studies

Reliability assessment (Normal, HALT, & HAST)

Competitive benchmarking studies

Statistical process control of design CFR performance

Critical Parameter Management (with capability growth index)

CDOV Phase 4B: Verification of Production

Phase 4B of the CDOV process examines capability production, assembly, and manufacturing process within the business as well as the extended supply chain and service organization. Figure 6–24 covers the major elements.

Build Initial Production Units Using Inspected Production Parts

Now that the organization has completed the verification of the final product design specifications, it is time to order actual production parts and build the initial production units that will bear revenue. This is not a serial process. Many parts, tools, and subassemblies will have long lead times. A good deal of judgment must be exercised to balance the risk of early order based on assumed verification versus waiting until the data is all in before ordering the "hard" tools and components.

Many design organizations get their part orders way out of "phase" with their performance and capability data. The designers keep designing before the testers finish testing! One way to

Start Date: ___	Build initial production units using inspected production parts	End Date: ___	Owner: _____
Start Date: ___	Assess capability of all CFRs and CTFs in production and assembly processes	End Date: ___	Owner: _____
Start Date: ___	Assess capability of all product level and subsystem level CFRs during assembly	End Date: ___	Owner: _____
Start Date: ___	Assess reliability of production units	End Date: ___	Owner: _____
Start Date: ___	Verify all requirements are being met across assembly processes	End Date: ___	Owner: _____
Start Date: ___	Verify all requirements are being met across production processes	End Date: ___	Owner: _____
Start Date: ___	Verify all service requirements are being met with service/support processes	End Date: ___	Owner: _____
Start Date: ___	Verify product, production, assembly, and service/support processes are ready for launch	End Date: ___	Owner: _____

Figure 6–24   Flow of Phase 4B of Product Commercialization

remedy this is to make the designer and tester the same person. At times it is good to have people so busy doing one thing they can't get anything else done! The real problems come when this phenomenon occurs habitually from the beginning of a new product development program. By the time the final production units are being built, virtually no one knows if or how well they will actually work.

The production parts need to be produced from and inspected against final, verify design specifications. A product design specification is the actual nominal specification and tolerance that the design can provide. The motivating forces behind the product design specifications are the system, subsystem, subassembly, component, and manufacturing process requirements. If the parts are not inspected during this earliest production build, you run the risk of not catching a number of human errors and mistakes that, once corrected, are likely to not be a problem ever again. If these mistake-proofing inspections do not occur, it is quite possible in any sufficiently complex product for these problems to go on for extended periods of time, particularly if the problem is intermittent.

Assess Capability of All CFRs and CTFs in Production and Assembly Processes

A thorough assessment of the Critical Parameter Management metrics that are transferring into steady-state production must be conducted at this point. This body of metrics is a relatively small, carefully selected group of CFRs and CTF specifications from the design CPM database. The production CPM database is a scaled-down version of the design CPM database. The design CPM database will be available to the production organization for the life of the product, but the specific production-oriented CFRs and CTF specifications will be the leading indicators that show that the manufacturing processes and the assembled parts coming from them are capable and under statistical control, assuring they are safe to use in steady-state production.

Assess Capability of All Product Level and Subsystem Level CFRs During Assembly

Within the actual product are a few CFRs that the assembly team tests as the subassemblies and subsystems are constructed. Once these are assembled into the product, the system level CFRs can be evaluated. There is no need for additional, special, or long-term burn-in style testing when Critical Parameter Management is systematically used across the phase and gates of the **CDOV** process.

The selection of these CFRs is somewhat analogous to the way your physician conducts your annual physical. Neither you or your doctor can afford to test everything that is important to your health. The few critical parameters that are the leading indicators of your health can efficiently and economically be evaluated. Once you pass these critical checks, out the door you go. This is essentially what must be done with products. Design in the quality, document and check the critical parameters, and get the product to customers so that they can be satisfied "on time" and your shareholders will see a great return on their investment in the business.

Assess Reliability of Production Units

An appropriately small number of the production units should be randomly selected for a complete reliability evaluation. This is done to complete the competitive benchmarking process and to assure that there are no lingering surprises in the initial production lot of system units. The data from this evaluation can be used as a baseline for your next product development program.

Once you are past the initial ramp-up of your production process, the data you need for reliability should come from the base of delivered products. A strong customer service and support process with a focused effort to establish data acquisition and transmission back to the design organization from a few strategic customers on CFRs and general reliability is essential. This is another way the Critical Parameter Management process is extended out into the steady-state use environment.

Verify All Requirements Are Being Met Across Assembly Processes

Just as you established Critical Parameter Management metrics out into the supply chain to quantify the capability of incoming component CTF specifications, you should also establish a series of CFR and CTF specification checks as the assembly process is conducted. Here we see a portion of the computer-aided data acquisition systems that were used in design transferred to the assembly team for continued use during steady-state production.

Verify All Requirements Are Being Met Across Production Processes

A comprehensive roll-up of all Critical Parameter Management data in the network of production processes must be gathered to demonstrate that all the critical manufacturing processes, components, assembled subassemblies, and subsystems are "on target with low variability," or Cpk measured. The ultimate requirements that matter the most are at the system or completed product level. All system level CFRs should be found capable on a routine basis. Assignable causes of variation should be identified using statistical process control. They should be removed under the scrutiny and balance of the CPM database.

Verify All Service Requirements Are Being Met with Service/Support Processes

All CFRs with critical adjustment parameters that the service community will use should be certified for "adjustability." The service engineering community will then be able to play its role in sustaining the Cpk values for the system and subsystem CFRs for which they are responsible.

The CPM database should be in the hands of the service engineering teams for diagnostic and maintenance purposes. One of the biggest payoffs in conducting CPM is the detailed technical performance information that can be passed on to the service organization, which can readily use the relational ideal/transfer function information resident in the CPM database to track down root causes of problems and take specific corrective action. This greatly increases the "fix it right the first time" expectation of modern customers.

Verify Product, Production, Assembly, and Service/Support Processes Are Ready for Launch

As a final Gate 4B deliverable, the entire launch readiness status, based on the capabilities that can be tracked in the CPM database, should be prepared for review. All the capability data across the organizations that are responsible for verifying launch readiness can be summarized and used to underwrite the integrity of the decision to launch or to postpone for corrective action.

Gate 4B Readiness

At Gate 4B we stop and assess the final, full set of deliverables and the summary data used to arrive at them. The tools and best practices of Phase 4B have been used and their key results can be summarized in a Gate 4B scorecard (Figure 6–25). A checklist of Phase 4B tools and their deliverables can be reviewed for completeness of results and corrective action in areas of risk.

Phase 4B Gate Review Topics

- Business/financial case development
- Technical/design development
- Manufacturing/assembly/materials management/supply chain development
- Regulatory, health, safety, environmental, and legal development
- Shipping/service/sales/support development

Phase 4B Verify

Deliverable	Owner	Original Due Date vs. Actual Deliv. Date	Completion Status Red Green Yellow			Performance vs. Criteria Red Green Yellow			Exit Recommendation Red Green Yellow			Notes
System CFR Nominal Performance Capability Indices			○	○	○	○	○	○	○	○	○	
System Sensitivity Analysis Results			○	○	○	○	○	○	○	○	○	
System Reliability Performance			○	○	○	○	○	○	○	○	○	
Subsystem, Subassembly, and Component Reliability and Life Test Performance			○	○	○	○	○	○	○	○	○	
Updated Critical Parameter Database and Scorecards			○	○	○	○	○	○	○	○	○	
Capability Growth Index Data			○	○	○	○	○	○	○	○	○	
Sustaining Production Engineering Support Plans			○	○	○	○	○	○	○	○	○	

Figure 6–25   General Scorecard for CDOV Phase 4B Deliverables

General Checklist of Phase 4B Tools and Best Practices

Measurement system analysis

Analytical tolerance design

    Worst case analysis

    Root sum of squares analysis

    Six Sigma analysis

    Monte Carlo simulation

Empirical tolerance design

    System level sensitivity analysis

Design of experiments

    ANOVA

    Regression

Multi-vari studies

Design capability studies

    Nominal system CFR design Cp studies

Reliability assessment (normal, HALT, & HAST)

Competitive benchmarking studies

Statistical process control of design CFR performance

Critical Parameter Management (with capability growth index)

References

Clausing, D. P. (1994). *Total quality development: Improved total development process.* New York: ASME Press.

Cohen, L. (1995). *Quality function deployment: How to make QFD work for you.* New York: Addison Wesley Longman.

Cooper, R. G. (2001). *Winning at new products: Accelerating the process from idea to launch* (3rd ed.). Reading, MA: Perseus Books.

Otto, K. N. & Wood, K. L. (2001). *Product design.* New Jersey: Prentice Hall.

Rechtin, Eberhardt. (1990). *Systems architecting: Creating and building complex systems.* New Jersey: Prentice Hall, Incorporated.

Ullman, D. G. (1997). *The mechanical design process.* New York: McGraw-Hill.

Ulrich, K. T. & Eppinger, S. D. (1995). *Product design and development.* New York: McGraw-Hill.

Wheelwright, S. C. & Clark, K. B. (1992). *Revolutionizing product development.* New York: The Free Press.

System Architecting, Engineering, and Integration Using Design For Six Sigma

Virtually every product made is some form of integrated system. So is every manufacturing and assembly process. All systems are built up from the subsystems, subassemblies, and/or components that are supplied from a network of internal and external manufacturing and assembly processes. Six Sigma methods are very much applicable to the development of new technology systems (commonly known as *platforms*), product systems, and production systems. A specific set of tools and best practices can be identified and applied by system architects and system integration engineers during the phases and gates of the I^2DOV and CDOV product development processes. This chapter is all about "what to do and when to do it" with respect to system architecting, engineering, and integration using the tools and best practices from technology development and Design For Six Sigma.

It has been our experience to find many of the companies that develop and design under the influence of Six Sigma methods do so largely independent of a system integration context. Most engineering teams tend to focus their efforts on subsystem, subassembly, component, or material development. They also tend to focus just on projects that correct, or react to, a preexisting problem that will have some form of rapid and reasonably significant cost reduction. Six Sigma projects have their historical roots grounded in rapid, strong financial returns. This, of course, is a great thing, but there exists a whole class of projects that develop their economic value over the course of time it takes to conduct the phases and gates of a product development process. Their financial value resides within the fulfillment of the new product's business case. This approach is not driven by reaction to problems but rather the prevention of problems. The ultimate economic value resides in the development, design, and delivery of a new system!

Developing new technologies, products, and production systems takes time. The development of a new system that fulfills a complete set of requirements can return enormous financial results. Ignoring or underinvesting in this class of applications for Six Sigma methods is starting to diminish—slowly. It is time for businesses that focus on revenue growth through the development and integration of systems to improve their results with the help of Six Sigma development and design tools and the process of Critical Parameter Management.

Just as in the previous two chapters, we will use a flow-down of actions to illustrate the major steps one can take to conduct system architecting, system engineering, and system integration for Six Sigma within the CDOV Phase/Gate process. This chapter includes the start date, end date, and owner blocks in the action flow-down diagrams. We do this to remind you that all these actions can and should be tracked using PERT charts for disciplined project management. The system architecting and engineering process for technology development is quite similar to the one used in product design, so we will use the CDOV process as a general approach that can be easily modified for use within the I^2DOV process.

System architecting, engineering, and integration teams typically reside within the technology, product, or production process development organizations. System architects are typically leaders of advanced product planning teams within the product development organization. They focus on the structuring of the system elements around the functions the system must perform to fulfill its requirements. They work with the system engineers to develop a complete set of system requirements and the network of measures that prove that the integrated subsystems, subassemblies, components, and manufacturing processes fulfill their system level requirements. System Architects are specialists who often work within a system engineering group. They guide the initial form, fit, and function structural definition work done in the concept phase of the CDOV process. System engineers and architects work together to manage the interfaces into, across, and out of the system. In some organizations they act as surrogates for field engineering, service, support, and assembly operations during the early phase of the development process. That is to say, they build and measure the performance of the first integrated units. They troubleshoot and repair them during testing. They maintain configuration control and document the bill of materials and initial product role-up costs. They transfer these important functions to the production and production support teams gradually during the later phases of the product development process (typically during the capability verification phase of the CDOV process).

The system engineering team owns the Critical Parameter Management process. It also forms the leadership core of a new multifunctional product development team called the *Quality Council*. It becomes the watchdog group for the program.

The system architects help initiate CPM but due to the nature of CPM as a verification and "certification of performance" tool, its proper place of ownership is within the system engineering team. The team documents all the product requirements and generates the computer-aided database that tracks the complex network of requirements and fulfilling capability metrics that verify that the system is ready for its gate reviews. The system engineering team evaluates both nominal and stressed performance (for Cp and Cpk). These teams are the new model for what used to be known as quality assurance groups. The old model was an organizationally independent team of inspectors who were quality conformance watchdogs. They owned nothing and were not held responsible for delivering anything except reports of how well or badly the system performed. They were largely reactive and rarely preventative with regard to the development of product quality. The new model, as prescribed in this text, is a new team that resides *within* the project or program. This team owns specific deliverables and is held directly accountable for the preventative development of system performance in terms of both cost and quality. These system engineers must perform their duties within the structured constraints of the CDOV process, so they also are held accountable for cycle-time. Old-line quality assurance organizations, because they only focused on policing the "suspicious behaviors" of the project team, usually added to cycle-time as they held the project hostage until "they" were forced to fix the problems. Now, the system engineers own the problems. With such responsibility comes a driving impulse to avoid the creation of problems in the first place. In fact, this approach encompasses specific tools and best practices and the time and resources to assure that quality is designed into the system in the earliest phases of the project.

We advocate the dissolution of old-line QA watchdog groups. Redeploy their talent into system engineering and integration teams and hold them accountable for the deliverables from the DFSS tools and best practices. As for the "independent and objective" policing work, this can be handled by making the system engineering manager a direct report to both the program manager and the corporate officer responsible for the business in which the program resides. In the majority of cases the conflict of interest is removed with this approach.

Phase 1: System Concept Development

The diagram in Figure 7–1 illustrates the major system architecting, engineering, and integration elements for the **CDOV** Process Phase 1: Develop a system concept based on market segmentation, the product line, and technology strategies.

The system architects and engineers participate heavily in Phase 1 inbound marketing activities. They serve in both consulting and leadership roles and provide a "feed forward" view of what is likely to be coming into the next commercialization project to the system engineering group. The divisional inbound marketing managers own the process and tools for voice of customer acquisition to create the database of "what is needed." The vice president or director of R&D owns the process and tools for voice of technology acquisition to develop the database of what technologies are available or need to be developed to meet the market needs. The system engineering manager owns the process and tools for VOC translation into system requirements, developing the CPM database to account for the hierarchy of requirements and metrics for the fulfillment of the VOC and the VOT.

The specific issues related to each Phase 1 action item (start date for each action item, the action item, end date for each action) will be reviewed in this section. Each of these action items can be conducted by using a tool or best practice. The flow of actions can be laid out in the form of a PERT chart for project management purposes.

Participate in Gathering the Voice of the Customer Data

Start Date: _____ *End Date:* _____ *Owner:* _____

Start Date: ___	Participate in gathering the voice of the customer data	End Date: ___	Owner: _____
Start Date: ___	Lead the process of refining and ranking of the voice of the customer using KJ analysis	End Date: ___	Owner: _____
Start Date: ___	Document the VOC in the Critical Parameter Management database**	End Date: ___	Owner: _____
Start Date: ___	Integrate and align the VOC with the voice of technology (Needs. . .Ideas. . .Technologies = Opportunities?)**	End Date: ___	Owner: _____
Start Date: ___	Lead in the creation of the product or system level House of Quality*	End Date: ___	Owner: _____
Start Date: ___	Conduct competitive assessments, investigations, benchmarking, and projections of vulnerabilities**	End Date: ___	Owner: _____
Start Date: ___	Generate product or system level requirements document	End Date: ___	Owner: _____
Start Date: ___	Lead the team in the definition of the functions that fulfill the system requirements	End Date: ___	Owner: _____
Start Date: ___	Oversee the generation of the system concept evaluation criteria	End Date: ___	Owner: _____
Start Date: ___	Work with inbound marketing team to generate, collect, analyze, refine, and categorize new "Blue Sky" product ideas*	End Date: ___	Owner: _____
Start Date: ___	Lead the team in the generation of system concepts that fulfill the functions	End Date: ___	Owner: _____
Start Date: ___	Lead the team in the evaluation of the system concepts	End Date: ___	Owner: _____
Start Date: ___	Lead the team in the selection of a superior system concept	End Date: ___	Owner: _____
Start Date: ___	Analyze, characterize, model, and predict nominal performance of the superior system concept	End Date: ___	Owner: _____
Start Date: ___	Identify new system and subsystem functions that will require new data acquisition equipment, calibration, and measurement processes	End Date: ___	Owner: _____
Start Date: ___	Develop reliability requirements, initial reliability model, and FMEA for the system	End Date: ___	Owner: _____
Start Date: ___	Calculate initial "*rough order of magnitude*" cost estimates for the superior concept	End Date: ___	Owner: _____

Figure 7–1 Flow of System Architecting, Engineering, and Integration Actions for Six Sigma in Phase 1 of Product Commercialization

Start Date: ___	Advise R&D on commercial system integration issues for new technology platforms and their integrated subsystems	End Date: ___	Owner: _____
Start Date: ___	Conduct risk assessment of superior concept from a system integration perspective	End Date: ___	Owner: _____
Start Date: ___	Create Phase 2 (design) system engineering and integration detailed project management plans	End Date: ___	Owner: _____
Start Date: ___	Assess performance and make changes in the current practice of system engineering and integration to assure state-of-the-art performance in the future	End Date: ___	Owner: _____

*(shared responsibility between system engineering and inbound product marketing group)
**(shared responsibility between system engineering, R&D, & inbound product marketing group)

Figure 7–1   *Continued*

System engineering participation during the VOC gathering activities is essential. Gathering the VOC breaks down into the following general areas:

1. Define who to visit, call, or survey by defining the market segments for the business.
2. Prepare a customer interview, call guide, or survey questionnaire.
3. Visit, call, or survey an appropriate number of customers.
4. Review and formally document the VOC data gathered to prepare it for KJ analysis.

If the system engineering team is not intimately involved in this activity, its ability to understand and lead the follow-on efforts of building the system House of Quality and creating the system requirements document will be weakened. In terms of Critical Parameter Management, the embryonic critical parameters are developed at this stage. They are the specific VOC needs that are new, unique, and difficult to fulfill. These new, unique, and difficult needs will ultimately be translated into the critical requirements and then developed into fulfilling critical parameters that the entire design engineering team will use to satisfy the new, unique, and difficult VOC needs. The deliverable from this activity is a complete, formally documented set of verbally and textually stated needs and use context images that are in the raw, unaltered "voice of the customer" vocabulary.

Lead the Process of Refining and Ranking of the Voice of the Customer Using KJ Analysis

Start Date: _____ *End Date:* _____ *Owner:* _____

The system architect responsible for producing the affinity diagram using KJ analysis takes a leadership role at this point. As the leader for this activity, the system architect must act in an unbiased, objective role as a KJ process facilitator. The multifunctional team conducting the KJ analysis should be from inbound marketing, engineering, service, sales, production, supply chain, and possibly even a few key customers. The team must process the VOC data into groups of needs that have an "affinity" to one another. The completed KJ analysis offers a diagram of hierarchical statements of need. It should look like a tree structure. This diagram contains the needs that are new, unique, and difficult. Needs that are not new, unique, or difficult are still important and need to be tracked to be sure their specific issue is directly covered by one or more of the system requirements that are soon to be documented. The primary reason we do an affinity diagram is to be sure we have comprehensively documented the most critical needs that the customer has articulated by word, text, or observed behavior. Two levels of KJ analysis can be conducted in the following order:

1. The **image KJ diagram** structures and ranks "images" the interview team formed during interview sessions and through opportunities to observe the physical behaviors customers exhibit by purely observing how they use or interact with the product or process being evaluated for new development.
2. The **requirements KJ diagram** contains the combined "imaged" needs and the actual stated needs of customers. The image KJ results are used to help create the requirements KJ diagram.

The requirements KJ is used to flow new, unique, and difficult customer needs into the system level House of Quality. The deliverable from this activity is a completed KJ diagram that contains the ranked and structured VOC needs (see Chapter 14 for details).

Document the VOC in the Critical Parameter Management Database

Start Date: _____ *End Date:* _____ *Owner:* _____

The system engineer responsible for the construction of the Critical Parameter Management relational database is now called into action. This individual enters the ranked and structured VOC data into the CPM database. This information is structured so it can be easily linked to the specific system requirements that are soon to be developed and entered in the next layer of the database. The Critical Parameter Management database must be a relational database so that each critical customer need can be linked to any and all system requirements to which it relates. This connectivity is extremely important as the system architecting team attempts to fully structure a system that can adequately balance the fulfillment of the critical customer needs.

Integrate and Align the VOC with the Voice of Technology (Needs. . .Ideas. . .Technologies = Opportunities?)

Start Date: _____ *End Date:* _____ *Owner:* _____

The system architectural team and system engineers must act as a balancing influence between the desire to give customers what they ask for and what current technology can reasonably provide. It is just as important for the system development teams to gather the voice of technology as it is the voice of the customer. The system engineering team becomes the clearinghouse for the inclusion of new technologies into the concept development phase. If a new technology is to be included in the product architecture, the system engineering team plays a lead role is the assessment of its contribution to program risk.

The voice of technology includes the following information:

1. New patents in the areas to be covered by candidate designs within the new product architecture
2. Benchmarking data on the current competitive products serving the target market
3. New trends in technology development related to the new product ideas
4. New technologies under development from R&D that are being readied for application

The system architecting and engineering teams will need this information to properly conduct quality function deployment. The House of Quality requires input from the voice of the customer as well as the voice of technology.

Lead in the Creation of the Product or System Level House of Quality

Start Date: _____ *End Date:* _____ *Owner:* _____

The system architecting team plays the lead role in the development and documentation of the system level House of Quality. The system engineering representative on the system architecting team typically has the responsibility for documenting the system House of Quality. The multifunctional team members from the business actually conduct QFD under the guidance of an experienced chief system architect.

The system House of Quality must be limited in what VOC needs are placed into it for translation into system requirements. Not all VOC needs are critical. Only those needs that are new, unique, and/or difficult should be translated and ranked in the system House of Quality. The system engineering team should provide assistance on what is considered new, unique, and/or difficult. All other VOC needs that are old, common, and easy should be directly translated and placed in the system requirements document. If a few of these types of needs are considered critical, they should be loaded into the system House of Quality.

Conduct Competitive Assessments, Investigations, Benchmarking, and Projections of Vulnerabilities

Start Date: _____ *End Date:* _____ *Owner:* _____

The system engineering team not only acts as the repository of customer needs and technology trend information but also conducts qualitative and quantitative competitive product assessments and benchmarking tests. These may include analytical assessments, reverse engineering and teardowns, baseline nominal and stress test evaluations, cost breakdowns, failure modes and effects analysis, life tests, service evaluations, and numerous other estimates of competitive performance characterizations. The system engineers provide benchmarking information to the system architecting team and to the subsystem teams as they begin their work in the concept design phase.

Generate Product or System Level Requirements Document

Start Date: _____ *End Date:* _____ *Owner:* _____

The system architectural team will need an extremely clear and well-worded system requirements document to drive the architecting process. It is risky to generate and evaluate system concepts in the absence of a well-structured set of system requirements.

The system requirements document contains all the requirements that must be fulfilled for the certification of the product design and its supporting hardware, software, services, and documentation. The system requirements will include both the critical new, unique, and difficult requirements along with the old, common, and easy requirements. Typically, the old, common, and easy requirements that are truly critical are few in number. All requirements—those that are critical to functional performance and those that are considered of nominal importance—must be included in the system requirements document.

It is especially important that software requirements be clearly defined and related to hardware requirements. The system architectural team needs to be sure that it has adequate representation from the software organization at the beginning of the conceptual design of the complete system architecture.

Lead the Team in the Definition of the Functions That Fulfill the System Requirements

Start Date: _____ *End Date:* _____ *Owner:* _____

Once the system architecting team is in possession of the system requirements, the system engineering representative can lead the team through the definition of the specific functions that the system must perform to fulfill the requirements. The list of system functions will begin the definition of the metrics for Critical Parameter Management. This activity often includes the refinement of functional performance targets and their units of measure. This is also the beginning of the definition of the requirements that will drive the development, design, and certification of a systemwide data acquisition system.

The documentation and linkage between the flow-down of needs to requirements to functions to concepts is essential to any Design For Six Sigma approach to product development.

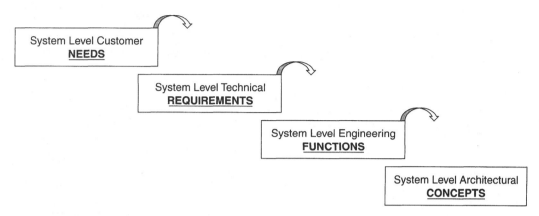

Figure 7–2 Linkage in DFSS Product Development

These four steps (Figure 7–2) are a core competency that you will need to develop if you have aspirations of being good at DFSS.

If you underinvest in any of this work, you can begin planning your schedule slips accordingly because the root causes of design mistakes are born here. Poor work here will result in design scrap and rework later in your product development process.

Oversee the Generation of the System Concept Evaluation Criteria

Start Date: _____ *End Date:* _____ *Owner:* _____

The system architecting team now uses the system requirements to help derive a set of system concept selection criteria. Concept selection criteria must evaluate two general characteristics:

1. Qualitative feasibility (as opposed to quantitative optimality)
2. Competitive vulnerability (against best-in-class benchmarks)

The system requirements are not the only components that drive the system concept selection criteria. Beyond the system requirements, these considerations drive concept evaluation criteria:

1. Legal and patent issues
2. Cost constraints in product development, production, and product delivery
3. Development cycle-time
4. Resource availability, allocation, and use

These considerations are the "custom" requirements that are unique to your business. They need to be considered for inclusion in the selection of a superior system concept. It is important that the architectural team initiate and lead a cross-functional effort to integrate system requirements with software architectural constraints and capabilities. They must ensure the software and subsystem teams are prepared to help in the generation and evaluation of system concepts as needed.

Work with Inbound Marketing Team to Generate, Collect, Analyze, Refine, and Categorize New "Blue Sky" Product Ideas

Start Date: _____ *End Date:* _____ *Owner:* _____

There are at least two major ways to get ideas for new products:

1. Obtain new product ideas from sources outside the system architectural team.
2. Generate new product ideas from within the system architecting team.

The system architectural team solicits new product ideas from any and all reasonable outside sources. These are often referred to as *blue sky* ideas. This approach is democratic and allows the internal organization and external influences to stimulate creativity during system concept development. A tyrannical or exclusionary approach to concept development is a bad idea. Organizations that fear the openness of "blue sky" ideation are limiting the potential of identifying concepts that have superior feasibility and low competitive vulnerability. If your organization has a few "concept tyrants," you need a "bloodless coupe." You cannot afford to miss a great idea from an individual who may be a "quiet genius."

Lead the Team in the Generation of System Concepts That Fulfill the Functions

Start Date: _____ *End Date:* _____ *Owner:* _____

With the "blue sky" ideas in hand, the system architectural team generates a reasonable set of system concepts. A reasonable number is a relative thing. Certainly it is at least two and probably would be better if it is six or more. The desire for numerous system concepts is justified by the fact that the team will use the best characteristics from the candidate concepts to synthesize a "super-concept," a hybrid built upon the innovations from across the group. One idea is not enough.

One person telling everyone else the way it's going to be is wrong. With many concepts come many options for system integration during concept selection and evaluation.

Lead the Team in the Evaluation of the System Concepts

Start Date: _____ *End Date:* _____ *Owner:* _____

The system engineering representative takes the lead (facilitation) role in conducting the Pugh concept evaluation and selection process. The role includes the assurance of the following elements:

1. A reasonable number of concepts to be evaluated
2. All concepts developed and illustrated to a common, equal level of representation
3. A complete set of evaluation criteria ready for use
4. A *best-in-class* benchmark datum concept identified for the comparison process
5. Adequate multifunctional team representation during the Pugh process
6. The participants all trained in the use of the Pugh process

System engineers act as an unbiased facilitator. Their role is to guide the initial run-through concept evaluation, subsequent rationalization, and hybridization of super-concepts and the selection of a new datum, if required. They lead the follow-on iterations of the hybrid concept evaluations and final selection of a superior concept relative to the final datum concept. The key is to be able to lead the system architecting team to a consensus based upon the criteria. This is a good point to remind the reader that the criteria should be representative of the voice of the customer, the voice of technology, and the voice of the business (let's not forget the shareholders!).

Lead the Team in the Selection of a Superior System Concept
Start Date: _____ *End Date:* _____ *Owner:* _____

As the system architecting team goes through the iterative cycles of the Pugh concept selection process (also called the *converging-diverging process* due to its iterative nature), a dominant concept emerges. This concept is almost never one of the original concepts. Usually it is a hybrid form of two or more of the initial candidate concepts that have been synthesized.

The team needs to reach two forms of consensus:

1. The selected concept is indeed superior to the datum concept.
2. The selected concept is rational and reasonable in comparison to the runner-up as they relate to the most critical of the selection criteria (ranking of the criteria is one way to settle this).

The second step of consensus is an important check and balance on the practicality of the outcome. From time to time the runner-up is actually better when the team takes the time to reflect on the top two concepts relative to the criteria that made them rank so high. Sometimes the top two concepts are codeveloped until the data is clearly supporting one over the other in terms of maturing toward the fulfillment of the critical system requirements.

Analyze, Characterize, Model, and Predict Nominal Performance of the Superior System
Start Date: _____ *End Date:* _____ *Owner:* _____

With a superior system concept selected and documented, it is now worth the time investment to analyze the design. The system engineers responsible for modeling system performance begin construction of the Critical Parameter Management database. This is done to prepare for in-depth analysis of the system performance. At this point the system is a black box of projected outputs and yet-to-be-defined internal relationships. In Critical Parameter Management terms the system is a collection of system level critical functional responses that need to be modeled and measured to see if they fulfill the system requirements. The system can be modeled in three ways:

1. **Analytical** model of system outputs based on inputs to system performance from external (user and environment) and internal (subsystem and subassembly) inputs
2. **Graphical** model of system's external space consumption and internal space use
3. **Physical** model of the system for functional and spatial performance characterization

These three methods of modeling characterize the estimated and actual form, fit, and function of the system. One cannot do all of these modeling activities within the concept phase! You need to have patience and conduct the right form and degree of modeling within the appropriate phase of the CDOV process.

The following suggestions align appropriate system modeling action with the proper phases in the CDOV process.

Phase 1 (concept design)

Analytical   At this point you should define the critical functional responses (Y variables) at the system level and link them to the system requirements they fulfill. These linked relationships should be documented in your CPM relational database. The formal $Y = f(x)$ models for all system CFRs will be estimated and verified in Phase 2 (design development). This limitation is due to the fact you have not yet defined the subsystem requirements, functions, and fulfilling architectures (form and fit). Do not be impatient. There will be plenty of time and opportunity to model the system level $Y=f(x)$ relationships after subsystem concept design is completed.

Graphical   The system architectural (form and fit) boundaries for external space consumption can be graphically modeled at this point. This initial cut at an external set of spatial boundaries certainly changes but now is the time to establish a baseline external configuration. This is particularly true if the system requirements have clear space constraints. Internal spatial layout can be estimated and allocated to each subsystem and subassembly. This will only be a preliminary allocation of internal space. The system engineers carefully manage the reapportionment and balancing of system space during the subsystem concept generation and selection process. Just in case you were wondering, yes, the system engineers are the "space police." Actually their jurisdiction is much broader than that. They are also the "interface police" and act much like the FBI. They trace the flow of variation in functional performance and space use across interface boundaries. They don't arrest and prosecute violators but they try to prevent as many violations against the system requirements as possible before they occur.

Physical   It is a major mistake to waste money and time prematurely slapping a system together in Phase 1 to evaluate functional performance at the system level! It is fine to create space models in prototype hardware for external and internal space allocation. If you rush off and integrate a system for performance and reliability evaluations this early, you have just created the following situations:

1. You have rushed past proper conceptual development of subsystem requirements and subsystem concept generation and selection (*poor concept design*).
2. You have prematurely integrated a pseudo-system with "designs" that are undermodeled and uncertified for nominal baseline performance (*poor nominal design*).
3. The subsystems are underdefined, immature, and overly sensitive to internal (part-to-part variability and deterioration), external, and system integration noise (*poor robust design*).

You need to slow down to speed up! Slow down and let the system requirements and system architecture drive the subsystem designs as you transfer into the design phase of the CDOV product development process. You need the right cycle-time—not the shortest (remember Chapter 4!).

Phase 2 (design development)

Analytical   Now the system engineering team can begin to construct and validate the numerous $Y = f(x)$ models that underwrite the critical relationships that exist between system critical functional responses (Y variables) and the subsystem critical functional parameters (x variables). It is typical early in the design phase to estimate the input (x) values and then predict how they will affect the system output (Y) variables. Monte Carlo simulations are often used to do this. These are exploratory evaluations that can help begin the balancing of subsystem baseline performance requirements early in the program. As the subsystem architectures and prototype designs are developed and modeled, the system models will fill out and be much more capable of assisting in iterative design trade-off studies.

Graphical   The system space allocation model will really begin to take shape as the subsystem teams develop their architectures (form and fit relationships). Just about all of the space-balancing activity will be done in this phase of the CDOV process. The system architecture will be adjusted and compromises will abound.

Physical   The subsystems are being defined, developed, and physically prototyped at this point. Once the subsystems have taken shape and possess stable, nominal behavior, then it is appropriate to integrate a subsystem into a "system mule" that enables the subsystem to begin to be evaluated for integration performance. A mule is some form of designed test fixture that either emulates enough of the system or an old system you have commandeered for the purposes of subsystem integration simulation. A mule is representative of the future system prototype hardware. We cannot overemphasize the point that now is *not* the time to integrate all the nominal subsystem designs into the actual system prototype! We recommend limited simulation of future integration hardware for the purpose of evaluating physical integration form, fit, and function for the baseline, nominal design. The subsystems are still in a nonrobust state and are not yet worthy of true system integration. If you break this rule and prematurely integrate at this point, you will get poor performance and will have invested in data that is just going to confirm the obvious—your system is not yet robust. Remain patient for one more phase.

Phase 3A (Subsystem robustness optimization)

Analytical   At this point in the CDOV process, the system models should be pretty mature for the $Y = f(x)$ relationships (ideal functions) that have been built on all the input critical functional parameters and critical-to-function specifications. These CFP and CTF specifications represent the (x) inputs from the various subsystems, subassemblies, and individual parts. Monte Carlo simulations are routinely performed to isolate system-to-subsystem and subsystem-to-subassembly-to-component sensitivity relationships that need to be "quieted" during the robustness activities within Phases 3A and 3B. The System Engineers expand and explore the ideal functions within and across the system to uncover critical sensitivities. The subsystem teams can be informed of the interface sensitivities associated with their designs so they can use Phase 3A to "desensitize" their design not only to their own set of noise factors but also to system integration noises.

Graphical   The system graphical models are now in a state of high maturity. The space allocation models are now used to provide guidance in the iterative changes that will come from mean value changes for numerous design parameters based on robust design data.

Physical   The subsystems are all going through robustness stress testing at this point in Phase 3. The mules will be used as part of the robustness evaluations. Interface noises can be emulated on the mules to identify robust set points for critical functional parameters and critical-to-function specifications on components that are estimated to affect system integration performance.

Phase 3B (system integration and stress testing)

Analytical   The system engineering team is now loaded with nominal set point information that has been derived from robustness stress testing data. These robust nominal set points can be evaluated in the sensitivity models in preparation for the design and execution of integrated system baseline and stress tests. The analytical models are used to help predict the many critical sensitivity points across the integrated system. These points of sensitivity need further empirical assessment and desensitization. Tolerance models are now developed to enable the system tolerance allocation to begin. The subsystem teams will use this information to begin to balance their tolerances with functional performance requirements.

Graphical   The space allocation models continue to be used to provide guidance in the iterative changes that will come from mean value changes. But now the system engineers shift over to become the "tolerance police." System tolerance allocations are projected down to the subsystems based on system integration stress test data. Monte Carlo simulations on geometric tolerances down through the system are now the focus.

Physical   This is the moment you have been waiting for! Now is the time to build the first true system integration prototype. The robust subsystems now contain enough stability and insensitivity to noise that it is worth investing in the integration effort. The data from system integration testing will now give the entire product development team the information they need to make key performance balancing decisions. Now you should only have to "balance a few bowling balls on the head of a pin" instead of many. System integration makes sense now. Three major tests are run by system engineering at this point:

1. System baseline performance and capability tests (under nominal conditions)
2. System stress tests (under system stress conditions using Taguchi noise factors)
3. System reliability test (under normal customer use conditions)

Phase 4A (product design verification)

Analytical   The system models are used to balance sensitivity and cost across the integrated set of critical parameters that are now fully documented in the critical Parameter Management database. Final production and service tolerance modeling is the dominant use at this time.

Graphical   System geometric tolerance allocations conclude as balance is attained down through the subsystems, subassembly, and components. Balance is measured through capability indices that are based on product design certification test data.

Physical   New, "production-like beta units" or product design certification models are built and evaluated for their capability in fulfilling the system requirements. The final product design specifications are derived from this set of test data.

Phase 4B (production process verification)

Analytical   The system analytical models are used to adjust the product design specifications as necessary during production certification.

Graphical   The system graphical models are used to adjust the product design specifications as necessary during production certification.

Physical   Production models are built and evaluated for their capability in fulfilling the product design specifications. The final production process specifications are derived from this set of test data.

Identify New System and Subsystem Functions That Will Require New Data Acquisition Equipment, Calibration, and Measurement Processes

Start Date: _____ *End Date:* _____ *Owner:* _____

The definition of the system functions that fulfill the system requirements may seem like a tedious thing to generate. They are anything but a set of tedious detail. They are essential to the proper planning and development of your data acquisition systems. The systems of transducers, meters, signal processing circuit boards, and computers will enable the collection of the data that actually makes the Critical Parameter Management process work. It is extremely important to carefully plan your data acquisition system designs, calibration processes, and application methods well in advance of when you have to use them. Part of the criteria for a superior system architecture is the ability to measure and verify the capability growth of all critical functional responses that integrate to fulfill the system requirements.

Develop Reliability Requirements, Initial Reliability Model, and FMEA for the System

Start Date: _____ *End Date:* _____ *Owner:* _____

A special effort may be warranted when your new product has new, unique, and difficult requirements for reliability performance. A set of specific, carefully worded reliability requirements is the first order of action in the concept design phase. Next we identify the system functions that have critical linkage to the system reliability requirements. This set of "Critical-to-reliability" functions must be measured both in terms of time or rate-based failure performance and function-based performance in the units of engineering scalars and vectors. We must keep two sets of "books" all the way through the CDOV phases and gates to account for all the mechanisms that enable the growth and certification of reliability.

Book 1: Time-Based Units of Measure

- Time to failure
- Rate of failure
- Percent surviving per unit time

Book 2: Function-Based Performance Units of Measure

- Mean of critical-to-reliability functional response (scalar or vector)
- Standard deviation of critical-to-reliability functional response (scalar or vector)
- S/N improvement of critical-to-reliability functional response (scalar or vector)
- Cp and Cpk of critical-to-reliability functional response (scalar or vector)

The idea here is to measure reliability directly and the ability to perform reliability critical functions with increasing levels of capability (growth of Cp) even in the presence of sources of variation that have a history of degrading reliability. This approach lends credibility to the idea that a robust design will help grow reliability. This is only true if the measure of robustness includes data that proves insensitivity to the noise factors that typically decrease reliability.

The system engineering team is responsible for building the system reliability prediction model and the system failure modes and effects analysis. Initially the system reliability model will be deterministic and based on the first estimates from the reliability block diagrams. As time goes on, in Phase 2, the model will shift to become a statistical model that is calculated using Monte Carlo simulations.

Calculate Initial "Rough Order of Magnitude" Cost Estimates for the Superior Concept

Start Date: _____ *End Date:* _____ *Owner:* _____

The system engineering team conducts the first "rough order of magnitude" cost estimates for the construction of the superior product concept, represented by hardware, firmware, and software prototypes. These cost models will eventually evolve into the actual costs associated with the bill of materials ordered to build the design certification model that is tested to produce the final product design specifications.

Advise R&D on Commercial System Integration Issues for New Technology Platforms and Their Integrated Subsystems

Start Date: _____ *End Date:* _____ *Owner:* _____

R&D is often working ahead of but in parallel with the product design team. It is often developing some form of new technology to transfer over to the design team, usually around the end of Phase 1 or early in Phase 2. System engineering must be a frequent visitor to and close partner of the R&D teams that will transfer their work into the design community. The system engineering personnel who work with R&D need to carefully communicate requirements and preventatively review system integration issues so that no surprises are delivered to the design team when the technologies are transferred.

Conduct Risk Assessment of Superior Concept from a System Integration Perspective

Start Date: _____ *End Date:* _____ *Owner:* _____

The system engineering team conducts a risk assessment for the system concept from a system integration and product design certification perspective. Any high-risk requirements are fully discussed and strategies put in place to deal with worst case scenarios that are likely to surface later in the program. These risk assessments focus on technical difficulties in terms of form, fit, and function. They also assess cost risks due to system complexity, assembleability, and serviceability. The risk assessment is underwritten by a formal system FMEA and system noise mapping.

Create Phase 2 (Design) System Engineering and Integration

Start Date: _____ *End Date:* _____ *Owner:* _____

The leader of the system engineering team creates the project management plans for the next phase. These plans are reviewed at the Phase 1 gate review. This is a preventative assessment to be sure all critical tasks are identified, resources prepared, and all responsibilities and deliverables clearly laid out. This is the point at which the system engineering team establishes the Phase 2 critical path for the program from a system development perspective.

Assess Performance and Make Changes in the Current Practice of System Engineering and Integration to Assure State-of-the-Art Performance in the Future

Start Date: _____ *End Date:* _____ *Owner:* _____

As a summary to Phase 1 in the CDOV process, the system engineering team should review its performance against its goals. The team should identify weaknesses in its ability to contribute to the gate deliverables in the concept phase. It should also assess tool use and execution and its ability to work within the multifunctional team. Future training plans can be formed and adjustments to the tasks and deliverables we have just discussed can be made. Gate reviews provide an excellent point to gather information for continuous improvement opportunities. See the score card for Phase 1 in Figure 7–3.

Phase 1 Goals

Assure that customer needs are adequately translated into a clear and readable system requirements document.

Assure generation and selection of superior system concept.

Assure potential subsystem concept "unknowns" are at an acceptable level of risk.

Assure that Critical Parameter Management database is established.

Provide a system engineering perspective and leadership in advanced product planning activities.

Act as a cross-functional team member primarily to R&D and inbound marketing.

Build strong awareness inside the system engineering team of incoming platform and product family plans.

Assist inbound marketing team in developing, evaluating, selecting, and modeling a superior system concept for Gate 1 approval.

Checklist of Phase 1 System Engineering Tools and Best Practices

VOC gathering methods

KJ analysis/affinity diagramming

QFD: House Of Quality #1 (system or product level)

Competitive benchmarking

Reverse engineering

System Engineering and Integration for Phase 1 Concept Development

Deliverable	Owner	Original Due Date vs. Actual Deliv. Date	Completion Status Red Green Yellow			Performance vs. Criteria Red Green Yellow			Exit Recommendation Red Green Yellow			Notes
Ranked and Documented VOC Database			O	O	O	O	O	O	O	O	O	
System Level House of Quality			O	O	O	O	O	O	O	O	O	
Competitive System Benchmarking Data			O	O	O	O	O	O	O	O	O	
System Level Requirements Documents			O	O	O	O	O	O	O	O	O	
Documented List of System Level Functions That Fulfill the System Requirements			O	O	O	O	O	O	O	O	O	
Documented Superior System Concept			O	O	O	O	O	O	O	O	O	
Portfolio of Candidate Technologies That Help Fulfill the System Functions & Requirements			O	O	O	O	O	O	O	O	O	
Documented Critical Parameter Database with Estimated System CFR Interaction Transfer Functions			O	O	O	O	O	O	O	O	O	
Documented Reliability Model and Predicted System Reliability Performance with Gap Analysis			O	O	O	O	O	O	O	O	O	
Estimated System Cost Projection			O	O	O	O	O	O	O	O	O	
Phase 2 (Design Development) Project Mgt. Plans			O	O	O	O	O	O	O	O	O	

Figure 7–3 General Score Card for CDOV Phase 1 System Engineering and Integration Deliverables

195

System architecting

Principles of modular/platform design

Knowledge-based engineering and modeling methods

Monte Carlo simulation

Concept generation methods (TRIZ, brainstorming, brain-writing)

Pugh concept evaluation and selection process

Noise diagramming and mapping

Data acquisition system design methods

Critical Parameter Management

Reliability prediction and allocation methods

Instrumentation and data acquisition system development

Cost estimating and accounting

Project management and planning methods

Phase 1 Major Deliverables

- Raw VOC data
- KJ analysis results (grouped and ranked VOC data)
- System level HOQ document
- Superior system concept (system architectural layout, analytical models, physical "prototype")
- System requirements document (product level requirements)
- System reliability requirements and allocation model
- System engineering input to the product level financial business case document
- Program risk assessment document (financial, material/human resources, and technical)
- Integrated program plan (IPP) . . . PERT and Gantt charts developed by the program manager
- Phase 2 system engineering project management plans (PERT and Gantt charts)

Phase 2: Subsystem, Subassembly, Component, and Manufacturing Concept Design

The diagram in Figure 7–4 illustrates the major system engineering and integration elements for CDOV Phase 2: The Design Phase of the CDOV Process. First, we will look at the unique set of design actions the system engineering team takes on to help its peers in a supporting role within the subsystem design teams.

The system engineering team also conducts a set of actions that continue the development of the system as an entity unto itself. You can see from the sets of actions in Figures 7–4 and 7–5 how the system engineering team has two roles: help the subsystem teams and integrate the results into system deliverables.

Phase 2: Develop Subsystem Concepts and Designs

Start Date: ___	Help create subsystem Houses of Quality	End Date: ___	Owner: _____
Start Date: ___	Help conduct subsystem benchmarking	End Date: ___	Owner: _____
Start Date: ___	Help generate subsystem requirements documents	End Date: ___	Owner: _____
Start Date: ___	Help develop functions that fulfill the subsystem requirements	End Date: ___	Owner: _____
Start Date: ___	Help generate subsystem concept evaluation criteria	End Date: ___	Owner: _____
Start Date: ___	Help generate subsystem concepts that fulfill the functions	End Date: ___	Owner: _____
Start Date: ___	Help evaluate subsystem concepts	End Date: ___	Owner: _____
Start Date: ___	Help select superior subsystem concepts	End Date: ___	Owner: _____
Start Date: ___	Refine system cost model based on subsystem cost models	End Date: ___	Owner: _____
Start Date: ___	Help develop reliability model and FMEA for each subsystem	End Date: ___	Owner: _____

Develop Subassembly, Component, and Manufacturing Process Concepts and Designs

Start Date: ___	Help create subassembly, component, and manufacturing process HOQs	End Date: ___	Owner: _____
Start Date: ___	Help generate subassembly, component, and manufacturing process requirements documents	End Date: ___	Owner: _____
Start Date: ___	Help evaluate subassembly, component, and manufacturing process concepts	End Date: ___	Owner: _____
Start Date: ___	Help select superior subassembly, component, and manufacturing process concepts	End Date: ___	Owner: _____
Start Date: ___	Refine system cost model based on subassembly and component cost models	End Date: ___	Owner: _____
Start Date: ___	Help develop reliability models and FMEAs for reliability critical subassemblies and components	End Date: ___	Owner: _____

Figure 7–4   Initial Flow of System Architecting, Engineering, and Integration Actions for Assisting Subsystem Teams in DFSS During Phase 2 of Product Commercialization

Phase 2: Design

Start Date: ___	Identify and resolve "conflicting requirement" problems by providing data from the VOC document and system requirements document	End Date: ___	Owner: _____
Start Date: ___	Help construct subsystem and subassembly noise diagrams	End Date: ___	Owner: _____
Start Date: ___	Refine system reliability model based on subsystem and subassembly and component reliability models	End Date: ___	Owner: _____
Start Date: ___	Generate the initial integrated system noise map	End Date: ___	Owner: _____
Start Date: ___	Generate system CAD spatial integration model	End Date: ___	Owner: _____
Start Date: ___	Generate system CAE functional performance model	End Date: ___	Owner: _____
Start Date: ___	Generate an estimated bill of material structure to prepare for configuration control and subsystem part inspection requirements	End Date: ___	Owner: _____
Start Date: ___	Generate system integration and evaluation prototype cost estimate	End Date: ___	Owner: _____
Start Date: ___	Create data acquisition system requirements	End Date: ___	Owner: _____
Start Date: ___	Conduct concept design for the systemwide data acquisition system	End Date: ___	Owner: _____
Start Date: ___	Continue to construct the relational database to document and track critical parameters	End Date: ___	Owner: _____
Start Date: ___	Initiate and manage a projectwide quality council to help manage and track critical parameters	End Date: ___	Owner: _____
Start Date: ___	Create Phase 3A system engineering and integration project plans	End Date: ___	Owner: _____
Start Date: ___	Assess performance and make changes in the current practice of system engineering and integration to assure state-of-the-art performance in the future	End Date: ___	Owner: _____

Figure 7–5   The Additional Flow of System Architecting, Engineering, and Integration Actions for Conducting DFSS During Phase 2 of Product Commercialization

System engineers in a technology development and Design For Six Sigma context play two key roles:

1. They work inside and between the subsystem teams to help with the tasks of subsystem development and design. In this role they fan out and "join" all the subsystem teams to gain intimate knowledge of the critical cross-boundary interface sensitivities and subsystem performance issues. They provide system integration information and carry back subsystem design and performance information to help plan for future system integration tests. Their key focus is on subsystem interface geometry and functional handoffs of mass, energy, and information (digital and analog logic and control signals).

2. They work as a team of objective, independent system integrators to build, evaluate, and verify the complete product design. In this role they own and are responsible for the integrated system.

We will not repeat the block diagram descriptions for the tasks that the system engineers help conduct with their peers on the subsystem design teams. Suffice it to say that the system engineers work with these teams and assist them as necessary. They literally help apply the tools and best practices of Technology and Design For Six Sigma. They should be evaluated and rated for their performance in this context with feedback to the system engineering manager. In this way they can be held accountable for their contributions to preventing design problems through these important teamwork tasks.

The system engineers have an important preintegration function to perform during the design phase of the CDOV process. Typically the system engineers divide up the subsystems between themselves and lead presystem integration design review meetings. This division of subsystems is not done on a random basis. The subsystem interface relationships are carefully examined and key groups of subsystems are "owned" by each one of the system engineers. Their job is to make sure each of the subsystem teams are "preventatively aware" of integration issues, variation transmission sensitivities, and potential failure modes. If there is a significant issue relative to system integration, it is the system engineer's responsibility to document it and evaluate solutions and work with the subsystem teams involved to coordinate a resolution to the problem. The system engineering team owns the integrated system and is responsible for its performance. The subsystem teams own their respective subsystems, subassemblies, components, and manufacturing/assembly process requirements. Critical Parameter Management responsibilities fall along these lines, but the CPM database is the responsibility of the system engineering team.

The specific issues related to each Phase 2 system engineering action item (start date for each action item, the action item, and end date for each action) will be reviewed in this section. Each of these action items can be conducted by using a tool or best practice. The flow of actions can be laid out in the form of a PERT chart for project management purposes.

Identify and Resolve "Conflicting Requirement" Problems by Providing Data from the VOC Document and System Requirements Document

Start Date: _____ *End Date:* _____ *Owner:* _____

We begin Phase 2 system engineering DFSS activities by working on clearing up any confusion or conflicts that come from the system requirements document. Two types of conflict issues generally surface at the beginning of Phase 2:

1. Incomplete, poorly worded requirements (a human communication problem)
2. Clearly worded requirements that are in conflict with one another (a VOC and VOT problem)

The system engineer responsible for the Critical Parameter Management database takes possession of the ranked voice of the customer data. It is this person's responsibility to use the VOC data to help correct poorly translated customer needs. Technical system requirements must reflect the customer needs in technical terms. The usual problem is that the technical vocabulary of the system requirement is either too ambiguous or too prescriptive. When the system requirement is too ambiguous, it needs to be rewritten with precise language, including units of measure and target values that everyone on the product development team can understand. When the system requirement is too prescriptive, it needs to be rewritten so that it does not imply a specific solution. System requirements should set a standard to be fulfilled with a unit of measure and a target for that measure. They should not state a solution, concept, or method of fulfillment. That is the realm of concept generation and selection. The system requirements must be stated in such a way that they can be used to evaluate numerous concepts or solutions. In this sense the system requirement implies a "function" or variable to be measured. You should be able to define many different "forms and fits" of architectural concepts that fulfill the function underwritten by the vocabulary of the system requirement.

When two or more well-written system requirements are in conflict with one another, the problem is much more serious. If we fulfill one system requirement, we absolutely disable our ability to fulfill another system requirement. This often happens when we have to add features and functions while simultaneously increasing reliability to historically unprecedented levels. About the only thing that can be done is to negotiate a balanced solution to this stand-off by going back and discussing the situation with the customers. A refined and compromised VOC need will usually result. New system requirements can then be derived from the new VOC input. The system House of Quality can also be of great help in this conflict resolution process. It contains a lot of ranking information to help make the required compromises between the conflicting system requirements.

Help Construct Subsystem and Subassembly Noise Diagrams

Start Date: _____ *End Date:* _____ *Owner:* _____

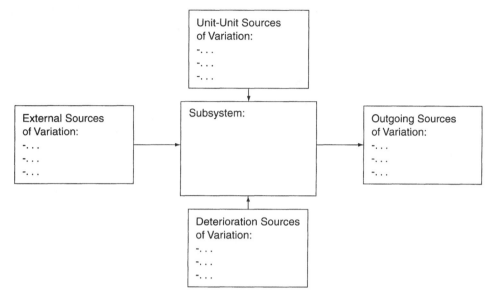

Figure 7–6   Subsystem Noise Diagram

One of the most critical pieces of information that the system integration team needs to properly evaluate the system is a clear understanding of what noises affect each of the subsystems. The best way for each system engineer to more fully understand the sources of variation within and between the subsystems they are responsible for integrating is to personally participate in the development of the subsystem noise diagrams. A noise diagram must be generated for each subsystem. The noise diagram contains the information depicted in Figure 7–6.

The **outgoing sources of variation** are important for proper stress test design prior to the system integration tests.

Generate the Initial Integrated System Noise Map

Start Date: _____ *End Date:* _____ *Owner:* _____

System noise maps can be generated by linking together all the **outgoing sources of variation** from the subsystem noise diagrams. A good system noise map will enable the system engineering team to develop credible system stress tests that expose the integrated set of subsystems to realistic forms of variation that will be encountered in the assembly, distribution, and use environment.

The system noise maps should be constructed in the design phase of the CDOV process so they can be reviewed by all of the subsystem teams to make sure they are reasonable and complete. The subsystem teams can also use them to structure their subsystem robustness test plans to include integration noise and all the input noises from the noise diagram. It is not enough to just make a subsystem robust to its own internal or within subsystem noise. When businesses launch products (integrated systems) and they have a lot of performance and reliability problems, it is usually because these integration noises were not identified and accounted for during subsystem development and integration. Developing subsystems for robustness to both internal and integration noise is a critical step in developing reliability into the system.

Refine System Reliability Model Based on Subsystem, Subassembly, and Component Reliability Models

Start Date: _____ *End Date:* _____ *Owner:* _____

Now that the subsystem noise diagrams, system noise map, and the system and subsystem FMEAs are in hand, refinements to the evolving reliability model can be made. It is also time to allocate reliability requirements and estimates down into the subsystems, subassemblies, and components. A fairly complete reliability prediction model should be taking shape for Gate 2 evaluation. Overall predicted system reliability performance will now begin to be adjusted by the new information down at the lower levels of the system architecture.

Generate System CAD Spatial Integration Model

Start Date: _____ *End Date:* _____ *Owner:* _____

Now that the subsystems are being designed, built, and evaluated for their baseline, nominal performance (ideal function modeling), the system engineering team can receive the geometric CAD files from the subsystem designers. All this form and fit information can be modeled within the system architecture for proper spatial relationships. Frequently "space wars" must be diplomatically solved and avoided. The system design leader, usually a skilled CAD operator, referees the allocation of space. A lot of analytical tolerance evaluation is conducted by the system CAD integration team. It is the system engineering team's responsibility to be the "space police." All subsystem space use requests must go through the system CAD integration leader so that interference issues can be evaluated prior to approving the use of the space. The system engineers worry about the functional interfaces while the system CAD integration leader worries about the form and fit interfaces. The system CAD integration leader will make frequent visits to the system integration meetings to review spatial issues across and between the numerous subsystem teams.

As you can see, system integration team members are characterized by what they worry about: *across* and *between* issues. Subsystem engineering teams tend to worry about *within* issues. The collaborative system integration meetings help assure the *across*, *within*, and *between* issues all get proper evaluation at the right time, using the right tools in the CDOV process.

Generate System CAE (Computer Aided Engineering) Functional Performance Model

Start Date: _____ *End Date:* _____ *Owner:* _____

The Critical Parameter Management database can be used to help model the functional performance of the system level critical functional responses. Each subsystem team should be developing ideal/transfer function models that are expressed in terms of $Y = f(x)$ relationships. These models can be linked together and used to model system performance. The models are started using models derived from "first principles." Later, in the design phase but mostly in the optimization phase, the models are empirically developed, refined, and confirmed.

Subsystem inputs that are linked to system outputs are often referred to as *transfer functions*. Once deterministic transfer functions are developed for the critical parametric relationships, statistical models in the form of Monte Carlo simulations can be developed. The system engineering team can run numerous what-if scenarios as the subsystems are changed as a result of design development, optimization, and verify certification as prescribed in the CDOV phases. These models will be used to perform analytical tolerance design prior to system integration stress tests in Phase 3B and product certification tests in Phase 4A.

It is important that the system engineering team build these critical parameter models and remain committed to their refinement and maintenance all the way through the CDOV process. The existence and refinement of such models is a strong sign that the system engineering team understands the critical parameters within the system.

Generate an Estimated Bill of Material Structure to Prepare for Configuration Control and Subsystem Part Inspection Requirements

Start Date: _____ *End Date:* _____ *Owner:* _____

The parts, subassemblies, and subsystems that comprise the system are maturing at a high rate in the design phase. Their physical architectures are defined and their earliest baseline set points are established. With this kind of information available, a preliminary part count and bill of material can be constructed. The system engineering team assembles the information through its system integration meetings with each subsystem team.

Generate System Integration and Evaluation Prototype Cost Estimate

Start Date: _____ *End Date:* _____ *Owner:* _____

The system engineering team can now bring in, from their work inside the subsystem integration meetings, the preliminary bill of material projections to enable the definition of a preliminary cost model. This cost model estimates the cost of the first system to be built for Phase 3B system integration tests. The model evolves with the designs and ultimately serves as the cost model for the actual product that will be certified at Gate 4A.

Create Data Acquisition System Requirements

Start Date: _____ *End Date:* _____ *Owner:* _____

The system interfaces are becoming clear as the subsystems and subassemblies take shape. The time is right to finalize a complete set of data acquisition system requirements. Each subsystem team works with its system engineering representative to define the metrology requirements for each of the critical functional responses they have created. The system engineering team specifies, purchases, and deploys a systemwide data acquisition system. Its job is to measure all system level critical functional responses and any subsystem or subassembly level critical functional responses that the subsystem team requests. The ability to adequately measure and document the many capability indices across the system rests at the foundation of a properly planned and designed data acquisition system.

Conduct Concept Design for the Systemwide Data Acquisition System

Start Date: _____ *End Date:* _____ *Owner:* _____

With a full set of data acquisition system requirements in hand, the process of designing the actual system can be undertaken. The process contains the following elements:

1. Define the specific functions to be measured (targets, discrimination, and extremes of data ranges likely to be encountered when exploring design space using designed experiments).
2. Generate at least two concepts for each critical functional response being measured.
3. Evaluate the concepts against a best-in-class benchmark (datum) metrology design against the specific data acquisition requirements and criteria.
4. Synthesize the best attributes from the proposed concepts and, if necessary, produce a hybrid data acquisition concept that surpasses the datum concept.
5. Develop, design, optimize, or purchase the superior concept; then verify the capability of the data acquisition equipment.

Continue to Construct the Relational Database to Document and Track Critical Parameters

Start Date: _____ *End Date:* _____ *Owner:* _____

The Critical Parameter Management database grows at a high rate during the design phase. The system, subsystem, subassembly, and component requirements are all initially defined at this point. The critical functional responses for the system, subsystems, and subassemblies are gathered and entered into the database by the system engineering team. Every subassembly CFR is linked to its requirement and traced up to the subsystem and system level CFRs and requirements they help fulfill. A many-to-many relational network of measured responses and their related requirements will begin to be developed in the design phase. All CFRs along with their initial baseline Cp performance measures should be documented at Gate 2.

Initiate and Manage a Projectwide Quality Council to Help Manage and Track Critical Parameters

Start Date: _____ *End Date:* _____ *Owner:* _____

A select set of engineers and professionals from the product development and operations teams should be formed into a Quality Council that oversees the execution of the Critical Parameter Management process. One experienced and highly respected individual from each of the following disciplines should be a member of the Quality Council:

1. System engineering (leader of the Quality Council)
2. Mechanical design engineering
3. Electrical design engineering
4. Software engineering
5. Materials engineering
6. Manufacturing engineering
7. Assembly operations
8. Supply chain management
9. Service and product support
10. Quality assurance
11. Program management

The purpose of the Quality Council is to promote and facilitate the deployment of the Critical Parameter Management process. Council members make sure all the people within their sphere of influence are trained and implementing the appropriate tools and best practices of DFSS to assure the continual generation and updating of the Critical Parameter Management database. If any one on the product development team has a problem or question, the Quality Council resolves the issue. Council members team up to help create solutions to exceptionally difficult metrology roadblocks. They refine and clarify definitions of requirements and critical parameter terms. The critical parameter database is an intricate and complex network of requirements and parametric terms and relationships. It takes discipline and hard work to build and maintain these databases. The Quality Council, under the leadership of system engineering, is responsible for helping everybody get it right. The system engineering team owns the CPM database, but the Quality Council is jointly responsible for its deployment. The Quality Council files a summary report at each gate review. This report informs the gate keepers of Critical Parameter Management issues, problems, and corrective actions undertaken as well as items for resolution prior to the next phase by program management.

In summary, the members of the Quality Council are CPM troubleshooters. They try to prevent problems as much as possible by communicating a clear approach to CPM. The problems they can't prevent they help solve. The problems they can't solve are forwarded up the chain of command in a time frame that is proportional to the nature and severity of the problem encountered. They typically meet once every week or two and also on an as-needed basis.

Create Phase 3A System Engineering and Integration Project Plans

Start Date: _____ *End Date:* _____ *Owner:* _____

The system engineering team meets to assess its required deliverables for the next phase and then structures detailed plans to deliver them in the allotted time. These plans are a requirement for passage of Gate 2.

Assess Performance and Make Changes in the Current Practice of System Engineering and Integration to Assure State-of-the-Art Performance in the Future

Start Date: _____ *End Date:* _____ *Owner:* _____

The Team meets to assess how they performed over Phase 2. They review their use of DFSS tools and best practices and the quality of their results. Future training plans and improvements are agreed to and included in the appropriate individuals performance plan.

In addition to the Phase 2 checklist and list of system engineering gate deliverables that follow, Figure 7–7 provides a Phase 2 scorecard.

Checklist of Phase 2 System Engineering and Integration Tools and Best Practices:

Competitive benchmarking (subsystem, subassembly, component)

VOC gathering methods

KJ analysis (relative to flow-down of system requirements to subsystem requirements)

QFD/Houses of Quality—Levels 2–5 (subsystem, subassembly, component, and manufacturing process levels)

Concept generation techniques (TRIZ, brainstorming, brain-writing)

Pugh concept evaluation and selection process

Design for manufacture and assembly

Value engineering and analysis

Design failure modes and effects analysis

Measurement systems analysis

Critical parameter management

Knowledge-based engineering methods

Math modeling (engineering analysis to define ideal functions and transfer functions)

Design of experiments (Full and Fractional Factorial designs and sequential experimentation)

Descriptive and inferential statistical analysis

ANOVA data analysis

Phase 2 Design Development

Deliverable	Owner	Original Due Date vs. Actual Deliv. Date	Completion Status Red Green Yellow	Performance vs. Criteria Red Green Yellow	Exit Recommendation Red Green Yellow	Notes
VOC Database Linked to Subsystem, Subassembly, Component, and Manufacturing Process Requirements			○ ○ ○	○ ○ ○	○ ○ ○	
Subsystem, Subassembly, Component, and Manufacturing Process Level Houses of Quality			○ ○ ○	○ ○ ○	○ ○ ○	
Competitive Subsystem, Subassembly, Component, and Manufacturing Process Benchmarking Data			○ ○ ○	○ ○ ○	○ ○ ○	
Subsystem, Subassembly, Component, and Manufacturing Process Level Requirements Documents			○ ○ ○	○ ○ ○	○ ○ ○	
Documented List of Subsystem, Subassembly, Component, and Manufacturing Process Level Functions That Fulfill Their Respective Requirements			○ ○ ○	○ ○ ○	○ ○ ○	
Documented Superior Subsystem, Subassembly, Component, and Manufacturing Process Concepts			○ ○ ○	○ ○ ○	○ ○ ○	
Portfolio of Candidate Technologies That Help Fulfill the System Functions and Requirements			○ ○ ○	○ ○ ○	○ ○ ○	
Documented Critical Parameter Database with Estimated Subsystem and Subassembly CFR Ideal Functions			○ ○ ○	○ ○ ○	○ ○ ○	

Figure 7-7 General Scorecard for CDOV Phase 2 System Engineering and Integration Deliverables: Phase 2 Design Development
(Continued)

Deliverable	Owner	Original Due Date vs. Actual Deliv. Date	Completion Status Red Green Yellow	Performance vs. Criteria Red Green Yellow	Exit Recommendation Red Green Yellow	Notes
Documented Critical Parameter Database with Estimated Subsystem and Subassembly CFR Interaction Transfer Functions (Between Subsystems and Subassemblies and Linkage up to the System CFRs)			○ ○ ○	○ ○ ○	○ ○ ○	
Documented Reliability Model and Predicted System Reliability Performance with Gap Analysis			○ ○ ○	○ ○ ○	○ ○ ○	
Estimated System, Subsystem, Subassembly, Component Cost Projections			○ ○ ○	○ ○ ○	○ ○ ○	
Documentation Supporting Design for Manufacturability and Assembleability			○ ○ ○	○ ○ ○	○ ○ ○	
Documented Design Failure Modes and Effects Analysis for Subsystems, Subassemblies, Reliability Critical Components, and Manufacturing Processes			○ ○ ○	○ ○ ○	○ ○ ○	
Descriptive and Inferential Statistics; Hypothesis Testing and Confidence Interval Analysis in Support of the Ideal Functions and Transfer Functions			○ ○ ○	○ ○ ○	○ ○ ○	
Sequential Design of Experiments (Initial identification of Subsystems, Subassembly, Component, and Manufacturing Process Critical Parameters)			○ ○ ○	○ ○ ○	○ ○ ○	
Analysis of Variance Data (ANOVA) That Underwrite Statistical Significance of Critical Parameters			○ ○ ○	○ ○ ○	○ ○ ○	
Documented Cp & Cpk Values for All Subsystem and Subassembly Design CFRs			○ ○ ○	○ ○ ○	○ ○ ○	
Initial Cp/Cpk Assessment for Candidate Critical-to-Function Specifications from Manufacturing Processes from Supply Chain			○ ○ ○	○ ○ ○	○ ○ ○	
Phase 3A (Optimization) Project Management Plans			○ ○ ○	○ ○ ○	○ ○ ○	

Figure 7–7 *Continued*

Regression and empirical modeling methods

Reliability modeling methods

Normal and accelerated life testing methods

Design capability studies

Multi-vari studies (for noise factor identification)

Statistical process control (for design stability and capability studies)

Gate 2 System Engineering Deliverables

Subsystem, subassembly, component, and manufacturing process level HOQ (assist and create link to system HOQ)

Subsystem, subassembly, component, and manufacturing process level requirement documents (assist and create links to system requirements document)

Superior subsystem, subassembly, component, and manufacturing process level concepts (assist and create links to system concept)

Software concept of operations, use cases, flow charts, block diagrams, sequence of events documents (assist and create links to SRD)

DFMA (assist and create links to system space allocation CAD model)

Subsystems, subassemblies, components, and manufacturing processes modeled $y = f(x)$ (assist and create links to system level CFRs in CPM database)

Subsystem DFMEAs (assist and use to refine system DFMEA)

Subsystem CFRs defined and linked to system CFRs (CPM database development)

Subsystem control factors, interactions, and linear/nonlinear model defined and documented in CPM database

Latest status of risks from new subsystems being transferred from R&D or being tracked from Phase 2 subsystem prototypes

Subsystem cost models (assist and create links to system cost model)

System noise map

System reliability model and latest subsystem allocation model

System reliability growth budget

Current bill of material structure

Status of early configuration control and problem reporting/resolution process superior design concept for the system level data acquisition system

Ongoing systemwide Critical Parameter Management database with inclusion of subsystem, subassembly, component, and manufacturing process level requirements and measured performance information (Cp, Cpk, etc.)

Status of Quality Council activities and agenda

Phase 3A system engineering and integration project plans and resource requirements

Phase 3A: Subsystem Robustness Optimization

The major system engineering and integration elements for CDOV: Phase 3 focuses on optimizing subsystems and the integrated system. For the system engineering team, Phase 3 contains differing roles within the two major parts:

- Phase 3A focuses on the robustness optimization of the individual subsystems and subassemblies within the system. Here the system engineers provide help in the subsystem optimization tasks.
- Phase 3B focuses on the integration of the robust subsystems, the nominal performance evaluation, robustness optimization, and initial reliability assessment of the complete system. System engineering "runs the show" in Phase 3B, and the subsystem teams help them in their integration tasks.

The diagram in Figure 7–8 outlines how system engineering helps the subsystem teams in Phase 3A.

The tasks outlined in Figure 7–9 are specific to the system engineering team. Many of the Phase 2 system engineering design tasks will continue on into Phase 3. System engineers have a lot to do to get ready for the Phase 3B integration tests. Much of what system engineers do breaks down into two focused areas:

1. Assist the subsystem teams in their Phase 3A robustness optimization tasks.
2. Get ready to conduct nominal and stressful system integration tests.

The aim of Phase 3A is the further development of subsystem and subassembly baseline designs resulting in the confirmation of their robustness optimization. Gate 3A assesses the subsystems and subassemblies relative to robustness optimization requirements. The aim of Phase 3B is system integration and stress testing, culminating in a gate review for system integration and performance certification.

The specific issues related to each Phase 3A system engineering action item (start date for each action item, the action item, and end date for each action) will be reviewed in this section. Each of these action items can be conducted by using a tool or best practice. The flow of actions can be laid out in the form of a PERT chart for project management purposes.

Continue to Resolve "Conflicting Requirement" Problems by Providing Data from the VOC Document and System Requirements Document

Start Date: _____ *End Date:* _____ *Owner:* _____

Even though we are starting Phase 3A, unsettled issues remain related to the flow-down of system to subsystem to subassembly to component to manufacturing process requirements. Some requirement issues can exist right to the bitter end. This is obviously not desirable but it does and

Start Date: ___	Help review and finalize the critical functional responses for each of the subsystems and subassemblies	End Date: ___	Owner: _____
Start Date: ___	Help develop subsystem noise diagrams and system noise map	End Date: ___	Owner: _____
Start Date: ___	Help conduct and assess noise factor experiments	End Date: ___	Owner: _____
Start Date: ___	Help define compounded noises for robustness experiments	End Date: ___	Owner: _____
Start Date: ___	Help define engineering control factors for robustness optimization experiments	End Date: ___	Owner: _____
Start Date: ___	Help design for additivity and run designed experiments or simulations	End Date: ___	Owner: _____
Start Date: ___	Help analyze data, build predictive additive model	End Date: ___	Owner: _____
Start Date: ___	Help run verification experiments to certify robustness	End Date: ___	Owner: _____
Start Date: ___	Help conduct response surface experiments on critical adjustment parameters	End Date: ___	Owner: _____
Start Date: ___	Help run verification experiments to certify tunability and robustness parameters for subsystems and subassemblies	End Date: ___	Owner: _____
Start Date: ___	Document critical functional parameter nominal set points, CAP and CFR relationships	End Date: ___	Owner: _____
Start Date: ___	Develop, conduct, and analyze reliability/capability evaluations for each subsystem and subassembly	End Date: ___	Owner: _____

Figure 7–8 Flow of Phase 3A of Product Commercialization

will continue to happen. The extent to which these conflicts remain unsettled is a function of how well the product development team conducts Phase 1 and Phase 2 concept engineering. If the tools and best practices of DFSS are properly deployed relative to requirement development and concept engineering, then there will be a low residual buildup of requirement conflicts. The remaining few conflicts are still managed and resolved by using the PM database, where the VOC and requirements hierarchy now resides.

Continue to Conduct Integration Meetings with Subsystem Teams

Start Date: _____ *End Date:* _____ *Owner:* _____

Start Date: ___	Continue to resolve "conflicting requirement" problems by providing data from VOC document and system requirements document (including all requirement allocations)	End Date: ___	Owner: _____
Start Date: ___	Conduct system integration meetings with subsystem teams	End Date: ___	Owner: _____
Start Date: ___	Help include system noises in subsystem and subassembly noise diagrams	End Date: ___	Owner: _____
Start Date: ___	Refine system reliability model based on subsystem, and subassembly, and component reliability models	End Date: ___	Owner: _____
Start Date: ___	Refine the integrated system noise map	End Date: ___	Owner: _____
Start Date: ___	Refine system integration and evaluation prototype cost estimate	End Date: ___	Owner: _____
Start Date: ___	Refine system CAD spatial integration model	End Date: ___	Owner: _____
Start Date: ___	Refine system CAE functional performance model	End Date: ___	Owner: _____
Start Date: ___	Refine an estimated bill of material structure to prepare for configuration control and subsystem part inspection requirements	End Date: ___	Owner: _____
Start Date: ___	Refine data acquisition system requirements	End Date: ___	Owner: _____
Start Date: ___	Refine concept design for the systemwide data acquisition system	End Date: ___	Owner: _____
Start Date: ___	Continue to construct the relational database to document and track critical parameters from Phase 3A experiments	End Date: ___	Owner: _____
Start Date: ___	Create system integration prototype construction, debug and test plans	End Date: ___	Owner: _____
Start Date: ___	Manage the projectwide quality council to help manage and track critical parameters	End Date: ___	Owner: _____
Start Date: ___	Create Phase 3B system engineering and integration project plans	End Date: ___	Owner: _____
Start Date: ___	Assess performance and make changes in the current practice of system engineering and integration to assure state-of-the-art performance in the future	End Date: ___	Owner: _____

Figure 7–9 Phase 3A Tasks for System Engineering Staff

The system integration meetings should be an established routine by now. The focus of these meetings in Phase 3A is on the development and execution of robustness optimization experiments. As the data comes in from the optimization experiments, the system team can update the CPM database. A pre- and post-robustness optimization Cp index for each subsystem CFR is required. The purpose of subsystem robustness optimization is to minimize the subsystem CFRs to statistically significant sources of noise. This is achieved by identifying significant interactions between the subsystem's control factors and noise factors. These unique interactions are typically quantified by using a series of comparisons of the signal-to-noise ratio.

Help Include System Noises in Subsystem and Subassembly Noise Diagrams

Start Date: _____ *End Date:* _____ *Owner:* _____

Each subsystem team refines and uses its respective noise diagrams to aid in the proper design of its noise experiments and the robustness optimization experiments. The system engineers assigned to respective subsystem teams need to be intimately involved in the robustness optimization experiments and resulting data. System engineering plays a key role in advising the subsystem teams on integration noises that will help the subsystem become robust to these unique noise factors. The goal here is to get the subsystems properly stress tested so that they are fairly easy to integrate and evaluate in the first integrated system. If the subsystems do not include integration noises in their experiments, there is a high degree of risk that the system integration tests will not go well and will likely overrun their time budget.

Refine the Integrated System Noise Map

Start Date: _____ *End Date:* _____ *Owner:* _____

The results of the subsystem and subassembly robustness evaluations may provide valuable information about integration noise factors that need to be included in the system noise map. The system has not been fully integrated yet but the system mules, in which the subsystems and subassemblies are evaluated, often yield new information about integration noise factors. Noise factors that were not obvious prior to the actual experimentation may even be discovered. The system noise map should be updated with any new information from all iterations of the robustness optimization experiments.

Refine System Reliability Model Based on Subsystem, Subassembly, and Component Reliability Models

Start Date: _____ *End Date:* _____ *Owner:* _____

The reliability information coming in from the various forms of subsystem, subassembly, and component life tests and accelerated life tests can be used to help refine the predicted reliability of the system. The reliability of the newly robust subsystems and subassemblies must be evaluated. Robust design has a proven track record of growing reliability when the robustness stress tests include the forms of noise that deteriorate reliability. Becoming insensitive to deteriorative sources of noise typically slows the rate of failures. Care should be taken to refer back to the FMEA data to be sure the noise diagrams and the FMEAs are correlated.

Refine System Integration and Evaluation Prototype Cost Estimate

Start Date: _____ *End Date:* _____ *Owner:* _____

The subsystems are maturing in part due to the changes in certain nominal set points resulting from the robustness work and in part due to changes for assembleability, cost, and any number of other requirement balancing issues. The cost estimates for the system integration prototypes can be refined to include Phase 3A improvements. Usually robustness improvements do not have a large impact on cost but other design changes do. Depending on the size and complexity of the system being developed, the cost to make and assemble the system prototype can be quite high. This may surprise managers if they are out of touch with the fact that you need to fabricate and build system prototypes that emulate the way your product will be stressed and deteriorated in the actual shipping and use environment. If you can't afford to proactively design robustness and durability into your product, you probably won't gain market share or enjoy the estimated revenues your business case projects. And be sure to open a sizable escrow account for warranty claims.

It is amazing how many managers tend to think only about the lowest cost and shortest cycle-time. What DFSS and the CDOV product development process are promoting is the development of the *right* cost and using the *appropriate* cycle-time that fulfills the system requirements—nothing more and certainly nothing less!

Refine System CAD Spatial Integration Model

Start Date: _____ *End Date:* _____ *Owner:* _____

Along with the set point changes that come from the results of the robustness optimization experiments come the inevitable spatial form and fit changes that affect the system architecture. The subsystem designers need to inform the system designers of the adjustments they are planning so they can be verified against the system space allocation model. Interferences, proximity noise inducement, and transmission (energy, mass, and signals where they are not supposed to be!) must be checked before the changes can be approved from a system performance perspective.

Refine System CAE Functional Performance Model

Start Date: _____ *End Date:* _____ *Owner:* _____

Functions are the main focus of Critical Parameter Management in DFSS. Any changes made for robustness optimization purposes are supposed to enhance functional integrity within and across the system. Just because you have improved the robustness of a subsystem does not necessarily mean you have helped the entire system perform better. The system engineers use the system model and system integration tests to help determine if changes in subsystem performance are harmonious with the maturation of system performance and the balanced fulfillment of the system requirements. Phase 3A uses the analytical and graphical models of the integrated system to check these issues. Phase 3B uses the physical system prototype model to evaluate the system level form, fit, and functional performance issues empirically.

Refine an Estimated Bill of Material Structure to Prepare for Configuration Control and Subsystem Part Inspection Requirements

Start Date: _____ *End Date:* _____ *Owner:* _____

Prior to Phase 3B, the bill of material must go through one more iteration to account for the changes from subsystem robustness optimization. The components that are fabricated and assembled into the system prototype model are inspected and placed under rigorous configuration control. This is done to assure that the system integration engineers know exactly what the critical parameter values at the component, subassembly, and subsystem level are prior to Phase 3B testing. If you do not inspect the critical-to-function specifications on the components prior to integration, you will experience performance problems that will consume valuable time to hunt down. You need to know what is being assembled into the system prototypes before you test them. History has shown conclusively that when robust subsystems and subassemblies are properly integrated, they can produce remarkable system performance during the initial system evaluations in Phase 3B.

Refine Data Acquisition System Requirements

Start Date: _____ *End Date:* _____ *Owner:* _____

After the subsystem teams master the methods of measuring their subsystem critical functional responses in Phase 3A, all their learning and improvements can be used to help the system integration team get ready to measure their CFRs during system performance evaluations in Phase 3B. Many things are likely to go wrong in Phase 3A subsystem CFR measurement. The system engineers help correct these problems and then transfer that knowledge into the planning for Phase 3B CFR measurement. It is a huge mistake for the system team to be "aloof" when the subsystem teams are setting up and taking data; team members should physically be there to learn how to get it right when it becomes their turn to measure functional performance.

Refine Concept Design for the Systemwide Data Acquisition System

Start Date: _____ *End Date:* _____ *Owner:* _____

The system data acquisition system architecture may need to change to accommodate changes in requirements and additions to what the subsystem team wants measured for them in Phase 3B.

Continue to Construct the Relational Database to Document and Track Critical Parameters from Phase 3A Experiments

Start Date: _____ *End Date:* _____ *Owner:* _____

All the results for subsystem and subassembly CFR capability growth need to be documented in the CPM database. The signal-to-noise improvements should also be documented to help estimate the trend in reliability growth based on increased robustness to deteriorative noise factors. The CPM hierarchy of critical relationships down through the system should be steadily growing. The critical adjustment parameters that will enable Cpk values to be adjusted to equal Cp should be completely documented for use in the Phase 3B integration and tolerance balancing evaluations. A rich set of data from the subsystem robustness and capability growth assessments should be in hand for Gate 3A review.

Manage the Projectwide Quality Council to Help Manage and Track Critical Parameters

Start Date: _____ *End Date:* _____ *Owner:* _____

The Quality Council should be actively engaged across the product development team. Council members should be solving measurement system problems and clarifying Critical Parameter Management requirements and responsibilities. They will go a long way to establishing a culture of rigor and discipline around taking data to assess the fulfillment of requirements across and within the system. Without their cross-functional work, the organization will drift and fractionate. The result is often that the left hand doesn't know what the right hand is doing. It also creates healthy peer pressure right where you want it—not settling for mediocre performance in gathering data and understanding relationships and sensitivities across the integrated system.

Create System Integration Prototype Construction, Debug and Test Plans

Start Date: _____ *End Date:* _____ *Owner:* _____

The system engineering team is heavily focused on planning system integration tests. These test plans are a required deliverable at Gate 3A. Team members are also busy making sure all the components, assembly procedures, and equipment are in place to build the system prototype models. The system data acquisition system has been purchased and is now being assembled, debugged, and assessed for measurement capability. They have communicated with the subsystem teams and made them aware that they will require all CFT specifications to be measured on all components prior to integration.

Create Phase 3B System Engineering and Integration Project Plans

Start Date: _____ *End Date:* _____ *Owner:* _____

The system engineering team meets to assess its required deliverables for the next phase (3B) and then structures detailed plans to deliver them in the allotted time. These plans are a requirement for passage of Gate 3A.

Assess Performance and Make Changes in the Current Practice of System Engineering and Integration to Assure State-of-the-Art Performance in the Future

Start Date: _____ *End Date:* _____ *Owner:* _____

The team meets to assess how it performed over Phase 3A. Team members review their use of DFSS tools and best practices and the quality of their results. Future training plans and improvements are agreed to and included in the appropriate individuals' performance plan. Figure 7–10 offers a scorecard of Phase 3A gate deliverables.

Phase 3A Goals

- Help assure selection of superior subsystem concepts.
- Help assure robust subsystems are ready for system integration tests.
- Assure that Critical Parameter Management is established and system data acquisition is in place.

Gate 3A System Engineering Deliverables

Subsystem HOQs (assist and create link to system HOQ)

Subsystem requirement documents (assist and create links to system requirements document)

Superior subsystem concepts (assist and create links to system concept)

Software concept of operations, use cases, flow charts, block diagrams, sequence of events

Documents (assist and create links to SRD)

DFMA (assist and create links to system space allocation CAD model)

Subsystem's modeled $y = f(x)$ (assist and create links to system level CFRs)

Subsystem DFMEAs (assist and use to refine system DFMEA)

Subsystem CFRs defined and linked to system CFRs

Subsystem control factors, interactions, and linear/nonlinear model defined and documented in CPM database

Phase 3A Optimization

Deliverable	Owner	Original Due Date vs. Actual Deliv. Date	Completion Status Red Green Yellow			Performance vs. Criteria Red Green Yellow			Exit Recommendation Red Green Yellow			Notes
Verification of Robustness and Tunability of the Subassemblies and Subsystems and Their Current Design Capability (Cp values) to Fulfill Their Respective Requirements			○	○	○	○	○	○	○	○		
Subsystem and Subassembly Noise Diagrams			○	○	○	○	○	○	○	○		
System Noise Map			○	○	○	○	○	○	○	○		
Complete Set of Statistically Significant Noise Vectors			○	○	○	○	○	○	○	○		
Complete Set of Pre-and Post-Robustness Optimization Subsystem and Subassembly CFR Cp & S/N Values			○	○	○	○	○	○	○	○		
Confirmation of the Additive S/N Model			○	○	○	○	○	○	○	○		
Complete Set of Critical Robustness Parameters and Their Nominal Set Points			○	○	○	○	○	○	○	○		
Complete Set of Critical Adjustment Parameters (incl. Their Ideal "Tuning" Functions)			○	○	○	○	○	○	○	○		
Updates to Critical Parameter Database and Scorecards			○	○	○	○	○	○	○	○		
Subsystem and Subassembly Reliability Test Plans and Current Data/Results Summary			○	○	○	○	○	○	○	○		
Phase 3B Project Management Plans			○	○	○	○	○	○	○	○		

Figure 7–10   General Score Card for CDOV Phase 3A System Engineering and Integration Deliverables

Latest status of integration noise transmission (risks from new subsystems being transferred from R&D or being tracked from Phase 2 subsystem prototypes)

Subsystem cost models (assist and create links to system cost model)

System noise map

System reliability model and latest subsystem allocation model

System reliability growth budget

Current bill of material structure

Status of early configuration control process superior design concept for the system level data acquisition system

Ongoing systemwide Critical Parameter Management database with inclusion of subsystem requirements & performance information

Status of Quality Council activities and agenda

System level risk analysis report based on summary of system test data

Phase 3B system engineering and integration project plans and resource requirements

General Checklist of Phase 3A System Engineering and Integration Tools and Best Practices

Subsystem and subassembly noise diagramming

System noise mapping

Measurement system analysis

Use of screening experiments for noise vector characterization

Analysis of means (ANOM) data analysis techniques

Analysis of variance (ANOVA) data analysis techniques

Baseline subsystem and subassembly CFR signal-to-noise characterization

Taguchi's methods for robust design

Design for additivity

Use of Full or Fractional Factorial experiments for robustness characterization

Generation of the additive S/N model

Response surface methods

Design capability studies

Critical Parameter Management

Reliability analysis

Life testing (normal, HALT, & HAST)

Project management and planning

Phase 3B: System Integration

The major elements of Phase 3B (see Figure 7–11) focus on system integration, nominal performance evaluation, robustness optimization, and initial system reliability assessment.

Start Date: ___	Integrate subsystems into system test units	End Date: ___	Owner: _____
Start Date: ___	Refine and complete the system requirements document	End Date: ___	Owner: _____
Start Date: ___	Verify capability of the systemwide data acquisition system	End Date: ___	Owner: _____
Start Date: ___	Conduct nominal performance evaluation on the system	End Date: ___	Owner: _____
Start Date: ___	Conduct system robustness stress tests	End Date: ___	Owner: _____
Start Date: ___	Document and track problem reports generated during tests	End Date: ___	Owner: _____
Start Date: ___	Calculate design capabilities for all system and subsystem CFRs from nominal performance and *stress tests*	End Date: ___	Owner: _____
Start Date: ___	Develop and document corrective action based on root cause analysis	End Date: ___	Owner: _____
Start Date: ___	Refine and update DFMEAs, noise diagrams, and noise maps	End Date: ___	Owner: _____
Start Date: ___	Conduct critical life and accelerated life tests	End Date: ___	Owner: _____
Start Date: ___	Refine subsystem set points to balance system performance	End Date: ___	Owner: _____
Start Date: ___	Refine and update critical parameter database and maps	End Date: ___	Owner: _____
Start Date: ___	Conduct initial system reliability assessment	End Date: ___	Owner: _____
Start Date: ___	Refine system and subsystem reliability models, predictions and budgets	End Date: ___	Owner: _____
Start Date: ___	Define production assembly and manufacturing CPM metrics for CFRs, CAPs, CFPs, and component CTF specifications	End Date: ___	Owner: _____
Start Date: ___	Assist in evaluating and documenting problems with system assembleability, serviceability, and maintainability	End Date: ___	Owner: _____
Start Date: ___	Verify system readiness for final product design capability development and assessment	End Date: ___	Owner: _____
Start Date: ___	Generate Phase 4A project management plans for product design verification	End Date: ___	Owner: _____
Start Date: ___	Assess performance and make changes in the current practice of system engineering and integration to assure state-of-the-art performance in the future	End Date: ___	Owner: _____

Figure 7–11   Flow of Phase 3B of Product Commercialization

The specific issues related to each Phase 3B System Engineering action item (start date for each action item, the action item, end date for each action) will be reviewed in this section. Each of these action items can be conducted by using a tool or best practice. The flow of actions can be laid out in the form of a PERT chart for project management purposes.

Integrate Subsystems into System Test Units

Start Date: _____ *End Date:* _____ *Owner:* _____

The long-awaited time to build the system integration prototype models has come. The system engineering team oversees the physical construction of the system models. The system models are usually built by technicians, assemblers from the production organization, and field service engineers. Since this is the first time anyone has actually integrated all the subsystems, sub-assemblies, and components into a functioning system, representatives from all the organizations that will end up assembling and servicing the end product are included on the build team. This gives them an early look and feel for the product and allows them to provide assembly and service critiques that inevitably will require adjustments and changes to the system architecture. The sooner these people get to know the system, the better they can help the design teams to iterate and optimize their designs.

Some businesses have the production organization completely conduct the first system model construction. This is usually due to tight resources and a feeling that "the system belongs to production so let them build it." The problem with this approach is that product organizations are usually ill equipped to provide the necessary technical feedback to the design organization. The intermediary role that a small, technically focused system engineering team provides is well suited to managing the interfaces and documenting both functional and spatial (form and fit) issues encountered in the initial system build process. The system engineering team has built strong communication links to each of the subsystem design teams, and it is quite easy for them to provide rapid and clear feedback on integration issues. The system engineering team will oversee the construction of all system test units through Phase 4A. At Phase 4B the production organization takes full ownership of the build process, and system engineering plays a supporting role.

Refine and Complete the System Requirements Document

Start Date: _____ *End Date:* _____ *Owner:* _____

A lot of learning takes place in the system integration phase. The system requirements will inevitably undergo some changes as the entire product development team assesses the data from the construction and evaluation of the system integration prototype models. The goal is to freeze as many of the system requirements as possible at the end of Phase 3B. Subsystem, subassembly, component, and manufacturing requirement balancing occur at a pretty heavy pace during this phase. Many things come to light that were not anticipated, and many of the lower level requirements need to be reallocated and adjusted for the sake of fulfilling the system requirements. Some people refer to Phases 3B and 4A as the tolerance balancing phases, "where we often have to rob Peter to pay Paul."

Verify Capability of the Systemwide Data Acquisition System
Start Date: _____ *End Date:* _____ *Owner:* _____

At the beginning of Phase 3B the system engineering team conducts measurement system analysis on system data acquisition to verify its capability to measure the many critical functional responses across and within the subsystems. Tens or even hundreds of high- and low-speed analog and digital signals may be sampled, stored, and analyzed down through the critical parameter hierarchy. Many forms of measurement are required to assure that the many-to-many functional relationships and sensitivities are fulfilling their respective requirements.

Conduct Nominal Performance Evaluation on the System
Start Date: _____ *End Date:* _____ *Owner:* _____

The first system evaluation that is conducted is a baseline performance capability evaluation. This test produces critical functional response data under nominal conditions with all critical parameters at their nominal target set points. The test results provide Cp indices for all CFRs at the system level and the selected CFRs down at the subsystem and subassembly level that were requested by the subsystem engineering teams.

This test is as easy as it is going to get for the newly integrated system. Just as there is no sense integrating the system until all subsystems are made robust, there is no sense running system stress tests until the system performance has been tested under stress-free, nominal conditions. If the system exhibits poor performance in the baseline tests, it will be a "basket case" when the system engineers hit it with the stressful Taguchi noise factors during the system stress tests. The Taguchi noise factors include external, unit-to-unit, and deterioration sources of variation. Don't even think about running a system reliability test until both the baseline capability tests and the system stress tests are completed and corrective actions implemented.

Conduct System Robustness Stress Tests
Start Date: _____ *End Date:* _____ *Owner:* _____

After the system successfully passes the required baseline evaluations, it will be set up to undergo rigorous Taguchi style stress tests. Here we purposefully induce realistic stress conditions in the system such as:

1. External sources of variation
   a. Contamination
   b. Temperature extremes

      c. Humidity extremes

      d. Vibrational extremes

      e. Inputs of energy and/or mass to which the system is believed to be sensitive

2. Unit-to-unit sources of variation

      a. Part-to-part

      b. Consumed materials in terms of batch-to-batch

      c. Assembly operators (adjustment variations)

      d. Inputs that are commonly associated with tolerances

3. Deterioration sources of variation

      a. Wear of surfaces (rate based)

      b. Depletion of bulk properties (rate based)

      c. Loss of mass

      d. Changes in physical properties due to applied loads, exposure to energy sources, and chemical reactions over time

The system stress test can be a series of evaluations as well as a single unified "blast of reality." Think of these tests in this way: Test the system as if you are the one who is going to be buying it for your business or personal use! You are obligated to systematically stress and evaluate the system the way your shipping partners and customers will, so that when they get it they are not the first to have applied such stress to the system. If you are not doing these system stress tests, you are using your customers as guinea pigs. This is no way to treat a customer you intend to keep for any length of time.

The CPM database is loaded with nominal, baseline Cp indices from the first system tests. Now we add Cp indices calculated from the CFR data under these stressful conditions. These Cp values will obviously be lower than the baseline values. But you need to see just how low! Some product development teams have established $Cp = 2$ standards for critical functional responses without stress and a standard of $Cp = 1$ with stress.

Document and Track Problem Reports Generated During Tests

Start Date: _____ *End Date:* _____ *Owner:* _____

During the baseline and stress tests, problems are encountered. Each problem should be recorded and logged into a problem reporting database. Each problem report is tracked for resolution. The only way to erase a problem report is to conduct corrective action and provide proof that the problem has been corrected.

Calculate Design Capabilities for All System and Subsystem CFRs from Nominal Performance and Stress Tests

Start Date: _____ *End Date:* _____ *Owner:* _____

After each system baseline and stress test, the mean and the standard deviation for each CFR can be used to calculate the Cp and Cpk values. These capability indices are the basis for Six Sigma performance measurement. They also are used to update the capability growth indices for each subsystem across the system. The capability growth index is explained in Chapter 12. The capability growth index is analogous to rolled throughput yield from Operations Six Sigma methods.

Develop and Document Corrective Action Based on Root Cause Analysis

Start Date: _____ *End Date:* _____ *Owner:* _____

The problem reports account for the things that go wrong during the system integration tests. Root cause analysis must be conducted on each problem. Once a corrective action is defined and applied, the system engineering team documents the completed problem resolution.

Refine and Update DFMEAs, Noise Diagrams, and Noise Maps

Start Date: _____ *End Date:* _____ *Owner:* _____

This disciplined effort to document and resolve problems also extends to the failure modes and effects analysis. Any and all failures are thoroughly investigated and documented. Life testing may be initiated on items that were not formerly identified as candidates for such evaluations. Additional engineering analysis can be conducted to identify the root cause of the failure. Failures are always due to some form of mass and energy transformation that, over time, resulted in a failure. Failure mechanisms can often be traced to a measurable parameter or response variable. The goal of the system and subsystem teams is to identify these variables so in the future they can be measured for "impending failures." In the copier industry, where paper jams are considered failures, paper position with respect to time is measured by LED and CCD sensors. In this way one can literally see paper beginning to change position; given enough time, the paper will mistrack to a point of jamming. This is what we mean by measuring impending failures.

Conduct Critical Life and Accelerated Life Tests

Start Date: _____ *End Date:* _____ *Owner:* _____

The system engineering team is routinely conducting regular and various forms of accelerated life testing. Numerous items are identified for life testing by FMEA methods. As just mentioned, design elements that fail during system evaluations may need to also undergo life testing now that their vulnerability has surfaced. These added candidates for life testing are analyzed and tested if the results warrant such action.

Refine Subsystem Set Points to Balance System Performance

Start Date: _____ *End Date:* _____ *Owner:* _____

The aggregated results from system baseline and stress testing are analyzed by the system and subsystem teams. Numerous system, subsystem, subassembly, and component set points are changed as a result of this analysis. The goal is to balance system performance against the system requirements document. These changes take time to complete and patience needs to be exercised before moving on to conduct the formal system reliability tests.

Refine and Update Critical Parameter Databases and Maps

Start Date: _____ *End Date:* _____ *Owner:* _____

After the system and subsystem teams have conducted their baseline and stress tests, all CFRs and CTF specifications with their unfolding sensitivity relationships can be documented in the Critical Parameter Management database. Once again, we focus on accounting for the capability growth of the network of critical functional responses throughout the integrated system.

Any new sources of noise transmission uncovered during the system evaluations can be documented in the system noise map. The system noise map is usually of great interest to the assembly and service organizations. They use these maps in conjunction with the CPM database to help them in mistake proofing, problem prevention, diagnostics, and troubleshooting.

Conduct Initial System Reliability Assessment

Start Date: _____ *End Date:* _____ *Owner:* _____

Once the corrective action changes are completed and system performance is taken to a new, more balanced level, the formal system reliability evaluations can begin. The system has evolved to a level worth evaluating in the context of a reliability test. Formal reliability evaluations on the system are run under the following conditions:

1. Nominal customer use conditions are established and maintained during the evaluations.
2. System failures and problems are corrected as if it was a real service call.
3. All components are set at their optimum, robust set points.
4. All assignable causes of variability are allowed to occur naturally (without intentional inducement) just as they would in the actual use environment.
5. All system and selected subsystem and subassembly critical functional responses are measured with the system data acquisition system after it has been checked for accuracy and repeatability (gage R&R capabilities and P/T ratios evaluated).

The initial reliability evaluation runs until it is clear that a settled down reliability or failure rate is reached. This is yet another excellent opportunity to conduct capability studies on the system.

The CFRs can be measured using the methods of statistical process control to establish their Cp and Cpk performance. Most engineers think SPC is only useful and applicable out in the supply chain where it is used during production operations. You can use SPC on subsystems starting in Phase 2 and continue to apply them to characterize stability of CFR performance all the way through Phases 3A, 3B, 4A, and finally 4B.

Refine System and Subsystem Reliability Models, Forecasts, and Budgets

Start Date: _____ *End Date:* _____ *Owner:* _____

We can refine the reliability prediction models and budgeted reliability allocations down through the system after we attain the service rate (mean time between service) and failure rate (mean time to failure) data from the reliability evaluation. This is the system team's first real look at objective data that represents the baseline reliability of the integrated system.

It is important to note that predicting reliability does not improve reliability. It only motivates one to action if the prediction is quite disparate with the perceived gap in current reliability. It is a waste of precious time to run reliability tests prematurely prior to subsystem robustness optimization. The FMEAs, noise diagrams, and noise map all point to subsystem liabilities and risks that must be the focus of our attention long before one should consider integrating and testing system reliability. Many say, "How do I know what to fix if I don't hurry up and build a 'system' and test it?" Translate this as, "Let me go create a bunch of problems so I can gain comfort from their existence and then I can get busy fixing them."

The DFSS approach using the CDOV process with the "vision" of Critical Parameter Management strategically slows the program relative to reliability tests. This is done for the sake of preventing integration problems by first making the subsystems and subassemblies robust to statistically significant noise factors. The noise factors are brought to our attention by conducting FMEA, noise diagramming, system noise mapping, and noise experiments in Phase 3A. You will be able to get through system performance and reliability evaluations much quicker if you slow down and take the time needed to develop robust subsystems in Phase 3A. If you don't do this, you will be hurrying up early in the program only to be greatly slowed down by the vicious results of all the noises to which subsystems remain sensitive after you have your disappointing reliability test data.

Define Production, Assembly, and Manufacturing CPM Metrics for CFRs, CAPs, CFPs, and Component CTF Specifications

Start Date: _____ *End Date:* _____ *Owner:* _____

Now that you have had a full-blown data acquisition event, you can stand back with the production and service experts and ask a few key questions:

1. What few critical functional responses should we continue to measure and track as this system approaches production and use in the field?
2. What critical adjustment parameters are we going to measure and track in production and service operations to keep Cpk = Cp?

3. What critical functional parameters within the subsystems should be measured as lower level CFRs during assembly and service procedures?
4. What component level critical-to-function specifications are worth tracking for Cp and Cpk performance as they are produced out in the supply chain?

This information will enable the production and service organizations to begin their portion of the Critical Parameter Management process.

Assist in Evaluating and Documenting Problems with System Assembleability, Serviceability, & Maintainability

Start Date: _____ *End Date:* _____ *Owner:* _____

During the system builds and evaluations, many assembly, service, and maintenance issues will become apparent. The system engineers need to take careful notes and file rigorous problem reports to be sure all types of issues are documented. It is not uncommon for resource limitations to prevent assembly and service personnel to be continuously present at all times during the initial system evaluations. It is the system engineering team's responsibility to act as the "*eyes and ears*" of these organizations until they can assume full responsibility for the system later in Phase 4B.

Verify System Readiness for Final Product Design Capability Development and Assessment

Start Date: _____ *End Date:* _____ *Owner:* _____

A thorough analysis and review of all the data from all three forms of system evaluation must be conducted prior to Gate 3B. The results are compared to the system requirements document. Many corrective actions and design improvements to balance cost and quality are under way. A risk analysis is conducted to account for the current state of the system. If the risks are deemed manageable, then the system engineering team can recommend passing Gate 3B. If the data and action items suggest too much risk remains, gate passage is denied and the program leaders manage the work-around strategy to keep the program as close to the schedule as possible. The whole idea behind DFSS tool, CDOV discipline, and CPM data requirements is to prevent schedule slips.

Generate Phase 4A Project Management Plans for Product Design Certification

Start Date: _____ *End Date:* _____ *Owner:* _____

The system engineering team meets to assess its required deliverables for the next phase and then structures detailed plans to deliver them in the allotted time. These plans are a requirement for passage of Gate 3B (see the scorecard in Figure 7–12).

Phase 3B Optimization

Deliverable	Owner	Original Due Date vs. Actual Deliv. Date	Completion Status Red Green Yellow	Performance vs. Criteria Red Green Yellow	Exit Recommendation Red Green Yellow	Notes
System CFR Nominal Performance Capability Indices			○ ○ ○	○ ○ ○	○ ○ ○	
System CFR Stressed Performance Capability Indices			○ ○ ○	○ ○ ○	○ ○ ○	
System Sensitivity Analysis Results			○ ○ ○	○ ○ ○	○ ○ ○	
System-to-Subsystem-to Subassembly Transfer Functions			○ ○ ○	○ ○ ○	○ ○ ○	
Refined Subsystem and Subassembly Ideal Functions			○ ○ ○	○ ○ ○	○ ○ ○	
System Reliability Performance			○ ○ ○	○ ○ ○	○ ○ ○	
Subsystem, Subassembly, and Component Reliability and Life Test Performance			○ ○ ○	○ ○ ○	○ ○ ○	
Updated Critical Parameter Database and Scorecards			○ ○ ○	○ ○ ○	○ ○ ○	
Phase 4A Project Management Plans			○ ○ ○	○ ○ ○	○ ○ ○	

Figure 7-12 General Score Card for CDOV Phase 3B System Engineering and Integration Deliverables

Assess Performance and Make Changes in the Current Practice of System Engineering and Integration to Assure State-of-the-Art Performance in the Future

Start Date: _____ *End Date:* _____ *Owner:* _____

The team meets to assess how it performed over Phase 3B. Team members review their use of DFSS tools and best practices and the quality of their results. Future training plans and improvements are agreed to and included in the appropriate individuals' performance plan.

Phase 3B Overall Goals

- Oversee the documentation of the product performance specifications as they are derived from the system requirements document and system performance/stress test data.
- Integrate subsystems to achieve full system performance with documented robustness and Cp performance for all CFRs.
- Prove, with data, that the system architecture fulfills all system requirements (with and without noise).

System Engineering Deliverables Gate 3B

Complete system requirements document

CFR Cp index "hurdle values"

First system prototypes

Risk assessment on nonintegrated subsystems and accessories

Certified computer-aided data acquisition system

Update on status of problem reports and corrective action

Nominal integrated system performance characterized

System sensitivities characterized

System CFRs and Cp values mapped

System reliability assessed

CPM database loaded with system test results

Updated DFMEAs, noise diagrams, and the system noise map

Status of life and accelerated life tests

Refined system and subsystem reliability models, allocations, and risk assessment

Reliability test results

Refined product cost estimates

Preliminary assembly documentation

Product design certification test plan

Engineering analysis reports on system sensitivities and integrated subsystem performance

Refined bill of materials

Documented system integration problem reports and corrective actions

System benchmarking test results and comparative assessment (your old product or a competitor) and risk analysis report for product design certification

Test results for environmental, health, safety, and regulatory compliance

Final component, subsystem, and system design specifications verified

CPM database updated from product certification test results

Documented problems with system assembleability, serviceability, and maintainability along with corrective actions completed

Construct, debug, deliver, and assist in testing of alpha systems

Documented results from regular and accelerated life tests

Updated DFMEAs, noise diagrams, and the system noise map from alpha systems testing

Updated Critical Parameter Management system-to-subsystem CFR maps and relational database

Preliminary service and maintenance documentation

Updated problem report and corrective action process and database

Refined bill of materials and configuration control database

System performance and integration risk assessment from all Phase 3 test data

Product performance specification document

All Phase 3 product design certification criteria met per the system requirements document

Project plans for Phase 4A

Phase 3B System Engineering Tools and Best Practices

QFD

Critical Parameter Management

Capability studies

Risk analysis

Engineering analysis

Baseline, stress, and reliability performance tests

Measurement system analysis

Excel database management

Design of experiments (screening and modeling)

ANOVA

Regression

Robust design

DFMEA

Subsystem noise diagramming

System noise mapping

Life testing

Accelerated life testing

HALT

HAST

Reliability modeling

Reliability allocation analysis

Reliability test planning

Project management

SPC

Statistical data analysis methods

Analytical tolerance design

Empirical tolerance design

Baseline performance tests

Competitive assessment

Regulatory and environmental performance test planning

Regulatory and environmental test capability studies

DFMA

Shipping performance test design

Performance, stress, and reliability testing

System–subsystem–subassembly transfer function development

System level sensitivity analysis

Health and safety testing

Risk analysis

Phase 4A: Final Product Design Certification

The major system engineering and integration elements for CDOV Phase 4 involve certification of final product design, production processes, and service capability. Phase 4 contains two major parts:

- Phase 4A focuses on the capability of the integrated system functional performance.
- Phase 4B focuses on the capability of production assembly and manufacturing processes within the business and the extended supply chain and service organization. This is also considered a part of the larger system and includes many elements that extend beyond the system we refer to as the actual functioning product.

Figure 7–13 diagrams the major elements of Phase 4A.

Start Date:___	Help conduct final tolerance design and cost balancing on components, subassemblies, and subsystems (Integration)	End Date: ___	Owner: _____
Start Date:___	Help all subsystem teams prepare summary competitive assessment documentation through tests, literature, and field evaluations	End Date: ___	Owner: _____
Start Date:___	Continue to conduct regular and accelerated life tests	End Date: ___	Owner: _____
Start Date:___	Assure all CTF components, CFRs from supply chain, and assembly operations are delivered and set at targeted nominals	End Date: ___	Owner: _____
Start Date:___	Build product design certification (alpha) units using production parts	End Date: ___	Owner: _____
Start Date:___	Evaluate system performance under nominal conditions	End Date: ___	Owner: _____
Start Date:___	Evaluate system performance under stress conditions	End Date: ___	Owner: _____
Start Date:___	Develop final regulatory requirements documents and conduct evaluations	End Date: ___	Owner: _____
Start Date:___	Develop, conduct, and analyze final health, safety, and environmental tests	End Date: ___	Owner: _____
Start Date:___	Evaluate system capability and reliability performance	End Date: ___	Owner: _____
Start Date:___	Complete and document corrective actions on problems	End Date: ___	Owner: _____
Start Date:___	Update all DFMEAs, noise diagrams, and the system noise map from product design certification (alpha) systems tests	End Date: ___	Owner: _____
Start Date:___	Continue to update and document the Critical Parameter Management map and relational database	End Date: ___	Owner: _____
Start Date:___	Verify product design meets all requirements	End Date: ___	Owner: _____
Start Date:___	Convert the system requirements document into the product performance specification document	End Date: ___	Owner: _____
Start Date:___	Generate system performance and integration risk assessment from all Phase 4A test data	End Date: ___	Owner: _____
Start Date:___	Generate Phase 4B system engineering plans	End Date: ___	Owner: _____
Start Date:___	Assess performance and make changes in the current practice of system engineering and integration to assure state-of-the-art performance in the future	End Date: ___	Owner: _____

Figure 7–13 Flow of Phase 4A of Product Commercialization

The specific issues related to each Phase 4A system engineering action item (start date for each action item, the action item, end date for each action) will be reviewed in this section. Each of these action items can be conducted by using a tool or best practice. The flow of actions can be laid out in the form of a PERT chart for project management purposes.

Help Conduct Final Tolerance Design and Cost Balancing on Components, Subassemblies, and Subsystems (Integration)

Start Date: _____ *End Date:* _____ *Owner:* _____

Just as the system engineers distributed themselves throughout the subsystem teams for the design and robustness optimization activities in Phases 2 and 3, they continue in their integration role for tolerance and cost balancing in Phase 4A. Numerous sensitivity relationships between manufacturing processes, components, subassemblies, and subsystems are fine-tuned in Phase 4A. The system engineers need firsthand knowledge of these changes to be sure they account for them in their product design certification test plans. They also need to be advising and negotiating with the subsystem teams to be sure their changes are not compromising system performance. Balancing a subsystem's cost and performance behavior at the expense of the system performance has to be carefully managed. This is the responsibility of the system engineer. The old watchdog role of the quality assurance organization shifts to the system engineering team in this context. The system engineers jealously guard the integrity of the system performance fulfillment of the system requirements. Seeking balance along with compromise is a common activity during Phase 4A.

Help All Subsystem Teams Prepare Summary Competitive Assessment Documentation Through Tests, Literature, and Field Evaluations

Start Date: _____ *End Date:* _____ *Owner:* _____

It is common for the system engineering team to "own" the purchased competitive equipment used for benchmarking. When competitive assessments are conducted at the system level, the system engineers run these evaluations. When competitive subsystem, subassembly, and component performance and architectural evaluations are conducted, they are usually performed by the respective subsystem teams with the assistance of their system engineering representatives. Other forms of competitive summary assessments can be conducted as required to help provide evidence that the product design has reached its goals of meeting or surpassing the appropriate requirements and benchmarks. Because it is the system engineering team's responsibility to prove the system is ready to pass Gate 4A requirements, team members must have time budgeted to assist in gathering the summary information in support of the busy subsystem teams.

Continue to Conduct Regular and Accelerated Life Tests

Start Date: _____ *End Date:* _____ *Owner:* _____

The system engineering team continues its role of conducting the various types of life tests required to verify the product design specifications. At this point, the life tests tend to become very aggressive. Highly accelerated stress tests (HAST) and highly accelerated life tests (HALT) are appropriate.

The system is nearing a point where it is about as good as it is going to get. Now is the time to take the system out to its real failure limits. The system and subsystem engineering teams must know where their critical failure limits are. This is particularly true for safety and high-cost reliability failure issues. Time is short in the overall program development cycle but not so short that design changes can't be made to help limit the risk of extreme failure conditions.

Assure All CTF Components and CFRs from Supply Chain and Assembly Operations Are Delivered and Set at Targeted Nominals

Start Date: _____ *End Date:* _____ *Owner:* _____

As the system engineering team gets ready to assemble and evaluate the product design certification (*alpha*) units, it needs proof that all critical-to-function specifications at the component level are made right on target nominal set points. The same is true for all subassembly and subsystem critical functional responses that are being delivered by the subsystem teams. It is against all that is good and pure to permit undefined and uninspected components, subassemblies, and subsystems to be integrated for these critical performance certification tests. This is where conformance to specifications is extremely important. Product configuration control must be rigorous at this point. We recommend a few hours in a "Six Sigma pain amplification chamber" for anyone caught trying to sneak in off-nominal components through configuration control for product design certification tests. Seriously, it is unacceptable for anyone to bypass the discipline of having the system "built to spec." The whole team is charged with verifying system performance at nominal specifications.

Build Product Design Certification (Alpha) Units Using Production Parts

Start Date: _____ *End Date:* _____ *Owner:* _____

The system engineering team oversees the physical assembly and adjustments of the components, subassemblies, and subsystems into the final product design certification units. Some people prefer to call these *alpha test units*. System engineering may have service and assembly personnel helping to do this assembly work. The service engineers definitely conduct any service or maintenance to these test units during the certification evaluations. All critical adjustment parameters and the CRFs they are adjusting are carefully monitored to be sure all CFR outputs are right on target—not just within specification limits.

Evaluate System Performance Under Nominal Conditions

Start Date: _____ *End Date:* _____ *Owner:* _____

The product design certification units are now set up and run at their nominal set points. This is the consummate test for everything being set at exactly what is called for in the design specifications. The results of this test are compared to the system requirements. All CFRs are measured using the systemwide data acquisition system. All capability indices are calculated and documented. This is truly a "noiseless" evaluation of the best the design can do under rigorously nominal conditions.

Evaluate System Performance Under Stress Conditions

Start Date: _____ *End Date:* _____ *Owner:* _____

The product design certification units are now exposed to the same variety of noises that were intentionally induced in Phase 3B. Sometimes the stress levels induced in Phase 3B are more aggressively stressful than those applied in Phase 4A. The stresses in Phase 4A certification testing must be a reasonable mixture of stresses that are likely to happen in the normal life cycle of the product. Phase 3B system stress tests must be more brutal than that because you are trying to force out the system weaknesses. Products rarely experience all noise factors occurring at high levels of intensity over continuous operating conditions. Phase 3B stress tests are harsh and are intended to flush out significant sensitivities and problems while time remains to correct them. Phase 4A stresses need to be metered out in more realistic doses of intensity. The mixtures of induced noises should mimic what your average customer group will experience in their environment and induce their use habits. Now, you are just seeing how far off-target your design means shift so you can emulate and document the value of Cpk for the system CFRs.

Each CFR is evaluated under these realistically stressful conditions. Their stressed (diminished) Cp and Cpk indices are calculated and then compared to the unstressed Cp and Cpk indices. It is helpful to make these comparisons at the Gate 4A review. This gives the gate keepers a clear picture of just how capable the product is across each of the CFRs. This paints a clear Six Sigma context view of the risks and accomplishments attained within the program up to this point. Credibility of certification is enhanced through the comparison of nominal and stressful data sets.

Develop Final Regulatory Requirements Documents and Conduct Evaluations

Start Date: _____ *End Date:* _____ *Owner:* _____

The system engineering team has an obligation to prove that all regulatory requirements have been fulfilled by the final product design specifications. We call particular attention to regulatory requirements here because without their fulfillment you probably are prohibited by law from selling the product.

This is a good place to remind the reader that requirements and specifications are often not the same thing. We wish they were a perfect match but they frequently are not. Specifications document what the design is capable of delivering. Requirements are what we hope the design can

fulfill. Specifications are the voice of final design and current process capability, while requirements are jointly representing the *voice of the customer and the business.*

Develop, Conduct, and Analyze Final Health, Safety, and Environmental Tests

Start Date: _____ *End Date:* _____ *Owner:* _____

Health, safety, and environmental requirements should not escape the scrutiny of Six Sigma standards for performance and capability. The system engineering team is responsible for evaluating
and providing certification data for these critical performance areas. There will undoubtably be
a set of critical functional responses related to the requirement for health, safety, and environmental requirements.

Evaluate System Capability and Reliability Performance

Start Date: _____ *End Date:* _____ *Owner:* _____

The product certification tests culminate in verifying system level performance that can be measured and reported in terms of Cp and Cpk. Several of the product design certification units are
dedicated, from the start, to run reliability certification evaluations. These evaluations are reported in mean time between service or mean time to failure metrics.

Complete and Document Corrective Actions on Problems

Start Date: _____ *End Date:* _____ *Owner:* _____

As is always the case for any system evaluation, there will be problems. The amount and severity of problems encountered in the product design certification evaluations should be few and minor, respectively. Nonetheless, any and all problems encountered during the certification tests
should be documented. Corrective actions can then be prescribed and results confirmed. All major corrective actions must be resolved to pass Gate 4A. Moderate and minor corrective actions
may be granted provisional passage at Gate 4A, but proof of their resolution must be provided in
a closely managed and tracked time frame.

Update All DFMEAs, Noise Diagrams, and the System Noise Map from Product Design Certification (Alpha) Systems Tests

Start Date: _____ *End Date:* _____ *Owner:* _____

All failure mode and effects analysis documents need to be updated after the design certification
tests are completed. New information about noise sensitivity within subsystems and across the
system should be added to the noise diagrams and the system noise map. All significant system

noises must be assessed and the status of their effects on the system CFRs documented in one of the following ways:

1. Noise has been evaluated and the effects are so serious that the noise has been removed or compensated for by adding some form of control system or scenario to the design at some added cost.

2. Noise has been evaluated and the effects have been safely diminished through robust design methods at minimal cost.

3. Noise has been evaluated and no known method of control or desensitization is available. The performance of the system against the system requirements is at risk.

    a. Warranty risk is high.

    b. Frequent service or maintenance will be required.

    c. Customer costs of ownership will be high.

Obviously if there is a rampant noise problem in the system, it will likely have been caught and dealt with during Phase 3A and 3B. It is extremely rare to see a system noise issue cause a delay in passing Gate 4A.

Continue to Update and Document the Critical Parameter Management Map and Relational Database

Start Date: _____ *End Date:* _____ *Owner:* _____

All new Cp and Cpk information from the design certification tests should be added to the Critical Parameter Management database. The database should be quite mature at this point and should have all sensitivity relationships down through the system hierarchy established, sensitivity coefficients quantified, with the transfer functions well modeled. These many-to-many relationships and math models should be confirmed, and the latest Cp and Cpk values entered for capability growth index reporting at Gate 4A.

Verify Product Design Meets All Requirements

Start Date: _____ *End Date:* _____ *Owner:* _____

A formal review of all measured CFRs and their capability indices within the product is conducted in comparison to the system, subsystem, subassembly, and component requirements documents. Formal certification occurs at Gate 4A. It is common to use the following hurdles for certification of design capability:

- Nominal CFR capability indices must be at or above a $Cp = 2$
- Stressed CFR capability indices must be at or above a $Cp = 1$

There is no hard and fast rule for what these hurdles need to be. That is up to your customers within the market segment you intend to serve.

Convert the System Requirements Document into the Product Performance Specification Document

Start Date: _____ *End Date:* _____ *Owner:* _____

The system engineering and subsystem engineering teams must document the final design specifications that completely identify all the set points for all the design parameters, those that are critical as well as all other parameters that define the product form, fit, and function. This aggregated group of defining parameters is referred to as the *product design specifications.* The product design specifications document contains what the product development team *was able to deliver* as opposed to what the system requirements document asked for. Gate 4A assesses whether these two documents are reasonably close to one another. If they are not, corrective action is definitely in order. If DFSS tools are used properly, with CPM actively tracking capability growth during the CDOV phases, the likelihood that the product design specifications will be grossly out of line with the system requirements is quite low. These specifications are used by the assembly, distribution, supply chain, service, and support organizations to build, ship, and maintain the product over its intended life cycle.

Generate System Performance and Integration Risk Assessment from all Phase 4A Test Data

Start Date: _____ *End Date:* _____ *Owner:* _____

The system engineering team constructs a risk assessment document based on the results or the lack thereof from Phase 4A activities.

Generate Phase 4B System Engineering Plans

Start Date: _____ *End Date:* _____ *Owner:* _____

The system engineering team meets to assess required deliverables for the next phase. Team members create detailed plans to deliver them in the allotted time. These plans are a requirement for passage of Gate 4A. See the scorecard in Figure 7–14.

Assess Performance and Make Changes in the Current Practice of System Engineering and Integration to Assure State-of-the-Art Performance in the Future

Start Date: _____ *End Date:* _____ *Owner:* _____

Phase 4A Verify

Deliverable	Owner	Original Due Date vs. Actual Deliv. Date	Completion Status Red Green Yellow	Performance vs. Criteria Red Green Yellow	Exit Recommendation Red Green Yellow	Notes
System CFR Nominal Performance Capability Indices			O O O	O O O	O O O	
System CFR Stressed Performance Capability Indices			O O O	O O O	O O O	
System Sensitivity Analysis Results			O O O	O O O	O O O	
System Reliability Performance			O O O	O O O	O O O	
Subsystem, Subassembly, and Component Reliability and Life Test Performance			O O O	O O O	O O O	
Updated Critical Parameter Database & Scorecards			O O O	O O O	O O O	
Capability Growth Index Data			O O O	O O O	O O O	
Phase 4B Project Management Plans			O O O	O O O	O O O	

Figure 7–14 General Scorecard for CDOV Phase 4A System Engineering and Integration Deliverables

The team meets to assess how it performed over Phase 4A. Team members review their use of DFSS tools and best practices and the quality of their results. Future training plans and improvements are agreed to and included in the appropriate individuals' performance plans.

Phase 4A Overall Goals

- Build and test the product design certification machines for overall product performance and robustness validation.
- Update and evaluate production certification (pilot) machines.
- Verify the product design specifications are fulfilled by measured system performance (with and without stress).
- Verify the product's serviceability and maintainability.
- Prove that the complete system is ready for production launch preparations.

General Checklist of Phase 4A System Engineering and Integration Tools and Best Practices:

Measurement system analysis

System noise mapping

Analytical tolerance design

    Worst case analysis

    Root sum of squares analysis

    Six Sigma analysis

    Monte Carlo simulation

Empirical Tolerance Design

    System level sensitivity analysis

Design of Experiments

    ANOVA

    Regression

Multi-vari studies

Design capability studies

    Nominal system CFR design Cp studies

    Stress case system CFR design Cp studies

Reliability assessment (normal, HALT, & HAST)

Competitive benchmarking studies

Statistical process control of design CFR performance

Critical Parameter Management (with capability growth index)

Nominal performance and stress testing

Statistical data analysis

Competitive benchmarking

Health, safety, regulatory, and environmental testing

Instrumentation and computer-aided data acquisition system capability (MSA)

Risk analysis

Shipping, handling, installation, safety, regulatory, and environmental testing and data analysis

Phase 4B: Production Verification

Phase 4B encompasses capability of production, assembly, and manufacturing processes within the business as well as the extended supply chain and service organization (see Figure 7–15). System engineering begins to play a secondary role as the certification efforts fully transition over to the production, assembly, service, support, and supply chain organizations.

The specific issues related to each Phase 4B system engineering action item (start date for each action item, the action item, end date for each action) will be reviewed in this section. Each of these action items can be conducted by using a tool or best practice. The flow of actions can be laid out in the form of a PERT chart for project management purposes.

Help Build Initial Production Units Using Inspected Production Parts

Start Date: _____ *End Date:* _____ *Owner:* _____

Phase 4B ushers in a major transition for the role played by the system engineering team. This is literally their ramp-down phase. The production, supply chain management, service, and support organizations take over the ownership of the system. In fact, the word *system* is rarely used in this later phase of product development. The vocabulary of Phase 4B is one of *product support* and *production operations.* The system engineering team plays a secondary role now, providing support to the new owners of the system.

Phase 4B has as one of its major goals the successful ramp-up of the production processes required to manufacture, assemble, evaluate, package, and distribute the product. System engineering had been leading the assembly and evaluation efforts in Phases 3B and 4A; now it transfers this responsibility. The resources in the system engineering organization act in an advisory role. The bill of material is transferred over to the production organization. The inspection requirements and process of initial production components, subassemblies, and subsystems are among the earliest responsibilities to be transitioned. System engineering participates in the early production builds where information regarding critical adjustment parameters and their sensitivity relationships with the critical functional responses are provided.

Help Conduct Shipping Tests and Packaging Assessments

Start Date: _____ *End Date:* _____ *Owner:* _____

Start Date:___	Help build initial production units using inspected production parts	End Date: ___	Owner: _____
Start Date:___	Help conduct shipping tests and packaging assessments	End Date: ___	Owner: _____
Start Date:___	Help assess capability of all CFRs and CFTs in production and assembly processes	End Date: ___	Owner: _____
Start Date:___	Help assess capability of all product level and subsystem level CFRs during assembly	End Date: ___	Owner: _____
Start Date:___	Assist production teams as they assume ownership of the product system and Critical Parameter Management process	End Date: ___	Owner: _____
Start Date:___	Help assess reliability of production units	End Date: ___	Owner: _____
Start Date:___	Help complete all shipping, handling, installation, safety, regulatory, and environmental tests	End Date: ___	Owner: _____
Start Date:___	Assist the production and service organizations in preparing, shipping, and installing the beta test units	End Date: ___	Owner: _____
Start Date:___	Help complete corrective action evaluations that are critical to shipping approval, and *document all remaining risks*	End Date: ___	Owner: _____
Start Date:___	Help verify all service requirements are being met with service and support processes	End Date: ___	Owner: _____
Start Date:___	Help verify product, production, assembly, and service/support processes are ready for launch	End Date: ___	Owner: _____
Start Date:___	Document lessons learned from Phase 1 through Phase 4	End Date: ___	Owner: _____
Start Date:___	Assess performance and make changes in the current practice of system engineering and integration to assure state-of-the-art performance in the future	End Date: ___	Owner: _____

Figure 7–15   Flow of Phase 4B of Product Commercialization

System engineers act in consulting roles as production engineers evaluate the product shipping process. The system engineers understand the structural elements that comprise the system architecture. They had the responsibility of developing and documenting the initial shipping requirements and conducting the tests to assess the system's ability to withstand shipping noises. The critical parameters and critical functional responses associated with the shipping requirements are transferred to the production team for additional evaluation in Phase 4B.

Help Assess the Capability of All CTFs in the Production Processes

Start Date: _____ *End Date:* _____ *Owner:* _____

The components that are produced out in the supply chain are a source of noise. The system engineering team transfers its knowledge of component quality issues that are still areas of risk from the product design certification evaluations.

Help Assess Capability of All Product Level and Subsystem Level CFRs During Assembly

Start Date: _____ *End Date:* _____ *Owner:* _____

Proceeding up the hierarchy of the system critical parameters, the system engineers can be helpful in the proper measurement of the capability of the subassembly and subsystem CFRs. They help assure no errors are made as these parameters are evaluated as the production processes ramp up to a steady state.

Assist Production Teams as They Assume Ownership of the Product System and Its Critical Parameter Management Process

Start Date: _____ *End Date:* _____ *Owner:* _____

System Engineering plays a key role in transfer of the Critical Parameter Management process and database. A representative from the sustaining engineering team in the production organization will take responsibility for the CPM tasks. The system engineer responsible for the CPM database must teach and assist the appropriate personnel in the production organization as they continue to measure and track the subset of critical parameters that will be used in production.

Help Assess Reliability of Production Units

Start Date: _____ *End Date:* _____ *Owner:* _____

When the first production units roll off the assembly line, they will need to be set up and evaluated for reliability. This is part of the formal certification of the manufacturing, assembly, and service processes. The system engineers created the reliability evaluation process for the product and play a key role in this Phase 4B reliability test. They have intimate knowledge of the details of running a disciplined reliability test and can greatly expedite the process in the production organization. This responsibility is not easily thrown over the wall.

Help Complete All Shipping, Handling, Installation, Safety, Regulatory, and Environmental Tests

Start Date: _____ *End Date:* _____ *Owner:* _____

Here, too, the system engineering team has great experience from Phase 4A testing. It has a strong experience base from which to draw as it helps the production, distribution, and service organizations complete the final certification evaluations in these areas prior to shipping approval.

Assist the Production and Service Organizations in Preparing, Shipping, and Installing the Beta Test Units

Start Date: _____ *End Date:* _____ *Owner:* _____

It is common for a small number of the earliest production units to be placed into beta test sites out in the global marketplace. The system engineers often participate in site selection and installation logistics as these units go out for early market testing. The system engineers usually partner with the service and support organization to take critical parameter data and solve technical problems.

Help Complete Corrective Action Evaluations That Are Critical to Shipping Approval, and Document All Remaining Risks

Start Date: _____ *End Date:* _____ *Owner:* _____

The system engineers have been keeping track of the corrective action database. New issues will arise from the production process certification tests that will have their resolution back in the design specifications. The system engineers can play a key role in maintaining the balance between costs and functional quality as the required design changes are made by the subsystem engineering teams.

Help Verify All Service Requirements Are Being Met with Service and Support Processes

Start Date: _____ *End Date:* _____ *Owner:* _____

During the beta tests (or external customer evaluations), the system engineers help the service and support teams make changes to critical adjustment parameters as necessary. This part of the production certification process is focused on assuring that as the product requires maintenance and CFR recentering, the service procedures are clear and reliable.

Help Verify Product, Production, Assembly, and Service/Support Processes Are Ready for Launch

Start Date: _____ *End Date:* _____ *Owner:* _____

The experience embedded in the system engineering team is a good check and balance to judge whether the production processes are ready and supportive of the formal product launch. It is not uncommon for members of the system engineering team to act as associate gate keepers for Gate 4B. See the Scorecard in Figure 7–16.

Phase 4B Verify

Deliverable	Owner	Original Due Date vs. Actual Deliv. Date	Completion Status Red Green Yellow	Performance vs. Criteria Red Green Yellow	Exit Recommendation Red Green Yellow	Notes
System CFR Nominal Performance Capability Indices			○ ○ ○	○ ○ ○	○ ○ ○	
System Sensitivity Analysis Results			○ ○ ○	○ ○ ○	○ ○ ○	
System Reliability Performance			○ ○ ○	○ ○ ○	○ ○ ○	
Subsystem, Subassembly, and Component Reliability and Life Test Performance			○ ○ ○	○ ○ ○	○ ○ ○	
Updated Critical Parameter Database and Scorecards			○ ○ ○	○ ○ ○	○ ○ ○	
Capability Growth Index Data			○ ○ ○	○ ○ ○	○ ○ ○	
Sustaining Production Engineering Support Plans			○ ○ ○	○ ○ ○	○ ○ ○	

Figure 7–16 General Scorecard for CDOV Phase 4B System Engineering and Integration Deliverables

Assess Performance and Make Changes in the Current Practice of System Engineering and Integration to Assure State-of-the-Art Performance in the Future

Start Date: _____ *End Date:* _____ *Owner:* _____

The team meets to assess how it performed over Phase 4B. Team members review their use of DFSS tools and best practices and the quality of their results. Future training plans and improvements are agreed to and included in the appropriate individuals' performance plans.

Document Lessons Learned from Phase 1 Through Phase 4

Start Date: _____ *End Date:* _____ *Owner:* _____

The system engineering team meets to document all the lessons learned during all the phases. These will include the successes the team attained and the failures it created, induced, and surpassed. This is part of the continuous improvement activity for the entire product development organization.

System Engineering Deliverables Phase 4B Gate

Product installation process validated

Customer quality metrics and machine performance assessed and corrective action identified/resolved

Reliability data documented

Software technical documentation validated (functionality/performance)

Validation of the complete manufacturing /assembly/ supply chain process

Phase 4B Overall Goals

- Help build initial production machines with production processes and test to production quality assurance test plans.
- Transfer CPM to production teams.
- Support production and service organizations as they attain shipping approval.

General Checklist of Phase 4B System Engineering and Integration Tools and Best Practices:

Measurement system analysis

Analytical tolerance design

    Worst case analysis

    Root sum of squares analysis

Six Sigma analysis

Monte Carlo simulation

Empirical tolerance design

System level sensitivity analysis

Design of experiments

ANOVA

Regression

Multi-vari studies

Design capability studies

Nominal system CFR design Cp studies

Reliability assessment (Normal, HALT, & HAST)

Competitive benchmarking studies

Statistical process control of design CFR performance

Critical Parameter Management (with capability growth index)

References

Blanchard, B. S. & Fabrycky, W. J. (1990). *Systems engineering and analysis.* New Jersey: Prentice Hall.

Lewis, J. W. (1994). *Modeling engineering systems.* Virginia: LLH Publishing.

Maier, M. W. & Rechtin, E. (2000). *The art of systems architecting* (2nd Ed.). Boca Ratan, Fla: CRC Press.

Oliver, D. W., Kelliher, T. P. & Keegan Jr., J. G. (1997). *Engineering complex systems with objects and models.* New York: McGraw-Hill.

Rechtin, E. (1991). *Systems architecting creating and building complex systems.* New Jersey: Prentice Hall.

Shetty, D., & Kolk, R. A. (1997). *Mechatronics system design.* Boston, Mass: PWS Publishing Co.

Stevens, R., Jackson, K., Brook, P., and Arnold, S. (1998). *Systems engineering coping with complexity.* Hertfordshire: Prentice Hall Europe.

P A R T   I I I

Introduction to the Use of Critical Parameter Management in Design For Six Sigma in Technology and Product Development

Few things focus a team on the most important customer needs, product requirements, design parameters, performance metrics, and deliverables for a product's success better than Critical Parameter Management. The notion of the existence of critical parameters and the fact that they must be rigorously managed is not new. What is new is the structured way in which DFSS guides the process of knowing what is critical, how critical parameters are measured, and how they are related to one another, documented, and tracked to verify the capability of new technologies and product designs.

Chapter 8 is a general introduction to the topic of Critical Parameter Management. Chapter 9 goes deeper into the definition of CPM by discussing the architecture of the CPM process. It also establishes how CPM fits within the structure of a product development organization. Chapter 10 defines the detailed process steps of conducting CPM. It examines the timing of what to do and when to do it in a Phase/Gate context. CPM scorecards are illustrated. Chapter 11 reviews the tools and best practices from DFSS that are directly applicable to CPM. The deliverables from the tools and best practices are also discussed.

Chapter 12 develops the metrics used within Critical Parameter Management. CPM is data intensive. This chapter demonstrates what is measured and how CPM data is used to document the required performance deliverables in a gate review. Chapter 13 ends the development and discussion of CPM by discussing how data acquisition systems and database architectures are used to enable the process of CPM.

Critical Parameter Management lies at the heart of DFSS. The data from CPM actions courses through the phases and gates of a commercialization process, enabling decisions to be made and risks to be managed. The data it produces is the lifeblood of DFSS. Without it an organization cannot keep track of the complex relationships that must be developed, designed, optimized, balanced, and made capable as the new product evolves toward certification for production, servicing, maintenance, and support.

CHAPTER 8

Introduction to Critical Parameter Management

Critical Parameter Management is rapidly becoming a strategic tool for improving new product development cycle-time. It focuses multifunctional technology development and product commercialization teams on the few things that matter the most relative to cost, quality, and time-to-market goals. It helps unify and integrate software and hardware engineering activities during the earliest phases of product development. Critical Parameter Management works well within the context of fulfilling critically ranked needs defined by the voice of the customer. It also delivers strong results when applied to the development of unique, breakthrough technologies that deliver functions that were not anticipated by customers.

Critical Parameter Management enhances the discipline and structure resident within organizations that are already following a formal technology development or product design process. Critical Parameter Management integrates technical processes, tools, and metrics that are used to develop and transfer technology, products, and manufacturing processes for efficient commercialization. An analogy can be used to help explain the unique nature of Critical Parameter Management.

Winning Strategies

We will focus on product commercialization as a race to be won in this example. Picture in your mind a racing crew from Harvard rowing a long narrow boat down the Charles River. Harvard has invested a lot of money in the sport of rowing and has a tradition of excellence that is important to the university, its alumni, and supporters. The President of Harvard wants assurance that this tradition of excellence will be maintained. He cares about the *quality metric* of wins vs. losses, or yield if you please. But the mechanics behind a win or a loss are not a matter of quality. They are a matter of physics, engineering, system integration, teamwork, and the fundamental measures of system performance.

A racing crew and its equipment make up a complex system that must efficiently transform and integrate the flow of energy, mass, and information. Each rower is putting energy into the system (the boat). Each rower delivers a unique, individual contribution to the speed and direction (vector) of the boat. One individual calls out a timing signal (software) to help integrate and co-ordinate the function of the rowers. The rowers must transform their potential energy into kinetic energy to transfer the mass of the boat as efficiently and harmoniously as possible. The boat is moving in an environment that is laden with sources of uncontrollable variation (water, wind drag, and lateral forces). The boat has its own internal sources of variability as well (worn parts, oar-to-oar differences). Crew members have an interesting problem: What are the *critical* factors that create on-target functions that fulfill their objective of winning races in spite of the "noise" of uncontrollable sources of variation? This is a system integration problem.

The type of responses they measure and the tools they use to optimize their individual and system performance in an environment rich in sources of uncontrollable variation are crucial to reducing the time it takes to attain repeatable and durable results. The level of quality that the crew attains is by the structure of the process used to optimize its functional performance. Crew members need to structure an approach for Critical Parameter Management.

Focus on System Performance

In the context of product or process commercialization, Critical Parameter Management considers the performance of the system as the highest priority. Subsystem and component performance are important but only within the context of system optimization. They are resources to be allocated to fulfill system requirements and goals. Critical Parameter Management gathers, integrates, and documents complex networks of critical performance relationships down from the voice of the customer to the system requirements, subsystem requirements, subassembly requirements, component requirements, and manufacturing process and material requirements. However, CPM goes well beyond the documentation of requirements as it also integrates and tracks what is supposed to happen vs. what is actually attained in terms of functional performance during the phases and gates of technology development and product commercialization processes.

Critical Parameter Management has as ultimate goals reduced time-to-market, high quality and low cost of production and ownership. Unfortunately one cannot easily use cycle-time, quality, or cost to measure the effect and efficiency of critical parameter relationships during technology, product, or manufacturing process development. These are *lagging* indicators of performance. By the time these metrics are backed up with real data, it is extremely expensive to take corrective action. These metrics foster a reactive "build-test-fix" mentality within engineering teams. When this mind-set governs organizational behavior, there is sure to be a need to do it over a second or third or fourth time—by design!

CPM measures the functions that are directly related to the laws of physics selected to control the transformation and flow of energy, mass, and information in technology, products, and manufacturing processes. These preventative measures are temporally efficient for the prevention of problems in cost, quality, and cycle-time. As a consequence, Critical Parameter Management processes are not born out of metrics found in Total Quality Management. Critical Parameter Management practitioners must measure functions, not quality. Metrology, in CPM, is grounded in measuring phenomena that are direct outcomes from the application of the principles of physical law. Critical parameter metrology and data analysis is governed by the engineering sciences with some assistance from the quality and applied statistical sciences. To be more specific, CPM performance is based on balancing system functions under the Law of Conservation of Energy and the Law of Conservation of Mass. CPM keeps two sets of "books," one in terms of physics and engineering metrics (**preventative** metrics) and one in terms of cost, quality, and time (**reactive** business, quality, and accounting metrics). The measured performance values gathered during product and manufacturing process development are stored, maintained, processed, and distributed as reports using a relational database, such as Access, Oracle, DOORS, the Six Sigma Cockpit, and others. This quantitative, interrelational information is used to make decisions, balance resources, and help manage the critical path of a product development project. In this context, CPM is highly supportive and synergistic with the methods of project management.

Critical Parameter Management is a process that encompasses methods and metrics conducted in a system engineering context. The process integrates and facilitates communication between managers and technologists: scientists, engineers, technicians, designers, production operations personnel, and field service and support personnel. Teams that succeed at Critical Parameter Management have leaders who excel at structuring and deploying engineering process mechanics, while the practitioners themselves are very good at engineering mechanics and design engineering best practices. Critical Parameter Management recognizes the value of a variety of metrics. There is a time and a place for measuring quality in CPM—after you've paid your engineering dues and actually have something functionally rational to measure. A dominant rule of CPM is to refrain from measuring attribute forms of quality that are distant from the function of the design parameters themselves.

Early in the development process, engineering scalars and vectors are the dominant values to be measured and evaluated in terms of means and standard deviations. Subsystems are optimized by evaluating the scalars and vectors as means, standard deviations, coefficients of variation, and signal-to-noise ratios. In this context they are viewed as leading indicators of insensitivity to noise (an uncontrollable source of variation) and reliability. Systems are balanced during subsystem integration and refinement by measuring the scalars and vectors in terms of short-term design and process capability indices (Cp). Tolerances are developed within subsystems and components measuring functional performance through the scalar and vector means, standard deviations, and Cp and Cpk indices. During steady-state production when Cpk values are routinely generated and evaluated, the use of Defects Per Unit and Defects Per Million Opportunities can legitimately be introduced to track the macro-performance of the finely tuned product and production process. Defect reduction and reliability improvement never come directly from measuring defects and reliability but from measuring the functions that cause the defects and poor reliability. Keeping the mean and standard deviations of critical functions on target lies at the heart of design defect reduction.

Data-Driven Process

Critical Parameter Management is data driven. Instrumentation, data acquisition, data analysis, and database management are some of the most important enabling best practices within CPM.

Organizations that deploy CPM are readily identified by their investment in computer-aided data acquisition systems linked to precision measurement instruments, efficient use of modern data analysis software, and the networked use of web-accessible relational databases. System performance knowledge is centralized, but its use is widely distributed to the cross-functional engineering teams across the project and extended supply chain.

Critical Parameter Management is a refinement that can be added to an existing product development process to make it better. By better we mean more structured, more disciplined, more concrete in quantifying the things that matter the most in satisfying the needs of the customer, the supply chain, and the business itself—its shareholders and employees. It forces critical issues to be covered purposefully early in a commercialization project while there is still time to prevent problems. It forces customer-focused quality, reliability, and low cost of ownership to be inherent in the evolving product.

Examples of key issues covered by Critical Parameter Management:

- Identification and documentation of customer needs
- Prioritization of customer needs into distinct groups of critical vs. general requirements
- Translation of the customer needs into technical terms that can be expressed as functions or physical characteristics that can be measured as continuous variables (engineering scalars and vectors) for mean, standard deviation, S/N, COV, Cp and Cpk evaluation, optimization within the product and its supporting network of manufacturing and service processes
- Identification and documentation of critical system, subsystem, and subassembly functional responses that fulfill the translated customer needs (technical requirements)
- Early identification and development of instrumentation and data acquisition systems to measure the critical functional responses
- Generation of design concepts that embody critical functional parameters that harmoniously control and fulfill the critical system and subsystem functions (concepts that are driven by measurement of functions and design for testability)
- Measurement of critical subassembly, subsystem, and system functions during the evaluation of competing design concepts (benchmarking) to see which concepts best fulfill the documented customer needs
- Selection of a superior product concept and attendant superior subsystem concepts that are high in inherent functional robustness and low in competitive vulnerability
- Development of a superior, integrated product and its subsystems into prototypes that are preengineered for ease of instrumentation so optimization of S/N and Cp indices can happen quickly (active and concurrent fulfillment of robustness and Six Sigma quantification goals)
- Creation of a relational, hierarchical database consisting of system requirements; system, subsystem, and component critical functional responses (CFRs); and the critical functional parameters (CFPs) that control the critical responses (including the documentation of CFP's sensitivity coefficients)

- Optimization of subsystem and subassembly critical functional responses in terms of S/N (robustness = insensitivity to noise) to identify nominal set points for component, subassembly, and subsystem critical functional parameters
- Integration of robust subsystems to define and balance system level sensitivities as measured by the system level critical functional responses using Cp metrics
- Definition of critical subsystem, subassembly, and component production tolerances to balance cost and quality for design certification (Cp metrics are proactively in use on the design and manufacturing supply chain prior to production process certification, balancing VOC with VOP)
- Transfer of metrology and data acquisition systems for manufacturing, assembly, and field service critical functional responses for ongoing quantification of Cp and Cpk values (to promote on-target performance) on into steady-state production and field use of the product

CPM uses design and manufacturing process capability indices [Cp ratios of criticality (USL–LSL) to sensitivity (6σ)] to track the maturing ability to fulfill the system and subsystem requirements that ultimately satisfy customer needs. As Jack Welch, former CEO of General Electric, says about Six Sigma results, "If the customer can't see and feel it . . . you haven't got it right yet" (Welch, 2001). Criticality has its definition rooted in the voice of the customer (targets and tolerances). Sensitivity is defined as the voice of the product and manufacturing processes (sample means and standard deviations). The voice of the customer cries out its demand, "This need is critical!" into the world of product and manufacturing process development. The echo back from the voice of the product and manufacturing processes is frequently the disappointing response, "I'm too sensitive to sources of variation!" Superior quality is the by-product of intentionally developed engineering functions within products and manufacturing processes that possess low levels of sensitivity. Yield and part per million quality are poor and indirect measures of sensitivity, while scalar and vector functions are continuous variables that provide measures of precision (variance or spread about the mean) and accuracy (mean in relation to target) needed to minimize sensitivity during development.

Critical Parameter Management uses a portfolio of best practices to reduce product and process sensitivity, or strength of contribution to variation, so there is wide latitude between critical tolerances and actual performance distributions (measured in sample standard deviations, s). In the world of Six Sigma, that latitude is expressed through short-term process capabilities equal to

$$2 \, (\text{Cp} = \text{USL} - \text{LSL})/6\sigma = 12\sigma/6\sigma$$

where σ really stands for a sample standard deviation, s.

Critical Parameter Management makes heavy use of best practices that provide rigorous quantification of intricate relationships between control parameter (independent variables we identify as critical functional parameters) and functional response sensitivities (dependent variables we identify as critical functional responses).

Balanced criticality and sensitivity relationships can be elusive commodities. This is especially true if one poorly translates and ranks the voice of the customer and then measures quality instead of physical functions. When the conservative laws of physics form the basis of our engineering measurement process, sensitivity is easier to measure and relate to criticality. In this context, the signal can be separated from the noise. We can quantify which engineering parameters have a dominant contribution to robustness and which have a dominant contribution to the adjustment of the mean. We can also quantify the percent contribution each parameter contributes to the standard deviation or mean of a critical functional response. With this information, the team can make enlightened technical and economic decisions up through the chain of criticality within any engineered system, right back to the original voice of the customer. If a project team does this across the board, fewer design mistakes are made, fewer things have to be reevaluated, and commercialization cycle-time is shortened.

CPM is particularly useful in preparing for and conducting software integration and evaluation. It plays a key role in the rapid identification of root causes of system and subsystem level performance relationships and problems. It helps prevent the system integration team from wasting time attributing problems to software when they are actually hardware or electrical problems and vice versa. CPM provides clear functional performance and timing paths (functional block or tree diagrams) for the software developers to follow as they generate and modify code. Software modules become better defined as they relate to functional trees that exist across subsystem boundaries. CPM helps software engineers understand the system and how they need to control and time the flow of information (logic and control signals) across the integrated product. The system's transformation and flow of energy, mass, and information are balanced under this kind of discipline.

The Best of Best Practices

Critical Parameter Management helps strategic, customer-focused Six Sigma initiatives spring to life. It provides a context to integrate numerous best practices to activate daily behaviors that gain results quickly and efficiently as organizations and teams seek to convert their energies into corporate cash flow.

In summary, Critical Parameter Management clarifies the process of generating, selecting, integrating, optimizing, and balancing product and production functions within a system context. It focuses the product development team on the key transformations and flows of energy, materials, and information. It amplifies time savings through rigor and discipline in what you measure, how you measure, and when you measure. World-class performance in quality, cost, and cycle-time flow from efficiencies gained in the disciplined balance of project management methods, active use of a Phase/Gate development process, and a rich and full portfolio of best practices. Critical Parameter Management thrives in such an environment.

Reference

Welch, J. F. (2001). *Jack: Straight from the gut.* New York: Warner Business Books.

The Architecture of the Critical Parameter Management Process

Building any form of architecture requires a rational starting point. Any foundation worth its salt comes from a sound understanding of fundamental principles and insight into basic definitions related to what is being built. For our purposes of defining an architecture for Critical Parameter Management we provide the following definitions and underlying principles.

Architecture is defined as "a style and method of design and construction" (American Heritage Dictionary, 1975). When we say we are going to discuss the architecture of a Critical Parameter Management process, we are talking about the methodology of designing and constructing a process to manage critical parameters across a technology and product commercialization process.

Critical is defined as that which is significant, influential, weighty, essential, primary, foremost, or principal. The act of being critical is accomplished through "characterization by careful evaluation" (American Heritage Dictionary, 1975).

For our purposes, criticality pertains to specific parameters being developed, measured, and evaluated during the phases and gates of a product development process. What is critical is derived from the many key deliverables that are associated with satisfying specific customer needs, business goals, governmental requirements, and societal needs. We take the time to define what is critical versus what is of general importance and focus on those few things that matter the most across the spectrum.

Parameter is defined broadly in our product development context as any of the following:

- A measurable functional response that is either a continuous or discrete variable (although we prefer it to be continuous)
- A controllable engineering factor that contributes to a measurable functional response
- An uncontrollable factor that contributes to a measurable functional response

Measurable functional responses are defined as *dependent variables* associated with a product or manufacturing, assembly, service, or maintenance process. Controllable or uncontrollable factors are defined as *independent variables* that are associated with a product or a manufacturing, assembly, service, or maintenance process.

We use these parameters to identify and manage critical functional relationships across the complete product or manufacturing, assembly, service, and maintenance processes. This is why Critical Parameter Management is considered a system engineering and integration tool.

We use the term **manage** to refer to a technical process associated with identifying, classifying, quantifying, regulating, constraining, governing, directing, controlling, influencing, dominating, guiding, steering, attenuating, balancing, overseeing, and documenting critical relationships between the parameters within product development processes.

Architecting a Critical Parameter Management process is heavily governed by these definitions and fundamental principles as well as the following overarching issues:

- System integration and evaluation
- Phases and gates employed in product development work flow
- Performance information developed through hierarchical relationships (serial and parallel networks of causal relationships)
- Traceability of ongoing measured results from maturing functional performance over Phase/Gate time
- Flexible application and reuse of the CPM process and its databases
- Cross-disciplinary use of a balanced portfolio of best practices
- Integrated and shared knowledge
- Limited scope that accounts for the critical few vs. the common many

Who Constructs the CPM Process?

The formal architecture of Critical Parameter Management is typically developed, documented, and maintained by a system engineering team or a system engineering center of excellence. This type of group has the ability and responsibility to integrate and manage total system performance. A subsystem team would not have the capacity or impetus to grapple with such a global set of tasks. The elements and flow of Critical Parameter Management are constructed through a multifunctional team from various parts of a business. Members should come from marketing, advanced product planning, technology development, product design, manufacturing, safety, health, environmental, and regulatory compliance, Supply Chain Management, distribution, and service/support organizations. The system engineering group acts in a leadership role as a unifying, integrating influence on the Critical Parameter Management architectural development process. The business owns the Critical Parameter Management process but the system engineering group acts as the steward.

Timing Structure of the CPM Process: Where and When Does CPM Begin?

Critical Parameter Management begins in an organization when the process of gathering and ranking the voice of the customer is initiated. If we define the right set of customers and query them properly, critical needs will emerge from their "voice." This is when criticality information is first available for processing. Critical Parameter Management ends the day the product is discontinued in the production and field service/support organizations. Ideally, CPM should be deployed within technology and platform building projects in R&D and then continued into specific product commercialization projects. Critical Parameter Management plays a key role in the technology transfer process. It finds its most economic application when an organization develops platforms of technology to be deployed in the generation of families of products. It greatly accelerates the commercialization of series of leveraged products because much of the work done in the first project is simply used over and over in the follow-on projects. Forward-looking companies that deploy a reuse strategy in the context of well-planned families of products gain a key advantage. Critical Parameter Management can be used on an individual project basis within companies that do not develop technology platforms. In these cases, CPM typically begins when the product development project is initiated.

What Are the Uses of CPM?

To gain insight into the construction of a Critical Parameter Management process, let's look at some of its general applications. From this pragmatic point of view we will see a rationale for how Critical Parameter Management should be structured as form follows function.

CPM provides discipline and structure for work processes that use best practices to produce critical parameter documentation and databases during technology and product design. The critical parameter documentation and databases are used by managers and engineering teams to drive preventative planning and corrective action within the development process. CPM databases allow comparisons to be made between requirements and actual measured performance at any level. The documentation and databases consist primarily of customer needs, technical requirements, functional performance measures, and criticality-to-sensitivity relationships that cover the entire span of a platform and product family, its production processes, and service/support architectures. Because Critical Parameter Management is archival in nature, it provides value in two directions besides its immediate use in technology, product, and manufacturing process design.

The first direction is to downstream steady-state production and service/support organizations. Here, key decisions on engineering changes for cost or quality are made on a continuing basis. CPM databases can provide the information required to prevent poor engineering decisions that could compromise functional performance during cost reduction, material/vendor changes, or troubleshooting projects. CPM databases contain complex hierarchical relationship trees that can be studied during engineering change management. One can think of these trees as "smart bills of materials." They offer detailed critical path information and relationships that provide quantitative answers to questions like: If I change this nominal set point, tolerance, or material, what will happen functionally throughout the product and ultimately to the customer's perception of performance, reliability, safety, and cost of use or ownership?

The second direction of value is to the upstream product planning and technology development organizations. They can use the existing CPM databases to perform what-if analysis as they generate and evaluate new platform, product system, or subsystem concepts. CPM databases can be quite helpful in identifying and ranking improvements for new projects. They also can be used to constrain the amount of feature and design creep that typically occurs in the vacuum of product planning processes that lack detailed historical data. If these teams have detailed information on criticality-to-sensitivity relationships within a current product, they can more readily match feasible technical improvements to evolving market opportunities. They can better rationalize what existing technologies should be leveraged into new designs vs. what designs should simply be reused as they currently exist. Typically this results in reduced cycle-times because it constrains how many new technologies or new designs are really needed to launch a new product. The reason it works is because the product planners can literally see function vs. requirement relationships in real numbers across the system for existing products. Far less decision making is left to the imagination, personal agendas, and political motivations. Data drives the process.

Real-time application of CPM within a technology or product design project has many benefits. The primary function of CPM is to provide information to help make technical and project management decisions in a preventative context during the earliest phases of technology development and product commercialization. The criticality information is designed to flow into a relational database for storage, retrieval, modification, and updating. The data is captured, analyzed, used, and then updated in rapid succession during each phase of product development. This helps promote rapid decisions and corrective actions relative to variation and critical sensitivities based on fresh information on what matters most. CPM provides a process to identify sensitivities and failures early so that engineering parameters can be inspected and adjusted early in the design process. This strategy helps to make critical functions as insensitive to sources of variation as early as possible. In this way critical system and subsystem functions are made robust, together by design—long before tooling is ordered for production.

General CPM applications include development, evaluation, and documentation.

For Technology Development

- Critical customer needs relative to technology building investment plans
- New or leveraged technology Houses of Quality
- Critical technology platform functional requirements and performance capabilities
- Relationships between critical responses and parameters in new or reused technologies
- Criticality ranking of engineering parameters in new or reused technologies
- Nominally robust baseline set points for parameters within the technologies (skeletal designs/drawings)

For Product and Manufacturing Process Development

- Critical customer needs relative to product development investment plans
- New or leveraged product/process Houses of Quality
- Critical product functional requirements and performance capabilities
- FMEA charts and noise diagrams/maps
- Relationship maps between critical responses and parameters in the product/processes

- Criticality ranking of engineering parameters in new or reused product/processes
- Robust nominal and tolerance set points of all critical parameters
- Design process capabilities and manufacturing process capabilities

The flow diagram in Figure 9–1 illustrates the structure of functions within Critical Parameter Management.

The CPM flow diagram illustrates, simultaneously from left to right and top to bottom, the nature of information development and integration as a general architectural construction. Once the critical information structures are developed, they can be linked to define detailed critical paths. One can trace criticality from many perspectives or starting points up or down the product architecture as necessary. Not only is the CPM database capable of tracing critical paths up and down the system architecture, it can also trace critical paths within and across subsystems. This demonstrates how CPM is highly supportive of a system engineering approach to technology and product development. Chapter 10 will fully expand on this general model.

The horizontal and vertical critical path structures in Figure 9–1 enable the engineering community to have a global view of interactivity and sensitivity. Without this kind of multidimensional architecture, the engineering teams easily lose sight of what is important across immensely complex product functions. This is particularly true of products with hardware, electronic, and electrical functions that are controlled and sequenced through complex networks of software modules. Software engineers can preventatively structure their code when they have access to critical path structures from the CPM database. This can greatly enhance coding productivity and accuracy as more and more contract software engineers are employed. It is inevitable that contract software engineers will be constrained by their lack of experience in a particular industry or product application. CPM provides a map for them to follow as they structure their code to facilitate product functions.

System engineering groups can use CPM data to rapidly identify root causes in evaluation tests, and they can strategically induce stressful noise conditions in system robustness and tolerance sensitivity experiments. Test planning, construction, and execution cycle-time can be greatly reduced when clear paths of critical functions are separated from those of nominal concern.

In summary a reasonable CPM architecture must provide the following functional attributes:

- Ability to trace critical functional relationships through all levels of a technology, product, or manufacturing process
- Accessibility so these critical functional relationships can be used by managers, engineers, and technical support personnel across an entire technology development and product commercialization team
- Durability for reuse over many product family offerings
- Flexibility so they can be used on small or large technology or product development projects
- Tunability for application across technology or product platforms

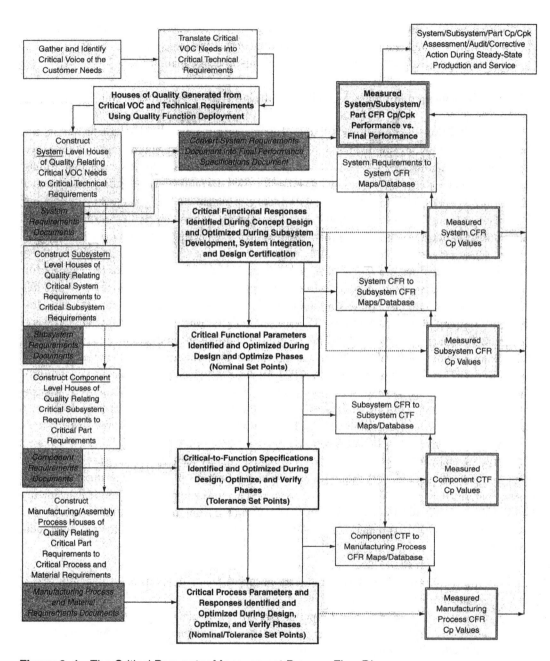

Figure 9–1 The Critical Parameter Management Process Flow Diagram

- Clarity to be understood across disciplines and organizational boundaries
- Configurability for rapid conversion to new applications
- Changeability for rapid addition or updating of new parameters, relationships, and data
- Compatibility with the output from a broad, modern portfolio of best practices
- Usability for ease of learning how to use the process, tools, and database
- Capacity to fulfill the critical information management needs of a technology or product development project

Reference

The American Heritage Dictionary of the English Language. (1975). Boston: American Heritage Publishing Co. & Houghton Mifflin Co.

The Process of Critical Parameter Management in Product Development

The steps of Critical Parameter Management can be mapped to a technology development or product commercialization process. CPM is laid out so it can account for any developing system of many-to-many critical parametric relationships. It helps structure a "designed" flow of tools and metrics on a phase-by-phase basis to manage risk. This fact alone requires CPM to be a process. At a more detailed level, CPM helps organize the specific tasks that reside within each development team's project management plan (PERT chart). The more you know about constructing project management plans within a technology development or product commercialization process, the more you can apply CPM toward your Six Sigma performance goals. Most importantly, CPM measures the results of the Technology Development and Design For Six Sigma tools that define and optimize the capability growth of all the critical functional responses in the system.

Critical Parameter Management establishes, tracks, and manages risk based upon the fundamental balancing model depicted in Figure 10–1.

Let's look at the big picture of how CPM relates to the general phases and gates processes for technology development and product commercialization. Recall that a general flow of phases

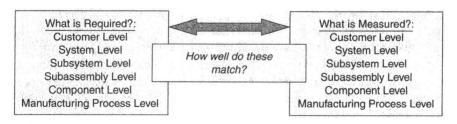

Figure 10–1   CPM Balancing Model

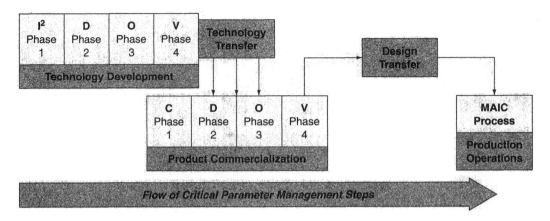

Figure 10–2   Product Development Flow

and gates for a complete product development process should look something like the diagram in Figure 10–2.

In Figure 10–3, we see a simplified, graphical model of the critical parameter flow-down that is structured across the phases of the I²DOV or CDOV Development Processes.

Critical Parameter Management can work in the context of a technology development process just as easily as it can work within the context of a product commercialization process. Many companies fail to recognize that this kind of discipline is applicable to R&D organizations, their work processes, and best practices. R&D organizations have traditionally been slow to adopt structured processes like CPM, often insisting that their world is one of free-flowing thought and tasks not amenable to such structure and discipline. One only need look at the number of failed commercialization projects due to the effects from underdeveloped, uncertified technology. These immature technologies typically create insurmountable functional problems during a product commercialization program.

Relative to what CPM does and when it does it, there is not much difference between technology building and commercialization cultures and processes. Both require gathering and translation of customer needs into technical specifications. Both require a clear set of system, subsystem, subassembly, component, and manufacturing process level requirements. Both need to assess sensitivities of concept models and prototypes to sources of variation that affect the robustness of their performance. Each must verify that the technology or design is safe to move forward into the commercialization or production process. CPM has the capacity to rally everyone on the development or design team around a focused set of critical requirements and their fulfilling critical functional parameters and responses. These metrics help assure the tasks of technology or design development, optimization, and verification progress on a predictable schedule.

In the context of cycle-time, we like to say these Design For Six Sigma processes take the right time to conduct—not necessarily the shortest. Usually when you have the shortest cycle-

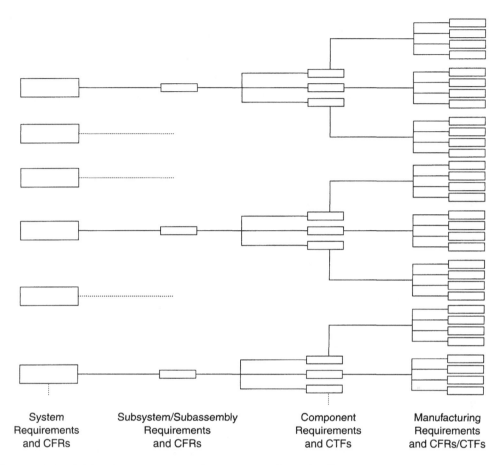

Figure 10–3 Critical Parameter Flow-Down Across CDOV or I²DOV Phases

time, you also have the maximum amount of shortcuts that actually represent the underutilization of the tools that help prevent design problems. The right cycle-time co-opts seemingly endless build-test-fix cycles that are characteristically associated with late product launches.

In a Six Sigma context, the Critical Parameter Management model relates requirements to measured performance using capability indices. The common symbols for capability are Cp and Cpk. Requirements are expressed in terms of upper and lower specification limits. Measured performance is documented in terms of means and standard deviations. Figure 10–4 depicts a CPM model in the Six Sigma context.

Product functional performance capability and manufacturing process capability can be related and tracked across technology development and product design using the model in Figure 10–4.

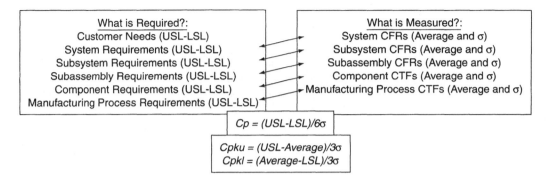

Figure 10–4 CPM Model in a Six Sigma Context

Definitions of Terms for Critical Parameter Management

Six basic terms are used to define the elements of Critical Parameter Management:

1. Critical customer **needs**
2. Critical technical **requirements**
3. Critical functional **responses**
4. Critical **adjustment parameters**
5. Critical **functional parameters**
6. Critical-to-function **specifications**

Critical implies that whatever is being measured and tracked is essential to fulfilling the highest priority needs that a majority of customers have expressed.

Functional conveys that a functional variable is being measured and tracked as opposed to some quality attribute. If the parameter is not directly measurable in units of some form of mass or energy, you are not conducting true CPM. When you think of the word *function,* think of physical performance.

Parameters can be represented as a flow-down of terms that are stored in a relational database, linked to one another, and evaluated in mathematical and empirical models referred to as *ideal/transfer functions.*

CPM elements flow down the system hierarchy as displayed in Figure 10–5.

The vocabulary of the flow-down of critical parameters is explained as follows.

System level critical functional responses (CFR): A continuous variable, typically either an engineering scalar or vector, that is a measure of a critical output from the system. System CFRs are directly sensed by the customer and can be directly compared to the new, unique, or difficult customer needs.

System level critical adjustment parameters (CAP): A continuous variable, typically either an engineering scalar or vector, that is specifically designed to adjust a system CFR.

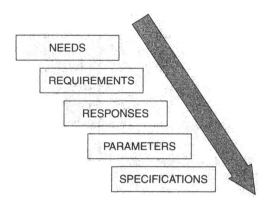

Figure 10–5 CPM Flow-Down

These variables are few in number and are often used by customers to adjust the mean output performance of the product.

Subsystem level critical functional responses (CFR): A continuous variable, typically either an engineering scalar or vector, that is a measure of a critical output from a subsystem. Subsystem CFRs are usually not directly sensed by the customer. When one or more subsystem CFRs are used as inputs to a system CFR, the name is changed to *Subsystem critical functional parameter (CFP)*. Subsystem CFRs are measured and documented for their capability to support system CFR capability.

Subsystem level critical adjustment parameters (CAP): A continuous variable, typically either an engineering scalar or vector, that is specifically designed to adjust a subsystem CFR. These variables are few in number and are sometimes available for customers to adjust but more often are used by assembly and service organizations.

Subassembly level critical functional responses (CFR): A continuous variable, typically either an engineering scalar or vector, that is a measure of a critical output from a subassembly. Subassembly CFRs are usually not directly sensed by the customer. When one or more subassembly CFRs are used as inputs to a subsystem CFR, the name is changed to *subassembly critical functional parameter (CFP)*. Subassembly CFRs are measured and documented for their capability to support subsystem CFR capability.

Subassembly level critical adjustment parameters (CAP): A continuous variable, typically either an engineering scalar or vector, that is specifically designed to adjust a subassembly CFR. These variables are also few in number and are often used by assembly and service organizations.

Component level critical-to-function specification (CTF spec): A continuous variable, typically an engineering scalar, that is a measure of a critical component dimension or feature. It may also be a characteristic relating to a surface or bulk property of a component or material. These CTF specifications are commonly communicated to the Supply Chain using "*spec sheets,*" CAD files, and component quality plans. They are usually measured under some form of variable data statistical process control as they are manufactured. Component CTF specs are designed, measured, and documented for their capability to support subassembly and subsystem CFR capability. This is the point where manufacturing process capability interfaces with and transitions into product or design level performance capability. Variation in the components affect the variation in the design functions.

Manufacturing process level critical-to-function specification (CTF spec): A continuous variable, typically an engineering scalar or vector, that is a measure of a critical manufacturing process set point. These variables are literally measured on a manufacturing process as it produces a component or material. One major use of manufacturing process capability is the measurement, control, and application of these CTF variables to fulfill component requirements. The manufacturing process capability index can also be used in reference to assembly processes as well. Manufacturing process requirements (USL-LSL) establish what is required of these variables. Manufacturing CTF specs are measured during production to ensure that what is required of the process is being provided. A manufacturing process requirement may be a pressure that is required to be at 25 psi +/− 1 psi. The measured manufacturing process CFT spec pressure might be measured at 24.3 psi in the morning and 25.1 in the afternoon. The CFT spec can be compared to the requirement and action taken if the manufacturing process has a control problem. The components are known to vary as the process pressure varies. CPM in a production context measures the in-process capability of the pressure. If the pressure and all other CTF specs are on target with low variability relative to the process requirements, the components need only limited measurement during production. During product development, manufacturing CTF specs are measured to document their capability to support component capability.

Critical Parameter Management helps develop three specific analytical models as it seeks to quantify nominal performance in Phase 2 (ideal/transfer functions), robust performance (additive models) in Phase 3, and capable performance (variance models) in Phase 4:

The Ideal/Transfer Function: $Y = f(x_A, x_B, \ldots x_N)$

   Developed in Phase 2 in I^2DOV and CDOV

The Additive Model: $S/n_{opt.} = S/n_{avg.} + [S/n_{A\ opt.} - S/n_{avg.}] + \ldots + [S/n_{N\ opt.} - S/n_{avg.}]$

   Developed in Phase 3 in I^2DOV and 3a in CDOV

The Variance Model: $\sigma^2_{total} = \sigma^2_A + \sigma^2_B \ldots + \sigma^2_N$

   Developed in Phase 4 in I^2DOV and Phases 2, 3b, and 4a in CDOV

Without a flow-down of critical parameter relationships underwritten by each of these three models, you really don't have control of your program and the data to credibly manage it in a Six Sigma context.

Critical Parameter Management in Technology Development and Product Design: Phase 1

The Critical Parameter Management process can be mapped against both a technology development process model (I^2DOV) and a product design process model (CDOV) using the block diagram structures, CPM flow-down models, and scorecards offered in this chapter. The intent of this chapter is to provide guidance on the construction of a critical parameter database. Because the I^2DOV and the CDOV processes both contain four phases and are focused on using relatively similar Six Sigma tools and best practices, we will present one CPM process that can be adapted to either process. Figure 10–6 charts the Phase 1 process steps for system development.

The example flow-down of VOC needs shown in Figure 10–7 illustrates the highest layer in the hierarchy of critical parameters. The voice of the customer is gathered and then processed using KJ methods. The processing includes:

1. Grouping of specific customer need statements into need categories.
2. Identification and segregation of the specific VOC needs into two types:
   a. Critical Needs (*): needs that are new, unique, or difficult
   b. Important Needs: needs that are old, common, or easy

Need categories align specific needs that are similar in nature into general categories. **Need types** separate needs into those that are termed critical and all others that are viewed as nominally important. A few needs are critical, but all needs are important. CPM tracks the critical needs, technical requirements, and the design and manufacturing parameters that fulfill them. All needs must be fulfilled to the extent that is economically feasible. Only the critical needs will be documented in the critical parameter database. They will be the needs that flow into the Houses of Quality. Only the needs that are new, unique, or difficult are worth the added expense of processing through QFD. A common mistake product development teams make is putting needs that are old, common, or easy through QFD. There is little or no reason to do this. These needs should be directly translated into technical requirements and placed into the appropriate requirements document. Critical needs go through the Critical Parameter Management process, which includes a trip through the Houses of Quality (Figure 10–8). Everything else is handled through the normal process of requirement documentation and certification.

The system House of Quality illustrates the linkage between the NUD VOC needs and the translated critical technical requirements. Notice, from the HOQ matrix in Figure 10–8, that numerous critical technical requirements can align with and help fulfill more than one NUD VOC need. This is one reason why the critical parameter database needs to be set up as a relational database. Depending on the size of the platform or system being developed, numerous many-to-many relationships may need to be linked within and down through the various layers of the design's architecture. This will be the case for all the lower level Houses of Quality to come.

Phase 1 delivers the first and upper-most layer of critical platform or system requirements and their fulfilling critical functional response metrics. After Phase 1, the Critical Parameter Management database will contain the general structure depicted in Figure 10–9.

System Development Process Steps	System Critical Parameter Management Process Steps
Gather and forecast customer needs and technology/product trends.	Isolate *critical* customer needs that reside within the general voice of the customer data and forecasted trends in technology and competitive products.
Define categories of the needs and establish them in a ranked order of importance.	Critical customer needs are derived from the prioritized categories of the voice of the customer data using the KJ method.
Translate the customer needs into system level requirements.	The critical customer needs are restated in the vocabulary of the technology or product development community and further ranked for importance. They are initially documented using the system House of Quality (*using the QFD method*).
Create a platform or system level technology or product requirements document from the list of technology or product requirements.	Critical technology or product requirements are documented in a relational CPM database (ACCESS/DOORS/Sigma Six Cockpit). It contains an integrated structure of critical technology or product requirements at the system or platform level.
Identify and characterize a set of measurable platform or system level functions that fulfill the technology or product requirement.	System level critical functional responses are developed and added to CPM database. CFRs are mapped to the technology requirements as measurable, continuous variables. The measured means and standard deviations from these system level CFRs will be used to define the system level technology or product capability indices: $Cp_{sys\ tech}$ and $Cpk_{sys\ tech}$. Platform or system level critical adjustment parameters are developed and modeled to assure they have a stable and linear relationship to the platform or system CFRs. This assures the technology or product is tunable. The CAPs are linked to their CFRs in the CPM database.
Generate numerous platform or system level technology or product concepts (architectures) that fulfill the system level functions.	Generate instrumentation and data acquisition system architectures that provide a clear measurement capacity and capability for the critical functional responses for each platform or system concept. "*If you can't measure it, you can't optimize it!*"

Gate 1: Platform or System Level Technology Invention/Innovation or Product Concept Development

Figure 10–6 Phase 1: Platform or System Level Technology Invention/Innovation or Product Concept Development

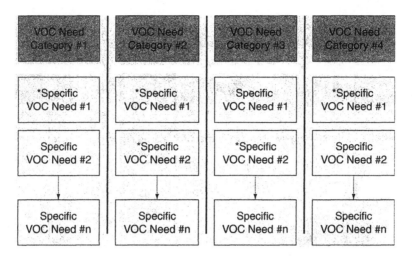

Figure 10–7 Example of the Flow-Down of Customer Need Categorization and Typing
*Note: The (*) symbol indicates a need that is critical because it is new, unique, or difficult (also referred to as a **NUD VOC Need**).*

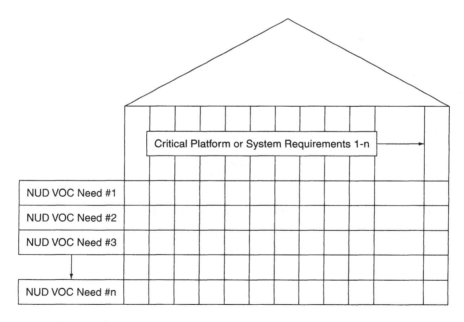

Figure 10–8 NUD VOC Needs-to-Platform or System Requirements HOQ

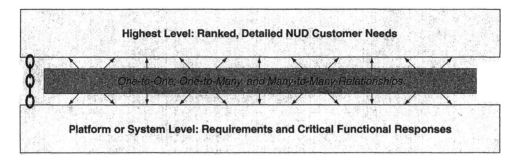

Figure 10–9   Structure of CPM Database After Phase 1

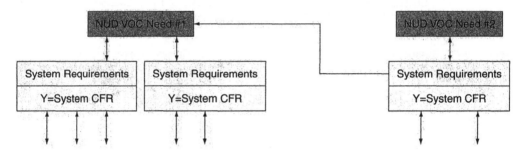

Figure 10–10   Flow-Down of Customer Needs to System Requirements

A more detailed view of a portion of the high-level flow-down of customer needs to system requirements and their measured CFRs would look like the diagram in Figure 10–10.

These parameters are documented in a relational database such as DOORS, ACCESS, or the Six Sigma Cockpit. The development team uses this relational database to continue the process of building a hierarchy of critical parameter relationships between the flow-down of needs, requirements, critical functional requirements, critical adjustment parameters, critical functional parameters, and critical-to-function specifications.

Critical Parameter Management Scorecard for System Level Performance in Phase 1

The system level CPM scorecard is typically documented as an Excel spreadsheet with the following information placed in its columns.

Desired Performance Based on VOC and VOT

System requirement

System requirement ID number

Sys. Reqts.	Sys. Reqt. #	Sys. CFR Name	Sys. CFR Target	Sys. CFR Tol.	Sys. CFR Mean	Sys. CFR Std. Dev.	Sys. CFR S/N Ratio	Sys. CFR Cp	Sys. CFR Cpk

Figure 10–11   Example System Level CPM Scorecard

System CFR name

System CFR target

System CFR tolerances

Actual Measured Performance

Measured system CFR mean

Measured system CFR standard deviation

Measured system CFR S/N ratio

Measured system CFR CP

Measured system CFR Cpk

These items are also stored in the CPM relational database along with the related transfer functions that link inputs from the subordinate subsystems, subassemblies, components, and manufacturing processes that drive the measured system CFR outputs.

Since this material is just documenting what information is actually available during Phase 1, no system level measured data available is yet available. In fact, the system CFRs won't be measured until Phases 3 and 4 during the integration evaluations. The only system information available at this point is the system requirements information, as noted in the scorecard in Figure 10–11.

System CFR:	Phase 1 CGI	Phase 2 CGI	Phase 3 CGI	Phase 4 CGI
	N/A	N/A		
	N/A	N/A		
	N/A	N/A		
	N/A	N/A		
	N/A	N/A		

Figure 10–12   Capability Growth Indices Available in Phases 3 and 4

The system requirements, targets, and their tolerances are due at Gate 1. This assures the gate keepers that the technology or product being developed is traceable back to a firm set of system level requirements. These system requirements will be used to generate two important items:

1. Subsystem technology requirements
2. System CFRs and their measurement system requirements

Normally, at the end of each phase, each CFR's capability growth index is calculated and documented, but the system level CFRs are only available to be measured in Phases 3 and 4, during the system integration and capability certification phases of technology or product development (see Figure 10–12).

The summary CPM tables and capability growth tables are essential communication tools at gate reviews. They summarize and present the critical parameter summary performance data required to run a product development program based on Six Sigma metrics, data, and standards.

We will end this section on Phase 1 with an example of a worksheet used to define a list of system requirements and their fulfilling set of critical functional responses. The worksheet on page 277 is used to help construct and provide orderly input to the CPM database.

Phase 2 in Technology Development or Product Design

Phase 2 begins at the system, subsystem, and subassembly level of technology development or product design (Figure 10–13) and works its way through the manufacturing process level.

The subsystem House of Quality is developed early in Phase 2. Its purpose in CPM is to identify the many-to-many relationships that exist between specific platform or system requirements and a specific subsystem (see Figure 10–14). For each given subsystem, specific system requirements are translated into specific subsystem requirements. This is also true for subassemblies that either align with system requirements or subsystem requirements.

Worksheet for Defining *System Level* Critical Parameters

List all *system level critical requirements* and the *system level critical functional responses* that fulfill each requirement:

Product/System Name: _____

System Level Critical Requirements: (Name and estimated nominal set point, and tolerance limits)	System Level Critical Functional Responses: (Name, measurement method, and units of measure)

Subsystem/Subassembly Development/Design Process	Subsystem/Subassembly Critical Parameter Management Process
Define rationalized categories of platform or system requirements and insert them into the appropriate **subsystem/subassembly** Houses of Quality.	Critical platform or system requirements are linked to critical subsystem/subassembly requirements. Critical subsystem/subassembly requirements are derived from the prioritized categories of system requirements using the KJ method.
For the Subsystem/ Subassembly HOQs, define subsystem/subassembly requirements that fulfill the incoming system requirements.	The critical subsystem/subassembly requirements are ranked relative to how well they fulfill system level requirements using the subsystem House of Quality (QFD).
Create a subsystem/ subassembly level requirements document from the list of subsystem/ subassembly requirements.	Critical subsystem/subassembly requirements are documented in the relational CPM database (ACCESS/ DOORS/Six Sigma Cockpit). It now contains an integrated structure of critical system-subsystem-subassembly requirements and CFR relationships.
Identify and characterize a set of measurable subsystem/ subassembly level functions that fulfill subsystem requirements.	Subsystem/subassembly level critical functional responses are developed and added to the CPM database. CFRs are mapped to the subsystem/subassembly requirements as measurable, continuous variables. The measured means and standard deviations of these subsystem/subassembly level CFRs will be used to define the subsystem/subassembly level design capability indices: **Cp and Cpk**. Subsystem/subassembly level critical adjustment parameters are developed and modeled to assure they have a stable and linear relationship to the subsystem/subassembly CFRs. This assures the technology or design is tunable.
Generate numerous subsystem/subassembly level design concepts (architectures) that fulfill the subsystem/subassembly level functions.	Generate instrumentation and data acquisition system architectures that provide a clear measurement capacity and capability for the subsystem/subassembly level critical functional responses for each subsystem/subassembly concept. *"If you can't measure it, you can't optimize it!"*

Figure 10–13   Charting Initial Phase 2 Process Steps

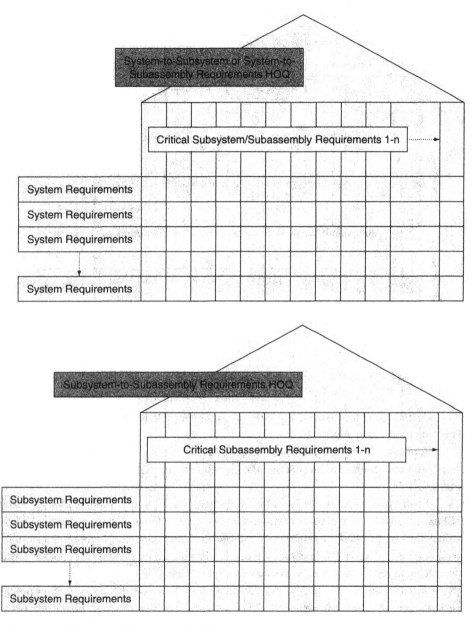

Figure 10–14 Subsystem Houses of Quality

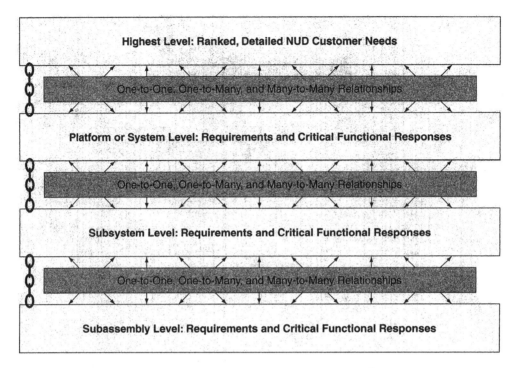

Figure 10–15   Structure of CPM Database at Phase 2

At this point in Phase 2 the Critical Parameter Management database now contains the structure depicted in Figure 10–15.

A detailed view of a portion of the high to middle level flow-down to subsystem requirements and their input critical functional parameters to the system CFRs looks like the diagram in Figure 10–16.

Each system level CFR is a (*Y*) variable controlled by the inputs from several subsystem critical functional parameters (*x* variables). The system level CFRs can have their mean output value adjusted by subsystem level critical adjustment parameters. A common example of such a parameter is the steering wheel on your car. CAPs that adjust system level performance are often accessible to the end user of the system. The subsystem CFPs and CAPs are the outputs from the transformations of mass, energy, and information from within the subsystems. An output from a subsystem is an input to the system! This is an excellent illustration of how every subsystem has a measurable and controllable output variable that can be referred to as a *critical functional response* with respect to the inputs to the subsystem. But as the subsystem, in turn, functions as an input to the system, its output is now referred to as a *critical functional parameter*. It now flows energy, mass, and information into the system along with other subsystems to enable specific system level CFR performance. This subsystem-to-system input-output relationship is quantified by a term called a *transfer function*. Transfer functions are set up as equations and are expressed in *Y* = *f(x)* terms. Transfer functions are either developed from analytical engineering models or by

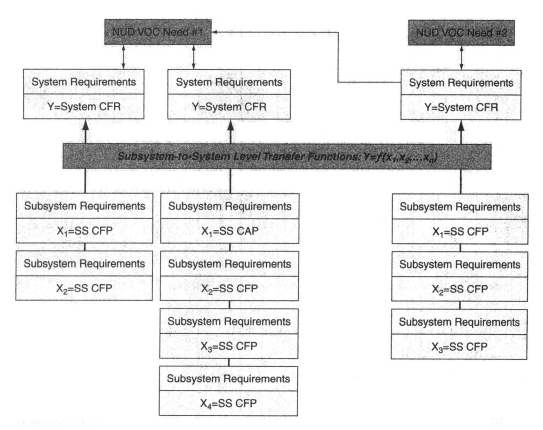

Figure 10–16   Flow-Down from Subsystem Requirements to Input Critical Functional
Parameters to System Critical Functional Responses

using designed experiments, analysis of variance, and regression techniques to derive the $Y = f(x)$
models. Transfer functions can be formulated at each level of the system flow-down structure.
The critical parameter management database is where all the requirements, critical parameters,
and their linked transfer functions reside.

One should be able to trace each subsystem requirement and critical functional response effect up through the platform or system level requirements and CFRs on up to their ability to help
fulfill specific NUD VOC need(s). Critical paths of sensitivity and fulfillment can be tracked and
managed with this approach to product development.

Critical Parameter Management Scorecard for Phase 2 for System-to-Subsystem Performance

The system-to-subsystem level CPM scorecard (Figure 10–17) is typically documented as an Excel spreadsheet with the following columns.

Sys. Reqt. #	Subsys. Reqt. #	Subsys. Reqts.	Subsys. CFR Name	Subsys. CFR Target	Subsys. CFR Tol.	Subsys. CFR Mean	Subsys. CFR Std. Dev.	Subsys. CFR S/N Ratio	Subsys. CFR Cp	Subsys. CFR Cpk	Subsys. Sensitivity Coefficient dY/dx

Figure 10–17   Sample System-to-Subsystem Level CPM Scorecard

Desired Performance Based on VOC and VOT

System requirement ID number

Subsystem requirement

Subsystem requirement ID number

Subsystem CFR name

Subsystem CFR target

Subsystem CFR tolerances

Actual Measured Performance

Measured subsystem CFR mean

Measured subsystem CFR standard deviation

Measured subsystem CFR S/N ratio

Measured subsystem CFR Cp

Measured subsystem CFR Cpk

Measured subsystem CFR sensitivity coefficient dY/dx

These items are also stored in the CPM relational database along with the related transfer functions that link inputs from subsystems, subassemblies, components, and manufacturing processes that drive the measured system CFR outputs.

At the end of Phase 2, each subsystem level CFR's baseline capability growth index is calculated and documented in a table like the one in Figure 10–18.

Subsystem CFR:	Phase 1 CGI	Phase 2 CGI	Phase 3 CGI	Phase 4 CGI

Figure 10–18 Documenting Baseline Capability Growth Indices of Subsystem Level CFRs

The system-to-subsystem summary tables are essential communication tools at Phase 2, 3, and 4 gate reviews. They summarize and present the system-to-subsystem critical parameter performance data. The worksheet on page 284 is helpful in setting up the CPM database at the subsystem level.

The detailed view in Figure 10–19 of the critical parameter flow-down from the middle to low level flow-down illustrates the addition of subassembly requirements and their measured critical parameters as inputs to the subsystems.

Notice in Figure 10–19 that the subsystem CFPs have now switched to being labeled CFRs. They were inputs to the system in the previous flow-down diagram, but now because they are receiving inputs from subassemblies, their output becomes a measurable critical functional response. Thus, subsystem and subassembly critical parameters can have two names:

1. **Critical functional response** (CFR) when they are being developed with subordinate level inputs coming into them.
2. **Critical functional parameter** (CFP) when they are acting as inputs to the next level up.

In this context, each system level CFR is a (Y) variable controlled by the inputs from several subsystem critical functional parameters (x variables). These subsystem CFPs are the outputs from the transformations of mass, energy, and information from within the subsystems. An output from a subsystem is an input to the system! This is an excellent illustration of how every subsystem has a measurable and controllable output variable that can be referred to as a *critical functional response* with respect to inputs to the subsystem. But as the subsystem, in turn, functions as an input to the system, it is now referred to as a *critical functional parameter*. It now flows energy, mass, and information into the system along with other subsystems to enable system CFR performance. A *transfer function* can quantify this subsystem-to-system input-output relationship. Subsystem-to-system transfer functions are set up as equations and are expressed in $Y = f(x)$ terms.

Worksheet for Defining *Subsystem Level* Critical Parameters

List all *subsystem level critical requirements* and the *subsystem level critical functional responses* that fulfill each requirement:

Associates System CFR

Name: _____

Subsystem Level Critical Requirements: (Name, and estimated nominal set point, and tolerance limits)	Subsystem Level Critical Functional Responses: (Name, measurement method, and units of measure)

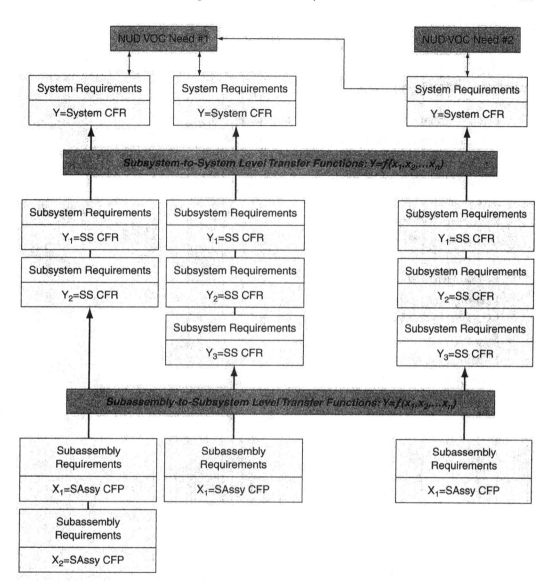

Figure 10–19   Critical Parameter Flow-Down from the Middle to Low Level

One should be able to trace each subsystem requirement and critical functional response effect up through to the platform or system level requirements and CFRs—on up to their ability to help fulfill specific NUD VOC need(s). Critical paths of sensitivity and requirement fulfillment can be tracked and managed with this approach.

Critical Parameter Management Scorecard for Phase 2
for Subsystem-to-Subassembly Performance

The subsystem-to-subassembly level CPM scorecard (Figure 10–20) is typically documented as an Excel spreadsheet with the following columns.

Desired Performance Based on VOC and VOT

Subsystem requirement ID number

Subassembly requirement

Subassembly requirement ID number

Subassembly CFR name

Subassembly CFR target

Subassembly CFR tolerances

Actual Measured Performance

Measured subassembly CFR mean

Measured subassembly CFR standard deviation

Measured subassembly CFR S/N ratio

Measured subassembly CFR Cp

Measured subassembly CFR Cpk

Measured subassembly CFR sensitivity coefficient dY/dx

Subsys. Reqt. #	Subassy. Reqt. #	Subassy. Reqts.	Subassy. CFR Name	Subassy. CFR Target	Subassy. CFR Tol.	Subassy. CFR Mean	Subassy. CFR Std. Dev.	Subassy. CFR S/N Ratio	Subassy. CFR Cp	Subassy. CFR Cpk	Subassy. Sensitivity Coefficient dY/dx

Figure 10–20   Sample Subsystem-to-Subassembly Level CPM Scorecard

Subassembly CFR:	Phase 1 CGI	Phase 2 CGI	Phase 3 CGI	Phase 4 CGI

Figure 10–21 Documenting Capability Growth Indices for Subassembly Level CFRs

These items are also stored in the CPM relational database along with the related transfer functions that link inputs from subsystems, subassemblies, components, and manufacturing processes that drive the measured system CFR outputs.

At the end of Phase 2, each subassembly level CFR's capability growth index is calculated and documented, as in Figure 10–21.

The two summary tables in Figures 10–20 and 10–21 are essential communication tools at Phase 2, 3 and 4 gate reviews. They summarize and present the subsystem-to-subassembly critical parameter performance data. The worksheet on page 288 is helpful in setting up the CPM database at the subassembly level.

The component Houses of Quality (Figure 10–23) are structured to establish relationships between critical subsystem and subassembly requirements and the newly defined component requirements.

Figure 10–24 offers an example of the structure for a House of Quality that links incoming critical subassembly requirements to critical component requirements.

At this point in Phase 2 the Critical Parameter Management database now contains the structure depicted in Figure 10–25.

Critical Parameter Management Scorecard for Phase 2 for Subsystem-to-Component Performance

The subsystem-to-component level CPM scorecard (Figure 10–26) is typically documented as an Excel spreadsheet with the following columns. The process for components CPM is illustrated in Figure 10–22.

Desired Performance Based on VOC and VOT

Subsystem requirement ID number

Component requirement

Component requirement ID number

Worksheet for Defining *Subassembly Level* Critical Parameters

List all *subassembly level critical requirements* and the *subassembly level critical functional responses* that fulfill each requirement:

Associates Subsystem & CFR

Name: _____

Subassembly Level Critical Requirements: (Name, and estimated nominal set point, and tolerance limits)	Subassembly Level Critical Functional Responses: (Name, measurement method, and units of measure)

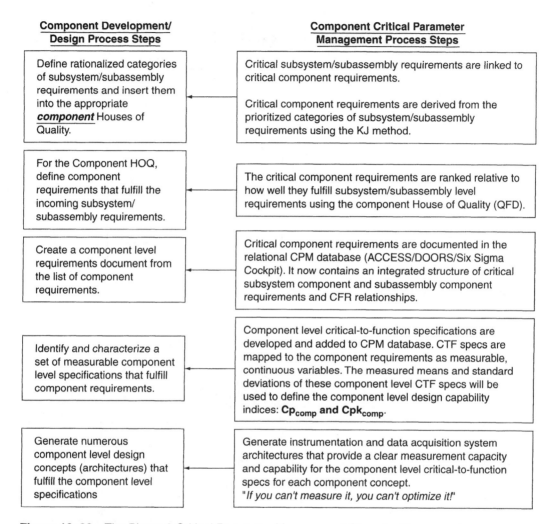

Component Development/ Design Process Steps	Component Critical Parameter Management Process Steps
Define rationalized categories of subsystem/subassembly requirements and insert them into the appropriate *component* Houses of Quality.	Critical subsystem/subassembly requirements are linked to critical component requirements. Critical component requirements are derived from the prioritized categories of subsystem/subassembly requirements using the KJ method.
For the Component HOQ, define component requirements that fulfill the incoming subsystem/ subassembly requirements.	The critical component requirements are ranked relative to how well they fulfill subsystem/subassembly level requirements using the component House of Quality (QFD).
Create a component level requirements document from the list of component requirements.	Critical component requirements are documented in the relational CPM database (ACCESS/DOORS/Six Sigma Cockpit). It now contains an integrated structure of critical subsystem component and subassembly component requirements and CFR relationships.
Identify and characterize a set of measurable component level specifications that fulfill component requirements.	Component level critical-to-function specifications are developed and added to CPM database. CTF specs are mapped to the component requirements as measurable, continuous variables. The measured means and standard deviations of these component level CTF specs will be used to define the component level design capability indices: Cp_{comp} and Cpk_{comp}.
Generate numerous component level design concepts (architectures) that fulfill the component level specifications	Generate instrumentation and data acquisition system architectures that provide a clear measurement capacity and capability for the component level critical-to-function specs for each component concept. *"If you can't measure it, you can't optimize it!"*

Figure 10–22 The Phase 2 Critical Parameter Management Steps for Component Development or Design

Component CTF name

Component CTF target

Component CTF tolerances

Actual Measured Performance

Measured component CTF mean

Measured component CTF standard deviation

Measured component CTF Cp

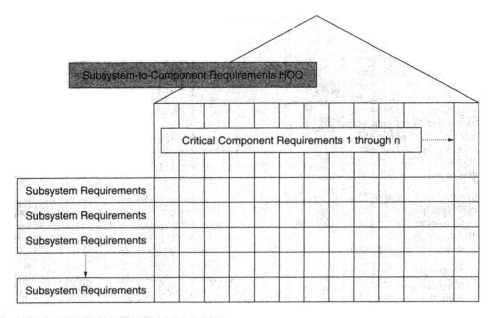

Figure 10–23 Component House of Quality

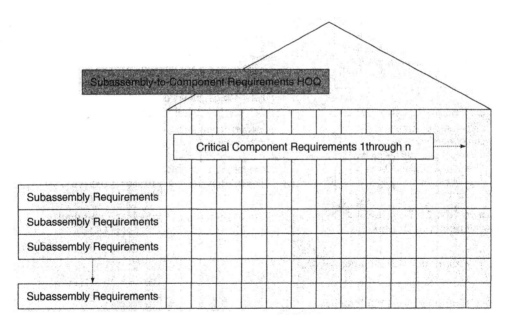

Figure 10–24 Linking Critical Subassembly Requirements to Critical Component Requirements in the House of Quality

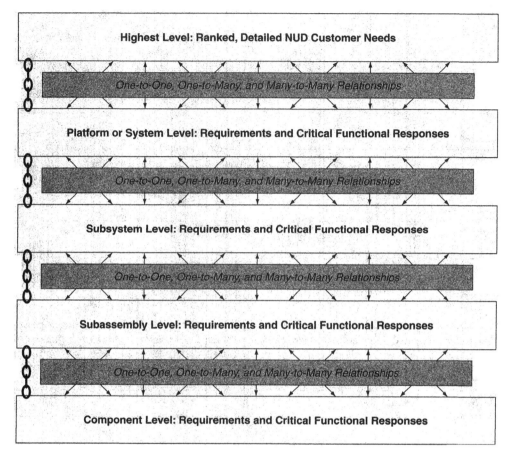

Highest Level: Ranked, Detailed NUD Customer Needs

One-to-One, One-to-Many, and Many-to-Many Relationships

Platform or System Level: Requirements and Critical Functional Responses

One-to-One, One-to-Many, and Many-to-Many Relationships

Subsystem Level: Requirements and Critical Functional Responses

One-to-One, One-to-Many, and Many-to-Many Relationships

Subassembly Level: Requirements and Critical Functional Responses

One-to-One, One-to-Many, and Many-to-Many Relationships

Component Level: Requirements and Critical Functional Responses

Figure 10–25 Adding Component Level Requirements and CFRs to the CPM Database

Measured component CTF Cpk

Measured component CTF sensitivity coefficient dY/dx

These items are also stored in the CPM relational database along with the related transfer functions that link inputs from subsystems, subassemblies, components, and manufacturing processes that drive the measured system CFR outputs.

At the end of Phase 2, each component level CTF's capability growth index is calculated and documented as in Figure 10–27.

The two summary tables in Figures 10–26 and 10–27 are essential communication tools at Phase 2, 3, and 4 gate reviews. They summarize and present the subsystem-to-component critical parameter performance. The worksheet on page 293 is helpful in setting up the CPM database at the subsystem-to-component level.

Subsys. Reqt. #	Comp. Reqt. #	Comp. Reqts.	Comp. CTF Name	Comp. CTF Target	Comp. CTF Tol.	Comp. CTF Mean	Comp. CTF Std. Dev.	Comp. CTF Cp	Comp. CTF Cpk	Comp. CTF dy/dx

Figure 10–26   Sample Subsystem-to-Component Level CPM Scorecard

Subsystem Related Component CTF Specification:	**Phase 1 CGI**	**Phase 2 CGI**	**Phase 3 CGI**	**Phase 4 CGI**

Figure 10–27   Documenting Capability Growth Indices for Subsystem Component CTF Specifications

Going on down to the next layer of inputs, we add the component requirements and their measurable critical-to-function specifications, as they relate to the subassemblies, to our growing illustration in Figure 10–28.

Critical Parameter Management Scorecard for Phase 2 of Technology Development for Subassembly-to-Component Performance

The subassembly-to-component level CPM scorecard (Figure 10–29) is typically documented as an Excel spreadsheet.

Worksheet for Defining *Component Level* Critical Parameters (for Subsystems)

List all *component level critical requirements* and the *component level critical-to-function specifications* that fulfill each requirement:

Associates Subsystem & CFR

Name: _____

Component Level Critical Requirements: (Name, and estimated nominal set point, and tolerance limits)	Component Level Critical-to-Function Specifications: (Name, measurement method, and units of measure)

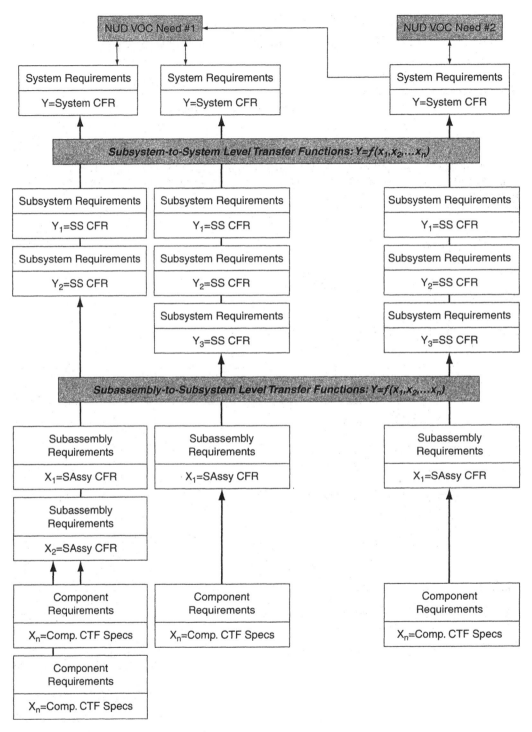

Figure 10–28 Adding Component Requirements and Their Critical-to-Function Specifications to the Flow-Down Chart

Subassy. Reqt. #	Comp. Reqt. #	Comp. Reqts.	Comp. CTF Name	Comp. CTF Target	Comp. CTF Tol.	Comp. CTF Mean	Comp. CFR Std. Dev.	Comp. CFR S/N Ratio	Comp. CTF Cp	Comp. CTF Cpk	Comp. CTF dY/dx

Figure 10–29   Sample Subassembly-to-Component Level CPM Scorecard

Desired Performance Based on VOC and VOT

Subassembly requirement ID number

Component requirement

Component requirement ID number

Component CTF name

Component CTF target

Component CTF tolerances

Actual Measured Performance

Measured component CTF mean

Measured component CTF standard deviation

Measured component CTF Cp

Measured component CTF Cpk

Measured component CTF sensitivity coefficient dY/dx

These items are also stored in the CPM relational database along with the related transfer functions that link inputs from subsystems, subassemblies, components, and manufacturing processes that drive the measured system CFR outputs.

At the end of Phase 2, each subassembly component level CTF specification's capability growth index is calculated and documented as in Figure 10–30.

Subassembly Related Component CTF Specification:	Phase 1 CGI	Phase 2 CGI	Phase 3 CGI	Phase 4 CGI

Figure 10–30   Documenting Capability Growth Indices for Subassembly Component CTF Specifications

The two summary tables in Figures 10–29 and 10–30 are essential communication tools at Phase 2, 3, and 4 gate reviews. They summarize and present the subassembly-to-component critical parameter summary performance data. The worksheet on page 297 is helpful in setting up the CPM database at the subassembly-to-component level.

The critical parameter management database adds many new layers of structure during Phase 2 (see Figure 10–31 on page 298). In fact, within Phase 2 the many-to-many relationships are first structured completely down through the system to subsystem to subassembly to component to manufacturing process architecture. There will be numerous Houses of Quality (see Figure 10–32 on page 299) developed for each layer of the system's architecture depending upon how many subsystems, subassemblies, components, and manufacturing processes exist.

At this point in Phase 2 the Critical Parameter Management database now reflects the structure depicted in Figure 10–33 on page 300.

One should be able to trace each of the manufacturing process requirements and critical-to-function specifications up to the component requirements and CFT specifications up to the subassembly requirements and critical functional responses. The next step is up through the subsystem requirements and CFRs or directly up to the platform or system level requirements and CFRs, on up to the specific NUD VOC need(s) that they help fulfill. The final example of the CPM flow-down model depicts these paths (see Figure 10–34 on page 301).

Phase 2 delivers the layers of critical subsystem, subassembly, component, and manufacturing requirements and critical parameter metrics. These are documented in the maturing relational database such as DOORS, ACCESS, or the Six Sigma Cockpit. The relational database will continue to be used to build and manage the hierarchy of critical parameter relationships between the system, subsystem, subassembly, component, and manufacturing process requirements, CFRs, CAPs, critical functional parameters, and critical-to-function component specifications.

The goals of CPM inside of Phase 2 activities are the baseline modeling and evaluation of all subsystem, subassembly, component, and manufacturing process CFRs and CTF specs. In

Worksheet for Defining *Component Level* Critical Parameters (for Subassemblies)

List all *component level critical requirements* and the *component level critical-to-function specifications* that fulfill each requirement:

Associates Subassembly & CFR

Name: _____

Component Level Critical Requirements: (Name, and estimated nominal set point, and tolerance limits)	Component Level Critical-to-Function Specifications: (Name, measurement method, and units of measure)

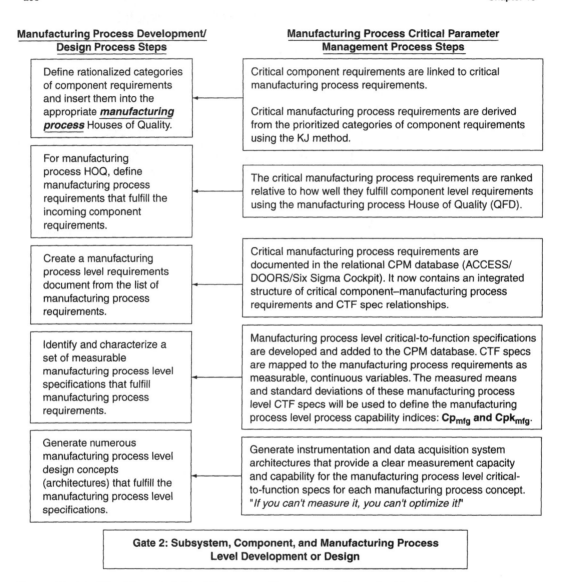

Manufacturing Process Development/ Design Process Steps

Define rationalized categories of component requirements and insert them into the appropriate **manufacturing process** Houses of Quality.

For manufacturing process HOQ, define manufacturing process requirements that fulfill the incoming component requirements.

Create a manufacturing process level requirements document from the list of manufacturing process requirements.

Identify and characterize a set of measurable manufacturing process level specifications that fulfill manufacturing process requirements.

Generate numerous manufacturing process level design concepts (architectures) that fulfill the manufacturing process level specifications.

Manufacturing Process Critical Parameter Management Process Steps

Critical component requirements are linked to critical manufacturing process requirements.

Critical manufacturing process requirements are derived from the prioritized categories of component requirements using the KJ method.

The critical manufacturing process requirements are ranked relative to how well they fulfill component level requirements using the manufacturing process House of Quality (QFD).

Critical manufacturing process requirements are documented in the relational CPM database (ACCESS/ DOORS/Six Sigma Cockpit). It now contains an integrated structure of critical component–manufacturing process requirements and CTF spec relationships.

Manufacturing process level critical-to-function specifications are developed and added to the CPM database. CTF specs are mapped to the manufacturing process requirements as measurable, continuous variables. The measured means and standard deviations of these manufacturing process level CTF specs will be used to define the manufacturing process level process capability indices: Cp_{mfg} **and** Cpk_{mfg}.

Generate instrumentation and data acquisition system architectures that provide a clear measurement capacity and capability for the manufacturing process level critical-to-function specs for each manufacturing process concept. "*If you can't measure it, you can't optimize it!*"

Gate 2: Subsystem, Component, and Manufacturing Process Level Development or Design

Figure 10–31 The Phase 2 Critical Parameter Management Steps for Manufacturing Process Development or Design

Phase 2 these values are measured under nominal conditions and compared to their requirements. Stressful sources of "noise" will be intentionally induced in Phase 3. These noises are carefully blocked out of Phase 2 modeling results. The $Y = f(x)$ relationships in Phase 2 are often called *ideal functions* due to the fact that the effects of statistically significant noise factors are intentionally minimized.

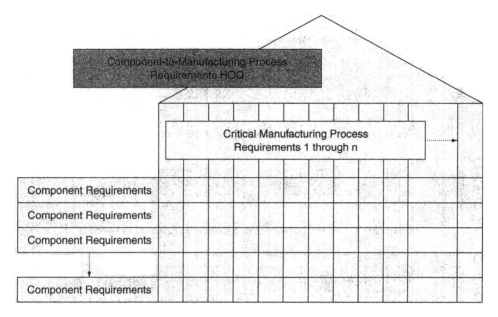

Figure 10–32 House of Quality for Component to Manufacturing Process Requirements

Critical Parameter Management Scorecard for Phase 2 for Component-to-Manufacturing Process Performance

The Component-to-Manufacturing process level CPM scorecard (Figure 10–35) is typically documented as an Excel spreadsheet. Figure 10–31 illustrates the manufacturing steps for CPM.

Desired Performance Based on VOC and VOT

Component requirement ID number

Manufacturing process requirement

Manufacturing process requirement ID number

Manufacturing process CFR name

Manufacturing process CFR target

Manufacturing process CFR tolerances

Actual Measured Performance

Measured Manufacturing process CFR mean

Measured Manufacturing process CFR standard deviation

Measured Manufacturing process CFR S/N ratio

Measured Manufacturing process CFR Cp

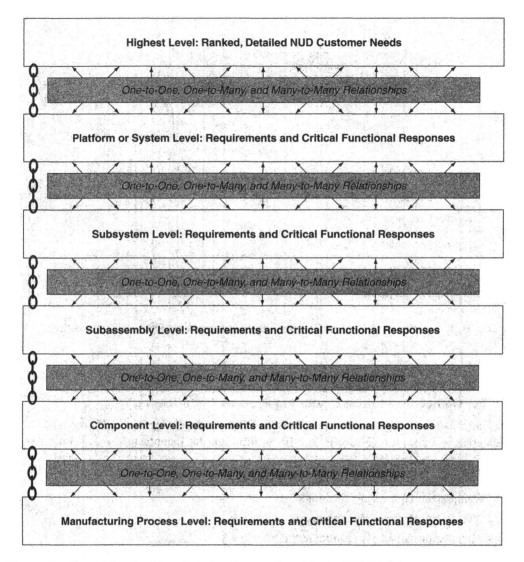

Figure 10–33   Adding the Manufacturing Process Level to the CPM Database

Measured Manufacturing process CFR Cpk

Measured Manufacturing process CTF specification sensitivity coefficient (dY/dx)

These items are also stored in the CPM relational database along with the related transfer functions that link inputs from subsystems, subassemblies, components, and manufacturing processes that drive the measured system CFR outputs.

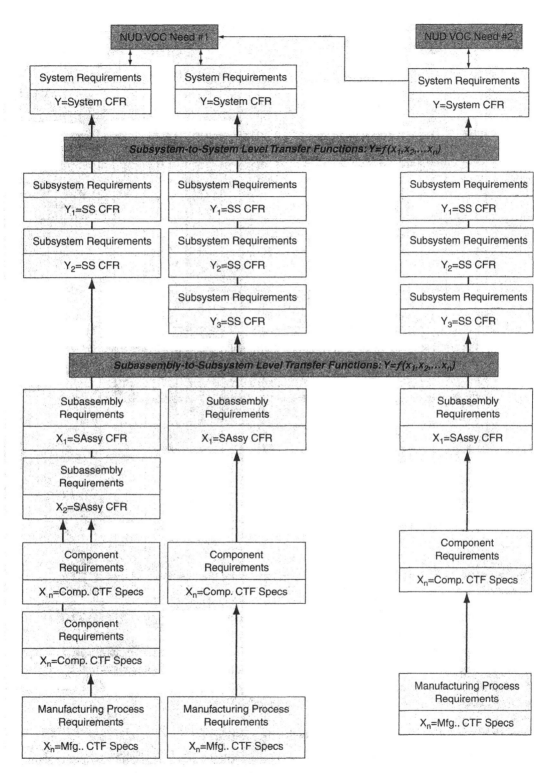

Figure 10–34 Final Phase 2 Flow-Down Chart

Comp. Reqt. #	Mfg. Reqt. #	Mfg. Reqts.	Mfg. CTF Name	Mfg. CTF Target	Mfg. CTF Tol.	Mfg. CTF Mean	Mfg. CTF Std. Dev.	Mfg. CTF S/N Ratio	Mfg. CTF Cp	Mfg. CTF Cpk	Mfg. CTF Sens. Coeff.

Figure 10–35   Sample Component-to-Manufacturing Process Level CPM Scorecard

Manufacturing Process CTF:	Phase 1 CGI	Phase 2 CGI	Phase 3 CGI	Phase 4 CGI

Figure 10–36   Documenting Capability Growth Indices for Manufacturing Process CTFs

At the end of Phase 2, each manufacturing process level CTF specification's capability growth index is calculated and documented, as in Figure 10–36.

The two summary tables in Figures 10–35 and 10–36 are essential communication tools at Phase 2, 3, and 4 gate reviews. They summarize and present the component-to-manufacturing process critical parameter summary performance data. The worksheet on page 303 is helpful in setting up the CPM database at the manufacturing process level.

Phases 3 and 4: Stress Testing and Integration

Phase 2 defines the requirements and critical metrics for all the subsystems, subassemblies, components, and manufacturing processes. The critical metrics are used in Phases 2, 3, and 4 to

Worksheet for Defining *Manufacturing Process Level* Critical Parameters

List all *manufacturing process level critical requirements* and the *manufacturing process level critical-to-function specifications* that fulfill each requirement: Associates Component CTF Specification

Name: _____

Manufacturing Process Level Critical Requirements: (Name, and estimated nominal set point, and tolerance limits)	Manufacturing Process Level Critical-To-Function Responses: (Name, measurement method, and units of measure)

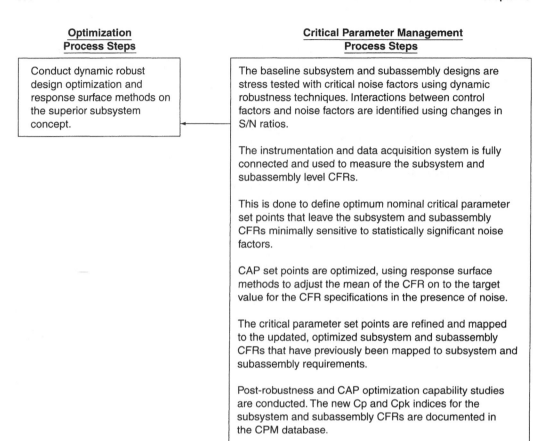

Optimization Process Steps	Critical Parameter Management Process Steps
Conduct dynamic robust design optimization and response surface methods on the superior subsystem concept.	The baseline subsystem and subassembly designs are stress tested with critical noise factors using dynamic robustness techniques. Interactions between control factors and noise factors are identified using changes in S/N ratios. The instrumentation and data acquisition system is fully connected and used to measure the subsystem and subassembly level CFRs. This is done to define optimum nominal critical parameter set points that leave the subsystem and subassembly CFRs minimally sensitive to statistically significant noise factors. CAP set points are optimized, using response surface methods to adjust the mean of the CFR on to the target value for the CFR specifications in the presence of noise. The critical parameter set points are refined and mapped to the updated, optimized subsystem and subassembly CFRs that have previously been mapped to subsystem and subassembly requirements. Post-robustness and CAP optimization capability studies are conducted. The new Cp and Cpk indices for the subsystem and subassembly CFRs are documented in the CPM database.

Gate 3: Subsystem and Subassembly Level Robustness Optimization

Figure 10–37   Phase 3: Subsystem and Subassembly Level Robustness Optimization

assess how well the evolving designs are maturing as they seek to fulfill the platform or system level requirements.

In Phase 3, the subsystems, subassemblies, components, and manufacturing processes are stress tested for their ability to contribute to robustness as stand-alone units of integrated technology (Figure 10–37). These stress tests include noises that are anticipated in the future Phase 4 integration environment. In Phase 4, the subsystems and subassemblies are integrated into existing or new platform or system structures where they will eventually reside. The Phase 4 capability, robustness, and tunability studies are conducted for baseline evaluations, robustness, and reliability stress testing at the platform or system level.

Phase 2 produced the Critical Parameter Management database flow-down structure for the nominal, baseline designs. Phase 3, the optimization phase, is focused on measuring sensitivity relationships between intentional changes to the established baseline critical parameters under the influence of intentionally induced noise factors. The following parameters are used to measure these sensitivity relationships:

1. Critical functional responses (these are *measured outputs*)
   a. subsystem level CFRs
   b. Subassembly level CFRs
2. Critical parameters at the subassembly and component level as they affect the subsystems (these are *measured and controlled inputs to a subsystem*)
   a. Critical functional parameters (engineered vectors or scalars)
   b. Critical-to-function specifications (component dimension or feature, usually a scalar)
3. Critical parameters at the component level as they affect the subassemblies (these are *measured and controlled inputs to a subassembly*)
   a. critical-to-function specifications (component dimension or feature, usually a scalar)
4. Noise factors (*measured and controlled inputs to both subsystems and subassemblies*)
   a. External sources of variation
   b. Unit-to-unit sources of variation
   c. Deterioration sources of variation

The goals for CPM in Phase 3 are to produce the following data:

1. Incoming baseline Cp indices (from Phase 2 to Phase 3) for all subsystem and subassembly CFRs. These Cp values are representative of the subsystems and subassemblies in the intentional absence of the effects of the noise factors.
2. Developing insensitivity to intentionally induced noise factors by measuring and documenting the interactions between critical parameter inputs and statistically significant noise factors. These interactions are quantified by using signal-to-noise ratio comparisons or by generating critical parameter × noise factor interaction plots.
3. Outgoing, post-robustness optimization Cp indices (from Phase 3 to Phase 4) for all subsystem and subassembly CFRs. It is common to present two forms of Cp indices as one exits Phase 3:
   a. Cp indices based on the absence of induced noise factors (the best the newly optimized CFRs can provide)
   b. Cp based on the inclusion of induced noise factors (the diminished performance of the newly optimized CFRs)

With subsystems and subassemblies that have been optimized for robust and tunable performance, it is now reasonable to integrate these designs into the platform or system structure for final technology transfer or product design certification.

Figure 10–38 presents a diagram of the processes of Phase 4 to verify platform or system level capability.

The goals for CPM in Phase 4 are to produce the following data:

1. Incoming Cp indices (from Phase 3 to Phase 4) for all system, subsystem, and sub-assembly CFRs. These Cp values are representative of the system, subsystems, and subassemblies after they have gone through system integration with performance adjusted and balanced in the presence of the statistically significant noise factors. They establish the capability of the product as it enters Phase 4. We need these to compare with the final capability indices at the end of Phase 4.

2. Assess the system's insensitivity to the transition over to production components, assembly, and service processes.

3. Production Cp indices (from Phase 4 to steady-state operations) for all system, subsystem, and subassembly CFRs and component and manufacturing process CTF specifications. It is also common to present two forms of Cp indices as one exits Phase 4:

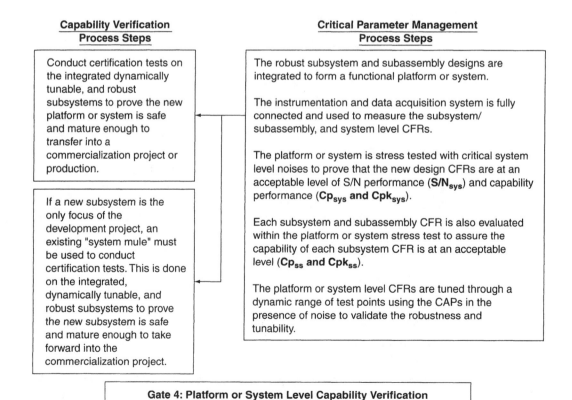

Capability Verification Process Steps

Conduct certification tests on the integrated dynamically tunable, and robust subsystems to prove the new platform or system is safe and mature enough to transfer into a commercialization project or production.

If a new subsystem is the only focus of the development project, an existing "system mule" must be used to conduct certification tests. This is done on the integrated, dynamically tunable, and robust subsystems to prove the new subsystem is safe and mature enough to take forward into the commercialization project.

Critical Parameter Management Process Steps

The robust subsystem and subassembly designs are integrated to form a functional platform or system.

The instrumentation and data acquisition system is fully connected and used to measure the subsystem/subassembly, and system level CFRs.

The platform or system is stress tested with critical system level noises to prove that the new design CFRs are at an acceptable level of S/N performance (S/N_{sys}) and capability performance (Cp_{sys} and Cpk_{sys}).

Each subsystem and subassembly CFR is also evaluated within the platform or system stress test to assure the capability of each subsystem CFR is at an acceptable level (Cp_{ss} and Cpk_{ss}).

The platform or system level CFRs are tuned through a dynamic range of test points using the CAPs in the presence of noise to validate the robustness and tunability.

Gate 4: Platform or System Level Capability Verification

Figure 10–38 Phase 4: Platform or System Level Capability Verification

    a. Cp indices based on the absence of induced noise factors (the best the production CFRs can provide)

    b. Cp based on the inclusion of induced noise factors on production units (the diminished performance of the production CFRs)

The main goal of Phase 4 in technology development is to bring new technology up to a point of maturity so that it can be safely inserted into the design phase (Phase 2) within the CDOV product commercialization process. The critical parameters are used to verify that the new technology is on equal footing with existing commercial designs that have already been through concept design, design development, and robust design in previous product commercialization projects.

It is not enough to optimize the robustness of a new subsystem or subassembly technology without integrating it into the system in which it will reside for baseline and stress testing prior to transferring it to a design program. Far too many underdeveloped technologies are prematurely transferred into product development programs. This approach to technology development prevents design problems from being transferred into the design organization. Certified critical parameter data and the system of data acquisition equipment are major deliverables at the end of technology certification.

The goal of Phase 4 in the CDOV product commercialization process is to verify the capability of both the product and the production process. The critical parameter management database and all appropriate data acquisition equipment are delivered to the production and service organizations so they can be used for change management, problem solving, and diagnostic purposes.

Capability Summaries and the Capability Growth Index

We have demonstrated that requirements flow from the system all the way down to the manufacturing processes, as in Figure 10–39.

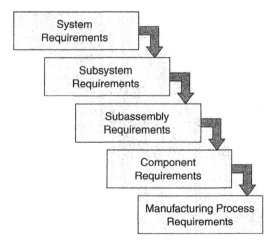

Figure 10–39 Requirement Flow-Down and Allocation

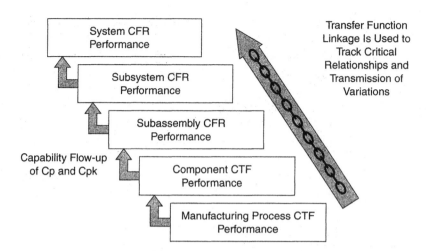

Figure 10–40 CPM Mapping from Manufacturing Process up to System Level Critical Functional Responses

CPM maps the variation transmission and capability flow up from the manufacturing processes to the system level critical functional responses (Figure 10–40).

Critical Parameter Management uses a summary statistic called the *capability growth index* to demonstrate the phase-to-phase growth of the capability of the critical parameters to the gate keepers. Critical parameter capability index status is used to help manage risk throughout the phases of the technology and product development processes. Cost and cycle-time are also important and must be equally tracked and measured along the way.

The capability growth index and the other metrics of Critical Parameter Management are discussed in the next chapter.

CHAPTER 11

The Tools and Best Practices of Critical Parameter Management

Critical Parameter Management is most effective when practiced under the discipline of a broad portfolio of Six Sigma tools and best practices. These tools and best practices focus on the development and documentation of a systemwide network of critical functional requirements that are fulfilled and validated by the measurement of critical functional responses and specifications (Figure 11–1). Without this network of requirement and response relationships, critical paths cannot be clearly defined, linked, optimized, tracked, and ultimately certified.

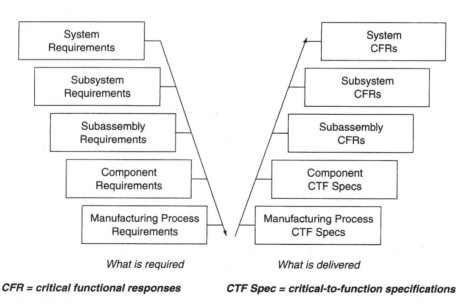

What is required   *What is delivered*

CFR = critical functional responses   **CTF Spec = critical-to-function specifications**

Figure 11–1 Six Sigma "Network" Employed in CPM

One of the most important deliverables from CPM comes from the use of DFSS tools in the development of *transfer functions*. Recall that a transfer function establishes a mathematical relationship between a dependent (Y) variable and any number of independent (x) variables. Many tools and best practices can be deployed in the development of Six Sigma performance during technology and product development. It is easy for us to define them—the challenge comes when project leaders try to get everyone on the development team to apply them rigorously in pursuit of the project deliverables. It is all about excellence in execution.

The term *best practice* is open to some debate. It has been the authors' experience that it is risky behavior to trust that every individual in the organization will naturally, of his or her own accord, do all the right things at the right times during product development. This simply does not happen well in industry. What usually happens is an unstable process of every person doing what he or she feels is "right" with respect to using product development tools and best practices. This invariably leads to random applications of limited and varying sets of tools. The end result of such anemic and inconsistent use of tools and best practices is mediocre quality and poor commercialization cycle-time. Poor product or process reliability is more often due to "death by a thousand cuts" than by any one glaring problem or defect in this scenario. The "cuts" come from poor tool use building up negative effects day by day.

Best practices are methods that are proven to work and deliver consistent results through rigorous application by companies that usually had no choice but to change how they conducted product development. If you want to find a tested set of best practices, go to companies that were on the verge of collapse but found a set of tools that brought them back to viability.

There is a set of tools that will consistently deliver 3 sigma quality. They are easy to find and define. Companies all over the world are currently using them and getting just what they invested in—mediocrity. A deeper and broader set of methods that we refer to as "best practices" consistently delivers 6 sigma functional quality. The companies that seem to learn this quickly are those at the end of their line, in a crisis, and very close to going out of business. In a word, they are motivated to apply rigor and discipline to tool use. Unfortunately, most reasonably healthy companies are complacent and rarely get serious about identifying and executing a modern set of best practices. No one tool or method can carry the burden of delivering 6 sigma functional performance and attain cycle-time reduction within budget. However, when properly linked and used at the right time with discipline and rigor, there are many that will do the job.

The Rewards of Deploying Proven Tools and Best Practices

The careful selection and integration of a group of tools and best practices within a Phase gate product commercialization process and detailed project management plan will outperform chaotic and random approaches to product design every time. If you buy a state-of-the art CAD (computer-aided design) system and focus on the rigorous use of the specification communication language of GD&T but ignore tolerance design tools, your product will likely require rework and encounter unforeseen reliability problems. If you design your technologies for nominal performance and focus on "design for patentability," you will likely experience technology transfer problems because the technology is not robust and tunable, nor is it likely to be readily manufacturable. Best practices create *complete* designs. Partial designs result from limited use of a few, favorite methods that leave critical design parameters underdeveloped. Frequently, development teams create a "new" design but don't bother to benchmark it or stress test it along with a competitive design. They then have the audacity to act surprised when it is eclipsed in the marketplace where it is exposed to real sources of variability. That which is real can humble that which is new!

A thorough set of best practices can be characterized by its ability to include and address all the realities and requirements that exist within technology development, product design, manufacturing, delivery, service, customer use, and life-cycle environments. Probably the biggest mistake we see being made by product development managers is the underfunding of time to execute the right number of tools and best practices to fulfill their program requirements. You may recall we had a bit to say about that in Chapter 4.

You should select a portfolio of tools and best practices that match your company's specific product development processes and goals. Every company is different, so no one company has the optimum recipe of best practices for anyone but themselves. Obviously the auto industry has a great deal of commonality with respect to best practices. This is where benchmarking of best practices can be helpful. Some companies consider their product development process and portfolio of best practices as proprietary. This is usually vanity more than a strategic maneuver. The value lies in the disciplined, active execution of the product development process and the deployment of a balanced set of tools and best practices that are designed to meet your objectives. The authors have seen lots of impressive product development process manuals with strong portfolios of best practices that get limited use if they are ever practiced at all. This is why your company should define the critical best practices that will meet your business and customer needs and then set out to do them better than anybody else. Having a product development process loaded with best practices and then actually doing them well is the secret to success.

Best practices can be broken down into many categories. For our focus on Critical Parameter Management, we will divide them into two broad areas: technology development and product commercialization. A lot of overlap and redundancy exists in the tool sets for these two processes. That is to be expected if you consider technology development a means of enabling brand new designs to catch up to the maturity of a reused or existing design that has already been out in the marketplace. Further categorization can take place around specific types of application. For technology development and product design we will focus on four common categories:

1. Invention, innovation, and concept design
2. Design development
3. Optimization
4. Capability verification

These four general areas have previously been shown to be readily structured into specific phases for both technology and design development.

Critical Parameter Management Best Practices for Technology Development

During technology development our focus is primarily limited to the conceptual design and parameter design of new or reused subsystem technologies or an integrated group of subsystems that are known as platforms. The technologies being developed need to be certified for the ability to merge with the rest of the design elements usually somewhere within Phase 2 of the CDOV product design process. We will continue our focus on phases and gates by illustrating which tools and best practices align with the phases and gates of the I^2DOV technology development process. We will also state the specific deliverable from each tool and best practice as it pertains to Critical Parameter Management.

Best Practices for Conceptual Design

Phase 1: Technology Concept Invention and Innovation

Gathering and processing the voice of the customer
Deliverable: Ranked and prioritized VOC data (new, unique, and difficult needs).

Quality function deployment and the Houses of Quality
Deliverable: Translated VOC, NUD technology requirements, additional ranking and prioritizing in terms of customer perception, and competitive benchmarking data. Target values for the technology requirements.

Competitive functional benchmarking
Deliverable: Data concerning measurable competitive product performance. This data helps substantiate the targets set to fulfill the technology requirements.

Technology requirements generation and documentation
Deliverable: Complete set of technology requirements document with target and tolerance values (including the NUD technology requirements).

Critical Parameter Management database development
Deliverable: Initial relational database constructed and loaded with the flow-down of the NUD customer needs and the technology requirements developed to fulfill them.

Best Practices for Baseline Development

Phase 2: Technology Development

Functional analysis and functional flow diagramming (diagramming the flow, transformation and state of energy, mass, and information within the technology concepts)

Deliverable: Candidate critical parameters that help begin the analytical and empirical modeling of the transfer functions, sometimes called *ideal functions* or *baseline performance models.*

Analytical Math Modeling (based on first principles, typically with CAE methods)
Deliverable: Initial transfer function models $[Y = f(x)]$.

Concept generation techniques (Form, fit, and function brainstorming, brain-writing, benchmarking, modeling, etc.)
Deliverable: Form, fit (architectural structuring), and functional parameter definition (transfer function model) to fulfill the technology requirements.

Failure modes and effects analysis (FMEA)
Deliverable: Initial assessment of the parameters that are critical to failure.

Fault tree analysis (FTA)
Deliverable: Initial assessment of the flow-down of parametric failure relationships within the functional architecture of the technology concepts.

Development and design of data acquisition systems (Instrumentation and measurement system analysis)
Deliverable: Definition and capability certification of the hardware, firmware, and software to be used to measure the critical functional responses and specifications for each critical parameter.

Empirical math modeling (Inferential statistics, sequential design of experiments, ANOVA, correlation, regression, response surface methods)
Deliverable: Transfer function models for critical parameters $[Y = f(x)]$.

Design for additivity and tunability (Synergistic functional interactivity of technology elements; platform design)
Deliverable: Initial assessment and quantification of the strength and directionality of all statistically significant (control factor $\times$ control factor) interactions; development and identification of critical adjustment parameters.

Preliminary DFx analysis (Testability, modularity, manufacturability, reliability, environment, etc.)
Deliverable: Identification of additional critical parameters for all performance areas required to meet the VOC needs.

Pugh concept evaluation and selection process
Deliverable: Superior technology concepts, including their critical parameters.

Critical parameter management database development
Deliverable: Relational database loaded with critical system, subsystem, subassembly, component, and manufacturing process parameters, their transfer functions, and their measured Cp performance as they are linked to the technology requirements.

Best Practices for Robustness Optimization

Phase 3: Technology Optimization

Failure modes, effects, and analysis (FMEA)
Deliverable: Additional FMEA documentation data from Phase 3 robustness evaluations and stress testing.

Fault tree analysis (FTA)
Deliverable: Additional data relative to failure flow-down relationships within the architectures of the maturing subsystems and subassemblies.

Generic noise diagramming and platform noise mapping (Product family noises)
Deliverable: Thorough documentation of the candidate noise factors for use in the robustness optimization experiments and analysis.

Development and design of data acquisition systems (Instrumentation and measurement system analysis)
Deliverable: Additional measurement system equipment and certification data prior to Phase 3 critical parameter optimization evaluations.

Noise experimentation and noise vector characterization
Deliverable: Statistically significant noises that will be used in robustness optimization experiments.

Dynamic robustness optimization experiments (Continuation of sequential design of experiments, additive model confirmation for critical parameters that underwrite robust and tunable technologies)
Deliverable: Set points for critical functional parameters (CFPs) and critical-to-function specifications (CTF specs) that leave the subsystem and subassembly CFRs minimally sensitive to the noise factors.

Critical Parameter Management database development
Deliverable: Relational database loaded with Cp indices for the subsystem and subassembly CFRs. Any new sensitivity linkage is mapped between critical parameters. Transfer functions are modified with new data.

Best Practices for Platform and Subsystem Certification

Phase 4: Technology Certification

Preliminary tolerance design and sensitivity analysis
Deliverable: Tolerance models built from analytical (Worst case, RSS, and Monte Carlo simulations) and empirical (Sequential DOE-based tolerance sensitivity) studies.

System integration and platform stress testing
Deliverable: CFR Cp and Cpk index data that includes the stressful effects from the statistically significant noise factors applied during the system/platform integration tests.

Technology capability assessment and characterization (Cp and Cpk)
Deliverable: Refined CFR Cp and Cpk indices after component, subassembly, and subsystem tolerance balancing and integration stress testing. Cp and Cpk data will be documented under nominal and stressful conditions for all CFRs.

Critical Parameter Management database development
Deliverable: Relational database loaded with all critical parameter relationships and transfer functions. The critical parameters are linked to the technology requirements to validate certification of the robustness and tunability of the technology.

The critical parameter database is a major part of what is transferred to the product design organization at the end of the technology development process. The data acquisition hardware, firmware, and software are also major deliverables to the design teams. The ability to measure new, unique, or difficult critical functional responses is a strategic advantage that can enable a commercialization team to stay on schedule as it integrates new technologies into the product design architecture. It is not uncommon to see major, new patents on data acquisition designs and technologies as a result of CPM in R&D organizations.

Critical Parameter Management Best Practices for Product Commercialization

The CDOV product commercialization process turns its focus on a broader set of best practices, mainly due to the fact that it must ensure that the integrated system (new technologies and reused designs) is ready for steady-state production and service processes.

R&D teams tend to use a smaller set of tools and best practices than product design teams. Their job is to get the technology to a mature enough state to transfer it into product commercialization. The product design teams must carefully plan what tools and best practices are appropriate for the product they are designing. The following set of tools and best practices is essential to the successful deployment of Critical Parameter Management.

Best Practices for Concept Design for Product and Manufacturing Process Commercialization

Phase 1: Concept Design

Gathering and processing the voice of the customer (Customer interviewing and KJ analysis)
Deliverable: Ranked and prioritized VOC data.

Quality function deployment and the Houses of Quality
Deliverable: Translated VOC, NUD system requirements, and additional ranking and prioritizing in terms of customer perception and competitive benchmarking data. Target values for the system requirements.

Competitive functional benchmarking
Deliverable: Data concerning measurable competitive product performance. This data helps substantiate the targets set to fulfill the system requirements.

System requirements generation and documentation
Deliverable: System requirements document with target and tolerance values. This includes the NUD system requirements from the House of Quality.

Critical Parameter Management database development
Deliverable: Initial relational database constructed and loaded with NUD customer needs that are linked to the system requirements.

Best Practices for Design Development

Phase 2: Design Development

Functional analysis and functional flow diagramming (diagramming the flow, transformation, and state of energy, mass, and information within the system, subsystem, and subassembly concepts)
Deliverable: Candidate critical parameters that help begin the analytical and empirical modeling of the transfer functions, sometimes called *ideal functions* or baseline performance models.

Analytical math modeling (based on first principles, typically with CAE methods)
Deliverable: Initial transfer function models $[Y = f(x)]$

Concept generation techniques (Form, fit, and function brainstorming, brain-writing, benchmarking, modeling, etc.)
Deliverable: Form, fit (architectural structuring), and functional parameter definition (transfer function model) to fulfill the product requirements.

Failure modes and effects analysis (FMEA)
Deliverable: Initial assessment of the parameters that are critical to failure.

Fault tree analysis (FTA)
Deliverable: Initial assessment of the flow-down of parametric failure relationships within the functional architecture of the technology concepts.

Development and design of data acquisition systems (Instrumentation and measurement system analysis)
Deliverable: Definition of the hardware, firmware, and software to be used to measure the critical functional responses and specifications for each critical parameter.

Empirical math modeling (inferential statistics, sequential design of experiments, ANOVA, correlation, regression, response surface methods)
Deliverable: Transfer function models for critical parameters $[Y = f(x)]$.

Design for additivity and tunability (Synergistic functional interactivity of design elements; platform deployment)

Deliverable: Initial assessment and quantification of the strength and directionality of all statistically significant (control factor $\times$ control factor) interactions. Development and identification of critical adjustment parameters.

Preliminary DFx Analysis (Testability, modularity, manufacturability, reliability, environment, etc.)
Deliverable: Identification of additional critical parameters for all performance areas required to meet the VOC needs.

Pugh concept evaluation and selection process
Deliverable: Superior technology concepts, including their critical parameters.

Critical Parameter Management database development
Deliverable: Relational database loaded with all critical system, subsystem, subassembly, component, and manufacturing process parameters, all transfer functions, and measured Cp performance values as they are linked to the system requirements. Baseline capability growth indices are documented for all subsystems.

Best Practices for Design Optimization

Phase 3A: Subsystem and Subassembly Robustness Optimization

Failure Modes, Effects, and Analysis (FMEA)
Deliverable: Additional data from robustness stress testing that is relevant to FMEA documentation and failure analysis.

Fault tree analysis (FTA)
Deliverable: Additional data relative to failure flow-down relationships within the maturing subsystems and subassemblies.

Specific noise diagramming and system noise mapping (Product specific noises)
Deliverable: Thorough documentation of the candidate noise factors for use in the robustness optimization experiments and analysis.

Development and design of data acquisition systems (Instrumentation and measurement system analysis)
Deliverable: Additional measurement system equipment and certification data prior to Phase 3A critical parameter evaluations.

Noise experimentation and noise vector characterization
Deliverable: Statistically significant noises that will be used in robustness optimization experiments.

Dynamic robustness optimization experiments [Continuation of sequential design of experiments: Evaluation of interactions between critical parameters (CFPs and CTF specs) and the statistically significant noise factors, additive model confirmation for robust and tunable subsystems and subassemblies]

Deliverable: Set points for critical functional parameters (CFPs) and critical-to-function specifications (CTF specs) that leave the subsystem and subassembly CFRs minimally sensitive to the noise factors. Confirmation data for the additive model.

Mean optimization experiments (Continuation of sequential design of experiments: Response surface methods applied to critical adjustment parameters to place the mean of the CFRs on to their VOC targets)
Deliverable: Set points for critical adjustment parameters (CAPs) that place the mean of the subsystem and subassembly CFRs on target.

Critical Parameter Management database development
Deliverable: Relational database loaded with Cp indices for the optimized subsystem and subassembly CFRs. Transfer functions and new sensitivity linkages are mapped between critical parameters. Transfer functions are modified with new data. The latest capability growth indices are documented for all subsystems.

Best Practices for System Integration and Stress Testing

Phase 3B: System Integration and Stress Testing

System integration stress testing (The integrated system is evaluated for Cp performance under nominal and stressful conditions)
Deliverable: System, subsystem, and subassembly CFR Cp and Cpk index data that includes the nominal performance and the stressful effects from the statistically significant noise factors applied during the system integration tests.

Product capability assessment and characterization (Cp and Cpk evaluations after the system, subsystems, and subassemblies have been balanced from the learning from the first system integration test data results)
Deliverable: Refined CFR Cp and Cpk indices after component, subassembly, and subsystem tolerance balancing and system integration stress testing. Cp and Cpk data will be documented under nominal and stressful conditions for all CFRs.

Analytical and empirical tolerance design and sensitivity analysis
Deliverable: Production tolerances (components, subassemblies, and subsystems) from tolerance models built from analytical (Worst case, RSS, and Monte Carlo simulations) and empirical (Sequential DOE-based tolerance and $+/-1\sigma$ sensitivity) studies.

Critical Parameter Management database development
Deliverable: Relational database loaded with all the latest critical parameter relationships. The transfer functions are updated and linked to the product requirements to validate the current nominal performance, robustness, and tunability of the system. Capability growth indices are documented for all subsystem and system level CFRs.

Best Practices for Product Design Certification

Phase 4A: Product Design Certification

Final, baseline component, subsystem, and system level capability assessment and characterization (Cp and Cpk)
Deliverable: Final product, subsystem, subassembly, component, and manufacturing process capability indices and capability growth indices.

Critical Parameter Management database development
Deliverable: Relational database loaded with all the latest critical parameter relationships. Final transfer functions are updated and linked to the product requirements to validate certification of the nominal performance, robustness, and tunability of the complete product. Capability growth indices are documented for all subsystem and system level CFRs.

Best Practices for Production Process Certification

Phase 4B: Production Process Certification

Once the product design is certified, the ramp-up to steady-state production can be evaluated and certified. A smaller but powerful set of best practices are required to sustain Critical Parameter Management on into production.

Transfer of data acquisition systems (Critical parameter instrumentation)
Deliverable: A select group of capable and certified transducers, meters, and computer-aided data acquisition equipment to continue measuring critical parameters during production.

Manufacturing process, component, subsystem, and system level product capability assessment and characterization (Cp and Cpk)
Deliverable: Ongoing statistical process control data to measure, analyze, improve, and control the capability of the critical parameters.

Critical Parameter Management database deployment
Deliverable: Relational database fully loaded with all the latest critical parameter relationships. Final transfer functions are linked to the product and production requirements to validate certification of the nominal performance, robustness, and tunability of the complete product and its production processes. Final capability growth indices are documented for all subsystem and system level CFRs.

As you can see, the best practices for CPM see their heaviest use early in the product development process. The more that is understood during the development of technology, the less time spent developing critical parameter knowledge during product commercialization. The best scenario is to build up a certified and reusable base of technologies, platforms, and designs that can be safely transferred and deployed into rapid commercialization projects. This approach enables a business strategy that is based on launching families of products.

Notice how few best practices are required once the product attains manufacturing certification and shipping approval. This is a clear sign that CPM has worked! Ideally one would like to produce products, subsystems, and components indefinitely with no corrective action at all during steady-state production. This, of course, is not realistic.

A few disciplined companies have attained a diminished need to redesign after they have launched a new product into the market. The alternative is the presence of special teams of "firefighters" from the design organization who reside in the production organization to literally finish developing the product while it is in production. These resources should be working on the next new product—not patching up all the shortcomings of one that was supposed to be completed. CPM tools and best practices go a long way in helping the product development team to get it done right the first time.

Metrics for Engineering and Project Management Within CPM

An unfortunate mismatch exists between the metrics traditionally used by management to measure product development process performance and the metrics engineering teams must use to conduct product development. Many of the current forms of product development metrics are rooted in Total Quality Management and standard business performance metrics. These metrics tend to focus on financial terms such as unit manufacturing costs (UMC), quality measures such as defects per million opportunities (DPMO), and time-based measures of reliability such as mean time to failure (MTTF). These are legitimate measures of business performance, but they are limited in scope and are *lagging* indicators of functional performance. Counting defects and measuring failures are considered inappropriate measures for Critical Parameter Management.

Critical Parameter Management requires *leading* indicators that are comprehensive in scope. A comprehensive scope includes functional performance measures that can be used throughout the phases and gates of product development—right on into the production operations and customer use environments. Engineering teams cannot efficiently attain on-target functional performance by measuring quality and reliability by reacting to such lagging indicators. Quality metrics such as yield (% good), percent defective, defects per million opportunities, defects per unit, go–no go attributes, and reliability metrics such as mean time between service calls and mean time to failure are all based on measuring the aftermath of physical functions that have moved off target. The key terms in these metrics are *defects, failure,* and *time.* A positive measure of improved performance in this TQM context typically focuses around acceptability of performance within tolerance limits.

The dominant paradigm is *acceptability,* which is tolerance centric. A CPM paradigm requires reduced variability around a target-centric system that measures critical physical functions that lead to quality and reliability. This is why we call critical functional responses *leading* indicators of quality and reliability. Critical Parameter Management must also operate beyond a nominal measurement context. The focus shifts from "How many defects does this product have under nominal, steady-state conditions?" to "How close to the desired targets do the critical functional responses remain when we expose the system to realistic sources of variation?" You can see how counting defects places an engineering team a long way from the measurable physics of function, energy transformations, flows of mass, and information (digital and analog logic and control signals). It is not reasonable or logical for managers to expect engineering teams to measure quality and reliability to optimize functional performance; they must measure the functions.

Key CPM Metrics

The proper metrics for Critical Parameter Management are as follows:

1. **Scalars and vectors** that are direct outcomes of energy and mass transformations, flows, or changes of state.
2. **Continuous variables** that are active over the continuum of initiating, sustaining, and concluding a function. Such metrics typically have an **absolute zero** start and end value. When energy is put into a system, the measured output starts from "absolute zero" and all measures of the function are relative to that "zero."

    Think of shooting a cannon at a target. The absolute zero is the distance from the cannon to where the projectile landed. A "relative zero" is the distance from the projectile to the actual target.

    These metrics also lend themselves well to plotting and characterizing relationships between the mean and standard deviation of a critical functional response with respect to the critical functional parameters and critical-to-function specifications (independent variables). This is especially true when optimizing transient performance.
3. Metrics that are **complete.** Such metrics cover the hierarchical network of ideal or transfer functions that are required by the flow-down of system, subsystem, or subassembly performance requirements.
4. Metrics that are **fundamental** measure functions in terms of the Law of Conservation of Energy and/or the Law of Conservation of Mass. These metrics quantitatively and additively account for the balances that are fundamental to physical performance.

The diagram in Figure 12–1 and models that follow are quite useful in helping determine what to measure in Critical Parameter Management.

The Ideal/Transfer Function: $Y = f(x_A, x_B, \ldots x_N)$

The Additive Model: $S/n_{opt.} = S/n_{avg.} + [S/n_{A\ opt.} - S/n_{avg.}] + [S/n_{B\ opt.} - S/n_{avg.}] + \ldots [S/n_{N\ opt.} - S/n_{avg.}]$

The Sensitivity or Variance Model: $\sigma^2_{total} = \sigma^2_A + \sigma^2_B \ldots + \sigma^2_N$

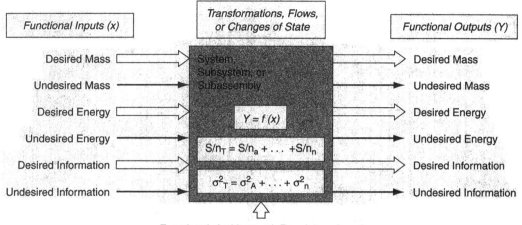

Figure 12–1 What to Measure in CPM

When these kinds of inputs and outputs are measured, functional interactivity contributions (transfer functions, additive models, and sensitivities) between engineering control factors can be documented with scientific accuracy. Real, functional interactivity relationships between critical parameters can be legitimately characterized on the basis of the conservative laws of physics. Then these functional contribution relationships (transfer functions and gains in robustness and sensitivities) can be designed for harmony, synergy, and additivity. You will encounter difficulties if you attempt to design for functional harmony and critical parameter additivity (remember the crew of rowers in Chapter 8!) if you measure quality attributes instead of functions that follow conservative physical law.

Recall, six basic terms are used to define the measured elements of Critical Parameter Management:

1. Critical customer **needs**
2. Critical **requirements**
3. Critical **functional responses (CFR)**
4. Critical **adjustment parameters (CAP)**
5. Critical **functional parameters (CFP)**
6. Critical-to-function **specifications (CTF spec)**

Critical implies that whatever is being measured and tracked is absolutely critical to fulfilling the highest priority needs that a majority of customers have expressed. When you think of the word *critical,* think of *customers.*

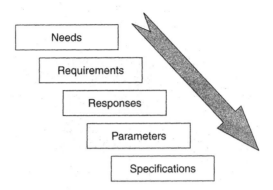

Figure 12–2 CPM System Hierarchy

Functional implies that a functional variable is being measured and tracked as opposed to some quality attribute. If the parameter is not directly measurable in units of some form of mass or energy, you are not conducting true CPM. When you think of the word *function,* think of *physics.*

The critical parameters can be represented as a flow-down of terms stored in a relational database, linked to one another, and evaluated in mathematical and empirical models, which are referred to as *transfer functions.* CPM terms flow-down the system hierarchy as in Figure 12–2.

The vocabulary of the flow-down of critical parameters is explained as follows:

System Level Critical Functional Responses (CFR):

A continuous variable, typically either an engineering scalar or vector, that is a measure of a critical output from the system. System CFRs are directly sensed by the customer and can be directly compared to the new, unique, or difficult customer needs.

System Level Critical Adjustment Parameters (CAP):

A continuous variable, typically either an engineering scalar or vector, that is specifically designed to adjust a system CFR. These variables are few in number and are often used by customers to adjust the performance of the product.

Subsystem Level Critical Functional Responses (CFR):

A continuous variable, typically either an engineering scalar or vector, that is a measure of a critical output from a subsystem. Subsystem CFRs are usually not directly sensed by the customer. When one or more subsystem CFRs are used as inputs to a system CFR, their name is changed to a *subsystem critical functional parameter (CFP).* Subsystem CFRs are measured and documented for their capability to support system CFR capability.

Subsystem Level Critical Adjustment Parameters (CAP):

A continuous variable, typically either an engineering scalar or vector, that is specifically designed to adjust a subsystem CFR. These variables are few in number and are sometimes available for customers to adjust but more often are used by assembly and service organizations.

Subassembly Level Critical Functional Responses (CFR):

A continuous variable, typically either an engineering scalar or vector, that is a measure of a critical output from a subassembly. Subassembly CFRs are usually not directly sensed by the customer. When one or more subassembly CFRs are used as inputs to a subsystem CFR, their name is changed to a *subassembly critical functional parameter.* Subassembly CFRs are measured and documented for their capability to support subsystem CFR capability.

Subassembly Level Critical Adjustment Parameters (CAP):

A continuous variable, typically either an engineering scalar or vector, that is specifically designed to adjust a subassembly CFR. These variables are also few in number and are often used by assembly and service organizations.

Component Level Critical-To-Function Specifications (CTF Spec):

A continuous variable, typically an engineering scalar, that is a measure of a critical component dimension or feature. It can also be a characteristic relating to a surface or bulk property of a component or material. These CTF specifications are commonly communicated to the supply chain using spec sheets, CAD files, and component quality plans. They are usually measured under some form of variable data statistical process control as they are manufactured. Component CTF specs are designed, measured, and documented for their capability to support subassembly and subsystem CFR capability. This is the point where manufacturing process capability interfaces with and transitions into product or design level performance capability.

Manufacturing Process Level Critical-To-Function Specifications (CTF Spec):

A continuous variable, typically an engineering scalar or vector, that is a measure of a critical manufacturing process set point. These variables are literally measured on a manufacturing process as it produces a component or material. One major use of manufacturing process capability is the measurement, control, and application of these CTF variables to fulfill component requirements.

The **manufacturing process capability index** (Cpm) can also be used in reference to assembly processes. Manufacturing process requirements (USL-LSL) establish what is required of these variables. Manufacturing CTF specs are measured during production to ensure that what is required of the process is being provided. A manufacturing process requirement may be a pressure that is required to be at 25 psi +/− 1 psi. The measured manufacturing process CFT spec pressure might be measured at 24.3 psi in the morning and 25.1 in the afternoon. The CFT spec can be compared to the requirement and action taken if the manufacturing process has a control problem. The components are known to vary as the process pressure varies. CPM in a production context measures the in-process capability of the pressure. If the pressure and all other CTF specs are on target with low variability relative to the process requirements, the components need limited measurement during production. During product development, manufacturing CTF specs are measured to document their capability to support component capability.

Statistical Metrics of CPM

Measuring quality or reliability attributes can lace the response values with "virtual" or false interactions that suggest codependencies (A × B) between critical parameters (CFPs and CTF specs) that physically do not exist in a fundamental context. Statisticians frequently provoke these artificial interactions that are little more than manifestations of a poor selection of a measured response. The engineer is then left to worry about all these "statistically significant" interactions based on quality attribute metrics. Most of these worrisome interactions don't even exist in the context of physical law. Products don't obey the statistics of quality attributes—they obey physical law! Products will tell you all about how they obey physical law if you measure the right things. To control physics, you must directly measure functions. Quality then comes from properly constrained physical functions. Statistics are superb quantifiers of physical performance, provided the mean and standard deviation are calculated on fundamental, complete, continuous variables that are scalar or vector quantities.

The following items are the basic statistical metrics used for Critical Parameter Management:

- The **mean or average** of a sample of data that are measures of scalar- or vector-based variables.

  Avg. $= (y_1 + y_2 + \ldots + y_n)/n$

- The **standard deviation** of a sample of data that are measures of scalar- or vector-based variables.

  $\sigma = $ Sqr. Root$[(\Sigma(y_i - \text{Avg.})^2)/n]$

- The **coefficient of variation** of a sample of data that are measures of scalar- or vector-based variables.

  COV $= \sigma/\text{avg}$

- The **signal-to-noise ratio** of a sample of data that are measures of scalar- or vector-based variables.

  S/N $= 10 \log (\beta^2/s^2)$ (dynamic form)

S/N = 10 log (avg^2/s^2) (static form: nominal the best type I also squared reciprocal of COV)

S/N = 10 log (1/s^2) (static form: nominal the Best Type II)

- The **short-term process capability index** of a short-term sample of data that are measures of scalar- or vector-based variables.

  Cp = (USL − LSL)/6s

- The **long-term process capability index** of a long-term sample of data that are measures of scalar- or vector-based variables.

  Cpk = Min[(USL − Avg.)/3s or (Avg. − LSL)/3s]

- The **capability growth index** (CGI)

Critical functional responses provide excellent high-visibility metrics to project managers from their subsystem engineering teams. To anchor the performance of the subsystem CFRs to the subsystem requirements document, the few critical subsystem level functional responses across an engineered subsystem can be aggregated together in the form of their short-term process capability indices (Cp). A summary capability statistic called the *design functional capability growth index* can be used to account for overall subsystem capability growth against a target of Cp = 2.

$$Cp = \frac{USL - LSL}{6s}$$

where **(Cp)** = short-term design capability index for a critical functional response

(USL − LSL) = difference between the upper spec limit and the lower spec limit for CFRs from the subsystem requirements document

(s) = sample standard deviation of the measured CFR

The capability growth index quantifies the aggregate functional capability maturity of a given subsystem (in this case each CFR is equally weighted).

$$CGI = \sum_{i}^{n}\left[\frac{100}{n\#CFR's}\left(\frac{Cp}{2}\right)_i\right]$$

The CGI is measured in units of percent. The percentage is the aggregated amount of subsystem CFRs that are at their goal of Cp = 2.

The value of the design functional CGI resides in its ability to provide an early, integrated, overarching measure of progress for Six Sigma performance at the subsystem level. From a project management perspective, each subsystem CGI could be aggregated into an overall system CGI. CGI metrics provide insight for human resource and hardware/material resource allocation and balancing. Simply put, one knows where the weak CFRs are and can assess their impact on the integrated project plan critical path. Quality and cycle-time can be more effectively managed.

CGI Example

Subsystem A has five critical functional responses being measured. Thus, $n = 5$. Each CFR has the following Cp and target comparison ratio (Cp target is 2) values:

CFR 1: Cp $= 1.3$ Target comparison ratio (Cp/2) $= 1.3/2 = 0.65$

CFR 2: Cp $= .84$ Target comparison ratio (Cp/2) $= .84/2 = 0.42$

CFR 3: Cp $= .71$ Target comparison ratio (Cp/2) $= .71/2 = 0.36$

CFR 4: Cp $= 2.3$ Target comparison ratio (Cp/2) $= 2.3/2 = 1.15 - Capped\ at\ 1.0$

CFR 5: Cp $= 1.62$ Target comparison ratio (Cp/2) $= 1.62/2 = 0.81$

Each target comparison ratio (Cp/2) value is converted to a percentage by multiplying by 100...

CFR 1: $0.65 \times 100 = 65\%$

CFR 2: $0.42 \times 100 = 42\%$

CFR 3: $0.36 \times 100 = 36\%$

CFR 4: $1.00 \times 100 = 100\%$ (*Capped at 1.0 because CFR 4 reached 100% of its growth goal*)

CFR 5: $0.81 \times 100 = 81\%$

These percentages give visibility to the individual CFR's current state of capability growth with respect to the capped target of Cp $= 2$. Any Cp value beyond 2 is obviously good but no credit is given to those CFRs with Cp >2. Doing so would unduly bias the aggregate subsystem Cp growth index upward while certain CFRs may potentially remain at unacceptably low Cp values. The intent is to motivate the subsystem teams to get all CFRs up to Cp $= 2$. Once all CFRs are safely at Cp $= 2$, the subsystem teams can focus their efforts on taking any CFR capabilities they wish to as high a Cp level as is economically feasible or helpful in delighting customers. Please note our focus on the overall Cp performance of the subsystem as opposed to the individual performance of an isolated CFR. This is good system engineering practice that emphasizes integrated teamwork instead of individual heroism.

Each CFR now has a percentage contribution calculated for overall subsystem capability growth. Each CFR is equally weighted (one could assign differing weights to CFRs as appropriate) as a contributor of one fifth (20 percent) of the overall capability growth. To quantify this contribution, we divide each CFR's Cp percent by the total number of CFRs in the subsystem (in this case $n = 5$).

CFR 1: 65%/5 = 13% contribution to overall subsystem capability (goal is 20%)

CFR 2: 42%/5 = 8.4% contribution to overall subsystem capability (goal is 20%)

CFR 3: 36%/5 = 7.2% contribution to overall subsystem capability (goal is 20%)

CFR 4: 100%/5 = 20% contribution to overall subsystem capability (goal is 20%)

CFR 5: 81%/5 = 16.2% contribution to overall subsystem capability (goal is 20%)

Summing the five CFR contribution values, we see that the subsystem CGI = 64.8%. The goal is 100%, so corrective action plans can now be designed and executed.

The Capability Growth Index and the Phases and Gates of Technology Development and Product Commercialization

The growth of capability accrues across the phases and gates process, as depicted in Figure 12–3.

The diagram in Figure 12–3 illustrates how functional capability is developed and grown from one phase to the next for *n* number of subsystem CFRs within a developing system. These bar charts are easy to generate and use in gate reviews to illustrate capability growth to the gate keepers. Senior managers can tell if CPM is successful at each gate. They can see how things are progressing on a phase-by-phase basis. If they don't see the growth in functional capability across the subsystems, then they can take action early to protect the business from a down-range catastrophe.

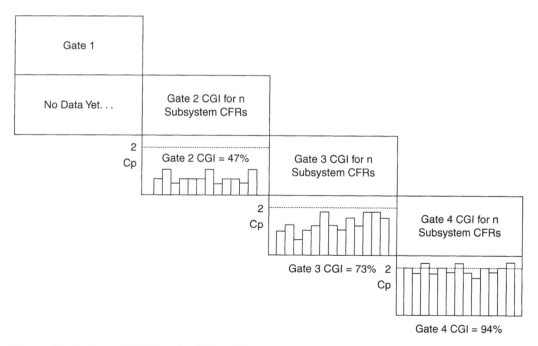

Figure 12–3   Accrual of Growth of Capability Across Phases and Gates

CHAPTER 13

Data Acquisition and Database Architectures in CPM

Data drives the key decisions that are made within the phases of the product development process. Data helps the gate keepers manage risk during gate reviews. Opinions are secondary in this context. Data provides the capacity to define and document criticality. Data is reality.

CPM starts with the gathered voice of the customer data. It progresses and enables capability growth through the measured voice of the product and its supporting network of manufacturing processes. The voice of the product and process is realized through the measurement of critical parameter data. Data is the mechanism used to validate the robustness and capability of the critical parameter relationships (ideal/transfer functions, additive models, and variance models). These are the models of Critical Parameter Management:

> The ideal/transfer function: $Y = f(x_A, x_B, \ldots x_N)$
>
> The additive model: $S/n_{opt.} = S/n_{avg.} + [S/n_{A\ opt.} - S/n_{avg.}] + [S/n_{B\ opt.} - S/n_{avg.}] + \ldots$
> $+ [S/n_{N\ opt.} - S/n_{avg.}]$
>
> The variance model: $\sigma^2_{total} = \sigma^2_A + \sigma^2_B \ldots + \sigma^2_N$

Data defines critical dependencies, sensitivities, and linkages all the way down through the system to the manufacturing process set points that constrain the fulfillment of critical-to-function requirements at the component level. Without the proper data, engineering team decisions are made at great risk.

Probably the single most important skill in CPM is the ability to define, develop, construct, assess, and use a capable data acquisition system. Without a capable data acquisition system your ability to trust your data will be limited.

Instrumentation, Data Acquisition, and Analysis in CPM

Every single critical functional response in an engineered system must be measured. This applies to products as well as manufacturing processes. One sure way for management to assess the health of a CPM program is to identify how many critical functional responses exist within the new product being developed *and then quantify how many of them are being actively measured.* This is usually quite a revelation to unsuspecting project managers. In Phase 2 (development or design) of either the technology development (I^2DOV) or product commercialization (CDOV) processes, the project manager should require that all CFRs be documented and measured by known data acquisition that can be calibrated.

Data Acquisition Systems

Consider the question, "What is the most fundamental act one can take to attain Six Sigma performance during the management of critical parameters?" The answer is all bound up in the act of taking data. Developing a new product with weaknesses and lapses in the proper acquisition of data is unacceptable in the context of Six Sigma.

Taking data is not always a simple matter during product development. As product complexity increases, so do the requirements and demands on data acquisition. In some situations measuring the product's functions is relatively straightforward and simple. This is the case when one is developing a popular product like the one-time-use camera. The complex production system that produces tens of millions of these cameras per year is a whole other matter. The data acquisition system for the camera is small, but the one for the production and assembly system is extremely complex. When one is developing complex products like a high-volume digital printer or an automobile, the data acquisition systems for the product itself become complicated. The data acquisition systems used in product development are often more involved than the ones used in production, assembly, and service.

Whatever the case, the data acquisition elements that make up your data acquisition systems should be in harmony with validating the requirements of the product and its production and service processes. The architecture of your data acquisition system will mirror the architecture of your requirements. An excellent way to audit a company for product development capability is to match how well its system, subsystem, subassembly, component, and manufacturing process requirements are being fulfilled in comparison to the structure of its data acquisition system. The data acquisition system elements that are being used to ensure the requirements are being properly fulfilled should pretty much tell the story. Of course the measurement capability of the data acquisition systems plays an integral role in the credibility of the data. We see from this that two important elements underwrite the integrity of any data acquisition system:

1. For every critical functional requirement, there must be a transducer that measures the CFR within the data acquisition system.
2. For every transducer measuring a CFR, there must be a measurement system analysis conducted to verify the capability of the measurement process (gage R&R and/or precision/tolerance study).

Many types of data acquisition systems are available on the commercial market. You can buy one designed and ready to use or you can build your own from scratch. If your measurement needs are relatively simple, buying the elements and integrating them yourself is probably the cheapest way to proceed. If you have a large network of CFRs ranging across a complex system architecture, you are probably better off partnering with a company that can help you design and integrate a computer-aided data acquisition system.

Transducer Selection and Calibration

Transducers are the elements within data acquisition systems that physically sense the critical functional responses within the system, subsystems, and subassemblies. Since there are no "reliability meters" or "yield meters," we must be realistic about what we measure. Literally dozens of transducer types and hundreds, if not thousands, of versions of transducers can be used in Critical Parameter Management.

One of the harsh realities that will quickly face anyone attempting CPM is the fact that some functions are difficult, sometimes even impossible, to measure directly. Sometimes you will have to invent new transducer technologies. Companies deploying Six Sigma in product development are often found to be filing for new patents for measurement technologies.

Other measurement cases will require you to measure a surrogate function that is close to the fundamental function. An example would be measuring a deflection that is related to a force you wish you could measure. The force is impossible to measure, but the deflection is proportional to the force so you measure the deflection and calculate the force. This functional compromise is still far better than measuring quality attributes! We recommend you define a full set of transducers that cover all the types of functions you need to measure. Where you have gaps or difficulties, define work-around or compromise strategies to handle the data acquisition requirements. Face these difficulties by thinking with an "out-of-the-box" mentality. Apply the DFSS tools to your transducer design problems. Invent or innovate, develop, optimize, and verify new transducer technologies for your business. Let transducer capability and performance become a strategic part of your product development capability in your business. This is a clear sign of growth and maturity in Six Sigma methods.

Data Processing and Storage

Once a function is acquired by a transducer, it must be processed and saved so it can be used to help in the management of critical parameters. A "signal" produced by a transducer can be collected in many ways. By far the most common form of signal is an analog voltage. Most transducers convert a function into a voltage. Once the voltage is produced in analog form, it can be further processed into a form that is useful to a digital computer. This is the well-known analog-to-digital (A/D) conversion of a measured signal. A/D conversion is a serious matter in data acquisition system design. Many commercial A/D conversion circuits or "boards" are available. We recommend you consult with an expert in data acquisition signal processing to guide you on the proper sizing and selection of the boards you should use. Whether you are gathering one or 1,000 signals, you need the right A/D board to get the data into your computer or meter.

Data will usually come in the following forms:

High-speed analog signals
Low-speed analog signals
High-speed digital signals
Low-speed digital signals

If you are not accustomed to working with modern data acquisition and processing hardware, firmware, and software, you need to become a student of this complex science. Any number of texts provide good places to start. Some recommendations are listed at the end of this chapter.

We also recommend computer-aided data acquisition and processing products such as LabView from National Instruments Corp.

Data Analysis

Data analysis has been greatly improved over the last few years. With the state of the art in modern PCs and data analysis software, it is not difficult to transfer data files between software programs. Statistical analysis tools such as Minitab have become popular software in use by the Six Sigma community at large. Minitab can help you evaluate data and get the statistical information you need to build and update the models used in Critical Parameter Management.

Minitab can receive data from most modern data acquisition programs. It can perform all of the following analysis (and much more) in support of CPM:

Capability analysis

Measurement system analysis

Design of experiments
- Full and Fractional Factorials (empirical modeling of transfer functions)
- Taguchi designs (additive modeling and variance modeling)
- Response surface methods (empirical modeling of transfer functions)
- Mixture experiments (empirical modeling of transfer functions)

ANOVA

Linear and nonlinear regression

Correlation

Inferential and descriptive statistics

Basic and advanced graphical data plotting

SPC charting

Many other statistical analysis software packages are available on the commercial market. Any package that can perform the analysis and plotting listed previously can be used for CPM.

Databases: Architectures and Use in CPM

Structuring, recording, manipulating, updating, and using the complex paths of critical functional relationships requires the use of modern software and computing technology. The best way of conducting Critical Parameter Management is to use a relational database. Relational databases enable the storage and tracking of the flowing types of relationships between critical parameters:

One-to-one relationships

One-to-many relationships

Many-to-one relationships

Many-to-many relationships

The biggest benefits from CPM come from the ability to identify and track many-to-many relationships across complex product and production systems. Product development organizations fall down on cycle-time, cost, and functional performance goals because they fail to understand interactions between complex parametric relationships. They underestimate the effects of complexity because they underinvest in the development of the knowledge they need to recognize critical relationships. They don't know about complex interactive relationships that characterize the sensitivities that dominate many of the problems they will inevitably face once the product is launched. In short, they don't know what they don't know.

However, when a product development team knows something of the complexity of relationships across, between, and within their system architectures, it can be quite preventative in its use of tools and best practices. Unfortunately, in the beginning stages of using Critical Parameter Management, the amount of information needed to do product development properly is a bit shocking to novice users of the CPM method. The good news is they quickly adjust their perspective, dig in, and get used to managing complexity instead of ignoring it. It is always amazing to see how human nature exhibits a "denial of the truth" attribute when confronted with complexity. The fact is we need to face complexity with a single-minded focus on being methodical, rigorous, and disciplined in breaking these problems down into manageable "bites." Critical parameter databases help us do this. It is interesting to see how a CPM database slowly but inexorably develops in the early phases of product development. Once people get used to methodically constructing such data structures, life gets relatively simple for those who have to track and verify the capability and robustness of the many complex relationships within the system. The hard part is to get started.

A few databases on the market can help in the performance of CPM:

Microsoft® Access: a capable tool for establishing a CPM database. Easy to learn and use to build your first CPM database. Macros must be written to perform the mathematical functions required to quantify capability calculations, transfer functions, perform sensitivity analysis, and achieve variational transmission tracking and modeling.

DOORS: a great requirements database. Staff must take short training classes to learn it. Advanced training is required to perform quantification of capability calculations, transfer functions, sensitivity analysis, and variational transmission tracking and modeling.

The Six Sigma Cockpit: a new relational database program based on object oriented programming. This software is designed to do everything CPM requires:

- Hierarchical requirements documentation
- Active linkage between all many-to-many critical parameter relationships
- Dynamic linking of all transfer functions for system variational modeling
- Easy input and storage of all associated documents for CPM

Six Sigma Cockpit can basically do everything required to conduct CPM on complex systems.

The choice of software to build a relational CPM database needs to be carefully considered. We recommend benchmarking these and other commercially available software packages to see which fits your short-term and long-term needs.

References

Beckwith, T. G., Marangoni, R. D., & Lienhard V., J. H. (1993). *Mechanical measurements.* Reading, Mass.: Addison-Wesley.

Doebelin, E. O. (1990). *Measurement systems application and design.* New York: McGraw-Hill.

Fraden, J. (1997). *Handbook of modern sensors* (2nd Ed.). Woodbury, NY: AIP Press.

Holman, J. P., & Gajda, Jr., W. J. (1989). *Experimental methods for engineers.* New York: McGraw-Hill.

Norton, H. N. (1989). *Handbook of transducers.* New Jersey: Prentice Hall.

Wheeler, A. J., & Ganji, A. R. (1996). *Introduction to engineering experimentation.* New Jersey: Prentice Hall.

Tools and Best Practices for Invention, Innovation, and Concept Development

The tools and best practices associated with Design For Six Sigma are what actually deliver results. That is only true if people use the tools and do so with discipline and rigor. The tools and best practices discussed in this section focus on deliverables required at gate reviews for the first phases in the I^2DOV and CDOV processes. These tools and best practices ensure the new product requirements are focused on fulfilling the voice of the customer. They also ensure that numerous concepts are generated and the superior product concept is selected. Part IV ends with a detailed explanation of the modeling process that DFSS requires to take the superior product concept through the phases and gates of product development.

Chapters 14–34 are structured so that the reader can get a strong sense of the process steps for applying the tools and best practices. The structure also facilitates the creation of PERT charts for project management and critical path management purposes. A checklist, a scorecard, and a list of deliverables are provided at the end of each tools and best practices chapter.

Each of these chapters follows a common format.

- A brief description of where the tool resides within the phases of the product development process:

Where Am I in the Process? Where Do I Apply the Tool Within the Roadmaps?

Linkage to the I²DOV Technology Development Roadmap:

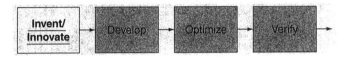

Linkage to the CDOV Product Development Roadmap:

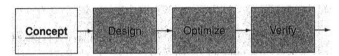

- A brief description of the purpose of the tool:

What Am I Doing in the Process? What Does the Tool Do at this Point in the Roadmaps?

The Purpose of the Tool in the I²DOV Technology Development Roadmap:

The Purpose of the Tool in the CDOV Product Development Roadmap:
- A brief description of the deliverables from using the tool:

What Output Do I Get at the Conclusion of this Phase of the Process? What are Key Deliverables From Gathering and Processing the Voice of the Customer at this Point in the Roadmaps?

- A flow-down of block diagrams that illustrate the major steps in using the tool:

Tool Process Flow Diagram

This visual reference of the detailed steps of the tool illustrates the flow of what to do.

- Text descriptions for each block diagram that explain the step-by-step use of the tool:

Verbal Descriptions for the Application of Each Block Diagram Within the Tool

This feature explains the specific details of what to do, including:

- Key detailed input(s) from preceding tools and best practices
- Specific steps and deliverables from each step to the next, within the process flow
- Final detailed outputs that flow into follow-on tools and best practices

- Illustration of example checklist and scorecard for the tool being discussed:

Tool Checklist and Scorecards

- A final review of the deliverables from the use of the tool:

What output do I get at the conclusion of this phase of the process? What are key deliverables from the tool at this point in the roadmaps?

- A list of recommended references on the tool.

The intent of the formatting is to show how the tools work, link together, and flow within the continuum of the phases of product development.

Gathering and Processing the Voice of the Customer: Customer Interviewing and the KJ Method

Where Am I in the Process? Where Do I Apply VOC Gathering and the KJ Method Within the Roadmaps?

Linkage to the I^2DOV Technology Development Roadmap:

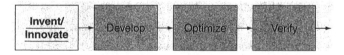

Gathering and processing the "long-term" voice of the customer is conducted within the **Invent/ Innovate** phase during technology development. It is applied to define, structure, and rank the voice of the customer for the development of new technology platforms and subsystems that will be integrated into a new platform or for integration into existing product designs.

Linkage to the CDOV Product Development Roadmap:

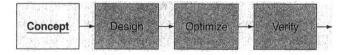

A second, more focused gathering and processing of the voice of the customer is conducted within the **Concept** phase during product development. Here we apply gathering and processing of the voice of the customer to a specific product within the family plan from the strategic product portfolio. This application of gathering and processing the voice of the customer tools ensures that the new, unique, and difficult needs are identified to help develop the critical requirements, functions, and architectures for a specific new product.

What Am I Doing in the Process? What Does VOC Gathering and Processing Using the KJ Method Do at this Point in the Roadmaps?

The Purpose of Gathering and Processing the Voice of the Customer in the I²DOV Technology Development Roadmap:

To set the stage for where this set of best practices fits into the grand scheme of product development, we show the strategic flow-down of how a business develops new technology and product concepts:

Define business strategy: *Profit goals and growth requirements*

Identify markets and market segments: *Value generation and requirements*

Gather long-term voice of customer and voice of technology trends

Develop product line strategy: *Family plan requirements*

Develop/deploy technology strategy: *Technology and subsystem platform requirements*

Gather product-specific VOC and VOT: *New, unique, and difficult needs*

Conduct KJ analysis: *Structure and rank the VOC*

Build system House of Quality: *Translate VOC*

Document system requirements: *New, unique, difficult, and important requirements*

Define the system functions: *Functions to be developed to fulfill requirements*

Generate candidate system architectures: *Form and fit to fulfill requirements*

Select the superior system concept: *High in feasibility, low in vulnerability*

In the Invent/Innovate phase of the I²DOV Process we gather the long-term voice of customer and voice of technology trends to obtain the data necessary to structure and rank the long-term customer needs that are new, unique, or difficult. The technology development team then conducts KJ analysis to structure and rank these needs. This data is used to define technology requirements capable of fulfilling the product line strategy. The technology requirements are used to define a set of adjustable functions. A unique set of architectural concepts (platforms) can then be developed to fulfill the various functions the family of products will eventually have to produce. After careful evaluation, one superior platform architecture is selected out of the numerous options to go through the rest of the I²DOV process.

The Purpose of Gathering and Processing the Voice of the Customer in the CDOV Product Development Roadmap:

In the CDOV process we gather product-specific VOC and VOT to focus on the new, unique, and difficult customer needs that are related to a specific product being developed out of the product family plan. The product design team conducts KJ analysis to structure and rank these product-specific needs. They are then used to define system (product level) requirements capable of fulfilling the product line strategy. The system requirements are used to define a set of product functions. Numerous system architectural concepts can then be developed to fulfill the product functions. After careful evaluation, one superior system architecture is selected out of the numerous options to go through the rest of the CDOV process. This product is but one of several that will be sequentially developed to fulfill the product line strategy.

What Output Do I Get at the Conclusion of this Phase of the Process? What Are Key Deliverables from Gathering and Processing the Voice of the Customer at this Point in the Roadmaps?

- A real set of actual customer need statements that forms the data that will be used to help define the system requirements
- A structured set of customer needs that are grouped by similarities (affinities) that relate like needs into rational sub-groups
- A ranked hierarchy of customer needs that begin to define criticality by drawing attention to the customer needs that are new, unique, and difficult.

Gathering and Processing the Voice of the Customer Process Flow Diagram (found on page 344)

This visual reference of the detailed steps of gathering and processing the voice of the customer illustrates the flow of what to do.

Verbal Descriptions for the Application of Each Block Diagram

This section describes each step in the diagram on VOC gathering and processing to explain specific details of:

- *Key detailed input(s)* from preceding tools and best practices
- *Specific steps and deliverables* from each step to the next, within the process flow
- *Final detailed outputs* that flow into follow-on tools and best practices

Key Detailed Inputs

> **Market segments defined—we understand who the customers are and their environment.**

With the market segments defined, the team knows exactly who their customers are and where to go to gather their needs.

INPUT . . .

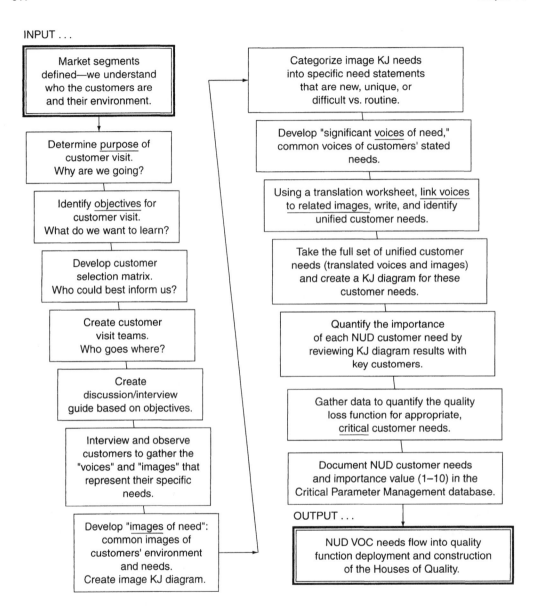

OUTPUT . . .

Specific Steps and Deliverables

Determine purpose of customer visit. Why are we going?

The purpose of visiting and interviewing a set of customers is to get data. You need to hear what they say and see what they do so that you develop a clear understanding of what their new product needs are. You want to explore and experience their world and come away with the data you need to develop the right product. The purpose of the visit should be worded in the context of a specific new product development initiative that the product strategy is suggesting. You may have new ideas and want to attain customer needs data to refine and focus your ideas for new products.

Identify objectives for customer visit. What do we want to learn?

The objectives of your customer visit are stated in terms of the detailed things you want to learn. In general terms, the critical things a product development team wants to learn are what needs are *new, unique,* and *difficult.* Customers usually don't want things that are old, common, and easy. If they do—fine, you should attempt to learn about those needs, too. The key is to be data driven in what you develop.

The details you want to learn are specific *customer needs.* These can fall into many different categories. Here are but a few of those categories from Garvin's eight "dimensions of quality" (1988)*:

1. **Performance**—primary operating characteristic of a product or service
2. **Features**—secondary characteristics of a product or service
3. **Reliability**—frequency with which a product or service fails
4. **Conformance or consistency**—match with specifications or standards
5. **Durability**—product life, resistance to problems and deterioration
6. **Serviceability**—speed, courtesy, and competence of repair or support
7. **Aesthetics**—fits and finishes
8. **Perceived quality**—reputation and life cycle ownership experience

Any visit to find out what customers need should explore details within and across these areas. You must be ready to take in whatever needs data is offered to you. You are not trying to "steer" their needs, but to be driven by their needs. If your customers are to drive your product development process, you need a fairly broad and detailed set of data from them.

Develop customer selection matrix. Who could best inform us?

To obtain a broad and detailed set of customer needs data, you must visit and interview a reasonable cross-section of the customers within your targeted market segments. A good way to ensure that this happens is to develop a customer selection matrix.

---

*Used with permission of Sigma Breakthrough Technologies, Inc.

Market Segment	Customer Name	Our Customer	Competitor Customer	Dual Customer	Competitive Technology	Lapsed Customer

Figure 14–1 Customer Selection Matrix

Source: Used with permission of Sigma Breakthrough Technologies, Inc.

A customer selection matrix defines the options of "who can best inform us." Figure 14–1 provides an example of a customer selection matrix.

The columns in this matrix make up a reasonable list of whom you should consider visiting to gather the VOC data to help define your technology and product development requirements.

> **Create customer visit teams. Who goes where?**

Once you know who to visit, you need to select a well-rounded set of customer interview teams. It is also important to know where you should go geographically and demographically so you can give some thought about exactly "who is going where."

It is common to send the interview teams out in groups of three:

1. A leader of the interview process
2. A "scribe" who writes the VOC need statements down exactly as they are given (no tainting or biasing of the VOC words)
3. A subordinate interviewer to help add balance and diversity of perspective to the discussions

Often the leader of the interview team is an expert from inbound marketing. The scribe is skilled in documenting verbal statements efficiently. The subordinate interviewer is typically an engineering team leader (often a DFSS "Blackbelt") at some level within the technology or product development organization.

> **Create discussion/interview guide based on objectives.**

The customer interview is not a sterile "yes–no" question and answer period. It is intended to be a flowing discussion influenced by a customer interview guide.

A customer interview guide needs to be developed and tested well in advance of the actual customer visits. A good customer interview guide contains these kinds of elements*:

1. *6–10 key topics/questions:*
   - Based on your objectives
   - Open-ended
   - Subtopics as needed
   - Possible multiple variations for different segments of your matrix (e.g., user vs. manager)
   - A guide, not a questionnaire
   - Evoke images of experiences and needs
2. *Past issues*
   - Weaknesses
   - Experiences
   - Problems
3. *Present issues*
   - Current considerations
   - Needs
   - Application
   - Technology
4. *Future issues*
   - Enhancements
   - Drivers
   - Competitive offerings
   - Direction
5. *Fitness to standard*
   - Problems with product
6. *Fitness to use*
   - Meets application requirements
7. *Value*
   - Comparison of performance vs. cost in a competitive environment
8. *Fitness to latent requirements*
   - Potential future enhancements

Your questions and discussions need to be oriented in several dimensions:

What—tend to focus conversation on events:

"*What* problems have you experienced?"

*Used with permission of Sigma Breakthrough Technologies, Inc.

How—tend to focus the discussion on a process:

> "*How* do you use. . .?"

Why—requires an explanation, but may elicit a defensive reaction:

> "*Why* is that?"

Could—queries are usually perceived as "gentle" and open:

> "*Could* you give us an example?"

The situation you want to avoid is verbal exchanges that are "closed ended" and result in yes or no answers. Your goal is to elicit responses or "word pictures" that are rich in descriptions of need.

These answers eventually end up being translated into technical requirements. The more you can get your customers to describe their needs in the form of "continuous" variables, the easier it is going to be for you to translate the VOC into technical requirements. Avoid "two-valued" language. Seek to have your customer communicate with you using "Multi-valued" language:

- The room is *hot or cold* vs. the room is *75 or 65 degrees.*
- The product needs to react *quickly* to my input vs. the product needs to react within *10 seconds.*

Think of getting the customer to talk to you in terms of the way a light dimmer changes the level of intensity of light in a room as opposed to the way a light switch goes from one extreme to another (on-off). Try to get targets and ranges as opposed to "more is better" or "less is better" kinds of need statements. Engineers use physical laws and design principles based on scalars and vectors that create continuous functions to fulfill product requirements. The closer you can get your customers to express their needs in this continuous variable context, the more likely the final product will satisfy the customer's originally stated needs. This is not the time to be vague.

Interview and observe customers to gather the "voices" and "images" that represent their specific needs.

Customer interviews can be extremely enjoyable events. It is quite literally an "open notebook–open mind" discovery process. Use your interview guide to establish a flow of conversation. Be intuitive, not coldly analytical. The engineers will take that tact later in the CDOV process where it belongs. You should go into the event with a strategy of *active listening*. Prime the interview process with stimulating questions and open-ended statements that are followed by listening, probing, and observation. Don't hesitate to ask if what someone just told you was what you actually heard. Go back and be sure you understood what was actually said. It is extremely easy to misunderstand people. Take the time to get it right. Above all else, be quiet and let your customers tell you their needs. This is not your forum.

Part of the interview process should include actual observation of customers using the product or process about which you are trying to learn. Watching customers, as opposed to talking with them, is called *contextual inquiry*. The data gathered in these sessions is often referred to as an *image*. The interviewer comes away with perceived images of customer use patterns, behaviors, and conditions. This book was structured, in part, from images we formed by watching how product development teams struggled to adequately inform themselves as they deployed DFSS.

> **Develop "images of need": common images of customers'
> environment and needs. Create image KJ diagram.**

Now that you have visited your selected set of customers, it is time to begin processing the VOC data. The first set of data is the transcribed images that were formed during interviews and contextual inquiry. Images can come from verbal customer statements that illustrate the environment and context of VOC needs. Mental images that your interview team formed during observed behaviors should also be documented and discussed as soon as possible after conducting the interviews. Don't wait — do this the evening after the interview was conducted, if possible. Be careful not to taint the VOC data with your personal biases and aspirations.

The team should create the first of the two forms of KJ diagrams (1. Image and 2. Requirements) at this point. An image KJ diagram is structured as follows:

1. Write down your specific images as "statements of need" on a Post-It note.
2. After all team members write down their images, adhere them to the wall. There may be many dozens of Post-It notes.

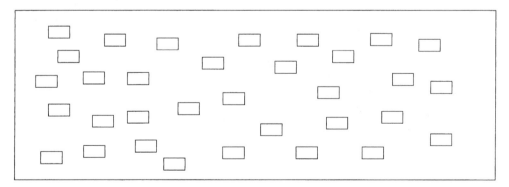

Many Detailed Imaged Needs on a Wall

3. Group the imaged needs into rational associations that are often referred to as *affinity groups*. In other words, take the Post-It notes that tend to express similar needs and group them together. You should end up with a number of grouped Post-It notes.

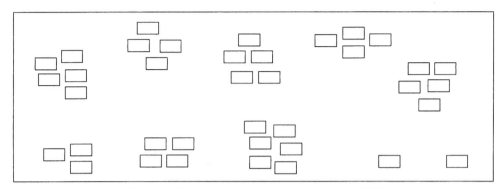

First Level Grouping of Imaged Needs on a Wall

4. For each group of Post-It notes, review the imaged need statements and reduce the number by throwing out redundant statements. If two or more are close to saying the same thing, accurately rewrite the imaged needs on a new Post-It note and throw out all the ones that were condensed into this one new statement.

5. For each group, write an overarching need statement (*shaded Post-Its in the figure that follows*) that summarizes the general topic expressed by the grouped set of imaged needs. This process is called "creating the ladder of abstraction." That is to say, you are creating a hierarchy or layered structure of customer needs by "laddered" categories. In the following figure, you have eight first level groups.

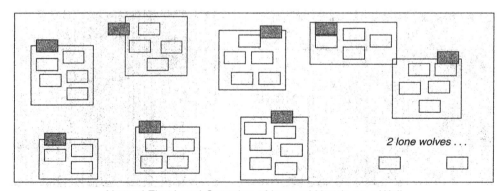

Named First Level Grouping of Imaged Needs on a Wall

6. With the many single imaged needs now grouped into their first level of need categories, review the wording of first level customer need statements (*shaded Post-Its*). Look for common themes between the first level groups and try to identify the next level of grouping. It is common to have 10 to 12 first level groups at this point. If a specific imaged need or first level group is not related to anything else, it is set aside for the time

being and labeled as a "lone wolf." Create the second level groups by grouping related first level groups. In the figure that follows, we see four second level groups.

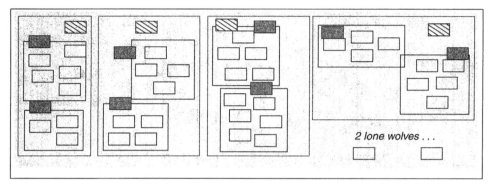

Named Second Level Grouping of Imaged Needs on a Wall

The customer need hierarchy at this point consists of just your imaged needs statements:

- 4 second level groups of imaged needs
- 8 first level groups of imaged needs
- 38 detailed level individual imaged needs (including two lone wolves)

7. Establish a form of criticality by ranking the importance of the eight first level groups of imaged needs. This is usually done by distributing three colored sticky dots to each team member. Each person gets one sticky dot of each color. The voting scenario follows this color scheme:
Red dot = 6 points
Blue dot = 3 points
Green dot = 1 point

The first level group that gets the most points is considered the most important set of imaged needs, the first level group that receives the second most points is considered the second most important group, and the third top vote getter ranks third. You are beginning to form a sense of what imaged needs are critical to those that are of nominal importance.

Relationships can be traced between the top three first level groups based on the following diagrams:

- If a first level group supports another group in a positive manner, the diagram at the top of page 352 shows an arrow pointing from the supporting group of needs to the "supported" group of needs.
- If a first level group is in contradiction to another group in a negative manner, the diagram shows a line pointing between the two groups with blocks on the ends.

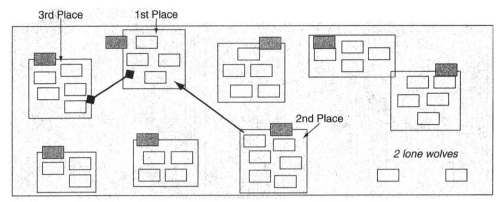

Ranking and Tracing Relationships Among First Level Groups

Categorize image KJ needs into specific need statements that are new, unique, or difficult vs. routine.

The final processing step on the imaged needs is the beginning of the Critical Parameter Management process. The Team looks at each detailed image need and assesses the need to see if it is new, unique, or difficult. In the following figure, we see an example of such needs highlighted in black.

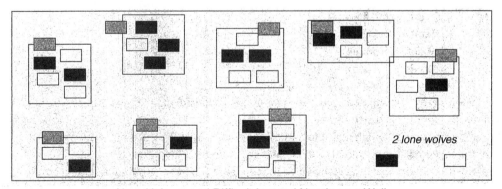

New, Unique, and Difficult Imaged Needs on a Wall

Develop "significant voices of need," common voices of customers' stated needs.

Now it is time to turn your attention to the raw VOC statements your team wrote down during the verbal interviewing sessions. The VOC statements are reviewed and redundancies removed to be sure they are concise and well understood. They can be converted to the Post-It note format during the review process just like the image needs from contextual inquiry. This step assures that the detailed VOC data is manageable.

> **Using a translation worksheet, link voices to related images, write, and identify unified customer needs.**

You now seek to merge the image needs Post-It notes with the stated VOC needs Post-It notes. You can use a translation worksheet to help accomplish this goal (see Figure 14–2).

Once you have merged the data, you can document that the unified customer needs are ready to go through the final KJ process. They are the needs you want to further process to potentially take into the Houses of Quality. Some unified customer needs will be directly translated into technical requirements and added directly to the system requirements document without going through the Houses of Quality. Putting nominally important customer needs (old, common, and easy needs) through QFD consumes valuable time and resources that are better spent on processing those critical needs that are new, unique, and difficult. You will fail at QFD if you try to put all VOC needs through the Houses of Quality.

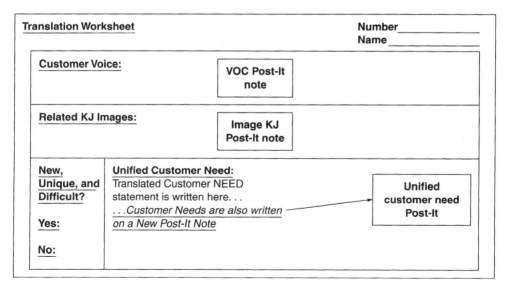

Figure 14–2 Sample Translation Worksheet
Source: Used with permission from Sigma Breakthrough Technologies, Inc.

> **Take the full set of unified customer needs (translated voices and images) and create a KJ diagram for these customer needs**

The team should create the second of the two forms of KJ diagrams at this point. A *unified customer needs* KJ diagram is structured as follows:

1. Adhere the unified customer needs to the wall. There may be dozens of such Post-It notes.

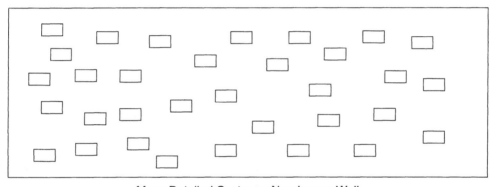

Many Detailed Customer Needs on a Wall

2. Group the needs into rational associations of affinity groups. Take the Post-It notes that tend to express similar needs and group them together. You should end up with a number of grouped Post-It notes.

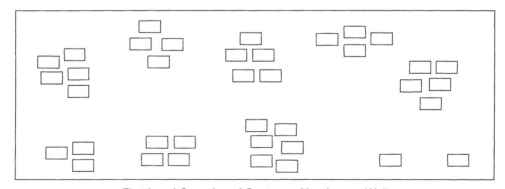

First Level Grouping of Customer Needs on a Wall

3. For each group of Post-It notes, review the unified customer need statements and reduce the number by throwing out redundant statements. If two or more are close to say-

ing the same thing, accurately rewrite the unified customer needs on a new Post-It note and throw out all the ones that were condensed into this one new statement.

4. For each group, write an overarching need statement (*shaded Post-Its in the following figure*) that summarizes the general topic expressed by the grouped set of unified customer needs. Again, you are creating the ladder of abstraction. This example shows eight first level groups.

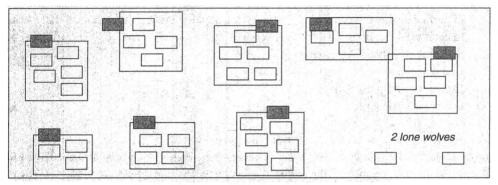

Named First Level Grouping of Customer Needs

5. With the many single unified customer needs now grouped into their first level of need categories, review the wording of first level unified customer need statements (*shaded Post-Its*). Look for common themes between the first level groups and try to identify the next level of grouping. It is common to have 10 to 12 first level groups at this point. If a specific unified customer need or first level group is not related to anything else, set it aside for the time being and label it as a *lone wolf*. Create the second level groups by grouping related first level groups. In the following figure, we see four second level groups.

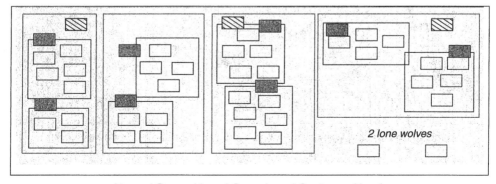

Named Second Level Grouping of Customer Needs

The customer need hierarchy at this point now consists of the unified customer needs statements:

- 4 second level groups of unified customer needs
- 8 first level groups of unified customer needs
- 38 detailed level individual unified customer needs (with two lone wolves)

6. Establish a form of criticality by ranking the importance of the eight first level groups of unified customer needs. This is usually done by distributing three colored sticky dots to each team member. Each person gets one sticky dot of each color. The voting scenario follows this color scheme:
Red dot = 6 points
Blue dot = 3 points
Green dot = 1 point

The first level group that gets the most points is considered the most important set of unified customer needs, the first level group that receives the second most points is considered the second most important group, and the group that receives the third most number of points is ranked third. Now you are beginning to form a sense of what unified customer needs are critical to those that are of nominal importance.

Relationships can be traced between the top three first level groups based on the following diagram:

- If a first level group supports another group in a positive manner, the diagram shows an arrow pointing from the supporting group of needs to the supported group of needs.

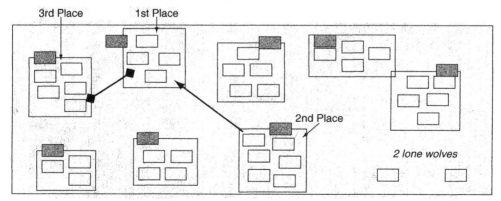

Ranking and Tracing Relations Among First Level Groups of Customer Needs

- If a first level group is in contradiction to another group in a negative manner, the diagram shows a line pointing between the two groups with blocks on the ends.

> **Quantify the importance of each NUD customer need
> by reviewing KJ diagram results with key customers.**

The final processing step on the unified customer needs is the beginning of the Critical Parameter Management process. Now the team looks at each detailed unified customer need and assesses the need to see if it is new, unique, or difficult. This time, though, you visit your key customers and review these results with them. The following example shows NUD customer needs highlighted in black.

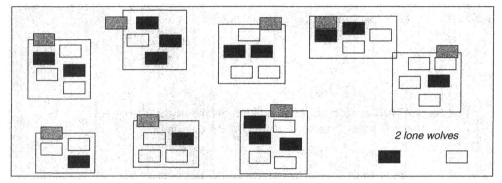

New, Unique, and Difficult Customer Needs

Get the customers to rank (on a scale of 1 to 10) the strength of importance for each need. Also, have the customer rank (on a scale of 1 to 5) the current fulfillment of these needs by your current product and their perception of how the competitive offerings on the market are fulfilling these needs. This data will be needed during QFD.

The outcome is a set of ranked, NUD unified customer needs that are worth taking into the QFD process. All other unified customer needs that are of nominal importance (old, common, and easy) are directly translated into technical requirements and placed in the system requirements document.

> **Gather data to quantify the quality loss function
> for appropriate, critical customer needs.**

While the team is visiting a selected set of key customers to validate and rank the KJ results, it is an excellent opportunity for one last interview. This one centers on gathering the data necessary to quantify Taguchi's **quality loss function:**

A_0: the dollar loss at the customer's "fail-to-satisfy" tolerance limits

Δ_0: the customer's "fail-to-satisfy" tolerance limits

The loss (A_0) is usually quantified in warranty dollars, cost of poor quality, and any form of cost that accrues due to the product performing off target at or beyond the fail-to-satisfy tolerance limits.

Some other typical situations that figure into repairing or replacing a product are:

- Loss due to lack of access to the product during repair
- The cost of parts and labor needed to make the repair
- The costs of transporting the product to a repair center

Regardless of who pays for the losses—the customer, the manufacturer, or a third party— all losses should be included in A_0. Substituting for k in the quality loss function $L(y) = k(y\text{-}m)^2$, we define the general form of the quality loss function as:

$$L(y) = A_0/\Delta^2{}_0\,(y\text{-}m)^2$$

Document NUD customer needs and importance value (1–10) in the Critical Parameter Management database.

We are at the end of the VOC gathering and processing tasks. The final act is to document the NUD customer needs in the Critical Parameter Management relational database. This data will soon be linked to the system, subsystem, subassembly, component, and manufacturing process requirements and critical parameters. It is essential for the product development teams to be able to trace their critical parameters back to the NUD VOC needs that they are fulfilling. These teams now know exactly what lies at the root of their requirements and the critical parameters they are developing and measuring to satisfy their customers.

Final Detailed Outputs

NUD VOC needs flow into quality function deployment and construction of the Houses of Quality.

The next process downstream of gathering and processing the VOC is quality function deployment. The results of this VOC process flow directly into the Houses of Quality, which are developed in the QFD process. Remember that we only take the NUD customer needs into QFD. If we try to take all customer needs through the Houses of Quality, it will be too much information and the team will bog down and waste its time. Only needs that are new, unique, and difficult are worth processing further during QFD.

VOC Gathering and Processing Checklist and Scorecard

This suggested format for a checklist can be used by a product planning team to help plan its work and aid in the detailed construction of a PERT or Gantt chart for project management of the tasks for the process of gathering and processing the VOC within the first phase of the CDOV or I^{2}DOV development processes. These items also serve as a guide to the key items for evaluation within a formal gate review.

Checklist for VOC Gathering and Processing

Actions:	Was this Action Deployed? YES or NO	Excellence Rating: 1 = Poor 10 = Excellent
1. Determine purpose of customer visit. Why are we going?	_____	_____
2. Identify objectives for customer visit. What do we want to learn?	_____	_____
3. Develop *customer selection matrix*. Who could best inform us?	_____	_____
4. Create *customer visit teams*. Who goes where?	_____	_____
5. Create *discussion/interview guide* based on objectives.	_____	_____
6. Interview and observe customers to gather their "voices" and "images" that represent their specific needs.	_____	_____
7. Develop "images of need," common image of customers' environment and needs. Create image KJ diagram.	_____	_____
8. Categorize image KJ needs into specific need statements that are new, unique, or difficult vs. routine.	_____	_____
9. Develop "significant voices of need," common voices of customers' stated needs.	_____	_____
10. Using a translation worksheet, link voices to related images, write, and identify unified customer needs.	_____	_____
11. Take the full set of unified customer needs (translated voices and images) and create a KJ diagram for these customer needs.	_____	_____
12. Quantify the importance of each NUD customer need by reviewing KJ diagram results with key customers.	_____	_____
13. Gather data to quantify the quality loss function for appropriate, critical customer needs.	_____	_____
14. Document NUD customer needs and importance value (1–10) in the Critical Parameter Management database.	_____	_____

Scorecard for VOC Gathering and Processing:

The definition of a good scorecard is that it should illustrate, in simple but quantitative terms, what gains were attained from the work. This suggested format for a scorecard can be used by an engineering team to help document the completeness of their work.

Number of Customers Interviewed: _____

Number of All Incoming Customer Needs from Interviews: _____

Number of NUD Customer Needs: _____

Table of NUD Customer Needs:

NUD Customer Need:	Customer Importance Rank:	Estimated Customer Satisfaction Metric (Target):	Estimated Quality Loss in Dollars (A_0) at Tolerance Limits:	Estimated Customer Tolerances ($+/- \Delta_0$):
1.	_____	_____	_____	_____
2.	_____	_____	_____	_____
3.	_____	_____	_____	_____

Let's review the final deliverables from VOC gathering and processing:

- *A real set of actual customer need statements* that forms the data that will be used to help define the system level and lower level technical requirements
- *A structured set of customer needs that are grouped by similarities (affinities)* that relate like needs into rational subgroups
- *A ranked hierarchy of customer needs* that begin to define criticality by drawing attention to the customer needs that are new, unique, and difficult
- *Quality loss functions based on data related to critical customer needs*

References

Clausing, D. (1994). *Total quality development.* New York: ASME Press.
Cohen, L. (1995). *Quality function deployment.* Reading, Mass.: Addison-Wesley Longman.
Garvin, D. A. (1988). *Managing Quality.* New York: The Free Press.

*We would like to thank Mr. Joseph Kasabula for his help in structuring this material.

Quality Function Deployment: The Houses of Quality

Where Am I in the Process? Where Do I Apply QFD Within the Roadmaps?

Linkage to the I²DOV Technology Development Roadmap:

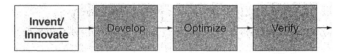

QFD is conducted within the Invent/Innovate phase during technology development. It is applied to translate, align, and quantitatively rank the relationships between the new, unique, and difficult voice of the customer needs and new, unique, and difficult technology requirements.

Linkage to the CDOV Product Development Roadmap:

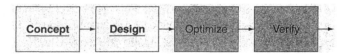

QFD is conducted within the **Concept** (at the system level) and **Design** (down at the subsystem, subassembly, component, and manufacturing process levels) phases during product development. Here we apply QFD to a specific product within the family plan from the strategic product portfolio. This application of QFD assures that the new, unique, and difficult needs are translated into the new, unique, and difficult requirements for the new product being developed.

What Am I Doing in the Process? What Does QFD Do at this Point in the Roadmaps?

The Purpose of QFD in the I²DOV Technology Development Roadmap:

Once again we need to show the strategic flow-down of how a business develops new product concepts:

Define business strategy: *Profit goals and requirements*

Identify markets and market segments: *Value generation and requirements*

Gather long-term voice of customer and voice of technology trends

Develop product line strategy: *Family plan requirements*

Develop/deploy technology strategy: *Technology and subsystem platform requirements*

Gather product-specific VOC and VOT: *New, unique, and difficult needs*

Conduct KJ analysis: *Structure and rank the NUD VOC needs*

Build system House of Quality: *Translate NUD VOC needs*

Document system requirements: *Create system requirements document*

Define the system functions: *Modeling*

Generate candidate system architectures: *Concept generation*

Select the superior system concept: *Pugh concept selection process*

In the Invent/Innovate Phase of the **I²DOV** process we **build the system House of Quality** to establish the requirements that are new, unique, or difficult. The technology development team then translates the new, unique, or difficult customer needs and technology requirements that flow down to subsystem, subassembly, components, and manufacturing processes Houses of Quality. Unlike the CDOV process, all Houses of Quality are derived in Phase 1 of the **I²DOV** process. The technology requirements that are defined in these Houses of Quality should be capable of fulfilling the product line strategy. The technology requirements are used to help define a set of adjustable or tunable functions. A unique set of architectural concepts known as *platforms* can then be developed to fulfill these functions over their range of projected product family plan applications. After careful evaluation, one superior platform architecture will be selected out of the numerous options to go through the rest of the **I²DOV** process. The value in such an approach is the investment of capital in a platform that can serve as a foundation for a family of products over a number of models and years. One investment—*many products!*

The Purpose of QFD in the CDOV Product Development Roadmap:

In the **CDOV** process we **build the system House of Quality** to translate the new, unique, and difficult customer and technology needs into new, unique, and difficult requirements that are related to the specific product being developed out of the product family plan. The product design team **builds the system House of Quality** to rank importance and set targets for these *product-*

specific requirements. The system (*product level*) requirements should be capable of fulfilling this specific product's contribution to the product line strategy. The system requirements are used to define a set of product functions. Numerous system architectural concepts can then be developed to fulfill the product functions. After careful evaluation, one superior system architecture (based on the underlying platform) is selected out of the numerous options to go through the rest of the **CDOV** process. This product is but one of several that will be sequentially developed to fulfill the product line strategy, based on the platform developed in the **I^{2}DOV** process. Subsequent Houses of Quality are developed for the subsystems, subassemblies, components, and manufacturing processes during Phase 2 (design) of the CDOV process. They define the NUD requirements at these respective levels within the system hierarchy.

What Output Do I Get at the Conclusion of this Phase of the Process? What Are Key Deliverables from QFD at this Point in the Roadmaps?

- **Translated VOC needs that are now documented as technology or system (*product level*) requirements** for what is considered new, unique, and difficult
- **NUD subsystem, subassembly, component, and manufacturing process requirements from their respective Houses of Quality**
- **Customer-based ranking of NUD needs** (down to all appropriate levels within the system)
- **Engineering team-based ranking of NUD requirements** (down to all appropriate levels within the system)
- **Critical Parameter Management data** that is documented in the CPM relational database:
  - Flow-down and linkage of all NUD requirements (down to all appropriate levels within the platform or system)
  - NUD requirement targets (with appropriate units of measure)
  - NUD requirement tolerances (estimated values from VOC data)
  - NUD requirement synergies and conflicts (from the "roof")
  - Customer and engineering benchmark values for NUD needs and requirements

QFD Process Flow Diagram (found on page 364)

This visual reference of the detailed steps of quality function deployment illustrates the flow of processes and responsibilities.

Verbal Descriptions for the Application of Each Block Diagram

This section details each step on the QFD process diagram:

- *Key detailed input(s)* from preceding tools and best practices
- *Specific steps and deliverables* from each step to the next within the process flow
- *Final detailed outputs* that flow into follow-on tools and best practices

INPUT . . .

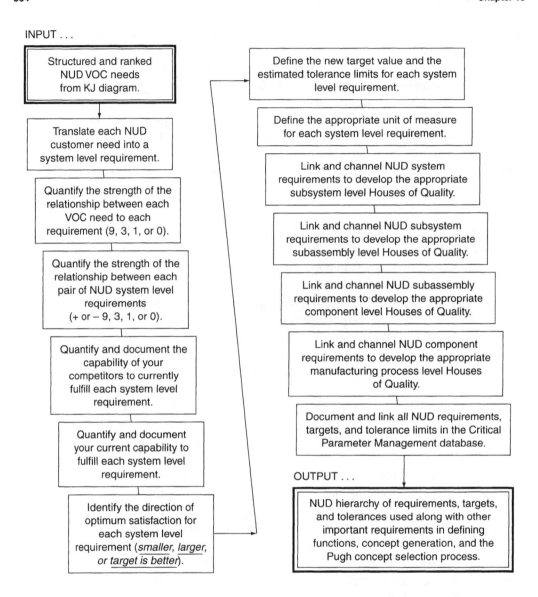

Key Detailed Inputs

Structured and ranked NUD VOC needs from KJ diagram

With the specific new, unique, and difficult voice of the customer needs defined, the team knows exactly *what* customers want. Now we need to translate the **whats** into **hows** that take the form of system level technical requirements.

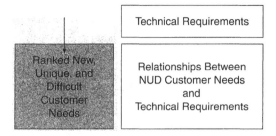

Specific Steps and Deliverables

> **Translate each NUD customer need into a system level requirement.**

We are ready to continue the Phase 1 process of developing customer needs into either technology or product-specific technical requirements. The term *translating* is commonly used when discussing how a need (what) is converted into a requirement (how). We must be careful when talking about requirements in terms of "how" a need is met. It is premature to be thinking about how a need is physically, architecturally, or conceptually met, but it is not too early to define how a need is measured in terms of fulfillment! So, don't think that we are talking about "how" in terms of design concepts and form/fit architectures; that comes later during the concept generation process. Requirements are specific terms that must be carefully structured to convey the right functional and feature performance information to the product design and technology development teams.

The biggest mistake one can make in stating a requirement is to structure the words in a manner that suggests or forces one specific solution or fulfilling concept. If your system level requirements state solutions, they are not correct. In fact, they are corrupting your ability to be creative. A properly worded requirement suggests a measurable function that can be fulfilled in many different ways. A good requirement will provoke many concepts or fulfilling architectures as candidate means to be measured as they fulfill the requirement! If your organization includes a "concept tyrant" who forces everyone to consider just one concept—his or hers—you need to declare a revolution toward a concept democracy. More on that issue later.

Here's an example of how a system level requirement can be poorly and properly worded:

- *A poor system requirement*: Car *acoustic damping materials* must be able to maintain internal acoustic noise level at or below 75dBA. (This is too **pre**scriptive; in fact, it is actually a component requirement!)
- *A proper system requirement*: Internal car *acoustic noise level* must not exceed 75dBA under any set of driving environmental conditions. (This is appropriately **de**scriptive, easy to see how to measure if you have met the requirement apart from any concept.)

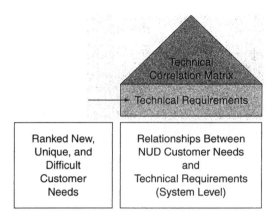

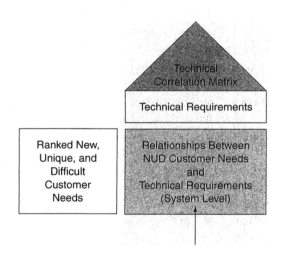

As one structures the rooms of the system level House of Quality (or any level of House of Quality for that matter), a matrix is used to relate incoming NUD VOC needs to the newly defined technical requirements.

The rows of the NUD VOC needs are run across the columns that contain the corresponding technical requirements. It is possible to have more than one technical requirement translated from a need. It is also possible that numerous technical requirements may have some form of fulfilling relationship to a NUD VOC need.

Once the needs are aligned with the requirements, numerical values are assigned to quantify the strength of the relationship that exists between the two statements:

9 = A strong fulfilling relationship between need and requirement

3 = A moderate fulfilling relationship between need and requirement

1 = A weak fulfilling relationship between need and requirement

0 = No relationship between need and requirement

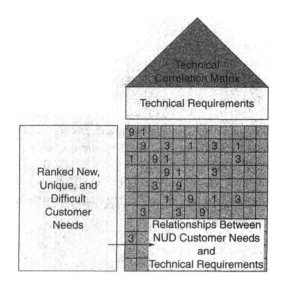

These values help establish which technical requirements are critical versus those that are of nominal importance. All NUD technical requirements must be fulfilled, but a few will emerge as the highest priority. Requirements that are particularly difficult to fulfill but are critical to customer satisfaction must be given the highest attention. They receive the resources necessary to assure they are fulfilled. Other requirements may need to yield.

> **Quantify the strength of the relationship between each pair of NUD system level requirements (+ or − 9, 3, 1 or 0).**

Every set of requirements developed to focus the application of tools, best practices, and an organization's material and human resources contains synergies and contradictions.

We turn our attention to the technical requirements themselves. You might compare these requirements to a family. Every family has its tensions, disagreements, and "black sheep." The roof of the House of Quality is used to numerically quantify both positive (+ 9, 3, and 1), supportive relationships (synergistic interactions) and negative (-9, -3, and -1) relationships (antisynergistic interactions). Since we are working with new, unique, and difficult (the black sheep) requirements, we need to carefully prescreen the requirements to assess where conflicts exist. Unintentional conflicts are frequently designed into new products. These functional conflicts are called *antisynergistic interactions* between the control factors within the design as well as antisynergistic interactions between control factors and noise factors. The problem stems from dysfunctional performance as two design elements attempt to provide their individual contribution to the design. Because of codependency or some form of physical linkage between the two engineering control factors, the measured function of the design is different than what is desired. This relationship is called *sensitivity* and can extend to include poorly behaved relationships between control and noise factors.

The reason we are concerned with these functional codependencies and sensitivities now is because this is where many of them are unintentionally created. When conflicting requirements are allowed to flow down into the concepts that comprise the detailed design elements, a design defect is born.

We need to address the sensitivities and conflicting relationships at the requirement level so that the designs developed to fulfill them are as synergistic as possible. As you may have heard, "a house divided against itself cannot stand."

The roof will identify the synergies and conflicts. Once they are brought to the team's attention, the hard work of creating a set of synergistic, harmonic requirements can then go forward.

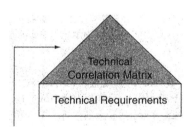

+/− 9 = A strong (+/−) relationship exists between the requirements
+/−3 = A moderate (+/−) relationship exists between the requirements
+/−1 = A weak (+/−) relationship exists between the requirements
0 = No relationship exists between the requirements

**Quantify and document the capability of your competitors
to currently fulfill each system level requirement.**

As part of the quantification of the importance or criticality of the NUD VOC needs, the team uses information gathered from the customers to assign ranking values (1 = highest fulfillment to 5 = lowest fulfillment) for how well your competitors are currently fulfilling each of the NUD VOC needs.

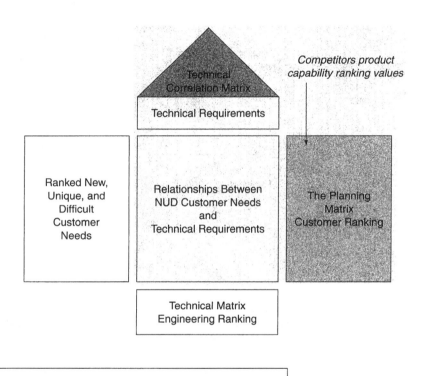

Quantify and document your current capability
to fulfill each system level requirement.

Right next to the competitive performance ranking for current fulfillment of NUD VOC needs, we document the ranking (1 = highest fulfillment to 5 = lowest fulfillment) of how well your current product is meeting those same needs (see the figure at top of page 370).

    With the ranking values in hand; one can perform a simple series of calculations to summarize the criticality value for each technical requirement (see the figure at the bottom of page 370).

    Here is an example:

    For each technical requirement, multiply the number in the relationship matrix by the ranking number for your current product from the planning matrix. Next add the products down for each technical requirement. For the first technical requirement the calculation is as follows: $[(9 \times 1) + (1 \times 3) + (3 \times 1)] = 15$. This value can then be compared to the other technical requirement values to give the team a sense of overall ranking of the importance of the technical

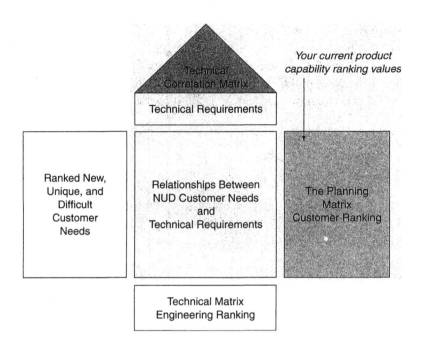

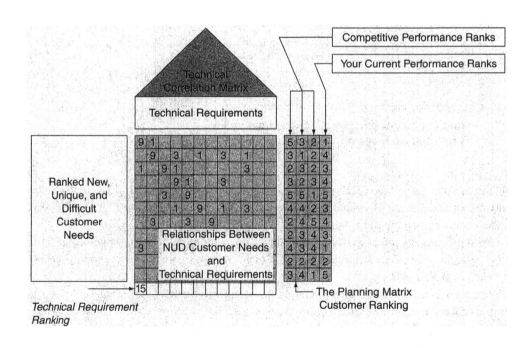

requirements. The strength of the ranks of the competition can be used to further motivate the team to consider what technical requirements are most critical.

> **Identify the direction of optimum satisfaction for each system level requirement (smaller, larger, or target is better).**

Just below each technical requirement a plus sign, a minus sign, or a capital T is placed to represent the direction of optimum fulfillment for each technical requirement.

$(+)$ = the **larger** the value of the measured variable for this requirement, the better the customer will be satisfied.

$(-)$ = the **smaller** the value of the measured variable for this requirement, the better the customer will be satisfied.

(T) = hitting the **targeted** value of the measured variable for this requirement best satisfies the customer.

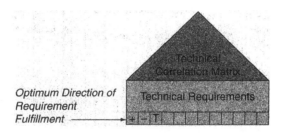

The preferred approach is to define as many target values as possible. Requirements with a target and tolerances contain the best information to verify that a design variable is meeting the requirement. "Smaller (or larger) is better" metrics are not as clear and descriptive but are used if a target metric is not possible.

> **Define the new target value and the estimated tolerance limits for each system level requirement.**

The fundamental elements used to define the new requirement targets and initial tolerances are as follows:

1. Current performance targets from your current product
2. Current performance targets from your competitor's product

3. Improvements desired by your customers (NUD needs)

4. New technologies that you have developed that can "delight" customers with unanticipated performance (that is, improved or advanced performance)

These four items can provide great insight as to what your requirement targets should be. They assure you reach for rational, balanced performance in a competitive context.

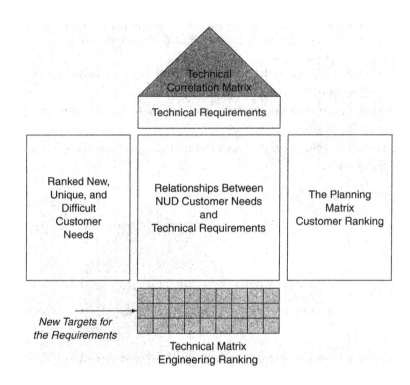

Define the appropriate unit of measure for each system level requirement.

Each technical requirement must have a well-defined unit of measure. It is preferable to express the units of measure as some form of continuous variable (a scalar or vector).

Link and channel NUD system requirements to develop the appropriate subsystem level Houses of Quality.

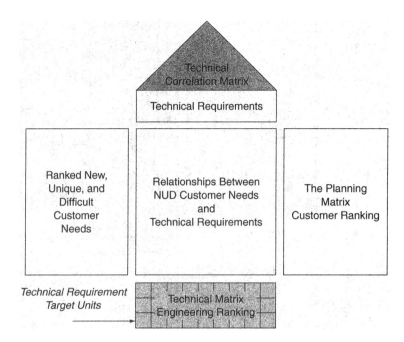

Once the system level House of Quality is defined, it is now possible to use the system level technical requirements as inputs to the various subsystem level Houses of Quality. Each subsystem level House of Quality illustrates and quantifies the relationships between the appropriate system level requirements and the subsystem requirements. NUD VOC needs that relate directly to a specific subsystem can also be used as inputs to the subsystem House of Quality.

The subsystem House of Quality is constructed in the same manner as a system House of Quality. The main difference is that the input is technical in nature and the challenge now is to translate system requirements into subsystem requirements, relationship and importance ranking values, as well as subsystem requirement targets. The planning matrix is typically not required down at the lower level Houses of Quality (see figure at the top of page 374).

> **Link and channel NUD Subsystem requirements to develop the appropriate subassembly level Houses of Quality.**

Once the subsystem level Houses of Quality are defined, it is now possible to use the subsystem level technical requirements as inputs to the various subassembly level Houses of Quality. Each subassembly level House of Quality illustrates and quantifies the relationships between the appropriate subsystem level requirements and the subassembly requirements. NUD VOC needs that relate directly to a specific subassembly can also be used as inputs to the subassembly House of

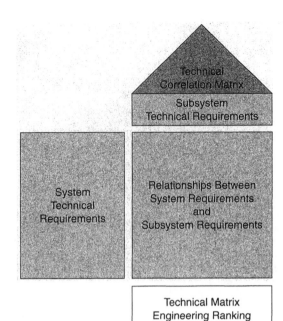

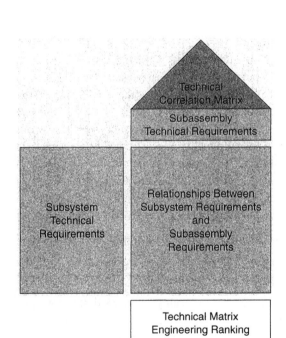

Quality. The subassembly House of Quality is constructed in the same manner as a subsystem House of Quality (see the figure at the bottom of page 374).

> **Link and channel NUD subassembly requirements to develop the appropriate component level Houses of Quality.**

Once the subassembly level Houses of Quality are defined, it is now possible to use the subassembly level technical requirements as inputs to the various component level Houses of Quality. Each component level House of Quality illustrates and quantifies the relationships between the appropriate subassembly level requirements and the component requirements. NUD VOC needs that relate directly to a specific component can also be used as inputs to the component House of Quality. The component House of Quality is constructed in the same manner as a subassembly House of Quality.

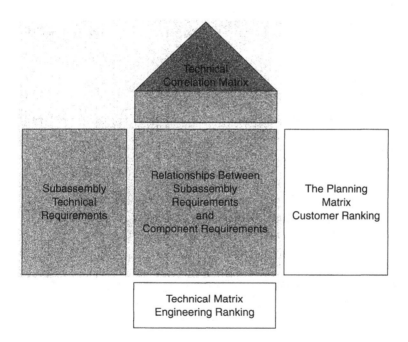

> **Link and channel NUD component requirements to develop the appropriate manufacturing process level Houses of Quality.**

Once the component level Houses of Quality are defined, it is now possible to use the component level technical requirements as inputs to the various manufacturing process level Houses of Quality. Each manufacturing process level House of Quality illustrates and quantifies the relationships between the appropriate component level requirements and the manufacturing process Requirements. The manufacturing process House of Quality is constructed in the same manner as a component House of Quality.

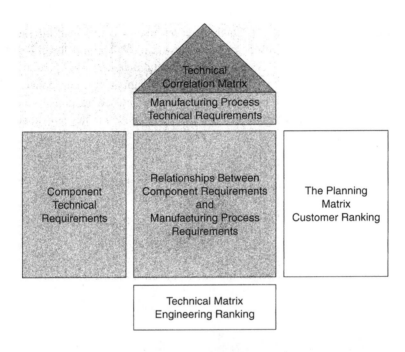

The Critical Parameter Management relational database can now have all the levels of requirements loaded into it. We now have the following structure defined, with the many-to-many relationships between the ranked requirements linked for traceability:

1. NUD VOC needs
2. System level requirements, targets, and tolerances
3. Subsystem level requirements, targets, and tolerances
4. Subassembly level requirements, targets, and tolerances
5. Component level requirements, targets, and tolerances
6. Manufacturing process level requirements, targets, and tolerances

The roofs and relationship matrices from the Houses tell us what requirements are to be linked in the CPM database.

The next set of entries for the Critical Parameter Management database will be the critical parameters in the form of CFRs, critical adjustment parameters, critical functional parameters, critical-to-function specifications and their hierarchy of linked transfer functions.

Final Detailed Outputs

> **NUD hierarchy of requirements, targets, and tolerances used along with other important requirements in defining functions, concept generation, and the Pugh concept selection process.**

The deliverables from the QFD process flow into the process of defining critical requirements, their fulfilling functions, generation of numerous concepts to fulfill the critical functions, and finally the Pugh concept selection process.

QFD Checklist and Scorecards

This suggested format for a checklist can be used by an engineering team to help plan its work and aid in the detailed construction of a PERT or Gantt chart for project management of the QFD process within the first two phases of the CDOV or the first phase of the I^{2}DOV development process. These items will also serve as a guide to the key items for evaluation within a formal gate review.

Checklist for QFD

Actions:	Was this Action Deployed? YES or NO	Excellence Rating: 1 = Poor 10 = Excellent
1. Structured and ranked NUD VOC needs from KJ diagram.	_____	_____
2. Translate each NUD customer need into a system level requirement.	_____	_____
3. Quantify the strength of the relationship between each VOC need to each requirement (9, 3, 1, or 0).	_____	_____
4. Quantify the strength of the relationship between each pair of NUD system level VOC requirements (+ or − 9, 3, 1 or 0) and technical requirements.	_____	_____
5. Quantify strength of relationships between the requirements.	_____	_____

(continued)

6. Quantify and document the capability of your competitors
   to currently fulfill each system level requirement.   _____   _____

7. Quantify and document your current capability to fulfill
   each system level requirement.   _____   _____

8. Identify the direction of optimum satisfaction for each
   system level requirement (smaller, larger, or target is better).   _____   _____

9. Define the new target value and the estimated tolerance
   limits for each system level requirement.   _____   _____

10. Define the appropriate unit of measure for each system level
    requirement.   _____   _____

11. Link and channel NUD system requirements to develop the
    appropriate subsystem level Houses of Quality.   _____   _____

12. Link and channel NUD subsystem requirements to develop
    the appropriate subassembly level Houses of Quality.   _____   _____

13. Link and channel NUD subassembly requirements to
    develop the appropriate component level Houses of Quality.   _____   _____

14. Link and channel NUD component requirements to develop
    the appropriate manufacturing process level Houses of Quality.   _____   _____

15. Document and link all NUD requirements, targets, and
    tolerance limits in the Critical Parameter Management database.   _____   _____

16. NUD hierarchy of requirements, targets, and tolerances used
    along with other important requirements in defining functions,
    concept generation, and the Pugh concept selection process.   _____   _____

Scorecard for QFD

The definition of a good QFD scorecard is that it should illustrate, in simple but quantitative terms, what gains were attained from the work.

This is a suggested format for a scorecard that can be used by a product planning team to help document the completeness of its work.

Total number of system level technical requirements: _____

Total number of subsystem level technical requirements: _____

Total number of subassembly level technical requirements: _____

Total number of component level technical requirements: _____

Total number of manufacturing process level technical requirements: _____

Number of NUD system level technical requirements: _____

Number of NUD subsystem level technical requirements: _____

Number of NUD subassembly level technical requirements: _____

Number of NUD component level technical requirements: _____

Number of NUD manufacturing process level technical requirements: _____

Table of NUD (Critical) System Level Requirements

System Requirement Name:	Target:	Estimated Tolerance: USL	Estimated Tolerance: LSL	Unit of Measure:
1.	_____	_____	_____	_____
2.	_____	_____	_____	_____
3.	_____	_____	_____	_____

Table of NUD (Critical) Subsystem Level Requirements:

Subsystem Requirement Name:	Target:	Estimated Tolerance: USL	Estimated Tolerance: LSL	Unit of Measure:
1.	_____	_____	_____	_____
2.	_____	_____	_____	_____
3.	_____	_____	_____	_____

Table of NUD (Critical) Subassembly Level Requirements:

Subassembly Requirement Name:	Target:	Estimated Tolerance: USL	Estimated Tolerance: LSL	Unit of Measure:
1.	_____	_____	_____	_____
2.	_____	_____	_____	_____
3.	_____	_____	_____	_____

Table of NUD (Critical) Component Level Requirements:

Component Requirement Name:	Target:	Estimated Tolerance: USL	Estimated Tolerance: LSL	Unit of Measure:
1.	_____	_____	_____	_____
2.	_____	_____	_____	_____
3.	_____	_____	_____	_____

Table of NUD (Critical) Manufacturing Process Level Requirements:

Manufacturing Process Requirement Name:	Target:	Estimated Tolerance: USL	Estimated Tolerance: LSL	Unit of Measure:
1.	_____	_____	_____	_____
2.	_____	_____	_____	_____
3.	_____	_____	_____	_____

We repeat the final deliverables from QFD:

- *Translated VOC needs that are now documented as technology or system (product level) requirements* for what is considered new, unique, and difficult.
- *NUD subsystem, subassembly, component, and manufacturing process requirements from their respective Houses of Quality*
- *Customer-based ranking of NUD needs* (down to all appropriate levels in the system)
- *Engineering team-based ranking of NUD requirements* (down to all appropriate levels in the system)
- *Critical Parameter Management data* that is documented in the CPM relational database:
  - Flow-down and linkage of all NUD requirements (down to all appropriate levels in the system)
  - NUD requirement targets (with appropriate units of measure)
  - NUD requirement tolerances (estimated)
  - NUD requirement synergies and conflicts (from the "roof")
  - Customer and engineering benchmark values for NUD needs and requirements

References

Clausing, D. (1994). *Total quality development.* New York: ASME Press.

Cohen, L. (1995). *Quality function deployment.* Reading, Mass.: Addison-Wesley Longman.

Concept Generation and Design for *x* Methods

Where Am I in the Process? Where Do I Apply Concept Generation and DFx Within the Roadmaps?

Linkage to the I²DOV Technology Development Roadmap:

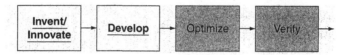

Concepts are generated within the Invent/Innovate phase while design for *x* (DFx) methods are applied within the Develop phase during technology development.

Linkage to the CDOV Product Development Roadmap:

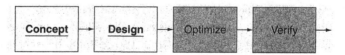

Concept generation is conducted within the Concept and Design phases during product development. Design for *x* methods are applied in the Design phase.

What Am I Doing in the Process? What Does Concept Generation and DFx Do at this Point in the Roadmaps?

The Purpose of Concept Generation and DFx in the I²DOV Technology Development Roadmap:

Once again we show the strategic flow-down of how a business develops new product concepts:

Define business strategy: *Profit goals and requirements*

Identify markets and market segments: *Value generation and requirements*

Gather long-term voice of customer and voice of technology trends

Develop product line strategy: *Family plan requirements*

Develop/deploy technology strategy: *Technology and subsystem platform requirements*

Gather product-specific VOC and VOT: *New, unique, and difficult needs*

Conduct KJ analysis: *Structure and rank the VOC*

Build system House of Quality: *Translate VOC*

Document system requirements

Define the system functions

Generate candidate system architectures

Select the superior system concept: *Pugh process*

In the Invent/Innovate phase of the I²DOV process we document system requirements, define the system functions, and generate candidate system architectures. The technology development team generates concepts that fulfill the new, unique, or difficult customer needs and technology requirements that flow-down to subsystem, subassembly, components, and manufacturing processes Houses of Quality. Unlike the CDOV process, all the technology concepts are derived in Phase 1 of the I²DOV process. The NUD technology requirements that are defined in the Houses of Quality, as well as other important requirements, are used to form the concept generation and selection criteria. The technology requirements are used to define a set of adjustable functions. A unique set of architectural concepts, or platforms, can then be developed to fulfill these functions over the projected range of product family application. After careful evaluation, one superior platform architecture is selected out of the numerous options to go through the rest of the I²DOV process. The value in such an approach is the investment of capital in a platform that can serve as the basis of your products over a number of models and years. One investment—many products!

DFx methods (general term for specific methods of design) are applied in Phase 2 of the I²DOV process. These methods help the team get an early start at making sure the platform is "designed for produceability" as opposed to just being "designed for patentability." Design for patentability is a common practice in R&D organizations today. The driving motive in technology building is to get a patent, not to help assure a family of products is developed and launched.

Technology is developed and patents attained, and then nothing comes of the designs. The usual problem is that the patented technologies are simply not produceable and thus never get commercialized.

The Purpose of Concept Generation and DFx in the CDOV Product Development Roadmap:

In the **CDOV** process we document system requirements, define the system functions, and generate candidate system architectures. Here the product development team generates system, subsystem, subassembly, component, and manufacturing process concepts. They are conceived to fulfill the new, unique, and difficult customer needs and technical requirements that are related to the specific product being developed out of the family plan.

The system requirements are used to define a set of product functions. Numerous system architectural concepts can then be developed to fulfill the product functions. After careful evaluation, one superior system architecture is selected out of the numerous options to go through the rest of the CDOV process.

DFx methods are applied in Phase 2 of the CDOV process. Design for *x* methods ensure that the design concepts are able to fulfill the following requirements:

Produceability (manufacturable)

Assembleability

Reliability

Serviceability (maintainable)

Environmentally friendly

Testability (measurable functions)

These attributes of a design are sometimes referred to as "design for the . . . ilities." It is extremely important to design in the attributes during the Concept and Design phases. If you try to add these late in the product development process, the costs will be exorbitant and you will definitely slip your schedule.

What Output Do I Get at the Conclusion of this Phase of the Process? What Are Key Deliverables from Concept Generation and DFx at this Point in the Roadmaps?

- *Documented candidate concepts at the system level* fulfilling what is considered new, unique, and difficult
- *Documented candidate concepts at the subsystem, subassembly, component, and manufacturing process levels* fulfilling what is considered new, unique, and difficult
- *Designs that are conceptually well rounded, both functionally and architecturally:*
  Documented concept feasibility for produceability (manufacturable)
  Documented concept feasibility for assembleability

Documented concept feasibility for reliability

Documented concept feasibility for serviceability (maintainable)

Documented concept feasibility for environmental friendliness

Documented concept feasibility for testability (measurable functions)

Concept Generation and DFx Process Flow Diagram

This visual reference of the detailed steps of concept generation and DFx illustrates the flow of processes and responsibilities at this stage.

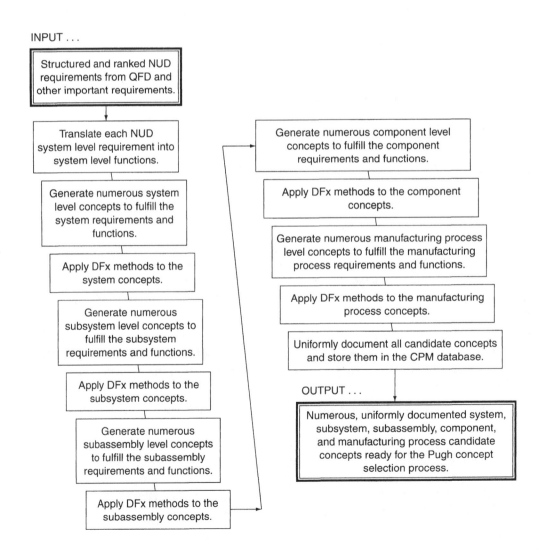

Verbal Descriptions for the Application of Each Block Diagram

In this section, we examine each step within the concept generation and DFx processes:

- *Key detailed input(s)* from preceding tools and best practices
- *Specific steps and deliverables* from each step to the next within the process flow
- *Final detailed outputs* that flow into follow-on tools and best practices

Key Detailed Inputs

> **Structured and ranked NUD requirements from QFD and other important requirements.**

With the specific, new, unique, and difficult requirements and other important technical and commercialization requirements defined, the team knows exactly how to measure the fulfillment of the requirements. Now we need to generate the form and fit architectures to fulfill the requirements.

1. Convert technical requirements to technical functions that must be engineered.
2. Construct input-output-constraint diagrams for each function down through the system hierarchy during concept generation.
3. Link all ideal functions/transfer functions using functional flow diagrams.

Specific Steps and Deliverables

> **Translate each NUD system level requirement into system level functions.**

System level technical requirements, when worded correctly, lead the team to define functions that must be engineered to fulfill the requirements. These functions are expressed in engineering units that can be measured with known transducers that can be calibrated. The result of this conversion process is the transformation of the system requirements into a complete set of measurable system level output functions (as shown in the figure on page 386).

The measurable system functions contain the subset of critical parameters we call *system level CFRs* (*c*ritical *f*unctional *r*esponses). These requirement-to-function relationships can be mapped for each level down within the system architecture (subsystems, subassemblies, components, and manufacturing processes).

> **Generate numerous system level concepts to fulfill the system requirements and functions.**

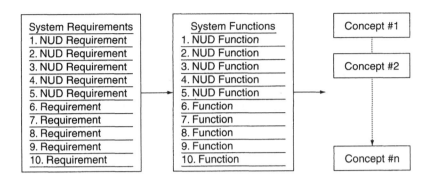

With the requirements and functions defined, it is now appropriate to develop a number of viable system level concepts that are high in feasibility and low in competitive vulnerability. It is not enough to develop one "strong" concept and run with it. The team must diligently construct (*architect* is perhaps a better term) a reasonable number of candidate system concepts. This allows the creative impulses across the system architecting team to flow freely. If a "concept tyrant" is active in your business, this social anomaly must be dealt with quickly and tactfully—but nonetheless this behavior cannot be allowed to continue. The more ideas from a variety of people the better. The idea is to get a lot of good elements that can be isolated from each individual concept. Later, hybrid concepts can be integrated from the suite of good ideas that everyone has contributed. Stifling creativity and input is a bad thing during concept generation, evaluation, and selection.

Concepts are generated in numerous ways:

1. Personal inspiration and vision (invention)
2. Benchmarking and reverse-engineering driving innovation based on form, fit, and function insights gained
3. Computer-aided invention and innovation using TRIZ (Theory of Inventive Problem Solving)
4. Brainstorming within a group of cross-functional team members
5. Brain-writing (ideas are written down and passed on for a team member to add to or change)
6. Patent studies
7. Literature searches
8. Consultant input and guidance (from professional and academic sources)

Concepts can be documented in several formats:

1. Math models (usually simple "first order" estimates)
2. Graphical models (typically hand sketches and simple CAD layouts)
3. Physical models (usually handmade "cardboard, duct tape, and hot-melt glue" models; stereolithography or other laser-sintered techniques)

The key rule to which all "concept architects" must adhere is that all candidate concepts should be developed and documented to approximately the same level of depth and scope. Problems arise during concept evaluation if one or more of the concepts are underdeveloped and underdocumented in comparison to the other concepts. All candidate concepts should be screened for consistency in this regard prior to conducting the Pugh concept selection process.

Remember that the requirements and functions should be fairly open-ended statements. They should be worded so that any number of concepts can be generated in an attempt to fulfill them. If your requirements are so specific that they essentially force you into a concept, they need to be reworded. If your customers have a specific architectural concept they are paying you to develop, then fine—go with the preselected concept. As a general rule, though, conceptual freedom is needed to generate alternatives.

> **Apply DFx methods to the system concepts.**

If you use the following design for *x* topics, the likelihood of your various concepts representing feasible architectural elements is quite high.

Produceability (Manufacturable)

Designs that are structured to be easy to manufacture during the conceptual stage of product development are much more likely to avoid a lot of redesign later when the system is being certified for production readiness. Probably the best way to ensure a concept is produceable is to have active involvement from the production and supply chain organizations during concept generation and selection. Any number of good texts on design for manufacture are available (see References at the end of this chapter). DFM is mainly concerned with materials and manufacturing processes. If your conceptual architectures suggest materials and processes that are expensive, fussy, or difficult to control, you may want to reconsider your form and fit alternatives as you attempt to fulfill the functions. Avoid exotic materials and processes unless you really need them.

Assembleability

Design for assembly is a close ally to design for manufacturability. This best practice concerns the ability of a design to be assembled either by hand or by some form of automation. Many design teams actually worry too much about assembleability to the detriment of functionality. It has been observed many times in industry that subsystem teams judge whether their design is ready for system integration by the simple check to see if all the parts actually go together. If they do, then the subsystem is ready for integration. This is only part of the story. If the nominal, "on-target" components assemble AND the measured functions of the subsystem are also on target with low sensitivity to noise factors—by design (robust design to be specific)—you have earned the right to integrate.

Assembleability issues typically arise in the following areas:

1. Ease of assembly motions
2. Number of assembly motions
3. Number of components
4. Complexity of component interfaces
5. Number of assembly interfaces
6. Type of fastening technology
7. Number of fasteners or fastening functions

Assembleability is very much related to nominal dimensions and their tolerances. Another enabling set of tools for assembleability analysis is analytical and empirical tolerance design.

Reliability

Design for reliability breaks down into two broad categories (Clausing, 1995).

1. Knowledge-based engineering is the science of designing things to bear applied loads or transfer heat—in a word, to transform, flow, and constrain the states of mass and energy according to accepted design engineering principles. If a concept is designed correctly in this context, it will contribute to reliable performance under nominal performance conditions. Properly sizing design elements for the performance required is a common notion in this context.

2. Variation control is the science of adequately accounting for the sources and effects of variation that can affect design functional performance. Variation in performance vs. requirements over some standard measure of time is measured or perceived as *reliability*. Designing to control the effects of variation breaks down into several areas:

a. **Robust designs** that are insensitive to both random and assignable cause noise factors. Here reliability is enhanced because the physics of noise is either slowed or minimized by exploiting the interactions that exist between control and noise factors.

b. **Compensation techniques** that actively correct a design's functional performance when disturbed by an assignable cause of variation. These are feedback or feed-forward control systems. Another form of compensation is redundancy. If one subsystem is critical to a product's performance, then two can be placed in the design so that if one fails to perform properly the second (or third or fourth as in the space shuttle's computers) takes over to maintain reliable performance.

c. **Noise suppression and elimination.** Here the noises themselves are acted upon to either diminish or remove their effect on functional performance. A noise factor is known to cause unacceptable functional performance; some means of isolating the functional elements of the design from the source of variation is used to ensure reliable performance (e.g., thermal insulation, seals on bearings, filtering). A source of variation is measured and acted upon to alter, cancel, or physically remove its effect.

Many tools and best practices in Design For Six Sigma work well to develop and grow reliability through variability control long before the reliability of the system can actually be measured! Here is a list of tools and best practices that help develop and grow reliability:

1. Reverse engineering highly reliable designs provide models of forms, fits, and functions that can be emulated.
2. Design for manufacture and assembly emphasizes the need for fewer parts, interfaces, and overly sensitive materials and processes.
3. Analytical and empirically derived ideal/function models provide $Y=f(x)$ relationships that can be measured to identify "impending failures." An example is to not count jams in a copier but to measure paper position (X translation, Y translation, and rotation about Z, the axis down through the center of the flat sheet of paper) as it flows through the machine. As the paper moves off target, you can "*see*" a jam coming. Don't measure reliability to develop it!
4. Failure modes and effects analysis identifies sources of failure and establishes a corrective action to reduce or eliminate the cause of failure.
5. Robust design desensitizes a design's critical functional responses to sources of variation by exploiting interactivity between control and noise factors.
6. Response surface methods are excellent for finding global optimum set points for placing CFR mean and standard deviation performance on a desired target for active control systems or critical adjustment parameters for assembly and service personnel.
7. Analytical tolerance design is useful in balancing assembly or functional tolerances to limit variation. It is most commonly applied to component tolerances and the constraint of part-to-part variation.
8. Empirical tolerance design is helpful in evaluating functional sensitivities inside complex, integrated systems. It is able to evaluate functional sensitivities due to all three major forms of noise: external, unit-to-unit, and deterioration sources of variation.
9. Capability studies are quite useful as prereliability evaluations that can act as leading indicators of functional performance at the subsystem and subassembly level long before they are integrated into a functioning system.
10. Multi-vari studies help uncover serious sources of assignable cause variation so that these noise factors can be dealt with properly.
11. Statistical process control also helps detect assignable causes of variation in design functions and manufacturing processes.

As you can see, there is no shortage of tools that can be proactively applied to develop and maintain reliability. DFSS concerns itself mainly with tools and best practices that can develop reliability before it can be measured.

Serviceability (Maintainable)

Designing concepts that are easy to service and maintain is extremely important up-front in product development. This topic has strong ties to design for assembly. Now the issues extend out into the service and maintenance process. It is a good idea to have the people who will maintain and service the new product participate in concept generation and selection.

A few key issues are accessibility, low fastener count, low tool count, one-step service functions, and predictable maintenance schedules. Extending the rate of deterioration is always helpful. Designing a component to fail predictably as opposed to randomly is also useful. If the service team knows when things need maintenance or replacement, they can plan for it. Unplanned service is costly and annoying to all concerned.

Environmentally Friendly

A modern concern is the design of environmentally friendly systems, subsystems, subassemblies, components, and manufacturing processes. Design elements need to be reusable as much as possible or be recyclable. There is decreasing tolerance for placing large amounts of industrial waste in a landfill.

Testability (Measurable Functions)

Probably the most significant attribute that one can design into a concept is active measurement of its ability to perform its ideal/transfer function. This means a viable data acquisition system must be included as an integral part of the concept design. It is best to measure continuous variables that are engineering scalars or vectors. Each concept should have these clearly identified along with the transducers and measurement systems that underwrite the integrity of the metrics.

These DFx items equally apply to all the sublevels of the product being developed and designed.

> **Generate numerous subsystem level concepts to fulfill the subsystem requirements and functions.**

One can take two approaches to subsystem concept generation:

1. After the system concepts are generated, they are evaluated and one superior system concept selected. Then we can move on to let the superior system architecture drive the subsystem concept generation process. Numerous subsystem concepts can be generated to fulfill the subsystem requirements.

2. After the system concepts are generated but before they are evaluated, some level of subsystem concept generation may be required to help define certain key elements of the system. This is not unusual and is actually done quite frequently. The system concepts may be fairly dependent on the architectural attributes of several key subsystems.

Here at the subsystem level we have the subsystem requirements driving the definition of the subsystem functions.

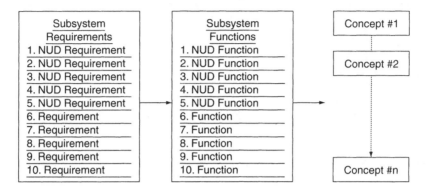

The measurable subsystem functions contain the subset of critical parameters we call *subsystem level CFRs*.

Many subsystem concepts can now be generated to fulfill these requirements and functions. They are generated to the same standards as were presented previously in the system level concept generation process.

The biggest difference now that we are down at the subsystem level of concept generation is that there can be many different subsystems. Numerous concepts should be developed for each distinct subsystem. This may seem time consuming but try going forward with a few bad subsystem concepts and you will find out what real-time consumption is all about when you have to fix bad concepts just prior to production launch! The few extra hours or days invested here pale in comparison.

> **Apply DFx methods to the subsystem concepts.**

The same issues apply to subsystem design for *x* issues as were reviewed in the system DFx section previously. The big difference is the level of detail now increases because we are getting down into a fair amount of internal subsystem detail. The breadth of complexity will be less because we are now typically looking at just one or two subsystems per team.

> **Generate numerous subassembly level concepts to fulfill the subassembly requirements and functions.**

Subassembly requirements are now driving the definition of the subassembly functions.

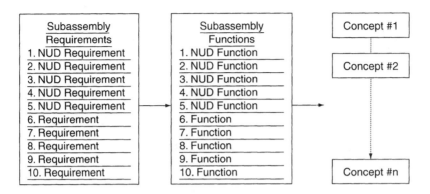

The measurable subassembly functions contain the subset of critical parameters we call *subassembly level CFRs* (*critical functional responses*).

Numerous subassembly concepts can now be generated to fulfill these requirements and functions. They are generated to the same standards as were presented in the system and subsystem level concept generation processes.

> **Apply DFx methods to the subassembly concepts.**

Design for *x* methods are applied just as they were for the subsystems and the system.

> **Generate numerous component level concepts to fulfill the component requirements and functions.**

Component requirements are now driving the definition of the component specifications.

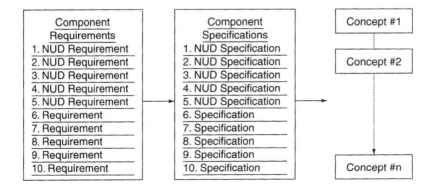

The measurable component specifications contain the subset of critical parameters we call *component level critical-to-function specifications* (CTF specs). The component requirements are what the component must provide to enable a subassembly or subsystem to assemble and function properly. A component specification is actually measured on a component to verify that it is fulfilling the requirement. An example is a shaft that has a required diameter of 2 inches +/− 0.0010 inches and a measured CTF spec of 2.0006 inches. Component specifications are used to measure dimensions, surface finishes, and bulk properties on or within components.

Numerous component concepts can now be generated to fulfill these requirements and functions. They are generated to the same standards as were presented in the previous levels of concept generation processes.

Apply DFx methods to the component concepts.

Design for *x* methods are applied just as they were for the higher levels of the system.

Generate numerous manufacturing process level concepts to fulfill the manufacturing process requirements and functions.

Manufacturing process requirements are now driving the definition of the manufacturing process specifications.

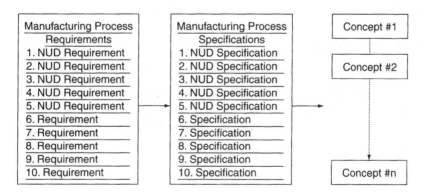

Manufacturing Process Requirements	Manufacturing Process Specifications	
1. NUD Requirement	1. NUD Specification	Concept #1
2. NUD Requirement	2. NUD Specification	Concept #2
3. NUD Requirement	3. NUD Specification	
4. NUD Requirement	4. NUD Specification	
5. NUD Requirement	5. NUD Specification	
6. Requirement	6. Specification	
7. Requirement	7. Specification	
8. Requirement	8. Specification	
9. Requirement	9. Specification	
10. Requirement	10. Specification	Concept #n

The measurable manufacturing process specifications contain the subset of critical parameters we call manufacturing process level CTF specs. The manufacturing process requirements are what the manufacturing process must provide to enable a component to be formed properly. If a manufacturing process set point is off target, the component dimensions, surface finishes, or bulk properties will be off-target, resulting in assemble and functional problems. A manufacturing process specification is measured on a manufacturing process to verify that it is fulfilling the requirement. An example is an injection molding machine that has a required cooling water inlet temperature of 60° F +/− 4° F and a measured CTF spec of 62° F. Manufacturing process specifications are used to measure process rates, temperatures, forces, pressures—any transformation, flow, or state of energy or mass that controls how components are produced.

> **Apply DFx methods to the manufacturing process concepts.**

Design for *x* methods have their role in the development of manufacturing processes. Typically the technology team is developing a new manufacturing process, while the design teams are simply selecting the best existing manufacturing process for their application to making the components they need.

> **Uniformly document all candidate concepts
> and store them in the CPM database.**

As we bring this process to a close, it is worth repeating the importance of developing and documenting any set of concepts, at any level within the system hierarchy, to a relatively consistent level. This helps assure that as the concepts are taken through the Pugh concept selection process, they all get a fair evaluation against the best-in-class benchmark datum.

The Six Sigma Cockpit database has the ability to store many kinds of documents. This is an excellent place to store the concept files for future reference.

Final Detailed Outputs

> **Numerous, uniformly documented system, subsystem,
> subassembly, component, and manufacturing process candidate
> concepts ready for the Pugh concept selection process.**

At the end of each concept generation process for each level of the system hierarchy the teams will have a well-developed set of candidate concepts that are ready to take into the Pugh concept selection process.

Concept Generation and DFx Checklist and Scorecards

This suggested format for a checklist can be used by an engineering team to help plan its work and aid in the detailed construction of a PERT or Gantt chart for project management of the modeling process within the first two phases of the CDOV or the first phase of the I²DOV development process. These items will also serve as a guide to the key items for evaluation within a formal gate review.

Checklist for Concept Generation and DFx

Actions:	Was this Action Deployed? YES or NO	Excellence Rating: 1 = Poor 10 = Excellent
1. Translate each NUD system level requirement into system level functions.	_____	_____
2. Generate numerous system level concepts to fulfill the system requirements and functions.	_____	_____
3. Apply DFx methods to the system concepts.	_____	_____
4. Generate numerous subsystem level concepts to fulfill the subsystem requirements and functions.	_____	_____
5. Apply DFx methods to the subsystem concepts.	_____	_____
6. Generate numerous subassembly level concepts to fulfill the subassembly requirements and functions.	_____	_____
7. Apply DFx methods to the subassembly concepts.	_____	_____
8. Generate numerous component level concepts to fulfill the component requirements and functions.	_____	_____
9. Apply DFx methods to the component concepts.	_____	_____
10. Generate numerous manufacturing process level concepts to fulfill the manufacturing process requirements and functions.	_____	_____
11. Apply DFx methods to the manufacturing process concepts.	_____	_____
12. Uniformly document all candidate concepts and store them in the CPM database.	_____	_____

Scorecard for Concept Generation and DFx

The definition of a good concept scorecard is that it should illustrate, in simple but quantitative terms, what gains were attained from the work. This suggested format for a scorecard can be used by a product planning team to help document the completeness of its work.

Total number of system level concepts developed: _____

Total number of subsystem level concepts developed: _____

Total number of subassembly level concepts developed: _____

Total number of component level concepts developed: _____

Total number of manufacturing process level concepts developed: _____

This forces teams to document numerous concepts and avoid just developing one concept that may be inferior or competitively vulnerable to a best-in-class benchmark or other concepts that were not developed.

Documented Concept Elements

System level concept documentation:

Math models _____

Graphical models _____

Physical models _____

Subsystem level concept documentation:

Math models _____

Graphical models _____

Physical models _____

Subassembly level concept documentation:

Math models _____

Graphical models _____

Physical models _____

Component level concept documentation:

Math models _____

Graphical models _____

Physical models _____

Manufacturing process level concept documentation

Math models _____

Graphical models _____

Physical models _____

We repeat the final deliverables from concept generation and DFx:

- Documented candidate concepts at the system level fulfilling what is considered new, unique, and difficult as well as other important requirements.
- Documented candidate concepts at the subsystem, subassembly, component, and manufacturing process levels fulfilling what is considered new, unique, and difficult as well as other important requirements.

- Designs that are conceptually well rounded, both functionally and architecturally:
  Documented concept feasibility for produceability (manufacturable)
  Documented concept feasibility for assembleability
  Documented concept feasibility for reliability
  Documented concept feasibility for serviceability (maintainable)
  Documented concept feasibility for environmental friendliness
  Documented concept feasibility for testability (measurable functions)

References

Boothroyd, G., Dewhurst, P., & Knight, W. (1994). *Product design for manufacture and assembly.* New York: Marcel Dekker.

Bralla, J. G. (1986). *Handbook of product design for manufacturing.* New York: McGraw-Hill.

Clausing, D. (1995). *Total quality development.* New York: ASME Press.

Otto, K., & Wood, K. (2001). *Product design.* Upper Saddle River, NJ: Prentice-Hall.

Rao, S. S. (1992). *Reliability-based design.* New York: McGraw-Hill.

Trucks, H. E. (1976). *Designing for economical production.* Dearborn, Mich.: SME Press.

Ullman, D. G. (1997). *The mechanical design process.* New York: McGraw-Hill.

Ulrich, K. T., & Eppinger, S. D. (1995). *Product design and development.* New York: McGraw-Hill.

The Pugh Concept Evaluation and Selection Process

Where Am I in the Process? Where Do I Apply the Pugh Concept Selection Process Within the Roadmaps?

Linkage to the I²DOV Technology Development Roadmap:

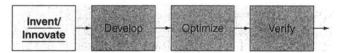

The Pugh concept selection process is conducted within the Invent/Innovate phase during technology development. It is applied to evaluate and select superior platform and subsystem concepts. Superior concepts are those that score the highest in feasibility in satisfying the voice of the customer and the corresponding technology requirements for the new technology platforms and subsystems.

Linkage to the CDOV Product Development Roadmap:

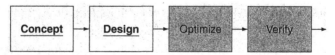

The Pugh concept selection process is conducted within the Concept and Design phases during product development. System level concept selection is conducted in the Concept phase. All lower level concepts are selected during the Design phase. Here we apply the Pugh concept selection process to a specific product within the family plan from the product portfolio. This application of the Pugh process ensures that superior concepts for the system, subsystems, subassemblies, components, and manufacturing processes are evaluated and selected to fulfill the hierarchy of technical requirements for the new product being developed.

What Am I Doing in the Process? What Does the Pugh Concept Selection Process Do at this Point in the Roadmaps?

The Purpose of the Pugh Concept Selection Process in the I²DOV Technology Development Roadmap:

Once again we show the strategic flow-down of how a business develops new product concepts:

Define business strategy: *Profit goals and requirements*

Identify markets and market segments: *Value generation and requirements*

Gather long-term voice of customer and voice of technology trends

Develop product line strategy: *Family plan requirements*

Develop/deploy technology strategy: *Technology and subsystem platform requirements*

Gather product-specific VOC and VOT: *New, unique, and difficult needs*

Conduct KJ analysis: *Structure and rank the VOC*

Build system House of Quality: *Translate VOC*

Document system requirements

Define the system functions

Generate candidate system architectures

Select the superior system concept: *Pugh process*

In the Invent/Innovate phase of the I²DOV process we document the platform and/or subsystem requirements, define the platform and/or subsystem functions, generate candidate platform and/or subsystem architectures, and select the superior platform and/or subsystem concepts.

The hierarchy of technology requirements is used to carefully evaluate and select superior concepts for and within the platform architecture. The evaluation process is conducted against a best-in-class benchmark datum concept. The superior concepts will go through the rest of the I²DOV process.

The Purpose of the Pugh Concept Selection Process in the CDOV Product Development Roadmap:

In the CDOV process we document the system, subsystem, subassembly, component, and manufacturing process requirements; define the system, subsystem, subassembly, component, and manufacturing process functions; generate candidate system, subsystem, subassembly, component, and manufacturing process architectures (concepts); and select the superior system, subsystem, subassembly, component, and manufacturing process concepts.

The hierarchy of system requirements are used to carefully evaluate and select superior concepts for and within the system architecture. The evaluation process is conducted against best-in-class benchmark datum concepts. The superior concepts will go through the rest of the CDOV process.

What Output Do I Get at the Conclusion of this Phase of the Process? What Are Key Deliverables from the Pugh Concept Selection Process at this Point in the Roadmaps?

- **Superior platform or system level concept** to take on into technology development or product design:
  Superior system level concept
  Superior subsystem level concepts
  Superior subassembly concepts
  Superior component concepts
  Superior manufacturing process concepts
- **Superiority is documented in terms of concepts that are:**
  Low in competitive vulnerability.
  High in functional feasibility and capability to fulfill requirements.
  Well documented in their ability to have their functional performance measured in terms of CFRs.
  Low in inherent sensitivity to noise factors.
  Underwritten by three basic models generated to a common level of completeness for each candidate concept:
  > Analytical (hand calculations or CAE model)
  > Graphical (hand sketch or CAD model)
  > Physical (early hardware prototype)

The Pugh Concept Selection Process Flow Diagram (found on page 402)

This visual reference of the detailed steps of the Pugh concept selection process illustrates the flow of tasks and responsibilities.

Verbal Descriptions for the Application of Each Block Diagram

This section *explains the specific details of* the Pugh concept selection process:

- *Key detailed input(s)* from preceding tools and best practices
- *Specific steps and deliverables* from each step to the next within the process flow
- *Final detailed outputs* that flow into follow-on tools and best practices

Key Detailed Input:

> **Requirements, functions, and numerous concepts from concept generation.**

INPUT . . .

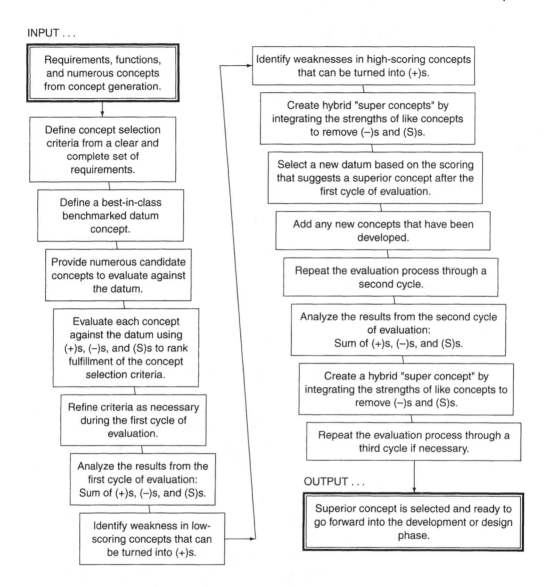

Requirements, functions, and numerous concepts from concept generation.

Define concept selection criteria from a clear and complete set of requirements.

Define a best-in-class benchmarked datum concept.

Provide numerous candidate concepts to evaluate against the datum.

Evaluate each concept against the datum using (+)s, (−)s, and (S)s to rank fulfillment of the concept selection criteria.

Refine criteria as necessary during the first cycle of evaluation.

Analyze the results from the first cycle of evaluation: Sum of (+)s, (−)s, and (S)s.

Identify weakness in low-scoring concepts that can be turned into (+)s.

Identify weaknesses in high-scoring concepts that can be turned into (+)s.

Create hybrid "super concepts" by integrating the strengths of like concepts to remove (−)s and (S)s.

Select a new datum based on the scoring that suggests a superior concept after the first cycle of evaluation.

Add any new concepts that have been developed.

Repeat the evaluation process through a second cycle.

Analyze the results from the second cycle of evaluation: Sum of (+)s, (−)s, and (S)s.

Create a hybrid "super concept" by integrating the strengths of like concepts to remove (−)s and (S)s.

Repeat the evaluation process through a third cycle if necessary.

OUTPUT . . .

Superior concept is selected and ready to go forward into the development or design phase.

With the specific concepts defined, the product planning team uses the requirements to define the criteria to be used to evaluate and select superior concepts. Concept selection is made by evaluating the numerous candidate concepts against a best-in-class benchmark datum.

Specific Steps and Deliverables

> **Define concept selection criteria from a clear and complete set of requirements.**

The numerous concepts must be evaluated to select the one concept that best fulfills the rational subset of the complete set of requirements. The requirements at any level within the system contain new, unique, and difficult requirements as well as those that are of nominal importance. It is common to select an essential group of concept evaluation criteria from these requirements. It is unrealistic to use every single requirement to evaluate the feasibility of the candidate concepts. Every single requirement will be used to verify the product as it evolves through the phases of product development. One must be careful to recognize that concept selection is not equal to product design, optimization, and capability certification. We are testing feasibility, not optimality! The key issues to evaluate will certainly be embodied within the NUD requirements plus a few select requirements that help prove the feasibility of the concepts in terms of competitive vulnerability.

> **Define a best-in-class benchmarked datum concept.**

To start the concept selection process, we refer back to the benchmarking information from the QFD process. From customer interviewing and technical benchmarking, the teams should have a good idea about what competitive or internal product or design element is the best in class. The best-in-class design is called the *datum concept* that is to be equaled or surpassed. This can be a product, subsystem, subassembly, component, or manufacturing process.

> **Provide numerous candidate concepts to evaluate against the datum.**

The set of candidate concepts, generated in the previous process, can be brought into the Pugh concept selection process at this time. Any new concepts can also be added, providing they are properly documented by a reasonable level of modeling:

- Analytical (hand calculations or CAE model)
- Graphical (hand sketch or CAD model)
- Physical (early hardware prototype)

It is good to remember these are just candidates at this point. None of these concepts is likely to be the winner yet!

Criteria:	Concept A	Concept B	Concept C	Concept n	DATUM
Criteria 1					
Criteria 2					
Criteria 3					
Criteria 4					
Criteria 5					
Criteria 6					
Criteria 7					
Criteria 8					
Criteria 9					
Criteria 10					
Criteria n					
Sum of (+)					
Sum of (−)					
Sum of (S)					

Figure 17–1   The Pugh Concept Evaluation Matrix

A matrix (Figure 17–1) is used to align each concept to the criteria.
The datum is not ranked. It serves as the concept against which all candidate concepts are evaluated.

> **Evaluate each concept against the datum using (+)s, (−)s,
> and (S)s to rank fulfillment of the concept selection criteria.**

Each candidate concept is now evaluated against the datum for each concept evaluation criteria. The team uses a simple evaluation technique consisting of a (+) for a concept that is better at fulfilling the criteria than the datum. A (−) is assigned to a concept that is worse than the datum, and (S) denotes a concept that is essentially the same as the datum.

Criteria:	Concept A	Concept B	Concept C	Concept n	DATUM
Criteria 1	+	+	−	S	
Criteria 2	+	−	+	S	
Criteria 3	−	+	S	+	
Criteria 4	+	S	−	+	
Criteria 5	+	−	−	−	
Criteria 6	S	+	S	+	
Criteria 7	+	−	S	+	
Criteria 8	−	−	S	+	
Criteria 9	−	S	−	+	
Criteria 10	S	+	+	−	
Criteria n	+	−	+	+	
Sum of (+)	6	4	3	7	
Sum of (−)	3	5	4	2	
Sum of (S)	2	2	4	2	

Refine criteria as necessary during the first cycle of evaluation.

It is quite common for the concept selection criteria to be "downsized" when two or more criteria are found to be repetitive or redundant. If criteria are found to be unclear, they should be modified. Criteria must be specific and clear. Cost as a general classification is a poor criterion. The elements that drive cost are good criteria. The same goes for reliability. Lower part count is a representative criterion that helps identify a concept that is feasible for improving reliability. Try to word criteria so that you can answer the question, "Why, specifically, is this design more feasible than the datum?"

> **Analyze the results from the first cycle of evaluation: sum of (+)s, (−)s, and (S)s.**

The number of (+)s, (−)s, and (S)s can be summed for each concept. This indicates which concepts are tending toward being superior to the datum. There may actually be one or more concepts that are clearly superior to the datum. In any case, the team needs to go to work on making strong concepts stronger. There may even be opportunities to convert weaker concepts into strong concepts. The method of creating hybrid concepts is used to improve Round 1 candidate concepts.

> **Identify weakness in low-scoring concepts that can be turned into (+)s.**

The team first turns its attention to the weaker concepts to see if their weaknesses can be corrected by altering their architectures, perhaps by adding strengths borrowed from other concepts.

> **Identify weaknesses in high-scoring concepts that can be turned into (+)s.**

The strong concepts also can benefit from improvements in their architectures. Each strong concept should be scrutinized so that its (−)s can be turned into (+)s. It is common to search through the other concepts to see if one of their (+) attributes can be brought into the strong concept's structure.

> **Create hybrid "super concepts" by integrating the strengths of like concepts to remove (−)s and (S)s.**

From time to time a group of two or more concepts that are similar can be integrated with the resulting hybrid concept becoming a "super concept." A super concept may rise to the top and become a good replacement for the current datum.

> **Select a new datum based on the scoring that suggests a superior concept after the first cycle of evaluation.**

The team reviews the results of the first round of concept evaluations. If a clearly superior concept emerges, then the datum is replaced with this superior candidate concept.

> **Add any new concepts that have been developed.**

It is perfectly legitimate to generate new concepts and introduce them to the Pugh process at this time. After the first round of concept evaluation, many new and stimulating ideas tend to be provoked. These should be developed and documented to an equal level with the concepts that are about to go through Round 2.

> **Repeat the evaluation process through a second cycle.**

The set of new and revised candidate concepts are now taken through a new round of evaluation against the new datum (or the original datum if it was not surpassed). Additional adjustments to the criteria should be made prior to Round 2.

> **Analyze the results from the second cycle of evaluation: Sum of (+)s, (−)s, and (S)s.**

After a second round of concept evaluation, the (+)s, (−)s, and (S)s are summed to reveal which concepts are emerging as superior to the datum.

> **Create a hybrid "super concept" by integrating the strengths of like concepts to remove (−)s and (S)s.**

Another round of concept hybridization can be conducted if necessary to build one or more concepts that can surpass the datum. It is common to arrive at a superior concept after the second round of the Pugh process.

Repeat the evaluation process through a third cycle if necessary.

A third round of concept amendment and evaluation can be conducted if necessary.

Final Detailed Outputs

Superior concept is selected and ready to go forward into the development or design phase.

Once a superior concept is selected, the team must begin to mature the concept into a viable baseline design that is now worthy and able to be fully modeled.

Pugh Concept Selection Process Checklist and Scorecard

This suggested format for a checklist can be used by an engineering team to help plan its work and aid in the detailed construction of a PERT or Gantt chart for project management of the Pugh concept selection process within the Optimize phase of the CDOV or I^2DOV development process. These items will also serve as a guide to the key items for evaluation within a formal gate review.

Checklist for the Pugh Concept Selection Process

Actions:	Was this Action Deployed? YES or NO	Excellence Rating: 1 = Poor 10 = Excellent
1. Define concept selection criteria from a clear and complete set of requirements.	_____	_____
2. Define a best-in-class benchmarked datum concept.	_____	_____
3. Provide numerous candidate concepts to evaluate against the datum.	_____	_____
4. Evaluate each concept against the datum using $(+)$s, $(-)$s, and (S)s to rank fulfillment of the concept selection criteria.	_____	_____
5. Refine criteria as necessary during the first cycle of evaluation.	_____	_____
6. Analyze the results from the first cycle of evaluation: Sum of $(+)$s, $(-)$s, and (S)s.	_____	_____

 7. Identify weakness in low-scoring concepts that
    can be turned into (+)s.   _____   _____

 8. Identify weaknesses in high-scoring concepts that
    can be turned into (+)s.   _____   _____

 9. Create hybrid "super concepts" by integrating the strengths
    of like concepts to remove (−)s and (S)s.   _____   _____

10. Select a new datum based on the scoring that suggests a
    superior concept after the first cycle of evaluation.   _____   _____

11. Add any new concepts that have been developed.   _____   _____

12. Repeat the evaluation process through a second cycle.   _____   _____

13. Analyze the results from the second cycle of evaluation:
    Sum of (+)s, (−)s, and (S)s.   _____   _____

14. Create a hybrid "super concept" by integrating
    the strengths of like concepts to remove (−)s and (S)s.   _____   _____

15. Repeat the evaluation process through a third cycle if necessary.   _____   _____

Scorecard for Pugh Concept Selection Process

The definition of a good concept selection scorecard is that it should illustrate, in simple but quantitative terms, what gains were attained from the work. This suggested format for a scorecard can be used by an engineering team to help document the completeness of its work.

Number of candidate concepts evaluated: _____

Number of selection criteria used: _____

Number of (+)s for superior concept: _____

Number of (−)s for superior concept: _____

Number of (S)s for superior concept: _____

We repeat the final deliverables from the Pugh concept selection process:

- *Superior platform or system level concept* to take on into technology development or
  product design
  Superior system level concept
  Superior subsystem level concepts
  Superior subassembly concepts
  Superior component concepts
  Superior manufacturing process concepts
- *Superiority is documented in terms of concepts that are:*
  Low in competitive vulnerability.
  High in functional feasibility and capability to fulfill requirements.
  Well documented in their ability to have their functional performance measured in
      terms of CFRs.

Low in inherent sensitivity to noise factors.
Underwritten by three basic models generated to a common level of completeness for
each candidate concept:
Analytical (hand calculations or CAE model)
Graphical (hand sketch or CAD model)
Physical (early hardware prototype)

References

Clausing, D. (1994). *Total quality development.* New York: ASME Press.

Pugh, Stuart. (1995). *Total design.* Reading, Mass: Addison-Wesley.

Pugh, Stuart. (1996). *Creating innovative products using total design.* Reading, Mass: Addison-Wesley.

Modeling: Ideal/Transfer Functions, Robustness Additive Models, and the Variance Model

Where Am I in the Process? Where Do I Apply Modeling Within the Roadmaps?

Linkage to the I²DOV Technology Development Roadmap:

Modeling of technology concepts is conducted within the Invent/Innovate phase during technology development. Modeling of technology baseline performance is conducted in the Develop phase. Modeling of robustness and tunability of the mean of critical functional responses is done during the Optimize phase. Modeling of variational sensitivities across the integrated technology platform is conducted in the Verify phase.

Linkage to the CDOV Product Development Roadmap:

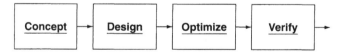

Modeling of product design concepts is conducted within the Concept and Design phases during product development. Modeling of the baseline performance is conducted in the later portion of the Design phase. Modeling of subsystem robustness and tunability of the mean of critical functional responses is done during the Optimize phase. Modeling of variational sensitivities across the integrated system is conducted in the Optimize and Verify phases.

What Am I Doing in the Process? What Does Modeling Do at this Point in the Roadmaps?

The Purpose of Modeling in the I²DOV Technology Development Roadmap:

Once again we begin by showing the strategic flow-down of how a business develops new product concepts:

Define business strategy: *Profit goals and requirements*

Identify markets and market segments: *Value generation and requirements*

Gather long-term voice of customer and voice of technology trends

Develop product line strategy: *Family plan requirements*

Develop/deploy technology strategy: *Technology and subsystem platform requirements*

Gather product-specific VOC and VOT: *New, unique, and difficult needs*

Conduct KJ analysis: *Structure and rank the VOC*

Build system House of Quality: *Translate VOC*

Document system requirements

Define the system functions

Generate candidate system architectures

Select the superior system concept: *Pugh process*

In the Invent/Innovate phase of the I²DOV process we define the platform functions and generate candidate platform architectures. The technology requirements will be used to define a set of adjustable or tunable functions. These functions require engineered technology concepts to transform, flow, and change the state of mass and energy. A unique set of architectural concepts, or platforms, are developed to fulfill these functions over an adjustable range of application. To help define the form and fit parameters that will fulfill the functions, initial modeling must be done. After careful evaluation, one superior platform architecture is selected out of the numerous options to go through the rest of the I²DOV process. Much more extensive modeling is then conducted on the superior concept during the Develop, Optimize, and Verify phases.

The Purpose of Modeling in the CDOV Product Development Roadmap:

In the CDOV process we define the system functions and generate candidate system architectures. The system requirements are used to define a set of product functions. Numerous system architectural concepts are then developed to fulfill the product functions. Modeling is used to help define and underwrite the feasibility of these concepts. After careful evaluation, one superior system architecture will be selected out of the numerous options to go through the rest of the CDOV process. When a superior system concept is selected, system modeling is initiated, soon to be followed by extensive model building down through the subsystems, subassemblies, components, and manufacturing processes. The models are used to define, develop, optimize, and verify the performance capability of the design's functions with and without stressful sources of noise applied to the system. Numerous forms of modeling are conducted in the Design, Optimize, and Verify Phases of the CDOV process.

What Output Do I Get at the Conclusion of this Phase of the Process? What Are Key Deliverables from Modeling Across the Roadmaps?

Models will be documented at all levels within the system:

System models

Subsystem models

Subassembly models

Component models

Manufacturing process models

Models will be documented in three forms:

Analytical models (equations based on first principles and empirical data)

Geometric or graphical models (CAD files, sketches, drawings, layouts)

Physical models (hardware, firmware, and software prototypes)

Invent/Innovate (I²DOV) and Concept (CDOV) Phase models:

- Rudimentary conceptual modeling conducted to underwrite conceptual feasibility
- General equations that express the underlying physical or chemical principles the concept will employ
- Hand sketches and CAD layouts that illustrate the form and fit of the concept
- Three-dimensional cardboard, styrofoam, stereo-lithography, or model-shop models that may possess basic functionality to demonstrate feasibility

Develop (I²DOV) and Design (CDOV) Phase models:

Baseline performance models will be documented internal to the system, subsystems, and subassemblies:

System Ideal Functions $[Y = f(x)]$

- Main effects of system engineering parameters
- Interactions between system engineering parameters
- Linear or nonlinear performance of system engineering parameters

Subsystem Ideal Functions [$Y = f(x)$]

- Main effects of subsystem engineering parameters
- Interactions between subsystem engineering parameters
- Linear or nonlinear performance of subsystem engineering parameters

Subassembly Ideal Functions [$Y = f(x)$]

- Main effects of subassembly engineering parameters
- Interactions between subassembly engineering parameters
- Linear or nonlinear performance of subassembly engineering parameters

Baseline performance models will be documented between the system, subsystems, and subassemblies:

System Transfer Functions [$Y = f(x)$]

- Main effects of system engineering parameters
- Interactions between system engineering parameters
- Linear or nonlinear performance of system engineering parameters

Subsystem Transfer Functions [$Y = f(x)$]

- Main effects of subsystem engineering parameters
- Interactions between subsystem engineering parameters
- Linear or nonlinear performance of subsystem engineering parameters

Subassembly Transfer Functions [$Y = f(x)$]

- Main effects of subassembly engineering parameters
- Interactions between subassembly engineering parameters
- Linear or nonlinear performance of subassembly engineering parameters

Capability Growth Models:

$CGI_{Subsystem} = Sum[(100/n) \times (Cp/2)]$ for n number of critical functional responses (CFRs)

Optimize Phase Models:

Robustness performance models will be documented internal to the system, subsystems, and subassemblies:

System Additive Models:

$S/N_{opt.} = S/N_{avg.} + (S/N_{A\ opt.} - S/N_{avg.}) + (S/N_{B\ opt.} - S/N_{avg.}) + \ldots + (S/N_{n\ opt.} - S/N_{avg.})$

- Main effects of system robustness parameters
- Interactions between system engineering parameters and noise factors

Subsystem Additive Models:

$S/N_{opt.} = S/N_{avg.} + (S/N_{A\ opt.} - S/N_{avg.}) + (S/N_{B\ opt.} - S/N_{avg.}) + \ldots + (S/N_{n\ opt.} - S/N_{avg.})$

- Main effects of subsystem robustness parameters
- Interactions between subsystem engineering parameters and noise factors

Subassembly Additive Models:

$S/N_{opt.} = S/N_{avg.} + (S/N_{A\ opt.} - S/N_{avg.}) + (S/N_{B\ opt.} - S/N_{avg.}) + \ldots + (S/N_{n\ opt.} - S/N_{avg.})$

- Main effects of subassembly robustness parameters
- Interactions between subassembly engineering parameters and noise factors

Critical Adjustment Parameter Models for the CFR Mean:

$Y = f(CAPs)$

- Main effects of subsystem and subassembly level CAPs on CFR mean
- Interactions between subsystem and subassembly level CAPs
- Linear or nonlinear performance of subsystem and subassembly level CAPs (response surface models on the CFR mean)

Capability Growth Models:

$CGI_{subsystem} = Sum[(100/n) \times (Cp/2)]$ for n number of critical functional responses (CFRs)

Capability Verification Phase Models:

Variance and tolerancing models will be documented internal to the system, subsystems, subassemblies, components, and manufacturing processes.

System Variance Models:

$\sigma^2_{total} = \sigma^2_A + \sigma^2_B + \ldots + \sigma^2_n$

- Sensitivities of system tolerance parameters
- Percent contribution of each parameter to variability of CFR performance (epsilon squared)

Subsystem Variance Models:

$$\sigma^2_{total} = \sigma^2_A + \sigma^2_B + \ldots + \sigma^2_n$$

- Sensitivities of subsystem tolerance parameters
- Percent contribution of each parameter to variability of CFR performance (epsilon squared)

Subassembly Variance Models:

$$\sigma^2_{total} = \sigma^2_A + \sigma^2_B + \ldots + \sigma^2_n$$

- Sensitivities of subassembly tolerance parameters
- Percent contribution of each parameter to variability of CFR performance (epsilon squared)

Component Variances:

$$\sigma^2_A, \sigma^2_B, \ldots \sigma^2_n$$

- Variability within each component tolerance parameter

Manufacturing Process Variance Models:

$$\sigma^2_{total} = \sigma^2_A + \sigma^2_B + \ldots + \sigma^2_n$$

- Sensitivities of manufacturing process tolerance parameters
- Percent contribution of each parameter to variability of CFR performance (epsilon squared)

Critical Adjustment Parameter Models for CFR Standard Deviations:

$$Y = f(CAPs)$$

- Main effects of subsystem and subassembly level CAPs on CFR σ
- Interactions between subsystem and subassembly level CAPs
- Linear or nonlinear performance of subsystem and subassembly level CAPs (response surface models for CFR σ)

Capability Growth Models:

$CGI_{subsystem} = Sum[(100/n) \times (Cp/2)]$ for n number of critical functional responses (CFRs)

Critical Parameter Management Data that Is Documented in the CPM Relational Database:

- Flow-down and linkage of all CFRs, CFPs, and CTF specifications (down to all appropriate levels within the system)
- Ideal/transfer functions documented
- Additive models documented
- Variance models documented
- Capability growth indices documented

When modeling is used extensively to develop, design, optimize, and verify a capable product, we can expect the following results.

1. Requirements that have been converted to functions that make it clear what must be engineered and developed
2. Construction of input-output-constraint diagrams for each function down through the system hierarchy during Concept and Design phases
3. Linkage of all ideal/transfer functions using functional flow diagrams
4. Definition of the flow and connectivity of models for each function:
   Math models
   Graphical models
   Physical models
5. Development of internal ideal function models for each subsystem and subassembly
6. Development of transfer function models between the system, subsystems, and subassemblies (the difference between ideal and transfer functions is subtle but worth noting)
7. Verification of all analytical models with empirical models
8. Documentation of the ideal and transfer functions and their linkages in the CPM database
9. Documentation of additive robustness models
10. Documentation of variance (tolerance) models

Modeling Process Flow Diagram (found on page 418)

The visual reference of the detailed steps of modeling illustrates the flow of these critical processes.

Verbal Descriptions for the Application of Each Block Diagram

This section *explains the specific details of* each step in the modeling process:

- *Key detailed input(s)* from preceding tools and best practices
- *Specific steps and deliverables* from each step to the next within the process flow
- *Final detailed outputs* that flow into follow-on tools and best practices

INPUT . . .

Requirement targets,
tolerances, functions,
and concepts.

Model ideal functions,
graphical architectures, and
physical prototypes at the
system concept level.

Model ideal functions,
graphical architectures, and
physical prototypes at the
subsystem concept level.

Model ideal functions,
graphical architectures, and
physical prototypes at the
subassembly concept level.

Model ideal functions,
graphical architectures, and
physical prototypes at the
component concept level.

Model ideal functions,
graphical architectures, and
physical prototypes at the
manufacturing process
concept level.

Develop capability growth
indices for all subsystem
CFRs in each appropriate
phase.

Develop and confirm robustness additive
models at the system level.

Develop and confirm robustness additive
models at the subsystem level.

Develop and confirm robustness additive
models at the subassembly level.

Develop and confirm robustness additive
models at the manufacturing process level.

Develop and confirm CAP models for
CFR means.

Develop and confirm variance models at
the system, subsystem, and subassembly levels.

Develop and confirm variance models
at the component and manufacturing
process levels.

Develop and confirm CAP models for
CFR standard deviations.

Document and link all CFRs, CFPs,
CTF specifications, and ideal/transfer
functions in the Critical Parameter
Management database.

OUTPUT . . .

Math, graphical, and hardware/firmware
models maturing into final product and
production specifications.

418

Key Detailed Inputs

> **Requirement targets, tolerances, functions, and concepts.**

With the specific new, unique, and difficult technical requirements defined, the team knows exactly how the requirements can be measured (targets/tolerances and units of measure) for fulfillment. We also use the defined functions, forms, and fits at the system, subsystem, subassembly, component, and manufacturing process levels.

Specific Steps and Deliverables

> **Model ideal functions, graphical architectures, and physical prototypes at the system concept level.**

The beginning of the modeling process always starts with a clear well-written set of requirements. This is true no matter whether you are working at the system level or are modeling down through the levels of the system hierarchy. At the system level, the first items the system architectural team determines are the functions that must be engineered to fulfill the system requirements. Once the functions are defined, the team can concern itself with defining numerous architectural designs (subsystems and interfaces) that can adequately generate the functions. This follows the old axiom of form following function. The flow-down of elements that drive the modeling process are:

Needs drive requirements
  Requirements drive functions
    Functions drive forms
      Form and function drive modeling

All this information is used to initiate the construction of math models, graphical models, and physical models. The first models that are done at the system level are graphical. Functional system models are difficult to construct until the subsystems, subassemblies, and components are defined. System math models are constructed later in the Develop (I^2DOV) or Design (CDOV) phases. The system math models take form as the critical parameter relational database is developed. In Phase 1 of both I^2DOV and CDOV processes, physical system models are restricted to architectural models of external forms and features (spatial models), functional system models are built and tested in the second part of the Optimize Phase (Phase 3B) in the CDOV process and in the third phase of the I^2DOV process. The premature integration of system hardware frequently causes program delays because of the extra effort it takes to desensitize all the "built-in" cross-boundary conflicts that the other forms of modeling at the lower levels would have prevented had they been conducted prior to system hardware integration.

The ideal or transfer functions at the system level are expressed as $Y = f(x)$ models with the following parameter definitions:

Y = a measured critical functional response

x_{CFP} = critical functional parameters (inputs from subsystems and subassemblies)

x_{CAP} = critical adjustment parameters (inputs from subsystems and subassemblies that are specifically designed to shift the mean of the CFR)

The system math models (ideal/transfer functions) are constructed from "n number of x" inputs from subsystems and subassemblies:

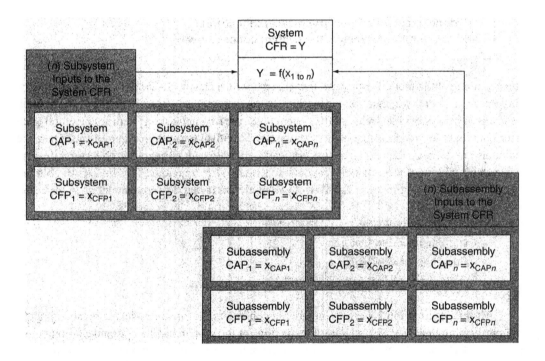

These models are derived either by "first principles" or from experimental data from designed experiments.

First principles modeling is a good place to begin to define the elements of a model and how those elements can be integrated to explain functional behavior. Your engineering books from your college or university days contain a wealth of first principles modeling methods and examples. If you sold them, go buy new ones.

A common approach to first principles modeling is the use of four analogs:

Electrical analogs

Mechanical analogs

Fluid analogs

Thermal analogs

These analogs provide a conceptual framework for any set of input–output relationships to be stated in a generalized block diagram format. This allows one to conceptually lay out the function (input–output relationships) in blocks representing elements from one of the four analog scenarios.

Once the conceptual block diagrams are laid out, the team can write down the mathematical forms of the functional elements representing the input–output relationships. With these in hand, differential equations can be structured and solved on a personal computer.

The process for first principles modeling can be structured as follows:

1. Construct block diagram models based on one of the analogs.
   - Review how work, power, and energy apply to your case.
   - Review how basic calculus applies to your case.
2. Set up first order differential equation based on block diagrams.
   - Solve closed or open form model on PC using Excel or Math/Solver software.
3. Set up second order differential equation based on block diagrams.
   - Solve closed or open form model on PC using Excel or Math/Solver software.
4. Set up higher order differential equation based on block diagrams as necessary.
   - Solve closed or open form model on PC using Excel or Math/Solver software.

An excellent text on the market explains the step-by-step process modeling of engineering systems. *Modeling Engineering Systems,* by Jack W. Lewis (1993, published by High Text), presents an easy-to-read explanation of math modeling. This book is a "must-have" to assist any engineer in modeling. This is especially true if you are a little rusty and need a gentle refresher.

Empirical modeling methods are covered in Chapters 27 and 29. The construction of the robustness additive models is covered in Chapter 28. Chapters 31 and 32 on analytical and empirical tolerancing explain how variance models are constructed. Chapter 10 explains how transfer functions are linked and used to do system modeling. The Six Sigma Cockpit software package from Cognition Software will enable you to do CPM-based system modeling. It will also enable you to create capability growth index roll-up models.

> **Model ideal functions, graphical architectures, and physical prototypes at the subsystem, subassembly, and component concept levels.**

Subsystem, subassembly, and component graphical models are first developed during the concepting activities of Phase 1 in the I^{2}DOV process and Phase 2 in the CDOV process. The graphical models for each subsystem, subassembly, and component are matured after superior concepts are selected. The graphical models are most commonly documented as CAD files:

- Hand sketches evolve into hand-drawn isometrics and integrated layouts.
- Hand-drawn conceptual architectures are then converted to CAD files.

The subsystem math models (ideal/transfer functions) are constructed from "n number of x" inputs from subassemblies and components:

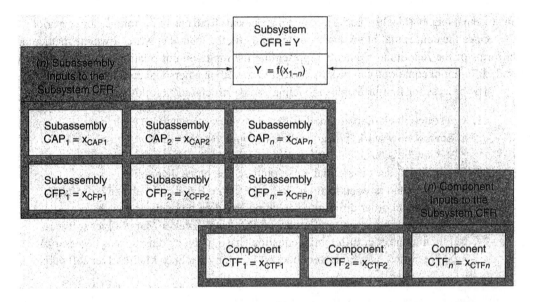

The subassembly math models (ideal or transfer functions) are constructed from "n number of x" inputs from components:

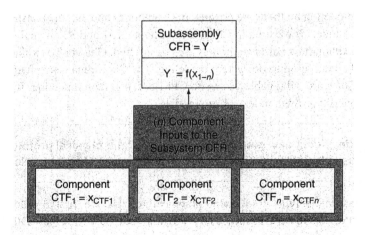

The component math models (ideal/transfer functions) are constructed from "n number of x" inputs from manufacturing process parameter set points:

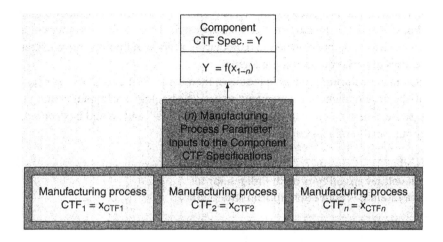

These models are derived either by first principles or from experimental data from designed experiments using physical models. They represent the baseline performance of the components without the debilitating effects of noise imposed upon them. That scenario will come later during the robustness and variance modeling in the Optimize and Verify Certification phases.

> **Model ideal functions, graphical architectures, and physical prototypes at the manufacturing process concept level.**

Manufacturing processes are usually strong sources of variation to components. All component models (math, graphical, and physical) will reflect the effects of manufacturing process capability.

Even when a manufacturing process is running "perfectly"—that is to say, when the only variation in the components is coming from random, natural sources—the component dimensions and its bulk and surface properties will vary. This "natural" variation must be evaluated to see if the design still functions properly when it is consistently present in the components. These kinds of models are called *baseline performance models*. They are evaluated in the Develop (I^2DOV) or Design (CDOV) phases.

Another set of models must be developed to account for both component assembleability and functionality. These models are usually graphical, mathematical, and eventually physical. The effect of manufacturing process set points on component assembleability and functionality can and should be modeled. If a manufacturing process is incapable of meeting component functional performance and assembleability requirements, the process itself may require redesign or a completely new technology.

The teams in charge of technology and design development must intentionally induce the significant forms of variation that are to be expected from the manufacturing processes into the components. This is often done through Monte Carlo simulations during analytical tolerance design (during Phase 2) and designed experiments (using $+/- 1\sigma$ values or the full tolerance limits) during Phase 3 and 4 using empirical tolerance design. These induced sources of component variation represent what happens when manufacturing process set points move off target due to assignable causes of variation during production.

If a new manufacturing process is required, then the I^2DOV and CDOV processes need to be applied to the development of a new manufacturing technology and manufacturing process design. In this sense, a new manufacturing process is a "product" and should be developed just like any other commercial product system.

> **Develop capability growth indices for all subsystem CFRs in each appropriate phase.**

At the end of the Develop/Design, Optimize, and Verify Certification phases, the capability growth indices for all subsystems should be calculated. This models the current state of capability for each critical functional response within each subsystem. The CGI keeps a running tally of the percentage of completion each subsystem possesses along the way of getting all CFRs up to a common capability of $Cp = 2$. When a subsystem has all its CFR Cp Indices equal to 2, its CGI will be 100 percent. This is the ideal goal for all subsystems. The CGI is explained in Chapter 12.

> **Develop and confirm robustness additive models at the system level.**

In Phase 4 of the I^2DOV process and Phase 3B of the CDOV process, the integrated platform or system can be evaluated for its robustness. Each CFR at the system level can be measured using a capable data acquisition system. The system is built up from subsystems, subassemblies, and components that are intentionally set up at levels that represent real variation that will be experienced when the product is in the customer's use environment. These sources of variation include:

- *External variation:* Run the system robustness tests in the presence of external sources of variation (contamination, temperature, humidity, altitude, or barometric pressure).

- *Subsystem and subassembly functional variation due to assembly, service critical adjustment parameter adjustment variation:* Intentionally set the critical adjustment parameters off target to simulate adjustment error by assembly or service personnel.

- *Component variation:* Have component dimensions, surface finishes, and bulk properties made off target to simulate variation coming in from your supply chain. These variations are usually at $+/- 1$ standard deviation of the distribution from the supply chain sample data.

• *Deterioration variation:* Physically alter components to look like they would after a reasonable period of use. Components or component interfaces are forced to deteriorate or to emulate a known amount of deteriorated condition based on time. This time is usually somewhere around 6-month to yearly intervals. What will the component or interface look like 6 months from now? How will it look 1, 2, or 3 years from now? Make it look that way! If you have no experience or historical knowledge of mass deformation, material depletion, or degradation, you will have to estimate it. Better to do that within reasonable bounds than to do nothing at all. If you do nothing at all, you are likely in for some big surprises later on after your customer uses your product.

The appropriate signal-to-noise metrics can be calculated from the mean and/or standard deviation from your sample of CFR data. Purposeful changes to critical parameters can be made at the subsystem, subassembly, and component level to improve the S/N values for the CFRs using designed experimental procedures for robustness optimization. The primary function of the S/N metrics in robust design is to quantify the strength of interactions between control factors and noise factors as they relate to a CFR. The additive model for summing the S/N gains takes the following general form:

$$S/N_{opt.} = S/N_{avg} + (S/N_{A\ opt.} - S/N_{avg}) + (S/N_{B\ opt.} - S/N_{avg}) + \ldots . + (S/N_{n\ opt.} - S/N_{avg})$$

Once an optimal level of system CFR robustness is attained, confirmation runs can be conducted to validate the additive model for the system CFR S/N values.

Develop and confirm robustness additive models at the subsystem level.

In Phase 3 of the I^2DOV process and Phase 3A of the CDOV process the individual subsystems can be evaluated for robustness. Each CFR at the subsystem level can be measured using a capable data acquisition system. The subsystem is built up from subassemblies and components that are intentionally set up at levels that represent real variation that will be experienced when the subsystem is in the customer's use environment as an integrated part of the system.

These sources of induced variation include:

External variation: Run the subsystem robustness tests in the presence of external sources of variation (contamination, temperature, humidity, altitude, or barometric pressure) as well as incoming noises from other subsystems or from the system environment that is immediately external to the subsystem. These system noises must be emulated and induced by special means since the system has not yet been assembled.

Subsystem and subassembly functional variation due to assembly, service critical adjustment parameter adjustment variation: Intentionally set the subsystem and subassembly critical adjustment parameters off-target to simulate adjustment error by assembly or service personnel.

Component variation: Have component dimensions, surface finishes, and bulk properties made off target to simulate variation coming in from your supply chain. These variations are usually at $+/-$ 1 standard deviation of the distribution from the supply chain sample data.

Deterioration Variation: Physically alter components and interfaces to look as they would after a reasonable period of use. Components or component interfaces are forced to deteriorate or to emulate a known amount of deteriorated condition based on time. This time is usually somewhere around 6-month to yearly intervals. What will the component or interface look like 6 months from now? How will it look 1, 2, or 3 years from now? Make it look that way! If you have no experience or historical knowledge of mass deformation or material depletion or degradation, you will have to estimate it. Better to do that within reasonable bounds than to do nothing at all. If you do nothing at all, you are likely in for some big surprises later on after you integrate the system.

The appropriate signal-to-noise metrics can be calculated from the mean and/or standard deviation from your sample of subsystem CFR data. Purposeful changes to critical parameters can be made at the subsystem, subassembly, and component level to improve the S/N values for the CFRs using designed experimental procedures for robustness optimization.

Once an optimal level of subsystem CFR robustness is attained, confirmation runs can be conducted to validate the additive model for the system CFR S/N values.

Develop and confirm robustness additive models at the subassembly level.

In Phase 3 of the I²DOV process and Phase 3A of the CDOV process the individual subassemblies can be evaluated for robustness. Each CFR at the subassembly level can be measured using a capable data acquisition system. The subassembly is built up from components that are intentionally set up at levels that represent real variation that will be experienced when the subsystem is in the customer's use environment as an integrated part of the system.

These sources of induced variation include:

External variation: Run the subassembly robustness tests in the presence of external sources of variation (contamination, temperature, humidity, altitude, or barometric pressure) as well as incoming noises from other subsystems or from the system environment that is immediately external to the subassembly. These system and subsystem noises must be emulated and induced by special means since the system or the subsystems have not yet been integrated.

Subassembly functional variation due to assembly, service critical adjustment parameter adjustment variation: Intentionally set the subassembly critical adjustment parameters off target to simulate adjustment error by assembly or service personnel.

Component variation: Have component dimensions, surface finishes, and bulk properties made off target to simulate variation coming in from your supply chain. These variations are usually at $+/-$ 1 standard deviation of the distribution from the supply chain sample data.

Deterioration variation: Physically alter components and interfaces to look as they would after a reasonable period of use. Components or component interfaces are forced to deteriorate or to emulate a known amount of deteriorated condition based on time. This time is usually somewhere around 6-month to yearly intervals. What will the component or interface look like 6 months from now? How will it look 1, 2, or 3 years from now? Make it look that way! If you have no experience or historical knowledge of mass deformation or material depletion or degradation, you will have to estimate it. Better to do that within reasonable bounds than to do nothing at all. If you do nothing, you are likely in for some big surprises later on after you integrate the subassembly into the subsystem.

The appropriate signal-to-noise metrics can be calculated from the mean and/or standard deviation from your sample of subassembly CFR data. Purposeful changes to critical parameters can be made at the subassembly and component level to improve the S/N values for the CFRs using designed experimental procedures for robustness optimization.

Once an optimal level of subassembly CFR robustness is attained, confirmation runs can be conducted to validate the additive model for the subassembly CFR S/N values.

> **Develop and confirm robustness additive models at the manufacturing process level.**

In Phase 3 of the I^2DOV process and Phase 3A of the CDOV process, the individual manufacturing processes can be evaluated for robustness. Each CFR at the manufacturing processes level can be measured using a capable data acquisition system. The manufacturing processes are built up from subsystems, subassemblies, and components that are intentionally set up at levels that represent real variation that will be experienced when the manufacturing processes are in the production environment.

These sources of induced variation include:

External variation: Run the manufacturing processes robustness tests in the presence of external sources of variation (contamination, temperature, humidity, altitude, or barometric pressure) as well as incoming noises from other inputs of unwanted mass or energy.

Manufacturing processes functional variation due to operator or control system critical adjustment parameter adjustment variation: Intentionally set the manufacturing processes critical adjustment parameters off target to simulate adjustment error by operations personnel or process controllers.

Raw material variation: Have raw material properties made off target to simulate variation coming in from your supply chain. These variations are usually at $+/-$ 1 standard deviation of the distribution from the supply chain sample data.

Deterioration variation: Physically alter manufacturing processes elements, tools, fixtures, and interfaces to look as they would after a reasonable period of use. Manufacturing process elements, tools, fixtures, and interfaces are forced to deteriorate or to emulate a known amount of deteriorated condition based on time. This time is usually somewhere around daily, weekly, monthly, to yearly intervals. What will the component or interface look like a few hours or days from now? How will it look 1, 2, or 3 months from now? Make it look that way! If you have no experience or historical knowledge of mass deformation or material depletion or degradation, then you will have to estimate it. Better to do that within reasonable bounds than to do nothing at all. If you do nothing, you are likely in for some big surprises later on after you apply the manufacturing process elements, tools, fixtures, and interfaces to make components.

The appropriate signal-to-noise metrics can be calculated from the mean and/or standard deviation from your sample of manufacturing process CFR data. Purposeful changes to critical parameters can be made at the subsystem, subassembly, and component level to improve the S/N values for the CFRs using designed experimental procedures for robustness optimization.

Once an optimal level of manufacturing process CFR robustness is attained, confirmation runs can be conducted to validate the additive model for the manufacturing process CFR S/N values.

Develop and confirm CAP models for CFR means.

All CFRs at the system level and down to the subsystem and subassembly levels are evaluated for the presence and need of critical adjustment parameters. Critical adjustment parameters are used to adjust the mean of a CFR back onto its desired target. They are often assembly or service adjustments down inside the subsystems or subassemblies. At the system level they are also typically available for the customer to use to adjust the output of the product or production process.

Critical adjustment parameter models $[Y = f(CAPs)]$ are initially developed during the Develop or Design phase and then optimized after robustness optimization is completed for the subsystems and subassemblies. Empirical DOE in the form of response surface methods is an excellent way to develop and confirm the optimized set points for critical adjustment parameters. If there is just one CAP controlling a CFR, linear or nonlinear regression can be used to optimize the CAP-CFR relationship.

Usually there are just a few (1–6) CAPs for any one CFR. Often the CAPs are actively controlled by feedback or feed-forward control systems. The RSM data can help clearly define a model for the controllers to use during adjustments.

Develop and confirm variance models at the system, subsystem, and subassembly levels.

Variance models are developed by technology development teams during Phases 2 and 4 of the I^2DOV process.

Variance models are developed by the system and subsystem design teams during Phases 2, 3B, and 4A of the CDOV process.

Analytical tolerance models are developed prior to empirical tolerance models. The analytical tolerance models are usually evaluated using Monte Carlo simulations on personal computers. A vector loop diagram is generated to model linear or nonlinear tolerance stack-up models. These models can be developed for many different kinds of variational build-up.

The following is the general form of variational models at the system, subsystem, and subassembly levels:

System Variance Models:

$$\sigma^2_{total} = \sigma^2_A + \sigma^2_B + \ldots + \sigma^2_n$$

- Sensitivities of system tolerance parameters
- Percent contribution of each parameter to variability of CFR performance (epsilon squared)

Subsystem Variance Models:

$$\sigma^2_{total} = \sigma^2_A + \sigma^2_B + \ldots + \sigma^2_n$$

- Sensitivities of subsystem tolerance parameters
- Percent contribution of each parameter to variability of CFR performance (epsilon squared)

Subassembly Variance Models:

$$\sigma^2_{total} = \sigma^2_A + \sigma^2_B + \ldots + \sigma^2_n$$

- Sensitivities of subassembly tolerance parameters
- Percent contribution of each parameter to variability of CFR performance (epsilon squared)

Empirical tolerance models are developed for complex design functional or assembly relationships that are not easily conducted through analytical modeling. Detailed information on the construction of analytical tolerance models can be found in Chapter 31 and in Chapter 32 for empirical tolerancing.

Develop and confirm variance models at the component and manufacturing process levels.

Component variances are important design parameters to understand and use in the development of higher-level functional and assembly capability.

Component Variances:

$\sigma^2_A, \sigma^2_B, \ldots \sigma^2_n$

- Variability within each component tolerance parameter

Manufacturing process variance models account for the build-up of variation in components due to variation in manufacturing process set points.

Manufacturing Process Variance Models:

$\sigma^2_{total} = \sigma^2_A + \sigma^2_B + \ldots + \sigma^2_n$

- Sensitivities of manufacturing process tolerance parameters
- Percent contribution of each parameter to variability of CFR performance (epsilon squared)

Develop and confirm CAP models for CFR standard deviations.

After careful screening experiments and Monte Carlo simulations are conducted on tolerance parameters at all levels in the system, a very few tolerance parameters that are most critical in limiting variation can be evaluated in a response surface experiment to optimize variance reduction. In this sense, a few tolerance parameters are used as critical adjustment parameters for the standard deviation of the CFRs.

**Document and link all CFRs, CFPs, CTF
specifications, and ideal/transfer functions
in the Critical Parameter Management database.**

After each stage of modeling is concluded, the critical functional responses, critical functional parameters, critical-to-function specifications, and their associated ideal/transfer functions should be documented in the Critical Parameter Management database. The ideal/transfer functions should also be linked to the appropriate inputs and outputs within the hierarchy of the CPM database. The ideal/transfer functions located inside the relational database can be exercised as required to evaluate the effects of induced variation in any term or set of terms. One or more x terms can be varied so that the transmission of variation can be traced up through the linked transfer functions—all the way to the system CFR level if desired.

Final Detailed Outputs

> **Math, graphical, and hardware/firmware models maturing into final product and production specifications.**

The results of modeling enable decisions to be made about improving predicted and measured functional and assembly performance. With the use of a balanced portfolio of modeling tools and best practices, timely, data-driven decisions can be made throughout the phases of product development.

Modeling Checklist and Scorecard

This suggested format for a checklist can be used by an engineering team to help plan its work and aid in the detailed construction of a PERT or Gantt chart for project management of the modeling process within the first two phases of the CDOV or the first phase of the I^2DOV development process. These items will also serve as a guide to the key items for evaluation within a formal gate review.

Checklist for Modeling

Actions:	Was this Action Deployed? YES or NO	Excellence Rating: 1 = Poor 10 = Excellent
1. Model ideal functions, graphical architectures, and physical prototypes at the system concept level.	_____	_____
2. Model ideal functions, graphical architectures, and physical prototypes at the subsystem concept level.	_____	_____
3. Model ideal functions, graphical architectures, and physical prototypes at the subassembly concept level.	_____	_____
4. Model ideal functions, graphical architectures, and physical prototypes at the component concept level.	_____	_____
5. Model ideal functions, graphical architectures, and physical prototypes at the manufacturing process concept level.	_____	_____
6. Develop capability growth indices for all subsystem CFRs in each appropriate phase.	_____	_____
7. Develop and confirm robustness additive models at the system level.	_____	_____
8. Develop and confirm robustness additive models at the subsystem level.	_____	_____

(continued)

9. Develop and confirm robustness additive models at the
subassembly level.   _____   _____

10. Develop and confirm robustness additive models at the
manufacturing process level.   _____   _____

11. Develop and confirm CAP models for CFR means.   _____   _____

12. Develop and confirm variance models at the system,
subsystem, and subassembly levels.   _____   _____

13. Develop and confirm variance models at the
component and manufacturing process levels.   _____   _____

14. Develop and confirm CAP models for CFR standard
deviations.   _____   _____

15. Document and link all CFRs, CFPs, CTF
specifications, and ideal/transfer functions in the
Critical Parameter Management database.   _____   _____

Scorecard for Modeling

The definition of a good modeling scorecard is that it should illustrate, in simple but quantitative terms, what gains were attained from the work. This suggested format for a scorecard can be used by a team to help document the completeness of its work.

Total number of subsystem level math models: _____

Total number of subsystem level graphical models: _____

Total number of subsystem level physical models: _____

Total number of subassembly level math models: _____

Total number of subassembly level graphical models: _____

Total number of subassembly level physical models: _____

Total number of component level math models: _____

Total number of component level graphical models: _____

Total number of component level physical models: _____

Total number of manufacturing process level math models: _____

Total number of manufacturing process level graphical models: _____

Total number of manufacturing process level physical models: _____

Table of System Level CFRs:

System CFR Name:	Target:	Estimated Tolerance: USL	Estimated Tolerance: LSL	Unit of Measure:
1.	___	___	___	___
2.	___	___	___	___
3.	___	___	___	___

Table of Subsystem Level CFRs:

Subsystem CFR Name:	Target:	Estimated Tolerance: USL	Estimated Tolerance: LSL	Unit of Measure:
1.	___	___	___	___
2.	___	___	___	___
3.	___	___	___	___

Table of Subassembly Level CFRs:

Subassembly CFR Name:	Target:	Estimated Tolerance: USL	Estimated Tolerance: LSL	Unit of Measure:
1.	___	___	___	___
2.	___	___	___	___
3.	___	___	___	___

Table of Component Level CTF Specifications:

Component CTF Spec. Name:	Target:	Estimated Tolerance: USL	Estimated Tolerance: LSL	Unit of Measure:
1.	___	___	___	___
2.	___	___	___	___
3.	___	___	___	___

Table of Manufacturing Process Level CTF Specifications:

Manufacturing Process CTF Spec. Name:	Target:	Estimated Tolerance: USL	Estimated Tolerance: LSL	Unit of Measure:
1.	_____	_____	_____	_____
2.	_____	_____	_____	_____
3.	_____	_____	_____	_____

Review now the output from modeling as outlined on pages 413-417 of this chapter.

References

Creveling, C. M. (1997). *Tolerance design.* Reading, Mass.: Addison-Wesley.

Fowlkes, W. Y., & Creveling, C. M. (1995). *Engineering methods for robust product design.* Reading, Mass.: Addison-Wesley.

Lewis, J. W. (1994). *Modeling engineering systems.* Eagle Rock, VA.: LLH Technology Publishing.

Otto, K., & Wood, K. (2001). *Product design.* Upper Saddle River, NJ: Prentice Hall.

Shetty, D., & Kolk, R. A. (1997). *Mechatronics system design.* Boston, Mass.: PWS Publishing Co.

PART V

Tools and Best Practices for Design Development

The tools and best practices discussed in this section focus on deliverables required at gate reviews for Phase 2 in the I^2DOV and CDOV processes.

The tools and best practices chapters are structured so that the reader can get a strong sense of the process steps for applying the tools and best practices. The structure also facilitates the creation of PERT charts for project management and critical path management purposes. A checklist, scorecard, and list of management questions end each chapter.

CHAPTER 19

Design Failure Modes
and Effects Analysis

Where Am I in the Process? Where Do I Apply DFMEA Within the Roadmaps?

Linkage to the I²DOV Technology Development Roadmap:

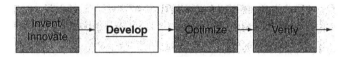

DFMEA is conducted within the Develop phase during technology development. Some technology development teams may decide to do a cursory version of DFMEA on each candidate technology concept prior to inclusion in the Pugh concept selection process. DFMEA is typically done on the superior concepts that emerge from the Pugh concept selection process.

Linkage to the CDOV Product Development Roadmap:

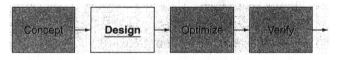

DFMEA is conducted within the Design phase during product development. Some product design teams also decide to do a cursory version on each of the candidate concepts prior to inclusion in the Pugh concept selection process. DFMEA is typically done on the superior concepts that emerge from the Pugh concept selection process.

Design FMEAs are applied to qualitatively define and quantitatively rank the failure modes for a system and its subsystems, subassemblies, and components. Process FMEAs are conducted on manufacturing processes.

What Am I Doing in the Process? What Does DFMEA Do at this Point in the Roadmaps?

The Purpose of DFMEA in the I²DOV Technology Development Roadmap:

DFMEA is applied to qualitatively define and quantitatively rank the failure modes and effects for new technology platforms and subsystems that are being developed for integration into a new platform or for integration into existing product designs. Preventative application of numerous tools and best practices can then be deployed to improve the resistance to failure modes and the strategic minimization of their effects.

The Purpose of DFMEA in the CDOV Product Development Roadmap:

Design FMEAs are applied to qualitatively define and quantitatively rank the failure modes for a system and its subsystems, subassemblies, and components. Preventative application of numerous tools and best practices can then be deployed to improve the resistance to failure modes and the strategic minimization of their effects.

What Output Do I Get at the Conclusion of this Phase of the Process? What Are Key Deliverables from DFMEA at this Point in the Roadmaps?

- Design failure modes identified
- Effects of a design failure mode identified
- Causes of a design failure identified
- Linkage of the following ratings to each cause:
  Severity rating
  Occurrence rating
  Detection rating
- Calculated risk priority numbers (RPN) for each cause
- Improved risk priority numbers (RPN) for each cause after corrective action

The DFMEA Flow Diagram (found on page 439)

This visual reference of the detailed steps of DFMEA illustrates the flow of tasks and processes employed at this stage.

Verbal Descriptions for the Application of Each Block Diagram

This section *explains the specific details of* each step of DFMEA:

- *Key detailed input(s)* from preceding tools and best practices
- *Specific steps and deliverables* from each step to the next within the process flow
- *Final detailed outputs* that flow into follow-on tools and best practices

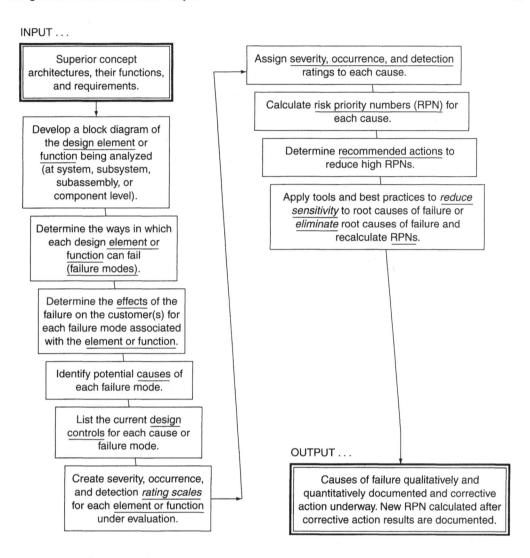

INPUT . . .

Superior concept architectures, their functions, and requirements.

Develop a block diagram of the design element or function being analyzed (at system, subsystem, subassembly, or component level).

Determine the ways in which each design element or function can fail (failure modes).

Determine the effects of the failure on the customer(s) for each failure mode associated with the element or function.

Identify potential causes of each failure mode.

List the current design controls for each cause or failure mode.

Create severity, occurrence, and detection *rating scales* for each element or function under evaluation.

Assign severity, occurrence, and detection ratings to each cause.

Calculate risk priority numbers (RPN) for each cause.

Determine recommended actions to reduce high RPNs.

Apply tools and best practices to *reduce sensitivity* to root causes of failure or *eliminate* root causes of failure and recalculate RPNs.

OUTPUT . . .

Causes of failure qualitatively and quantitatively documented and corrective action underway. New RPN calculated after corrective action results are documented.

Key Detailed Inputs

Superior concept architectures, their functions, and requirements.

The form, fit, and functions and the requirements they seek to fulfill at the system, subsystem, subassembly, and component levels are inputs to the DFMEA process.

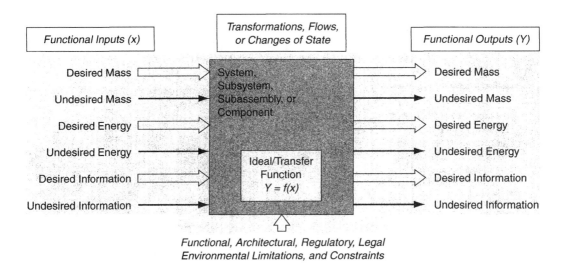

Figure 19–1   System Inputs and Outputs

Specific Steps and Deliverables

> **Develop a block diagram of the design element or function being analyzed (at system, subsystem, subassembly, or component level).**

A block diagram illustrates the input–output relationships at the system, subsystem, subassembly, or component level. In particular they illustrate the flow of mass, energy, and information into and out of a design element. For example, a hydraulic subassembly may have fluid (mass), forces (energy), and digital logic and control signals (information) flowing in some form that is associated with its ideal function.

The object of the block diagram in Figure 19–1 is to understand the input to the block, the process (function) performed within the block, and the output from the block. It also establishes the relationship of each item/function to all other associated system elements.

With the design element, its inputs, outputs, and ideal/transfer function defined, the team has the information it needs to begin to assess its potential failures modes.

> **Determine the ways in which each design element or function can fail (failure modes).**

Functions	Failure Modes	Effects	Causes	Design Controls

The team works together to identify and document all the ways the design can fail. These are referred to as *failure modes*. The failure modes include all forms of functional failures. Failures can be associated with how the design transforms mass, energy, and information. These three items cover everything that is associated with any design's function. A design can fail to flow, transform, or change the state of mass, energy, and information (digital or analog logic and control signals).

Function	**Failure Modes**	Effects	Causes	Design Controls

> **Determine the effects of the failure on the customer(s) for each failure mode associated with the element or function.**

The team traces the effects each failure mode has on the customers that use the product or process. Effects can span a wide range. Some examples include:

- Economic loss
- Physical safety
- Loss of desired function

No matter what the relative impact of the effect, every effect should be documented. Every effect should be traceable through its failure mode back to the ideal/transfer function that underwrites the fulfillment of a requirement. There should be clear linkage and traceability between failure effects, failure modes, ideal/transfer functions, and their requirements.

Function	Failure Modes	Effects	Causes	Design Controls

Identify potential causes of each failure mode.

A thorough root cause analysis should be performed on each failure mode. Root cause analysis is a detailed, fundamental engineering investigation into the physical causes for the failure. The ideal/transfer function for the design element in which the failure occurred is the starting point. Something associated with the transfer, flow, or change of state of the design's functions has gone wrong. Root cause analysis in this context rigorously documents the reasons why the failure occurred.

Function	Failure Modes	Effects	Causes	Design Controls

List the current design controls for each cause or failure mode.

After each root cause is unveiled, the team needs to identify the controls available to the design team to prevent, forecast, or detect the failure mode. Controls for the prevention of design defects (failure modes) or reducing the rate of occurrence include: mathematical modeling, engineering stress and failure analysis, reliability modeling, robust design, and tolerance design.

Forecasting or detection of either the causes or failure modes are not as desirable as prevention but may be necessary. Examples of forecasting or detection include performance testing, lab analysis, prototyping, stress testing, and alpha site testing.

Function	Failure Modes	Effects	Causes	Design Controls

Create severity, occurrence, and detection rating scales for each element or function under evaluation.

Three categories can be established to quantify the nature of failures modes and their detection:

1. *Severity:* importance of the effect on customer requirements; could also be concerned with safety and other risks if failure occurs.
2. *Occurrence:* frequency with which a given cause occurs and creates the failure mode.
3. *Detection:* (capability of current design controls) ability of current design control scheme to detect the causes or the effect of the failure modes.

Assign severity, occurrence, and detection ratings to each cause.

The three categories can now have numerical ratings assigned to quantify the nature of the failure modes and their detection:

Severity: importance of the effect on customer requirements; could also be concerned with safety and other risks if failure occurs.

1 = Not severe, 10 = Very severe

Occurrence: frequency with which a given cause occurs and creates the failure mode.

1 = Not likely, 10 = Very likely

Detection: (capability of current design controls) ability of current design control scheme to detect the causes or the effect of the failure modes.

1 = Likely to detect, 10 = Not likely at all to detect

The accompanying table illustrates a ranking scale (you can devise your own to suit your business).

Rating	Severity of Effect	Likelihood of Occurrence	Ability to Detect
10	Hazardous without warning	Very high: Failure is almost inevitable	Cannot detect
9	Hazardous with warning		Very remote chance of detection
8	Loss of primary function	High: Repeated failures	Remote chance of detection
7	Reduced primary function performance		Very low chance of detection
6	Loss of secondary function	Moderate: Occasional failures	Low chance of detection
5	Reduced secondary function performance		Moderate chance of detection
4	Minor defect noticed by most customers		Moderately high chance of detection
3	Minor defect noticed by some customers	Low: Relatively few failures	High chance of detection
2	Minor defect noticed by discriminating customers		Very high chance of detection
1	No effect	Remote: Failure is unlikely	Almost certain detection

Source: Used with permission from Sigma Breakthrough Technologies, Inc.

Calculate *risk priority numbers (RPN)* for each cause.

The quantified values associated with the severity, occurrence, and detection of failure modes and their causes can be used to calculate a summary value called the *risk priority number.* This assigns a value to the risk each failure mode represents.

RPN = (Severity) $\times$ (Occurrence) $\times$ (Detection)

Determine *recommended actions* to reduce high RPNs.

The team now establishes its plan of action to reduce the risk associated with the effect of each failure mode. The highest RPN items are the highest priority for action. The tools and best practices of DFSS are excellent means to correct these risks.

> **Apply tools and best practices to reduce sensitivity to root causes of failure or eliminate root causes of failure and recalculate RPNs.**

Once a plan of action is defined, the team members apply the appropriate tools and best practices that have been selected to correct the root causes of the failure modes.

Final Detailed Outputs

> **Causes of failure qualitatively and quantitatively documented and corrective action underway. New RPN calculated after corrective action results are documented.**

The final, corrective actions produce results that are documented in the FMEA database. New RPN numbers can now be calculated for each failure mode. The failure modes are still possible in some cases, but their probability of occurrence is greatly reduced.

DFMEA Checklist and Scorecard

This suggested format for a checklist can be used by an engineering team to help plan its work and aid in the detailed construction of a PERT or Gantt chart for project management of the DFMEA process within the Optimize phase of the CDOV or I^2DOV development process. These items also serve as a guide to the key items for evaluation within a formal gate review.

Checklist for DFMEA

Actions:	Was this Action Deployed? YES or NO	Excellence Rating: 1 = Poor 10 = Excellent
1. Develop a block diagram of the design element or function being analyzed (at system, subsystem, subassembly, or component level).	_____	_____
2. Determine the ways in which each design element or function can fail (*failure modes*).	_____	_____
3. Determine the effects of the failure on the customer(s) for each failure mode associated with the element or function.	_____	_____
4. Identify potential *causes* of each failure mode.		

(continued)

Checklist for DFMEA

Actions:	Was this Action Deployed? YES or NO	Excellence Rating: 1 = Poor 10 = Excellent
5. List the current *design controls* for each cause or failure mode.	_____	_____
6. Create severity, occurrence, and detection rating scales for each element or function under evaluation.	_____	_____
7. Assign severity, occurrence, and detection ratings to each cause.	_____	_____
8. Calculate risk priority numbers (RPN) for each cause.	_____	_____
9. Determine recommended actions to reduce high RPNs.	_____	_____
10. Apply tools and best practices to reduce sensitivity to root causes of failure or eliminate root causes of failure and recalculate RPNs.	_____	_____

Scorecard for DFMEA

The definition of a good DFMEA scorecard is that it should illustrate, in simple but quantitative terms, what gains were attained from the work. This suggested format for a scorecard can be used by an engineering team to help document the completeness of its work.

Number of failure items/functions in product: _____

Average RPN for the product prior to corrective action: _____

Number of items/functions under corrective action: _____

Average RPN for the product after corrective action: _____

Table of DFMEA Items/Functions :

Failure Item/ Function Name:	Failure Mode:	RPN Prior to Corrective Action:	RPN After Corrective Action:	Corrective Action:
1.	_____	_____	_____	_____
2.	_____	_____	_____	_____
3.	_____	_____	_____	_____

We repeat the final deliverables from DFMEA:

- Design failure modes identified
- Effects of a design failure mode identified
- Causes of a design failure identified
- Linkage of the following ratings to each cause:
  Severity rating
  Occurrence rating
  Detection rating
- Calculated risk priority numbers (RPN) for each cause
- Improved risk priority numbers (RPN) for each cause after corrective action

References

Potential failure mode and effects analysis (3rd ed.). (1995). AIAG (810-358-3003).

Stamatis, D. H. (1995). *Failure mode and effect analysis.* Milwaukee, Wisc.: ASQC Quality Press.

Reliability Prediction

This chapter was written by Edward Hume, a seasoned veteran in Six Sigma methods. Concurrent with the introduction of Motorola's Six Sigma initiative, Ted led the development and operation of his business unit's first Six Sigma manufacturing operation. He practiced Six Sigma in design and management positions over much of his career. Ted is an electrical engineer specializing in the design and reliability sciences.

Where am I in the Process? Where Do I Apply Reliability Prediction Within the Roadmaps?

Linkage to the I²DOV Technology Development Roadmap:

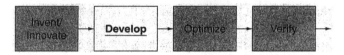

Reliability predictions are conducted within the Develop phase during technology development.

Linkage to the CDOV Product Development Roadmap:

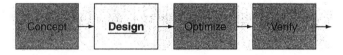

Reliability predictions are conducted within the Design phase during product development.

What Am I Doing in the Process? What Does Reliability Prediction Do at this Point in the Roadmaps?

The Purpose of Reliability Prediction in the I²DOV Technology Development Roadmap:

Reliability prediction is applied to qualitatively define and quantitatively predict the baseline reliability of new technology platforms and subsystems that are being developed for integration into a new platform or for integration into existing product designs. Preventative application of numerous tools and best practices can then be deployed to improve reliability, as required, prior to the technology being transferred into a commercialization program.

The Purpose of Reliability Prediction in the CDOV Product Development Roadmap:

Reliability predictions are first applied to predict the reliability of subsystems, subassemblies, and components and then to predict the overall integrated system reliability. Preventative application of numerous tools and best practices can also be deployed to improve the reliability as required prior to design certification.

What Output Do I Get at the Conclusion of this Phase of the Process? What Are Key Deliverables from Reliability Prediction at this Point in the Roadmaps?

- Predicted reliability performance of subsystems, subassemblies, and components
- Predicted reliability performance of the system
- Component, subassembly, and subsystem reliability performance improvement activities
  Required deliverables from specific tools and best practices that improve reliability by reducing variability and sensitivity to deterioration and external noise factors

The Reliability Prediction Flow Diagram (found on page 451)

This visual reference traces the detailed steps of the reliability prediction process.

Applying Each Block Diagram Within the Reliability Prediction Process

This section steps through the specifics of:

- *Key detailed input(s)* from preceding tools and best practices
- *Specific steps and deliverables* from each step to the next within the process flow
- *Final detailed outputs* that flow into follow-on tools and best practices

Key Detailed Inputs

> **Reliability requirements for system, historic reliability data, and DFMEA inputs.**

INPUT . . .

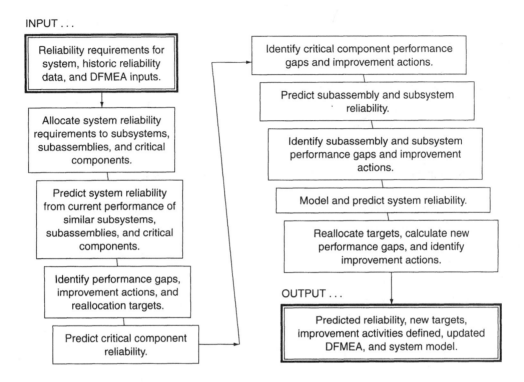

System Reliability Requirement Inputs

- Voice of the customer (concept engineering)
- Competitor comparisons, best within industry (benchmarking and QFD)
- World class, best of all industries, benchmarking
- Business unit improvement plans
- Government agencies
- Reliability data on similar current products
- Lessons learned and design rules

Concept Gate Review Question

- What was the basis for the reliability requirements?

Requirements Content

Requirements must be complete and consistent. Otherwise, a great reliability effort can fail to satisfy customer and/or business needs. Some key items should be in the requirements.

- Measurable definitions of reliability requirements:
  - The time to reach 1%, 10%, etc., failures (often referred to as B1, B10, etc., life)
  - MTTF—mean time to failure

- MTBF—mean time between failures, used to describe repairable equipment with constant failure rates
- % per month or year failure rate
- Environmental conditions:
  - Where is it used: in an office, factory, vehicle, or aircraft?
  - Is it air-conditioned, subject to ambient temperatures, or subject to raised temperatures because of its enclosure?
  - Is it exposed to dust, rain, or ice?
  - Is it dropped? How is it shipped?
  - Is it subject to spills, chemicals, or being cleaned with a hose?
  - If it uses electrical power, will it be used in developed and/or undeveloped countries?
- Governmental agency requirements: All agency requirements from all countries of intended use must be identified.

Concept Gate Review Questions

- Are reliability requirements measurable?
- Can reliability requirements be achieved within the scope of the project?
- Are environmental requirements completely described for current and derivative products?
- Are governmental agency requirements completely known for current and derivative products?

These inputs must be used to generate a detailed environmental specification to guide the design and testing. One of the most common causes of significant reliability problems is not designing for the appropriate environmental conditions.

Historical Warranty and Reliability Data Inputs

Many companies do not have good historical reliability data. For example:

- No data is collected beyond the warranty period.
- Time to failure is not recorded or is difficult to derive because data is kept primarily to support standard monthly reports with percent failure rates and/or warranty cost.
- Cause of failure is not determined or, in many cases, what is reported is a symptom rather than a root cause.

As an example of symptom vs. cause, let's look at a cell phone that no longer works properly. The symptom is that the battery won't hold its charge. Repair includes replacing a circuit board that controls battery charging operations, the antenna assembly, and the display. All replaced items have been damaged by what must have been a severe shock. What was the cause of the failures? The phone was dropped on concrete producing a stress (high deceleration) that exceeded the strength of solder joints, display, and connections in the antenna assembly. The information that often gets reported is that the phone did not hold a charge and sometimes that the circuit board, display, and antenna assembly were replaced. To learn from the repair we need to understand the stress that caused the damage (when possible), all the items that were damaged, and the nature and the location of the damage.

A system that provides information on time to failure and cause of failure over the expected life of the product is a plus when starting the Develop/Design phase. Many companies maintain records of all failures that occur in the warranty period. To add cause of failure to all repair records and extend them to cover the life of the product would be expensive. It would also be difficult to achieve because of the requirement to train a large number of people on proper failure analysis. Only a representative sample of the repair data is necessary or desirable. As in all Six Sigma efforts, small manageable sets of data are used for learning and improvement.

Collection of Field Failure Data

A best practice is to establish a partnership with a few key customers and/or repair depots and train them to provide needed field failure information. Those selected should provide a good cross-section of the customer base. Field failure information should be collected over the entire design life of the product.

Lessons Learned and Design Rules

Another best practice is to turn the experience from each product development and field problem into lessons learned. Turning the experience into design rules ensures that the lessons will be used in future designs. Those companies that have a process to learn from each experience clearly will have steeper learning curves than those companies that don't.

Concept Gate Review Questions

- Do we have historic reliability data, including the cause of failures, to use as a starting point for prediction efforts?
- Do we have a database of lessons learned and has it been reviewed?
- Do we turn lessons learned into design rules?

DFMEA Inputs

From a customer system FMEA, failure modes or a combination of failure modes that have a severe effect (high effect score) on the customer's system constitute DFMEA inputs. In system, subsystem, and subassembly DFMEAs, a high RPN number requires special attention during the design process, reliability prediction, and reliability testing.

Concept Gate Review Questions

- Has the initial DFMEA been performed on the selected concept?
- Has a customer systems FMEA been performed?

Specific Steps and Deliverables

> **Allocate system reliability requirements to subsystems, subassemblies, and critical components.**

As soon as the final concept has been determined, the overall system reliability requirements should be allocated to each of the subsystems, subassemblies, and critical components to create targets for reliability development. This initial allocation is based on current knowledge of the reliability of all the items in the design concept.

> **Predict system reliability from current performance of similar subsystems, subassemblies, and critical components.**

When the design concept is selected and the targets are identified, it is important to do a first pass prediction to identify performance gaps. Since the design is in its infancy, detailed component predictions are not possible. However, the performance of similar subsystems, subassemblies, and critical components can be used to provide a reasonably accurate estimate of reliability performance. The reliability performance of similar subsystems, subassemblies, and critical components are adjusted to those in the current design concept by comparing the complexity, inherent stress levels, external stress levels, expected length of life, and base reliability of critical components.

Develop a reliability block diagram (RBD) as described in the last step, "Model and predict system reliability," to use as the model to contain the estimated information.

If you lack historical information on some of the subsystems, subassemblies, and critical components, put together an estimate of the bill of materials, then follow the detailed process of prediction in the steps that follow. In the case of electronics, Military Handbook 217F (MIL-HDBK-217F) offers a part count method of predicting electronics reliability. This is a reasonable method of initially establishing relative reliability. Don't wait for the design to be complete before doing predictions. Once a design is complete, it is too late to prevent the development of a product that doesn't meet reliability requirements.

> **Identify performance gaps, reliability improvement actions, and needed target reallocation.**

Gaps between the predictions and targets can lead to a number of actions:

- Reevaluation of the system concept
- Reallocation of targets
- Prioritized reliability development actions

> **Predict critical component reliability.**

Most products have individual components that, by themselves, are a significant part of the overall system's reliability. These components are typically involved in cost or performance vs. reliability trade-off and require significant reliability development efforts.

Examples of such components are the following:

Bearings

Fans

Capacitors

Batteries

Structural members that are flexed during operation

Semiconductor power devices

Complex semiconductor devices such as microprocessors

Sources of data to make predictions will include:

- Supplier information
- Reference databases:
  MIL-HDBK-217F—Available through prediction programs
  Telcordia (Bellcore 6)—Available through prediction programs
  Prism, Electronic Parts Reliability Data (EPRD), and Non-Electronic Parts Reliability
     Data (NPRD)—Reliability Analysis Center (RAC), raciitri.org, 888-722-8737
- Field reliability data
- Test data

Supplier information and reference databases usually specify reliability in terms of a constant base failure rate for each part. Part failure rate in your application is a function of the base failure rate modified by stress, quality, and environmental factors. Reliability/survival or cumulative failure probability are calculated from the part failure rate.

Field reliability and test data are fit to known statistical distributions. Reliability/survival or cumulative failure probabilities are calculated from the fitted distribution.

Before we proceed with critical part reliability prediction, let's consider some basic reliability information. Four functions are used to describe reliability performance:

Probability density function—Plot of amount of failures vs. time (or cycles, miles, etc.), $f(t)$. This function would fit a histogram of your data. The most common example of a probability density function is a standard normal curve.

Cumulative Distribution Function—For any given time, the probability that the part is not working is $F(t) = \int f(t)dt$.

Reliability/Survival Function—For any given time, the probability that the part is still working is $R(t) = 1 - F(t)$

Hazard Function—Rate of failure at any given time is $h(t) = f(t)/R(t)$. This failure rate is represented in a bathtub curve, as in the following example.

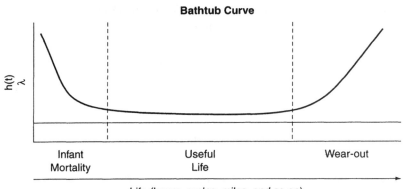

Bathtub Curve

Infant Mortality   Useful Life   Wear-out

Life (hours, cycles, miles, and so on)

Known statistical distributions are used to fit data from field and test data. Some commonly used distributions are the exponential, Weibull, normal, and lognormal. The pertinent functions for the exponential distribution are shown in this equation:

$$f(t) = le^{-tl} \ (t > 0)$$
$$F(t) = 1 - e^{-tl}$$
$$R(t) = e^{-tl}$$
$$h(t) = 1$$

Note that the exponential distribution has some unique properties. R(t) and F(t) are easy to calculate with a calculator. Since the hazard function is a constant (λ), failure rate is not a function of time. This makes it natural for modeling the useful life, constant failure rate, portion of the bathtub curve. Mean time between failure (MTBF) is symbolized by θ and for the exponential distribution only, $\theta = 1/\lambda$. The pertinent functions for the Weibull distribution are shown in the following equation.

$$f(t) = \frac{\beta}{\alpha^{\beta}} t^{(\beta-1)} e^{-(t/\alpha)^{\beta}} \ (t \geq 0)$$
$$F(t) = 1 - e^{-(t/\alpha)^{\beta}}$$
$$R(t) = e^{-(t/\alpha)^{\beta}}$$
$$h(t) = \left(\frac{\beta}{\alpha}\right)\left(\frac{t}{\alpha}\right)^{(\beta-1)}$$

The Weibull distribution also has some interesting properties. It can fit data sets that represent all portions of the bathtub curve. The shape/slope parameter β controls the shape of the curve. When $\beta < 1$, the failure rate decreases with time. When $\beta = 1$, the Weibull distribution becomes an exponential distribution and failure rates are constant. When $\beta > 1$, failure rate increases with time, modeling the wear-out portion of the bathtub curve. The scale parameter alpha (α), also called characteristic life eta (η), represents the point where 63.2% of the units have failed. The pertinent functions for the normal distribution are shown in this equation:

$$f(y) = (2\pi\sigma^2)^{-1/2} \exp[-(y - \mu)^2/(2\sigma^2)], -\infty \le y \le \infty$$

$$F(y) = \int_{-\infty}^{y} (2\pi\sigma^2)^{-1/2} \exp[-(y - \mu)^2/(2\sigma^2)]$$

$$F(y) = \phi[(y - \mu)/\sigma], -\infty \le y \le \infty$$

The location parameter μ is also the mean time to failure (MTTF) and the mean of the data set. Scale parameter σ is the standard deviation of the data set. The pertinent functions for the lognormal distribution are shown as

$$f(t) = \{.4343/[(2\pi)^{1/2}t\sigma]\} \exp\{-[\log(t) - \mu]^2/(2\sigma^2)\}, t \ge 0$$

$$F(t) = \Phi\{[\log(t) - \mu]/\sigma\}, t \ge 0$$

The location parameter μ is also the MTTF and the mean of the log of the data set. It is commonly called the *log mean*. Scale parameter σ is the standard deviation of the log of the data set. It is commonly called the *log standard deviation*. The lognormal distribution forms a number of different shapes with variation of the scale parameter σ. Shapes include both the infant mortality and wear-out portions of the bathtub curve, making it quite useful in reliability work.

Prediction with Supplier Information and Reference Databases

The MIL-HDBK-217 is used to give an example of the process used to make predictions from supplier information and reference databases. Each part type has a base failure rate reported in failures per million hours of operation. This is modified by a number of π factors to come up with the part failure rate in the proposed application.

$$\lambda_p = \lambda_b \pi_T \pi_A \pi_R \pi_S \pi_C \pi_{CV} \pi_Q \pi_E$$

λ_p = Part failure rate

λ_b = Base part failure Rate

π_T = Temperature factor

π_A = Applications factor

π_R = Resistance factor, power rating

π_S = Voltage applied/voltage rated factor

π_C = Construction factor

π_{CV} = Capacitance factor

π_Q = Quality factor

π_E = Environment factor

Each part is described by some or all of these factors as specified by MIL-HDBK-217. Each of these failure rates is a constant and represents the useful life portion of the bathtub curve. The exponential distribution has a constant failure rate and is the one we turn to for calculating reliability.

An aluminum electrolytic capacitor will be used as an example: $\lambda_p = \lambda_b{}^*\pi_{CV}{}^*\pi_Q{}^*\pi_E$.

λ_b is a function of voltage and temperature stress. The capacitor is rated at 400µf, 85°C, and 25v. Operating conditions will be 60°C and 17.5v.

$\lambda_b = .33 \times 10^{-6}$ failures per hour

$\pi_{CV} = 1, \pi_Q = 1, \pi_E = 2$

$\lambda_b = (33 \times 10^{-6}) \times 1 \times 1 \times 2 = .66 \times 10^{-6}$ failures per hour

Exponential distribution: $f(t) = \lambda e^{-t\lambda}$ $(t > 0)$, $F(t) = 1 - e^{-t\lambda}$, $R(t) = e^{-t\lambda}$, $h(t) = \lambda$. Using a calculator we can determine $F(t)$ and $R(t)$ at 1 year (8,760 hours) and 10 years (87,600 hours).

$R(t) = e^{-t\lambda} = e^{-8700 \times (.00000066)} = .994$, probability for survival of 1 year is 99%.

$R(t) = e^{-t\lambda} = e^{-87000 \times (.00000066)} = .944$, probability for survival of 10 years is 94%.

$F(t) = 1 - R(t) = 1 - .994 = .006$, $1 - .944 = .056$, probability for failure at 1 and 10 years, respectively, is 1% and 6%.

A similar process can be followed for supplier information and other reference databases.

Tool Summary

Limitations

- Failure rates in the database are not collected under identical conditions to those to which your product will be exposed.
- Failure rates in the database are collected on parts whose construction and manufacturing processes may have changed for the parts used in your product.
- Failure rates in the database often do not describe infant mortality and wear-out failures.

Benefits

- Causes a review of the quality level of selected parts.
- Causes a review of all pertinent stress levels on the part in question.

- Provides a first order approximation of part reliability.
- Provides input to reliability allocation.

Recommendations

- Use to provide reliability comparisons to known performance of similar products.
- Use for reliability allocation.
- Use to provide a basis for reliability improvement efforts.
- Use to review stress levels applied and inherent quality performance of key parts.
- Verify predictions through testing.

Prediction with Field and Test Data

The Minitab program is used to show how a three-step process will take you from data to predicted reliability with confidence intervals. The steps are the following:

1. Fit a known distribution to the data.
2. Review the shape of statistical functions $f(t)$, $R(t)$, and $h(t)$ and observe the distribution parameters.
3. Project B1, 10, etc., life and project reliability/survival probabilities at times such as 10 years, 100,000 miles, etc.

We will use the results of a test data from 40 incandescent light bulbs as an example.

Step 1. Fit a known distribution to the data. Open Minitab and enter time to failure for each light bulb into one column labeled "N hours." Sample data follows.

3666	6823	4489	7707	5591
6579	8248	5565	4743	5591
6671	5506	2356	1786	4786
5525	7675	7979	1426	4672
7742	3513	3290	5194	6225
4840	3804	803	6627	5267
6243	5990	5729	6466	6019
3949	2806	6282	6466	6019

Using pull-down menus, follow this path—*Stat, Reliability/Survival, Distribution ID Plot-Right Cens.* Load "N Hours" into the dialog box and run the program. Graphical results showing the best fit are shown in Figure 20–1.

Use visual feedback showing how well the data points fit the proposed distribution and the Anderson-Darling (adj.) number to determine which distribution provides a better fit. The Anderson-Darling (adj.) number is a goodness of fit measurement, in which smaller is better.

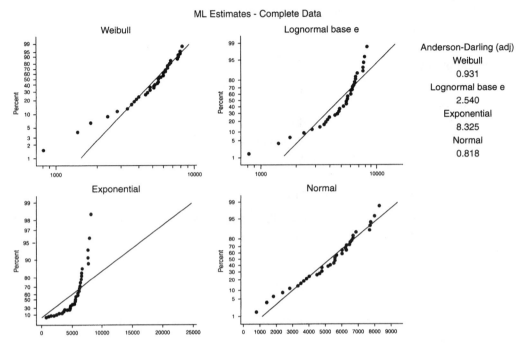

Figure 20–1   Four-Way Probability Plot for N Hours

Both approaches show the normal distribution is a slightly better fit than the Weibull distribution. The exponential and lognormal distributions are not good enough fits to be considered. Since the normal distribution fits the best, it will be used for further analysis.

Step 2. Using pull-down menus, follow this path—*Stat, Reliability/Survival, Distribution Overview Plot-Right Cens.* Load "N Hours" into the dialog box and select the normal distribution. Graphical results are shown in Figure 20–2.

The probability density function, *f(t)*, is the normal curve we are used to seeing. Survival function, *R(t)*, depicts the probability of survival as the hours of use increase. Hazard function, *h(t)*, shows the increasing failure rate of a wear-out function. The location parameter, MTTF, or mean is 5,266.5 hours. Scale parameter or standard deviation is 1,778.7 hours. Some important information found on this graph is the shape of *f(t)*/probability density function and the shape of *h(t)*. From *h(t)* we learn if we have infant mortality, constant failures, or wear-out. This analysis is done one failure mode at a time. Data sets with more than one failure mode are split apart and analyzed separately.

Step 3. Using pull-down menus, follow this path—*Stat, Reliability/Survival, Parametric Dist Analysis Plot-Right Cens.* Load "N Hours" into the dialog box and select the normal distribution. Select the Estimate button and load times for which you desire survival probabilities. Session window output results follow.

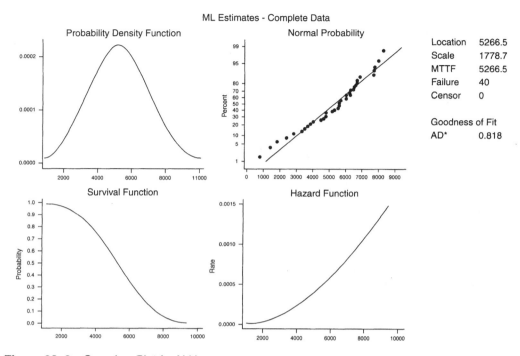

Figure 20–2   Overview Plot for N Hours

Table of Percentiles

Percent	Percentile	Standard Error	95.0% Lower	Normal CI Upper
1	1128.514	541.4131	67.364	2189.664
10	2986.921	379.5390	2243.038	3730.803
50	5266.450	281.2413	4715.227	5817.673
90	7545.979	379.5390	6802.097	8289.862

Table of Survival Probabilities

Time	Probability	95.0% Lower	Normal CI Upper
2000.000	0.9669	0.9080	0.9905
3000.000	0.8987	0.8043	0.9546
3500.000	0.8397	0.7305	0.9149
4000.000	0.7618	0.6425	0.8552
4500.000	0.6667	0.5426	0.7748
5000.000	0.5595	0.4357	0.6778
5500.000	0.4478	0.3291	0.5714

Tool Summary

Limitations

- Requires significant investment in field data collection and/or testing facilities

Benefits

- Time to 1%, 10%, etc., failures is projected with confidence limits
- Survival probability at selected times is projected with confidence limits
- The hazard function *h(t)* is defined graphically, providing input to how failure rates vary over life

Recommendations

- Use to decide on advertised specifications of life, such as a 4,000 hour light bulb
- Use to set warranty policy and replacement intervals
- Use to predict reliability and time to failure
- Use for reliability allocation
- Use to provide a basis for reliability improvement efforts

**Identify critical component performance gaps
and reliability improvement actions.**

Reliability performance in critical components can drive a number of improvement actions:

- Reviewing the basic product concept to find ways of reducing stress on the components
- Adding components to reduce stress or provide redundancy
- Improving the quality of the component

Predict subassembly and subsystem reliability.

Predictions based on supplier information, reference databases, or other source of constant failure rate information are made using a unique property of the exponential distribution: failure rates are additive. The failure rate of the subassembly or subsystem is simply the sum of all the individual failure rates. Once the total failure rate is calculated, *F(t)* and *R(t)* are found for the subassembly or subsystem using the same equations used in the previous example:

$$F(t) = 1 - e^{-t\lambda}, R(t) = e^{-t\lambda}$$

A number of programs are available that contain one or more reference databases, are capable of importing bills of materials, and can calculate individual failure rates and system or subsystem reliability.

Prediction Software

- Relex Software, *www.relexsoftware.com,* 724-836-8800
- Reliability Analysis Center RAC "Prism," *rac.iitri.org/,* 888-722-8737
- RelCalc for Windows, *www.t-cubed.com,* 818-991-0057
- Item Software (USA) Inc., *www.itemsoft.com/,* 714-935 2900

Tool Summary

Limitations

- The same as reported at a component level

Benefits

The same as reported at a component level plus:

- Software tools available make tabulating and storing subsystem and subassembly reliability relatively easy.
- What-if calculations help to quickly define improvement actions.

Recommendations

- The same as reported at a component level

> **Identify subassembly and subsystem performance gaps and reliability improvement actions.**

Reliability performance in subsystems and subassemblies can drive a number of improvement actions:

- Reviewing the subsystem or subassembly design to find ways of reducing stress on the components
- Reducing the number of components in the designs
- Improving the quality of the components

Model and Predict System Reliability

System reliability is modeled using a reliability block diagram (RBD). Subsystems, subassemblies, and major components are arranged into series and parallel orientation. All items whose failure will

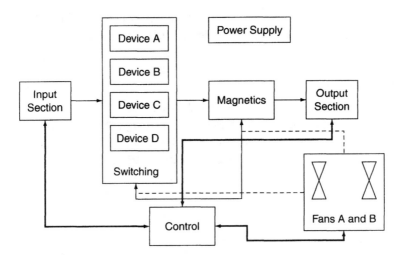

Figure 20–3 Sample Diagram of Power Supply

cause the system to fail are oriented in series. All items that have a degree of redundancy, where more than one item is required to fail for the system to fail, are oriented in parallel. Some of the subsystems, subassemblies, and major components may have competing failure modes. In this case, the competing failure modes are given separate series blocks. The reliability survival function is used to calculate probability of survival for all subsystems, subassemblies, and major components. Figure 20–3 is an example of a functional block diagram for a power supply.

Translating the power supply into an RBD gives us the following:

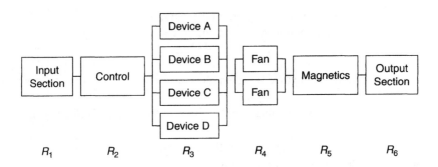

Six blocks are labeled R, each representing the reliability/survival probability of the block. Blocks $R_{3,4}$ are made up of critical components that have redundancy. The reliability of the system is equal to the product of each of the blocks.

$$R_S = R_1 *R_2 *R_3 *R_4 *R_5 *R_6$$

If blocks R_{1-6} have a 99 percent probability of surviving for 10 years then the reliability of the system would be: $R_S = (.99)\,(.99)\,(.99)\,(.99)\,(.99)\,(.99) = (.99)^6 = 0.941$.

Reliability of blocks R_3 and R_4 is determined by calculating the probability of survival of at least those items required for successful operation. There are two possible outcomes, survival R and failure Q. Therefore the probability of $R + Q = 1$. If only one fan is required, then the possible outcomes of both fans are $R_A R_B + R_A Q_B + Q_A R_B + Q_A Q_B$. The probability of $R_A R_B + R_A Q_B + Q_A R_B + Q_A Q_B = 1$. At least one fan is required for survival, so outcomes $R_A R_B + R_A Q_B + Q_A R_B$ represent survival, and outcome $Q_A Q_B$ represents failure. If the probability of survival of both fans is 90 percent, we can calculate the probability of survival of block R_4 as follows:

$$R_4 = R_A R_B + R_A Q_B + Q_A R_B = 1 - Q_A Q_B = 1 - (1 - .90)(1 - .90)$$
$$R_4 = .99$$

Using the assumption that each of the switching devices has an equal probability of survival, the equation for all possible outcomes is:

$$R^4 + 4R^3 Q + 6R^2 Q^2 + 4R Q^3 + Q^4 = 1$$

If three devices are required for operation, $R^4 + 4R^3 Q$ represents the successful outcomes. If the probability of survival of each device is 95 percent, we can calculate the probability of success for R_3:

$$R^4 + 4R^3 Q = 1 - (6R^2 Q^2 + 4R Q^3 + Q^4) = 1 - ((6^*(.95)^{2*}(.05)^2) + 4^*(.95)^*(.05)^3$$
$$+ (.05)^4)$$
$$R_3 = .986$$

Tool Use Summary

Benefits

- All sources of reliability data can be turned into a reliability/survival probability and used in the model.
- Creating an RBD helps identify the critical subsystems, subassemblies, and components that contribute most to lack of success.

Recommendations

- Use of this tool should be initiated after the design concept is selected and updated continuously.
- Reliability performance of the system, subsystems, subassemblies, and critical components predicted by the model should be tracked against the target.
- Consider using prediction software that creates and maintains complex reliability block diagrams (such as Relex Software, *www.relexsoftware.com,* 724-836-8800).

> **Reallocate targets, calculate new performance gaps,
> and identify improvement actions.**

Now system, subsystem, subassembly, and critical component reliability predictions have been completed. Performance gaps have been identified. Significant contributors to loss of reliability have been identified. The following summary lists actions that should be taken:

- Reallocate targets that add up to meeting reliability requirements.
- Search for additional supplier data, database, field data, and test data that will confirm failure rates or reliability of critical components.
- Design improvement activities to:
  Reduce stress on critical parts
  Select higher reliability parts
  Reduce complexity
- Plan reliability testing to confirm predictions and improvements.

Final Detailed Outputs

> **Predicted reliability, new targets, improvement
> activities defined, updated DFMEA, and system model.**

Outputs of the prediction process should include the following:

1. Targets for the system, subsystems, subassemblies, and critical components
2. Predictions for the system, subsystems, subassemblies, and critical components
3. Reliability performance gaps
4. Improvement activities planned for performance gaps
5. Reliability testing plans

Reliability Prediction Checklist and Scorecard

This suggested format for a checklist can be used by an engineering team to help plan its work. The list will also aid in the detailed construction of a PERT or Gantt chart for project management of the reliability prediction process within the Optimize and Capability Verification Phases of the CDOV or I^2DOV development process. These items also serve as a guide to the key items for evaluation within a formal gate review.

Checklist for Reliability Predictions

Actions:	Was this Action Deployed? YES or NO	Excellence Rating: 1 = Poor 10 = Excellent
1. Allocate system reliability requirements to subsystems, subassemblies, and critical components.	_____	_____
2. Predict system reliability from current performance of similar subsystems, subassemblies, and critical components.	_____	_____
3. Identify performance gaps, improvement actions, and reallocation targets	_____	_____
4. Predict critical component reliability.	_____	_____
5. Identify critical component performance gaps and improvement actions.	_____	_____
6. Predict subassembly and subsystem reliability.	_____	_____
7. Identify subassembly and subsystem performance gaps and reliability improvement actions.	_____	_____
8. Model and predict system reliability.	_____	_____
9. Reallocate targets, calculate new performance gaps, and identify improvement actions.	_____	_____

Scorecard for Reliability Predictions

The definition of a good reliability prediction scorecard is that it should illustrate, in simple but quantitative terms, what gains were attained from the work. This suggested format for a scorecard can be used by an engineering team to help document the completeness of its work.

Table of Documented Reliability Failures and Corrective Actions

System, Subsystem, Subassembly, or Critical Component Name:	Target Reliability:	Predicted Reliability:	Gap:	Improvement Action and Tools:
1.	_____	_____	_____	_____
2.	_____	_____	_____	_____
3.	_____	_____	_____	_____

Leadership Review Guidelines

A manager should ask these questions at gate reviews to promote problem prevention and forward-looking support issues to foster a problem prevention culture during technology and product development. Questions for the Concept or Invent/Innovate gate review presented earlier in this chapter are listed again here:

- What was the basis for the reliability requirements?
- Are reliability requirements measurable?
- Can reliability requirements be achieved within the scope of the project?
- Are environmental requirements completely described for current and derivative products?
- Are governmental agency requirements completely known/described for current and derivative products?
- Do we have historic reliability data, including the cause of failures, to use as a starting point for prediction efforts?
- Do we have a database of lessons learned and has it been reviewed?
- Do we turn lessons learned into design rules?
- Has the initial DFMEA been performed on the selected concept?
- Has a customer systems FMEA been performed?

Let's review the best practices again as well:

Collection of Field Failure Data

A best practice is to establish a partnership with a few key customers and/or repair depots and train them to provide needed field failure information. Those selected should provide a good cross-section of the customer base. Field failure information should be collected over the entire design life of the product.

Lessons Learned and Design Rules

A best practice is to turn the experience from each product development and field problem into lessons learned. Turning the experience into design rules ensures that the lessons will be used in future designs. Those companies that have a process to learn from each experience clearly have steeper learning curves than those companies that don't.

Asking a lot of questions about the results of reliability predictions at the end of the phases of development is not useful. By then it is simply too late to do much about what was not attained or done! Thus, we worry about the questions to be asked *prior* to conducting reliability predictions, so that goes as smoothly as possible. Everything you need to know as an output from reliability predictions was stated back in the final step in the reliability predictions process flow diagram:

Failure rates are qualitatively and quantitatively predicted, improvement action and tools are identified, and DFMEAs are updated.

Finally, we repeat the final deliverables from reliability predictions:

1. Targets for the system, subsystems, subassemblies, and critical components
2. Predictions for the system, subsystems, subassemblies, and critical components
3. Reliability performance gaps
4. Improvement activities planned for performance gaps
5. Reliability testing plans

References

Condra, L. W. (1993). *Reliability improvement with design of experiments.* New York: Marcel Dekker.

Elsayed, E. A. (1996). *Reliability engineering.* Reading, Mass.: Addison-Wesley.

Leitch, R. D. (1995). *Reliability analysis for engineers.* Oxford: Oxford University Press.

O'Connor, P. D. T. (1991). *Practical reliability engineering* (3rd ed.). Chichester: J. Wiley & Sons.

Military Handbook (MIL-HDBK-217F).

Introduction to Descriptive Statistics

Where Am I in the Process? Where Do I Apply Descriptive Statistics Within the Roadmaps?

Linkage to the I^2DOV Technology Development Roadmap:

Descriptive statistics are used across all the phases of I^2DOV, perhaps to a lesser degree during the Invent/Innovate phase.

Linkage to the CDOV Product Development Roadmap:

Descriptive statistics are used across all the phases of **CDOV,** perhaps to a lesser degree during the Concept phase.

The Purpose of Descriptive Statistics in the I^2DOV Technology Development Roadmap:

The purpose of descriptive statistics in I^2DOV is to describe and predict technology performance over the latitude of system parameters and stressful generic noises that it is expected to experience during commercialization and end use.

The Purpose of Descriptive Statistics in the CDOV Product Development Roadmap:

The purpose of descriptive statistics in CDOV is to describe and predict product performance during manufacture and end use based on the performance of the few prototypes available during CDOV.

What Output Do I Get from Using Descriptive Statistics?

- Probability distributions and models of critical responses
- Descriptive statistics providing estimates of mean performance and variation
- Predictions of performance based on the descriptive statistics
- Information needed to calculate capability

What Am I Doing in the Process?

Employing descriptive statistics, we can take measurements of our system's critical responses to estimate their distributions, means, and variances. We calculate the mean and variance from the sample data. And we use these calculations and distributions to predict the probability of producing values above or below the specification limits.

Descriptive Statistics Review and Tools

The Mean

As design engineers, we are faced with many challenges during product development. Two major challenges are described in these questions:

- How can we quantify and control variation in our critical responses?
- How can we predict the performance of a "population" of products based on the few prototypes available during design?

Statistics provide the tools to answer these questions. Let's begin our review of statistics with the average. Several types of averages are employed in product design and development. The equation for an important one, the arithmetic mean, is given in these equations:

$$\overline{X} = \frac{\sum_{i=1}^{n} x_i}{n}$$

$$\mu = \frac{\sum_{i=1}^{N} x_i}{N}$$

The first equation is the familiar formula we use to find our grade point average in school. Add up all the grades and divide by the number of grades. The first equation applies to the sample mean $\overline{X}$, and the second to the population mean μ. A population may be of infinite size while a sample consists of a finite number of observations, say n, taken from the much larger population. The mean provides an estimate of central tendency, center of gravity, or point of balance. We use the mean to determine one aspect of robustness: Is our critical response on target?

The mean is sensitive to outliers. An outlier is an observation that is unusual in magnitude compared to the rest of the observations in a sample. For example, the mean of 1, 2, 3, 4, and 5 is 3. The mean of 1, 2, 3, 4 and 100 is 22. A single outlying observation has changed the mean by a large amount. We use the mean because we want to detect outliers. An outlier generally has a negative connotation, but it may indicate a condition of extremely good or extremely poor performance. We want to know about such a condition so we can stay away from it or strive to operate at it.

Two other types of averages are the median and mode. The median is the middle observation when the sample has been put in numeric order. The mode is the most frequently occurring number. These types of averages are less sensitive to outliers.

The Variance and Standard Deviation

Next, we discuss the variance and standard deviation. The formulae for the sample variance s^2 and the sample standard deviation s are given in the first of the two following equations. The population variance σ^2 and standard deviation σ are given in the second equation.

$$s^2 = \frac{\sum_{i=1}^{n} (x_i - \overline{X})^2}{n - 1} \text{ and } s = \sqrt{s^2}$$

$$\sigma^2 = \frac{\sum_{i=1}^{N} (x_i - \mu)^2}{N} \text{ and } \sigma = \sqrt{\sigma^2}$$

The variance tells us whether our observations cluster near the mean or spread far away from the mean. As the variance increases, the observations spread out farther from the mean. In general, we want a small variance. The square root of the variance is the standard deviation, which has the advantage of being in the same units as the measured quantity. We use the variance to determine another aspect of robustness: Is the variation in our critical response adequately insensitive to noise?

Let's look at the components of the sample variance equation. The quantity in parenthesis, $(x_i - \overline{X})$ the deviation, is the distance of each observation from the mean. This is calculated for all n observations. The next step is to square the deviates. Why square them?

If we were to sum up all of the deviations, the sum would be zero. Try it using any set of numbers. We could sum the absolute value of the deviations to avoid this but absolute values are difficult for linear operations such as differentiation. The decision to square the deviations addresses these two issues and has the advantage of "penalizing" more heavily as observations move farther from the mean, as shown in Figure 21–1.

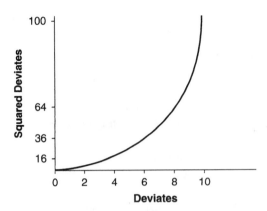

Figure 21-1   Relationship of Deviates to Squared Deviates

The next step is to calculate the average squared deviation by dividing the sum of all the deviations by $n - 1$. Why $n - 1$ and not n? One reason has to do with information. The concept of information is critical in statistics, so it is worth spending some time on it now. As we noted earlier, we use transfer functions to model the effects our critical parameters have on our critical responses. Each term in the transfer function estimates a parameter's effect and is considered a "piece of information." It is assigned a **degree of freedom** (dof). Information and dof are synonymous.

Consider the following examples. Imagine a container, held horizontally, that can only hold three tennis balls. At step one, I am free to put a ball in any of the three open positions. At step two, I am free to put a ball in any of the two remaining positions. But, at step three, I have no freedom as to where I put the last ball. It must go in the only remaining open position. Using three balls, I have two choices or "two degrees of freedom."

To consider another example of dof, imagine that I have five numbers in mind and I tell you four of them. Say, 1, 2, 3, 4 and ?. I also tell you that the average of the five numbers is 3. There is no freedom in the choice of the fifth number. It must be 5. Out of n numbers, I have '$n - 1$' dof. The mean contains information about the response value we expect to get most of the time and it "costs" one dof to get that information. The mean is contained in the equation used to calculate the variance. Therefore, one dof has been consumed, so we divide by $n - 1$, not n.

Another reason for $n - 1$ comes from estimation theory. The sample variance is an estimate of the population parameter, σ^2. The $n - 1$ is required to make s^2 an unbiased estimator of σ^2.

Thus, we have two statistics we can use to quantify the variation in, and the average performance of, our critical responses. These are called *descriptive statistics,* and they help us with our challenge to quantify variation. How about predicting performance? What does Statistics offer here?

As we stated earlier, the sample variance is an unbiased estimator of the population parameter, σ^2. Let's investigate this a bit. A sample is made up of the measurements we make of our design's critical responses, and they come from a much larger population of possible measurements. The population is characterized by certain parameters, two of which are the mean μ and the variance σ^2. The population may be finite or it may be infinite, and we may never know the values of μ and σ^2. The purpose of taking a sample is to estimate them.

Statistical theory tells us that the expected value of the sample mean and sample variance equals the population mean and variance, respectively. This is shown in the following equation for the random variable X.

$$E(\overline{X}) = \mu_X \text{ and } E(s_X^2) = \sigma_X{}^2$$

This is true in the long run. However, we usually can't wait that long and are content to use the sample mean and variance as estimates of the population mean and variance. This is shown in the following equation, where the "hat" or "roof" is used to indicate an estimate.

$$\hat{\sigma}_X^2 = s_X{}^2 \text{ and } \hat{\mu}_X = \overline{X}$$

If our sample adequately represents the population, we can estimate the population parameters by the sample statistics. Herein lies the solution to the challenge of predicting performance. If we can estimate the population's parameters and its distribution from our sample, we can predict the performance of a population of product from our sample. But what is a distribution?

The term *distribution* is synonymous with the shape of the data when plotted as a histogram. The histogram provides a graphical view of how the data distribute themselves in terms of relative frequency. Sometimes, as with the normal distribution, we have an equation, called a *probability density function* (pdf), that describes the distribution mathematically.

The distribution has a total area of one. The area under the curve that is below a particular value is equal to the probability of getting numbers that are less than or equal to that value. The area under the curve that is above a particular value is equal to the probability of getting numbers that are greater than or equal to that value. Figure 21–2 gives an example of this using the normal distribution.

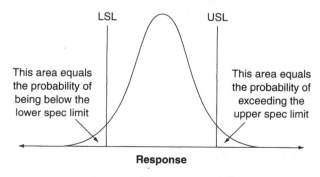

Figure 21–2   Example of Normal Distribution

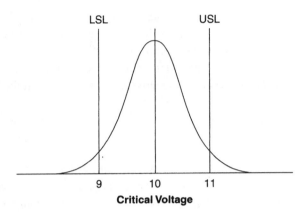

Figure 21–3   Critical Voltage

Probabilities have been calculated for the important distributions and can be found in the back of most statistics books. Luckily, software packages, such as Minitab, can calculate these probabilities for us.

Let's try a simple example. Say we have a critical voltage with the following characteristics:

- A target value of 10V
- An historic standard deviation of 1V
- Upper and lower specification limits of 11V and 9V, respectively
- Adequately described by the normal distribution

What are the chances of producing voltages above and below the specification limits? The situation is shown in Figure 21–3.

We need to determine the area total under the normal distribution, which is below 9 and above 11. Let's use Minitab.

We use the Minitab pull-down menu to go to Calc > Probability Distributions > Normal, and we get the window shown on page 477. We enter the mean, sd (standard deviation), and the lower specification limit. The output is given in Minitab's session window, shown on opposite page.

Cumulative Distribution Function

```
Normal with mean = 10.0000 and standard deviation = 1.00000
x        P(X <= x)
9.0000   0.1587
```

Minitab has calculated the probability of getting numbers from a normal distribution that are less than or equal to 9. This is the area to the left of the LSL. The area to the right of the USL is the same since the limits and the normal distribution are both symmetric about the mean.

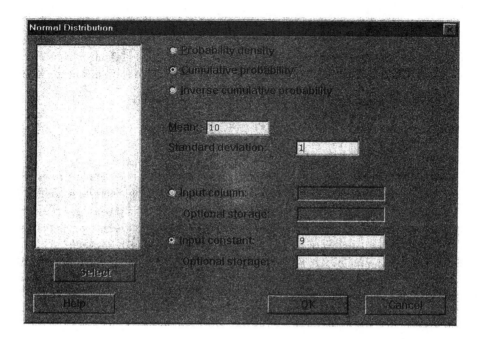

Thus, the probability of not being within the limits is $0.1587 + 0.1587 = 0.3174$, or 31.74%. The probability of being within the limits is $1 - 0.3174 = 0.6826$ or 68.26%. This, of course, is the familiar area, from a normal distribution, that falls within one standard deviation above and below the mean. We can now predict that, on average, our product will produce "out of spec" voltages 31.74 percent of the time, or 317,400 parts per million. We have met the design challenge to predict the performance of a population of products based on the few prototypes available during design.

Before we conclude our discussion, we need to review a few additional concepts from basic statistics regarding distributions and the averaging of random samples.

A special case of the normal distribution is the standard normal distribution, also known as the *z-distribution*. The z-distribution has the same shape as the normal, but its mean is zero and its standard deviation is one, $\mu_z = 0$ and $\sigma_z = 1$. Therefore, values from the z-distribution are in units of standard deviations. A normal distribution can easily be converted to the standard normal using the following equation, where X is a vector of normally distributed values and Z is a vector of z-distributed values.

$$Z = \frac{X - \mu}{\sigma}$$

As we noted earlier, we do not know the parameters of our population. But we must know μ and σ to use the z-distribution. In cases where we do not know σ but we do know μ, such as a target value, we use a different distribution called the *t-distribution*. The t-distribution has the same shape as the normal but its tail areas are a bit larger than the normal. As the sample size increases, the t-distribution approaches the normal distribution. For a normal random variable X, the random variable t_ν is t-distributed and is given by the following equation, where s is the sample standard deviation.

$$t_\nu = \frac{X - \mu}{s}$$

The t-distribution is determined by its dof, labeled ν, which is the sample size minus one.

Our last fact addresses the variance of the average of a random sample. A common technique in signal processing is to take many observations of a signal and process the average of the many observations instead of the individual observations. The reason for this is that the variation of the average is smaller than that of the individual observations. In fact, it is reduced by the sample size. If a random sample of size n is taken from a population whose variance is σ^2, the variance of the average of the sample will be σ^2/n. This is shown in the following equation where X is a random variable and n is the sample size.

$$\sigma_{\bar{X}}^2 = \frac{\sigma_X^2}{n}$$

Consequently, the sample averages of a random variable X are t-distributed given by this equation:

$$t_{\bar{X},\nu} = \frac{\bar{X} - \mu}{\dfrac{s}{\sqrt{n}}}$$

We use this fact in confidence intervals and t-tests.

We have reviewed the descriptive statistics that enable us to succeed at one of the challenges discussed at the beginning of this section, "How can we predict the performance of a 'population' of products based on the few prototypes available during design?" But, we have not addressed how to identify and control variation in our critical responses. The next chapter provides a review of the inferential statistics needed for this crucial aspect of DFSS.

Introduction to Inferential Statistics

Where Am I in the Process? Where Do I Apply Inferential Statistics Within the Roadmaps?

Linkage to the I²DOV Technology Development Roadmap:

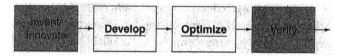

Inferential statistics are mainly used during the Develop and Optimize phases of I²DOV.

Linkage to the CDOV Product Development Roadmap:

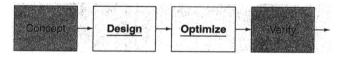

Inferential statistics are mainly used during the Design and Optimize phases of CDOV.

The Purpose of Inferential Statistics in the I²DOV Technology Development Roadmap:

Inferential Statistics in I²DOV provide the statistical theory required to determine whether various system factors have an effect on the responses of our technology's critical-to-function responses.

The Purpose of Inferential Statistics in the CDOV Product Development Roadmap:

Inferential statistics in CDOV provide the statistical theory required to determine whether various design factors have an effect on the critical-to-function responses of our product.

What Am I Doing in the Process?

Employing inferential statistics allows us to take data from the system and analyze the data to determine, with statistical rigor, whether certain factors of a design have an effect on the design's response.

What Output Do I get from Using Inferential Statistics?

Statistical tests and indicators are generated upon deciding which various factors of a design have an effect on the design's response.

Inferential Statistics Review and Tools

Describing and quantifying variation in a critical response is necessary but not sufficient. We must also identify the sources of the variation and quantify how much variation each source contributes. Identifying the sources of variation and quantifying their relative contribution is a critical skill that the DFSS practitioner must master. Let's review the inferential statistics upon which this is based.

Assume we measured the heights of 200 people. Half were young children and half were adults. Our goal is to determine if age affects average height. This is a simple experiment, so let's pose the problem in the terminology of experimentation. We have one *factor,* called "person." Our factor has been varied between two *levels,* "child" and "adult." Our critical *response* is height in inches. We have two samples of 100 observations each. Let's look at the data that was collected in the histogram shown in Figure 22–1.

The effect seems clear. The samples cluster into two distinct groups based on age. It is unlikely that we would be wrong if, assuming a proper experiment was run, we concluded that age does affect height. The histogram indicates this because the graphs have no overlap and their

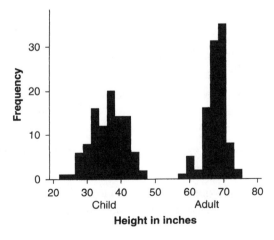

Figure 22–1   Histogram of Age Relationship to Height

means are far apart. It is very unlikely that an observation from the child sample could have actually come from the adult sample. Statistics is not really needed here to tell if age has an effect on height. Now, let's run another experiment.

Once again, we measure the heights of 200 people. This time, they are all adults. Half are 40 years old and half are 50 years old. Our goal is to determine if age affects height. The histogram is shown in Figure 22–2.

Now the problem is not so easy. We do not have two distinct clusters.

Our sample distributions overlap. Determining if two samples are different becomes more and more difficult as the amount of overlap increases. Overlap is affected by the mean and variance of the samples. If there were no variation, there would be no overlap, regardless whether the means were close or far apart. But there is always some amount of variation. How do we determine if our samples are from two different populations when there is significant overlap?

Our approach to this problem is to start with a practical problem stated as a practical hypothesis, convert it into a statistical hypothesis, draw a statistically valid conclusion from data, and then convert the statistical conclusion back into a practical conclusion.

This approach is common practice in DFSS.

Hypotheses and Risk

Let's pose the practical hypothesis that age does not affect height. This is called the *null hypothesis* and is denoted as H_0. Our statistical null hypothesis is that the means of our two samples are equal, written H_0: $\mu_1 = \mu_2$. The alternative to this simple hypothesis, or H_a, is that age does have an effect on height. Our statistical alternate hypothesis is that the means are not equal, written H_a: $\mu_1 \neq \mu_2$.

We have converted our practical problem into a statistical one. We must now gather evidence upon which to reject, or fail to reject, the null hypothesis. The evidence we seek is derived from measured data. Once we have the data, we calculate a test statistic and use that statistic to

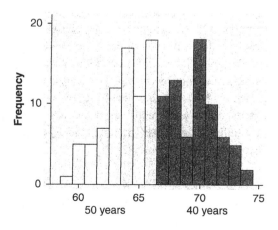

Figure 22–2   Comparing Heights of 40- and 50-year-olds

determine whether we should reject or fail to reject the null hypothesis. Keep in mind that fail to reject is not the same as "accept." We can never prove that the null is true. We can only seek evidence to prove that the null is not true. One instance of "not true" is sufficient to reject the null. However, one, two, or three instances in which the null is found to be true does not mean it is always true. We cannot investigate all possible instances. Nevertheless, in practice when we fail to reject, we act as though the null is true and has been accepted. We even get sloppy and sometimes use the phrase "accept the null hypothesis."

Sounds simple? Well, it gets more complicated. Our decision to accept or reject the null is based on our samples, and we can make two correct decisions or we can make two wrong decisions. The correct decisions are to accept or reject the null when we should. The incorrect decisions are to accept or reject the null when we shouldn't. Let's look at the two types of incorrect decisions starting with incorrectly accepting the null.

The error of incorrectly accepting the null is called a *type II error.* For our child/adult height example, it is like saying, "Age has no effect on height" when in fact it does. It is also like saying, "Based on the absence of complaints, my customers are happy." This type of error has been the death of some companies. Some customers do not complain. They quietly switch to the competitor, and the loss of market share is the lagging indicator of the switch. The risk of making a type II error is sometimes called the *consumer's risk.* Imagine a company that tests pacemakers. If the null hypothesis is "the device under test is good" and a type II error is made, then a bad unit gets shipped. In the worst case, the consumer gets to test it out in his or her chest.

In short, a type II error is the mistake of saying the null should be accepted when it really should not. To be certain we have not made this error, we need to continue testing until we finally find evidence to reject the null. This may take a long time and a lot of money. The market will wait only so long. Instead, we allow a certain risk of making this error. The risk or probability of making the error is called β *risk.*

There's another error we can make. What if we conclude that we should not accept the null when in fact we should? This is a *type I error,* and the probability of making this error is termed α *risk.* Alpha risk is sometimes called the *manufacturer's risk.* If our pacemaker company makes this error, it unnecessarily reworks or scraps a device. Typical values for α and β are 5 percent and 20 percent, respectively; however, like "n-sigma," each company must determine its own values of α and β based on the safety, legal, and economic consequences of each type of error.

What about the two correct decisions we can make? We can correctly reject the null and we can correctly fail to reject the null. These have probabilities of $(1-\beta)$ and $(1-\alpha)$, respectively. The statistical power of a test is its ability to correctly reject the null and has a probability of $(1-\beta)$. The probability of correctly failing to reject the null is $(1-\alpha)$. This discussion, in the context of the pacemaker manufacturer, is shown in Figure 22–3.

The t-Test

How do we use our hypotheses to determine if age has an effect on height? If the null is true, age does not have an effect on height, and both samples are from the same distribution, their means are equal and the difference between the two population means should be zero. Thus, the null can also be written as H_0: $\mu_1 - \mu_2 = 0$. If the null is not true, age does have an effect on height, and

Truth	Decision	
	Device is good	Device is bad
Device is good	Correctly accept the null $(1 - \alpha)$ risk	Incorrectly reject the null Type I Error α risk
Device is bad	Incorrectly accept the null Type II Error β risk	Correctly reject the null $(1 - \beta)$ risk

H_0: the device is good
H_a: the device is bad

Figure 22–3   Facing α and β Risk Errors

the samples are from different distributions, their means are not equal, and the difference between the two population means should not be zero.

The alternative hypothesis can be written as H_a: $\mu_1 - \mu_2 \neq 0$. We can compare the difference with zero to test our hypotheses. This must be a statistical comparison because each sample mean is a random variable and so their difference is also a random variable.

Now let's choose a value for alpha, the risk of a type I error. Remember, this is saying the means are different when they are not. In general, this can be true in two ways. The younger mean can be greater than the older mean, or it can be less than the older. The difference $\mu_1 - \mu_2$ can be positive, or it can be negative. So, let's split α in half, 2.5 percent above and 2.5 percent below zero. This is called a *two-tailed test* and is shown in Figure 22–4. We use a t-distribution because we don't know σ.

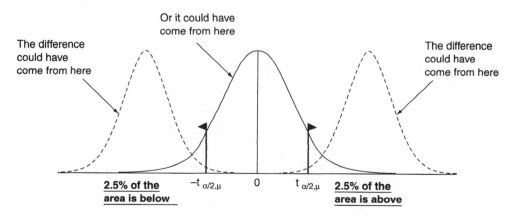

Figure 22–4   Sample "Two-Tailed" Test

The test statistic for zero difference between two normally distributed random variables when σ is unknown is t-distributed and is given in the following equation.

$$t = \frac{\bar{x}_1 - \bar{x}_2}{\sqrt{\dfrac{s_1^2}{n_1} + \dfrac{s_2^2}{n_2}}}$$

If the population variances are equal ($\sigma^2 = \sigma_1{}^2 = \sigma_2{}^2$), and there are $(100 - 1) + (100 - 1) = 198$ dof, we use this equation:

$$t = \frac{\bar{x}_1 - \bar{x}_2}{s_p\sqrt{\dfrac{1}{n_1} + \dfrac{1}{n_2}}}$$

where the pooled sd is $s_p = \sqrt{\dfrac{(n_1 - 1)s_1^2 + (n_2 - 1)s_2^2}{n_1 + n_2 - 2}}$

The test with equal variances is slightly more powerful than the test with unequal variances.

So far, we have our hypothesis, our risk level, our data, and our test statistic. Now we need a number with which to compare our test statistic, to determine if we should reject or fail to reject the null. We do this by comparing the calculated value of the test statistic to the numbers from a t-distribution that have 2.5 percent of the area below it and 2.5 percent of the area above it.

This is a two-tailed test. The magnitude of these two numbers is the same since the t-distribution is symmetric and is termed $t_{(\alpha/2, v)}$. In our case dof = 198. We can use Minitab to determine $t_{(\alpha/2, v)} = t_{(0.05/2, 198)}$. We use the Minitab pull-down menu to go to Calc > Probability Distributions > t Distribution, and we get the window shown here.

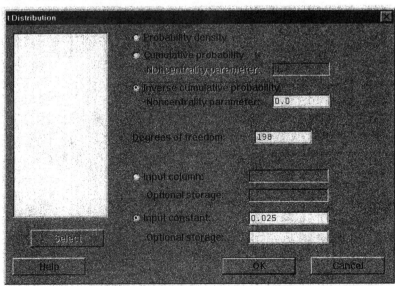

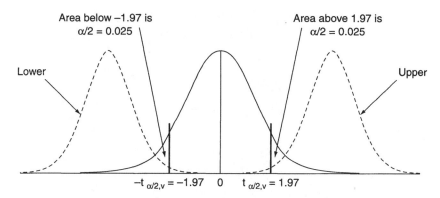

Figure 22–5 Histogram of a Two-Tailed Test

We check inverse cumulative probability, choose 198 dof, and input the constant 0.025 as our area. The output is given in Minitab's session window:

Inverse Cumulative Distribution Function

```
Student's t distribution with 198 DF
P(X<=x)            x
  0.0250        -1.9720
```

This tells us that, for our t-distribution, 2.5 percent of the total area is to the left of -1.97 and 2.5 percent is to the right of $+1.97$. We have found a range of values, the interval from -1.97 to $+1.97$, which bounds 95 percent of the area and occurs with a cumulative probability of 95 percent, as shown in Figure 22–5.

If the difference between our two sample means (in t units) falls outside this interval, we reject the null. Values that are outside this interval occur with a cumulative probability of only 5 percent. So, we decide it is more likely that our samples came from different populations and that age does affect height. Of course the difference does have up to a 5 percent chance of actually having come from the t-distribution with a mean of zero. But we have decided, by our choice of alpha, to allow a 5 percent chance of making this error.

This is a good place to introduce the concept of a *p-value*. The p-value is equal to the area that is beyond the value of the test statistic we have calculated. In this case, our test statistic is the difference between two means and is t-distributed. The p-value associated with 1.97 is 0.025. If we calculated a t-value greater than 1.97, its p-value would be less than 0.025. If a test statistic's p-value is less than alpha, we reject the null. If it is greater than alpha, we fail to reject the null. We will use p-values in t-tests and ANOVA.

This is the statistical basis upon which we decide if a sample is different from a given value or if two samples are different from each other. This is called a t-test. Luckily, all of this is done for us in statistical software packages like Minitab. Let's use Minitab to test whether age affects height in our first experiment. Use Stat > Basic Statistics > 2-sample t.

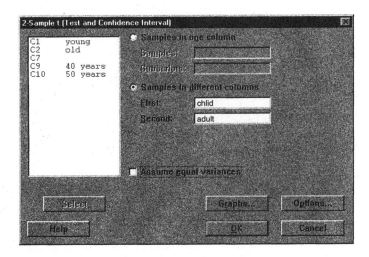

Our samples are in different columns and we assume unequal variances since we don't know if they're equal. Note that the dof is now 165, not 198.

Choose the alternative hypothesis of "not equal." The session window output follows:

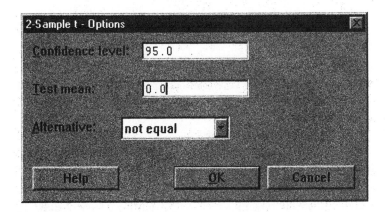

Two-Sample T-Test and CI: young, old

Two-sample T for young vs old

	N	Mean	StDev	SE Mean
young	100	36.57	5.19	0.52
old	100	67.88	3.22	0.32

Difference = mu young - mu old
Estimate for difference: -31.312

```
95% CI for difference: (-32.517, -30.107)
T-Test of difference = 0 (vs not =): T-Value = -51.31
P-Value = 0.000 DF = 165
```

A t-value of -51.31 is a very large negative value and is way out in the left tail of the t-distribution. Its p-value, reported as "P-value=0.000," is much less than our alpha of 0.05. This tells us to reject the null. Our practical conclusion is that age does affect height among the people we sampled in this experiment.

Now we'll use Minitab to test whether age affects height in our second experiment. We use the same Minitab utilities and choose 40 years and 50 years for the column names.

The session window output follows:

Two-Sample T-Test and CI: 40 years, 50 years

```
Two-sample T for 40 years vs 50 years
```

	N	Mean	StDev	SE Mean
40 years	100	68.11	3.02	0.30
50 years	100	65.17	2.93	0.29

```
Difference = mu 40 years - mu 50 years
Estimate for difference: 2.944
95% CI for difference: (2.114, 3.773)
T-Test of difference = 0 (vs not =): T-Value = 7.00
P-Value = 0.000 DF = 197
```

The p-value is much less than alpha and so we reject the null. Our two sample t-test enabled us to decide, with a 5 percent chance of being wrong, that age does affect height among the people we sampled in the experiment.

Let's compare the "old" group with the "40 year" group. The null hypothesis is that the average heights of the two age groups are the same. That is, age does not affect average height. The session window output follows:

Two-Sample T-Test and CI: 40 years, old

```
Two-sample T for 40 years vs old
```

	N	Mean	StDev	SE Mean
40 years	100	68.11	3.02	0.30
old	100	67.88	3.22	0.32

```
Difference = mu 40 years - mu old
Estimate for difference: 0.229
```

```
95% CI of difference: (-0.641, 1.099)
T-Test of difference = 0 (vs not =): T-Value = 0.52
P-Value = 0.604 DF = 197
```

The t-value is relatively small (0.52) and its p-value (0.604) is larger than alpha. We fail to reject the null and conclude that age does not affect height between these two groups.

Analysis of Variance

The t-test is only useful when comparing at most two samples. The test loses statistical power as the number of paired samples to be tested increases. What do we do if our factor has more than two levels or we have many factors, each at many levels? How do we compare the means without losing power? The answer is ANOVA, or *analysis of variance.*

ANOVA is used to test whether the means of many samples differ but it does so using variation instead of the mean. ANOVA can be thought of, as Professor Tom Barker at the Rochester Institute of Technology would say, as "analysis of the means by way of the variance (Barker 1994)." ANOVA compares the amount of variation within the samples to the amount of variation between the means of the samples. If the "between" variation is significantly larger than the "within" variation, we conclude that the mean of our response has changed.

Let's see how ANOVA is used to analyze the result of the experiments we run. The process consists of:

- Posing a practical and statistical null and alternative hypothesis
- Selecting risk levels
- Running an appropriate experiment to generate sample data
- Calculating a test statistic from the data
- Using the probability distribution of the test statistic to determine if we should reject or fail to reject the null

This sounds like the steps in a t-test. They are! The difference is that we use ANOVA to generate a test statistic that is based on variation unlike the t-test, which is based on the mean.

Let us pose a null and alternative hypothesis. What we want to determine in DFSS is whether x affects y. That is, do changes in x have an effect on the mean value of y? Our critical-to-function response is y, and x is a critical parameter or factor that we believe affects y. We pose the practical null hypothesis that, "No, x does not have an effect on y." The alternative hypothesis is that x does have an effect on y. We'll assume a 5 percent alpha risk. But we need a statistical hypothesis that allows a data-driven decision to reject or accept the null. The data generated from an experiment is tightly coupled to the statistical hypothesis to be tested. That is why the study of DOE is necessary. Let's say we ran an experiment involving one factor x and our critical response y. We tested over many levels of x, say n levels, and we took m observations at each of those n levels. The observation, $y_{i,j}$ is the jth observation at the ith level of x.

We can think of each observation as having come from one single distribution with a mean of μ plus some error, written as $y_{i,j} = \mu + \delta_{i,j}$. But, we can also think of each group of m measurements, taken at a particular level of x, as having come from their own distribution with a mean of μ_i plus some error, written as $y_{i,j} = \mu_i + \varepsilon_{ij}$. There's a subtle, but profound difference between these two equations. If the first equation is correct, x has no effect on y since μ is not a function of x. If the second equation is correct x has an effect on y, since μ_i is a function of x. This is shown in Figure 22–6.

The statistical null hypothesis is H_0: $\mu_1 = \mu_2 = \mu_i = \ldots \mu_n$. All the individual means are the same regardless of the levels of x. The statistical alternative hypothesis is H_a: $\mu_i \neq \mu_q$ for at least one $i \neq q$. At least one mean is different from the others.

These hypotheses can be stated in another form. The effect we are looking for is the amount by which each μ_t has been moved away from μ. This effect, for the ith level of x, is called τ_i. This is also shown in Figure 22–6. We can now rewrite $y_{i,j} = \mu + \delta_{i,j}$ to get this equation:

$$y_{i,j} = \mu + \tau_i + \varepsilon_{ij}$$

By the way, if all the effects are zero then x has no effect on y. So, the null can also be written as H_0: $\tau_1 = \tau_2 = \tau_i = \ldots \tau_n = 0$ and the alternative written as H_a: $\tau_i \neq 0$ for at least one i.

Now let's calculate a test statistic from the data. We'll use ANOVA to do this. Our best estimate of μ is $\bar{\bar{y}}$, which is the grand average of all the observations from our experiment. Our best estimate of μ_1 is $\bar{y}_i$, the average of all m observations taken at the ith level of x. Our best estimate of τ_i is $(\bar{y}_i - \bar{\bar{y}})$. And finally, an estimate for the error ε_{ij} is the distance of the jth observation taken at the ith level of x from its local mean, given by $\varepsilon_{ij} = y_{ij} - \bar{y}_i$.

We can substitute these into the previous equation, subtract $\bar{\bar{y}}$ from both sides, and get $(y_{ij} - \bar{\bar{y}}) = (\bar{y}_i - \bar{\bar{y}}) + (y_{ij} - \bar{y}_i)$. See Figure 22–7 for a graphical view of this equation.

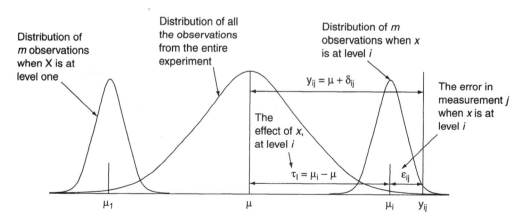

Figure 22–6 ANOVA Representation of Possible Effects

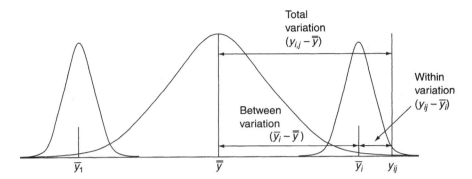

Figure 22–7 ANOVA Representation of Experimental Data

Now, this equation applies to one particular observation. We want an equation that uses all the observations. If we were to sum this equation over all the observations for all the levels of *x*, we would end up with zero as we did with the variance. So, we need to first square both sides and then take the sum. We then obtain this equation:

$$\sum_{i=1}^{n} \sum_{j=1}^{m} (y_{ij} - \overline{\overline{y}})^2 = m \sum_{i=1}^{n} (\overline{y}_i - \overline{\overline{y}})^2 + \sum_{i=1}^{n} \sum_{j=1}^{m} (y_{ij} - \overline{y}_i)^2$$

This is an extremely important relationship, so let's understand it term by term.

The term on the left is the sum of the squared distances of each observation from the grand mean, treating all the observations in the experiment as though they came from one large population. This term is related to the variance of all the observations, taken as a whole. It is called the **total sum of squares,** SS_{total}. It estimates the total variation seen in the entire experiment.

The second term on the right is also a sum of squares. Each of the inner sums provides an estimate of the variation seen within each level of *x*. This is considered to be error. The "within" variation or error should be due to all sources of variation, *except for changes in x*, which cause the response value to vary. One source of this noise is the measurement system itself. If the measurement system is too noisy, we may attribute the variation we see to the system under test when most of it is due to the measurement system alone. This is why it is crucial that a measurement systems analysis (MSA) is done *before* data are taken and used to draw conclusions. Another source of this noise is variation in "factors" that affect our response but have not been included in the experiment either knowingly or unknowingly. Ambient temperature, humidity, and altitude are examples of this. The within variation serves as an estimate of the error in our experiment. We expect the error to be relatively constant and small, and it should have very little to do with the levels of *x* or the magnitude of *y* over which we experimented.

The outer sum adds the within variation for all the levels of x to get a "pooled" estimate of overall experimental error, σ^2.

This term is called *sum of squares due to error,* SS_{error}. If we divide SS_{error} by its dof, we get the mean sum of squares, MS_{error}.

The middle term estimates how different each "local" mean is from the overall mean. This is the factor effect we are trying to detect. We sum them all up to get an estimate of how much the variation in x has caused the local means to vary compared to the error. This term is called *sum of squares due to treatment,* SS_{trt}. If we divide SS_{trt} by its dof, we get the mean sum of squares, MS_{trt}. The term *treatment* is a carryover from the origins of DOE. DOE has its roots in the field of agriculture (pun intended). The soil was treated with various nutrients to determine their effects on plant growth. The term *treatment* has persisted. A **treatment combination** is used to describe a unique combination of levels of several factors in a multifactor experiment. For example, 20 percent nitrogen, 10 percent phosphorus, and 5 percent potash is a treatment combination.

Back to the story. If MS_{trt} is larger than MS_{error}, we suspect that x has an effect on y and that the null hypothesis can be rejected. If MS_{trt} is less than or equal to MS_{error}, we suspect that x has no effect on y and we fail to reject the null. We are making decisions about means based on variation. This is why ANOVA can be thought of as analysis of the means by way of the variance.

If MS_{trt} is large compared to MS_{error}, we suspect that x has caused our response to change by more than chance or experimental error alone. How large should this ratio be? Statistics provides the answer.

We must impose some restrictions on y if we are to use statistics to obtain the answer. We require that the error for any observation, ε_{ij}, is normally distributed with a mean of zero and a variance of σ^2. This can be written as $\varepsilon_{ij} \sim N(0, \sigma^2)$. If this condition is true, y is also a normal random variable. If the null is false, then $y_{ij} \sim N(\mu_j, \sigma^2)$. Now we can use a statistically rigorous test to determine if MS_{trt} is large enough.

We need a few facts to proceed:

- The mean of a normal random variable is also a normal random variable.
- If a standard normal random variable, y, is squared, the new variable y^2 is chi-square distributed.
- The ratio of two chi-square distributed random variables is F-distributed.

Please refer to *John E. Freund's Mathematical Statistics (Sixth Edition).*

The chi-square and F-distribution are special distributions that arise in random sampling and play a key role in our statistical test.

Now, both terms, MS_{error} and MS_{trt}, involve squaring a normal random variable. Each of these terms is chi-square distributed. Their ratio is therefore F-distributed.

I can hear you saying, "So what?" Well, now we have a test statistic, called the F-ratio given by

$$Fratio = \frac{MS_{trt}}{MS_{error}}$$

whose distribution we know. Since we know its distribution, we can determine the probability of getting any value from this distribution. For the F-distribution, increasing F values occur with decreasing probabilities. We can now use the F-ratio to determine if we should reject or accept the null hypothesis by comparing its p-value to alpha, as we did in the t-test. If the p-value of the calculated F-ratio is less than the alpha risk, we reject the null.

Remember, the null is "x has no effect on y." If the null is true, MS_{trt} will be close to MS_{error}, the F-ratio will be close to one, and its p-value will be large. If the p-value is greater than alpha, we accept the null and conclude that x does not affect y. If the null is false, MS_{trt} will be larger than MS_{error}, the F-ratio will be larger than one, and its p-value will be small. If the p-value is less than alpha, we reject the null and conclude that x does affect y.

We are long overdue for an example. Let's go back to the two experiments we ran on age versus height. We'll use Minitab to determine if age affects height using ANOVA.

We use Stat > ANOVA > One-way (Unstacked) since the data happen to be in separate columns.

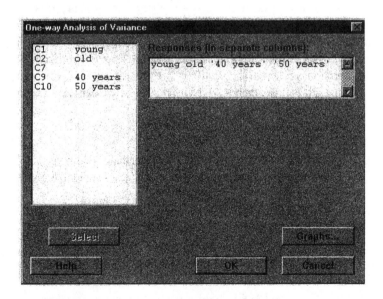

We select our four response columns and click OK. Remember ANOVA now allows us to compare more than two samples. The session window output follows. The reader is encouraged to use Minitab to calculate the total and error sums of squares (SS).

One-Way ANOVA: Young, Old, 40 Years, 50 Years

```
Analysis of Variance
Source      DF        SS        MS        F        P
Factor       3    70229.4   23409.8  1704.23   0.000
Error      396     5439.6      13.7
Total      399    75669.0
                                      Individual 95% CIs For Mean
                                      Based on Pooled StDev
Level        N      Mean      StDev   -----+------+------+-----+-
young      100    36.572      5.186   (*
old        100    67.884      3.217                            (*)
40 years   100    68.113      3.019                            (*)
50 years   100    65.169      2.931                        (*)
                                      -----+------+------+-----+-
Pooled StDev =     3.706               40      50      60      70
```

The F-ratio is large and its p-value is much smaller than our alpha of 0.05. We therefore reject the null hypothesis and conclude that age does affect height.

We don't know which level of age has driven the height difference. We only know that the mean height of at least one age group is different from the others. A multiple comparisons test can tell us which is different and is available in the one-way ANOVA utility.

We have accomplished our goal. We have identified a source of variation with statistical rigor, 95 percent statistical confidence. We can quantify the amount of variation that the source contributes to our critical response, called **epsilon squared,** using this equation:

$$\varepsilon^2 = \frac{SS_{source}}{SS_{total}} = \frac{70,229}{75,669} = 0.93 \quad (0.5)$$

For our example, 93 percent of the variation in height is due to age. This makes age a critical parameter of height.

Now that we have identified and quantified the sources of variation, we want an equation or transfer function that describes the relationship. We use regression to obtain the transfer function, but we need a few more building blocks first. In Chapter 23, we explore measurement systems analysis.

References

Barker, Thomas (1994). *Quality by experimental design* (2nd ed.). New York: Marcel Dekker, Inc.

Miller, I., Miller, M., & Freund, J. E. (1999). *John E. Freund's mathematical statistics* (6th ed.). New Jersey: Prentice Hall.

Measurement Systems Analysis

Where Am I in the Process? Where Do I Apply Measurement Systems Analysis Within the Roadmaps?

Linkage to the I²DOV Technology Development Roadmap:

A measurement systems analysis is conducted prior to the use of any measurement system during any phase of I²DOV. It is less likely to be used during the Invent/Innovate phase.

Linkage to the CDOV Product Development Roadmap:

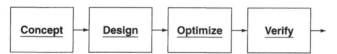

A measurement systems analysis is conducted prior to the use of any measurement system during any phase of CDOV. It is less likely to be used during the Concept phase.

What Am I Doing in the Process? What Does a Measurement Systems Analysis Do at this Point in the Roadmap?

The Purpose of MSA in the I²DOV Technology Development and CDOV Product Development Roadmap:

The purpose of a measurement systems analysis, whether in I²DOV or CDOV, is to qualify a measurement system for the acquisition of data. It is imperative that this is done to help ensure that the conclusions drawn from the data are valid. During an MSA, we run an experiment consisting of repeated and replicated measures of responses from a number of test items.

The replicated measures should contain variation due to different operators or environmental conditions and the repeated measures should contain variation that is mainly due to the measuring instrument.

What Output Do I get at the Conclusion of Measurement Systems Analysis? What Are Key Deliverables from MSA at this Point in the Roadmap?

- *Gage R&R* to assess repeatability and reproducibility
- *Performance to tolerance ratio* to compare measurement system error to the tolerance
- *Number of distinct categories* to assess the number of categories into which the measurement system is capable of "batching" the measurements

MSA Process Flow Diagram

This visual reference of the detailed steps of MSA illustrates the flow of tasks at this stage.

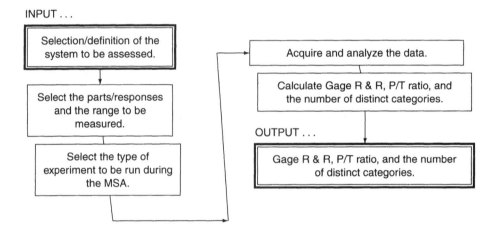

Verbal Descriptions for the Application of Each Block Diagram

This section explains the specific details of measurement systems analysis:

- *Key detailed input(s)* from preceding tools and best practices
- *Specific steps and deliverables* from each step to the next within the process flow
- *Final detailed outputs* that flow into follow-on tools and best practices

Key Detailed Inputs

> **Selection/definition of the system to be assessed.**

A measurement systems analysis (MSA) is an experimental method used to assess the performance of a measurement system. A measurement system consists of the equipment, fixtures, procedures, gages, instruments, software, environment, and personnel that together make it possible to assign a number to a measured characteristic or response. It is important to realize that the instrument used to make the measurement, (for example, a voltmeter), is but one component of the overall measurement system.

An MSA must be done regardless of the precision, accuracy, or cost of the measurement instrument since it is only one component of the entire measurement system. Do not draw conclusions from or take action based on data obtained from a measurement system for which an MSA has not been performed! You cannot detect effects if the variation in the measurement system is approximately equal to the variation in the response. This sounds like an obvious statement. However, this is a common mistake made by many Black Belts when experimentally investigating effects.

Specific Steps and Deliverables

> **Select the parts/responses and the range to be measured.**

A measurement system possesses several characteristics: accuracy, repeatability, reproducibility, linearity, and stability. Accuracy is with reference to a known standard value and is assessed through calibration. The measurement system is used to measure a primary, secondary, or tertiary standard. The difference between the known value of the standard and the average of repeated measurements is a measure of accuracy called *bias*. Linearity refers to how a measurement system's accuracy or precision changes as the magnitude of the measured quantity changes. Stability refers to how accuracy and precision change over time.

Repeatability and reproducibility pertain to precision. The variation in the measured value comes from two sources, the variation due to the response or part being measured and the variation due to the measurement system itself, as shown in this equation.

$$\sigma_{total}^2 = \sigma_{part}^2 + \sigma_{ms}^2$$

Measurement system variation, σ_{ms}^2, can be further broken down into variation due to repeatability, estimated by σ_{rpt}^2, and variation due to reproducibility, σ_{rpd}^2:

$$\sigma_{ms}^2 = \sigma_{rpt}^2 + \sigma_{rpd}^2$$

Repeatability pertains to variation when the same person measures the same test item with the same instrument many times. Reproducibility allows more variation to enter into the measured value.

Reproducibility pertains to measurements made by different people, at different times on the same test item. Reproducibility is therefore a measure of variation from operator to operator. Reproducibility can be broken down into two components, one due to the various operators and another component due to an interaction between an operator and the part being measured, as shown in this equation:

$$\sigma_{rpd}^2 = \sigma_{op}^2 + \sigma_{op \times part}^2$$

Thus, total measurement variation consists of four variance components:

$$\sigma^2_{total} = \sigma^2_{part} + \sigma^2_{op} + \sigma^2_{op \times part} + \sigma^2_{rpt}$$

The goal of an MSA is to estimate the four variance components shown in this equation. The estimates are then used to calculate three critical measures of measurement system performance:

Gage R&R

Performance-to-tolerance ratio (P/T ratio)

Number of distinct categories

Gage R&R, called "% Study Variation" by Minitab, determines what percentage of the total variation is taken up by measurement system variation. The P/T ratio, called "% Tolerance" by Minitab, determines the fraction of the tolerance that is taken up by measurement system variation. The formulae are

$$gage\ R\&R = \frac{\sigma_{ms}}{\sigma_{total}} * 100\%$$

$$P/T = \frac{5.15 * \sigma_{ms}}{Tolerance}$$

The 5.15 multiplier in the P/T ratio is used to span 99 percent of the measurement system variation, assuming it is normally distributed.

The third metric considered is the number of distinct categories. This is sometimes called a *discrimination ratio*. It identifies how many distinct, 97 percent nonoverlapping normal distributions the measurement system is capable of grouping into measurements. Two distinct categories is a go–no-go gage. The formula is as follows:

$$distinct\_cat = round\left(\frac{1.41 * \sigma_{part}}{\sigma_{ms}}\right)$$

Select the type of experiment to be run during the MSA.

We need to estimate the variances shown in the total measurement variation equation to calculate these three metrics. This is accomplished by running a hierarchical experiment. The experiment is designed so that several operators make repeated measurements on a set of parts. A nondestructive test is one in which the test item is unchanged after the test has been performed and the feature can be remeasured. The same physical part can be used throughout the test. A destructive test is one in which the test item is changed by performing the test, so the feature cannot be remeasured. In this case, a group of nearly identical parts must be available to simulate repeated measurements. A nondestructive experiment is called *crossed,* while a destructive case is *nested.*

The typical hierarchical experimental structure follows for i operators measuring j parts, k times each.

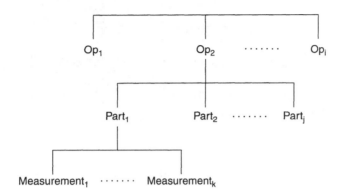

The operators should not know which part they are measuring during the experiment nor should they know the values the other operators have obtained. This reduces the chance of data corruption due to an operator's desire to be consistent within a part or when compared to an operator who may be favored or an expert.

The range of the response in the experiment must represent the range expected when the measurement system is placed into service. If the range is smaller than expected, measurement system variation may be a large portion of the total variation and we may erroneously conclude that the measurement system is not capable.

Acquire and analyze the data.

Once the data have been obtained, our estimates of the variance components in the total measurement variation equation are obtained from the sums of squares and degrees of freedom of each source of the variation. It is worth spending some time on the analysis of the data since it is not, as they say, intuitively obvious to the casual observer. The necessary information is obtained from Minitab by running a Gage R&R study using the ANOVA method. The Xbar and R method cannot estimate the operator × part (op × part) interaction and therefore is not preferred. We will use a file provided by Minitab to explain the data analysis. The file is "Gageaiag.mtw" and can be found in Minitab's data folder.

We begin the analysis using the Minitab pull-down menu to go to Stat > Quality Tools > Gage R&R (crossed) for nondestructive and get the window that follows. We select Part, Operator, and Response and the default ANOVA method of analysis. Add a tolerance, 1 in this example, under Options.

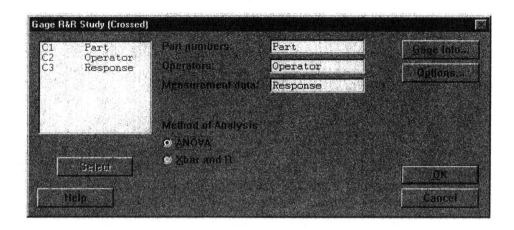

Minitab produces the following information. Let's see how the Gage R&R numbers are obtained.

Gage R&R Study–ANOVA Method

Gage R&R for Response

Two-Way ANOVA Table With Interaction

Source	DF	SS	MS	F	P
Part	9	2.05871	0.228745	39.7178	0.00000
Operator	2	0.04800	0.024000	4.1672	0.03256
Operator*Part	18	0.10367	0.005759	4.4588	0.00016
Repeatability	30	0.03875	0.001292		
Total	59	2.24912			

Gage R&R

Source	VarComp	% Contribution (of VarComp)
Total Gage R&R	0.004437	10.67
Repeatability	0.001292	3.10
Reproducibility	0.003146	7.56
Operator	0.000912	2.19
Operator*Part	0.002234	5.37
Part-To-Part	0.037164	89.33
Total Variation	0.041602	100.00

Source	StdDev (SD)	Study Var (5.15*SD)	%Study Var (%SV)	%Tolerance (SV/Toler)
Total Gage R&R	0.066615	0.34306	32.66	34.31
Repeatability	0.035940	0.18509	17.62	18.51
Reproducibility	0.056088	0.28885	27.50	28.89
Operator	0.030200	0.15553	14.81	15.55
Operator*Part	0.047263	0.24340	1.17	24.34
Part-To-Part	0.192781	0.99282	94.52	99.28
Total Variation	0.203965	1.05042	100.00	105.04

Number of Distinct Categories = 4

Total variability is estimated from the total sums of squares. This is found using all k observations made on all j parts by all i operators. The formula is given in the following equation where Y_{obs} are the observations and $\overline{Y}$ is the mean of all the $i \times j \times k$ observations.

$$SS_{total} = \sum_{obs=1}^{ijk} (Y_{obs} - \overline{Y})^2 = 2.24912$$

Variation due to part is found from the variation of the average measurement made by all the operators on a part from the overall average. This is then summed over all the parts. We want this variation to be the largest part of total variation since we care about variation from part to part. The formula for **sums of squares** and mean sums of squares is given in the following equation block where $\overline{Y}_{part}$ is the average of all the measurements made by all the operators on a particular part.

$$SS_{part} = ik \sum_{part=1}^{j} (\overline{Y}_{part} - \overline{Y})^2 = 2.05871$$

$$MS_{part} = \frac{SS_{part}}{dof_{part}} = 0.22875 \qquad (0.9)$$

$$dof_{part} = j - 1$$

Variation due to the operators is found from the variation of the average of the measurements made by each operator on all the parts from the overall average. This is then summed over all the operators. This should be small since we do not want measurement precision to depend on who makes the measurement. The formula for sums of squares and mean sums of squares is given in the following equation block where $\overline{Y}_{op}$ is the average of all the measurements made on all the parts by a particular operator.

$$SS_{op} = jk \sum_{op=1}^{i} (\overline{Y}_{op} - \overline{Y})^2 = 0.0480$$

$$MS_{op} = \frac{SS_{op}}{dof_{op}} = 0.0240$$

$$dof_{op} = i - 1$$

Variation due to the operator $\times$ part or interaction is due to a difference in how an operator measures one part compared to how the same operator measures another part. There should be very little operator $\times$ part interaction. We want parts to be measured equally well independent of who is measuring what part. The formula for sums of squares and mean sums of squares is given in the following equation block where $\overline{Y}_{part,op}$ is the average of the k measurements each operator makes on each part.

$$SS_{op \times part} = k \sum_{part=1}^{j} \sum_{op=1}^{i} ((\overline{Y}_{part,op} - \overline{Y}_{op}) - (\overline{Y}_{part} - \overline{Y}))^2 = 0.10367$$

$$MS_{op \times part} = \frac{SS_{op \times part}}{dof_{op \times part}} = 0.005759$$

$$dof_{op \times part} = (i - 1)*(j - 1) = (3 - 1)*(10 - 1) = 18$$

Repeatability, also known as *error*, is obtained from the repeated measurements each operator makes on the part. We can estimate repeatability by subtracting total variability from all the remaining sources of variability. This is shown in the following equation block, along with the mean sum of squares for repeatability.

$$SS_{rpt} = SS_{total} - (SS_{part} + SS_{op} + SS_{op \times part}) = 0.03874$$

$$MS_{rpt} = \frac{SS_{rpt}}{dof_{rpt}} = 0.001291 \qquad (0.12)$$

$$dof_{rpt} = dof_{total} - dof_{op} - dof_{part} - dof_{op \times part}$$

In summary:

Our best estimate of repeatability is $\hat{\sigma}_{rpt}^2 = MS_{rpt} = 0.00129$.

Our best estimate of op $\times$ part variation is $\sigma_{op \times part}^2 = \dfrac{MS_{op \times part} - \hat{\sigma}_{rpt}^2}{k} = 0.002234$

Our estimate of operator variation is $\hat{\sigma}_{op}^2 = \dfrac{MS_{op} - k*\hat{\sigma}_{op \times part}^2 - \hat{\sigma}_{rpt}^2}{j*k} = 0.000912$

Our estimate of reproducibility is $\hat{\sigma}_{rpd}^2 = \hat{\sigma}_{op}^2 + \hat{\sigma}_{op \times part}^2 = 0.003146$

Our estimate of part variation is $\widehat{\sigma}^2_{part} = \dfrac{MS_{part} - k*\widehat{\sigma}^2_{op \times part} - \widehat{\sigma}^2_{rpt}}{i*k} = 0.037164$

Our estimate of total variation is $\widehat{\sigma}^2_{total} = \widehat{\sigma}^2_{part} + \widehat{\sigma}^2_{rpt} + \widehat{\sigma}^2_{rpd} = 0.041602$

We now have our estimates of the variance components. This is listed under the column (VarComp) in Minitab's output table.

The %Contribution of VarComp is found by dividing each VarComp by the total variation and multiplying by 100%.

The standard deviation (StdDev column in Minitab) is the square root of VarComp.

Study Variation (Study Var in Minitab) is 5.15*StdDev. This spans 99% of each variance component.

Percent study variation listed as % Study Var in Minitab.

%Study Var = (Study Var)/(Total Study Var)

Calculate Gage R&R, P/T ratio, and the number of distinct categories.

So, Gage R&R = %Study Var for Total Gage R&R = 32.66%. This tells us that 32.66 percent of the total variation is due to the measurement system. This is too large. Gage R&R should be less than 20 percent.

P/T = %Tolerance = Study Var/Tolerance = 34.31%. Remember we set the tolerance to unity. This tells us that 34.31 percent of the tolerance is taken up by measurement system variation. This is also too large. P/T should be less than 20 percent.

The number of distinct categories is four. This tells us that the measurement system is capable of dividing the measurements into four distinct, 97 percent nonoverlapping normal distributions, as shown in Figure 23–1. At least five distinct categories are recommended.

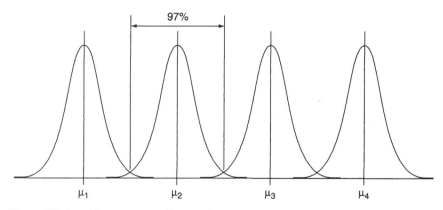

Figure 23–1 Distinct Measurement Categories

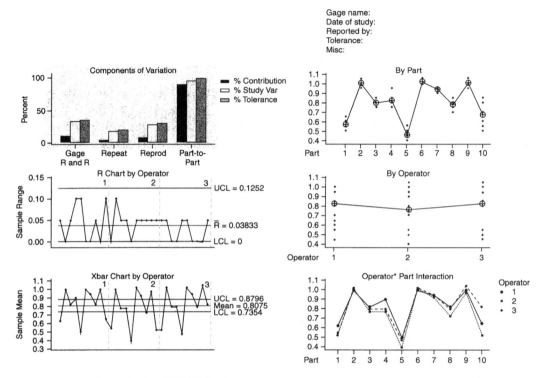

Figure 23–2 Gage R&R (ANOVA) for Response

Minitab also provides a graphical representation of this information given above. This is shown in Figure 23–2. The upper left graph shows three bars located above the label 'Gage R&R'. They are the percent contribution of measurement system variation, the Gage R&R, and P/T, from left to right, respectively.

Final Detailed Outputs

> **Gage R&R, P/T ratio, and the number of distinct categories.**

Once we have assessed our measurement system and found it to be capable, it is safe to use it to make measurements from which to draw conclusions and take actions.

MSA Checklist and Scorecard

This suggested format for a checklist can be used by an engineering team to help plan its work and aid in the detailed construction of a PERT or Gantt chart for project management. These items also serve as a guide to identify the key items for evaluation within a formal gate review.

Checklist for MSA

Actions:	Was this Action Deployed? YES or NO	Excellence Rating: 1 = Poor 10 = Excellent
Definition of the system to be assessed.	_____	_____
Select the parts/responses and the range to be measured.	_____	_____
Select the type of experiment to be run during the MSA.	_____	_____
Acquire and analyze the data.	_____	_____
Calculate Gage R&R, P/T ratio, and the number of distinct categories.	_____	_____

Scorecard for MSA

The definition of a good scorecard is that it should illustrate, in simple but quantitative terms, what gains were attained from the work. This suggested format for a scorecard can be used by a product planning team to help document the completeness of its work.

Table of MSA Requirements

Measurement System Name:	Gage R&R:	Part/Response Tolerance:	P/T Ratio:	# of Distinct Categories:
1.	_____	_____	_____	_____
2.	_____	_____	_____	_____
3.	_____	_____	_____	_____

We repeat the final deliverables from an MSA:

- *Gage R&R* to assess repeatability and reproducibility
- *Performance to tolerance ratio* to compare measurement system error to the tolerance
- *Number of distinct categories* to assess the number of categories into which the measurement system is capable of "batching" the measurements

References

Automotive Industry Action Group (AIAG) and the American Society for Quality Control (ASQC). (1998). *Measurement systems analysis.* AIAG.

Montgomery, D. C. (1997). *Introduction to statistical process control.* New York: Wiley & Sons.

Wheeler, D. H., & Lyday, R. W. (1989). *Evaluating the measurement process.* Knoxville: SPC Press.

Capability Studies

Where Am I in the Process? Where Do I Apply Capability Studies Within the Roadmaps?

Linkage to the I²DOV Technology Development Roadmap:

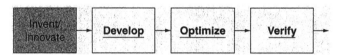

Capability studies are used during the Develop, Optimize, and Verify phases of I²DOV. A capability study is less likely to be conducted during the Invent/Innovate phase.

Linkage to the CDOV Product Development Roadmap:

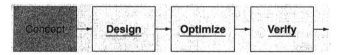

Capability studies are used during the Design, Optimize, and Verify phases of CDOV. A capability study is less likely to be conducted during the Concept phase. However, as various concepts are proposed, the capability of the potential manufacturing processes required by each concept should be accessed.

What Am I Doing in the Process? What Does a Capability Study Do at this Point in the Roadmap?

The Purpose of Capability Studies in the I²DOV Technology Development and CDOV Product Development Roadmaps:

The development team is measuring the system's critical responses in the presence of stressful noise. The measurements are used to calculate the mean and standard deviation of the response.

The mean, standard deviation, together with the specification limits, are then used to calculate a capability index. The capability index is used to track improvement in the technology's or product's insensitivity to noise and its readiness for use in the next phase of development.

What Output Do I Get at the Conclusion of a Capability Study? What Are Key Deliverables from a Capability Study at this Point in the Roadmap?

- Assessment of the stability of the critical-to-function responses
- Estimates of the mean and standard deviation of the critical-to-function responses
- Estimate of the capability index C_p and C_{pk} used to help decide if the system is ready for the next phase of development

Capability Study Process Flow Diagram

This visual reference of the detailed steps of a capability study illustrates the flow of tasks and responsibilities.

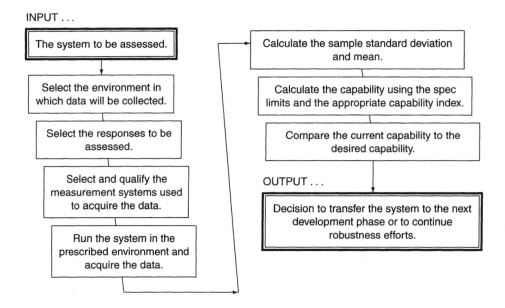

Verbal Descriptions of the Application of Each Block Diagram

This section explains the specific details of what to do in a capability study:

- *Key detailed input(s)* from preceding tools and best practices
- *Specific steps and deliverables* from each step to the next within the process flow
- *Final detailed outputs* that flow into follow-on tools and best practices

Capability Study Overview

A capability study provides a snapshot of the capability of a system's critical responses. It is used to assess how robust a response is to noise and therefore how mature or ready the design is for the next phase in the product development process. The primary technical goal of DFSS is to produce a product that is robust to noise—that is to say, acceptably insensitive to the sources of variation that will degrade the product's performance throughout its life. The phrase "acceptably insensitive" is used to remind the development team that the required level of robustness is a balance between quality, the cost to achieve it, and the consequences of a nonconforming product.

Robustness is measured by two quantities, average performance quantified by the sample mean $\overline{X}$ and variation quantified by the sample standard deviation s. This is the voice of the design (VOD). A product is robust if the mean and standard deviation of its critical responses do not vary beyond what the customer would perceive as degraded quality. The customer, however, does not speak in terms of mean and standard deviation. The voice of the customer (VOC) is in terms of quality. The engineer's task is to translate the VOC into the upper specification limit (USL) and lower specification limit (LSL) of the engineering functions that will satisfy the customer's needs.

Taken literally, the goal of DFSS is to design a product so that, for each of its critical responses, the USL is 6 standard deviations above its mean and the LSL is 6 standard deviations below its mean—thus "Design For Six Sigma." This is shown in Figure 24–1. This condition must exist in the presence of the stressful noises to be found in the product's intended-use environment. We assume here that the response has a desired target, which is centered between its specification limits. We further assume that the response is normally distributed and that its mean is on target.

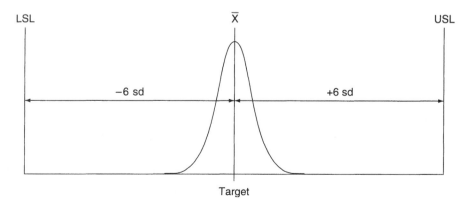

Figure 24–1 Defining Design For Six Sigma

These assumptions, combined with some basic statistics, allows the engineer to predict how a population of products will perform based on the performance of a smaller sample. Statistics tells us that:

1. The normal distribution has a density function that exists from negative infinity to positive infinity.
2. The total area under the distribution is unity.
3. Area is equivalent to probability.

Based on this, we conclude that:

1. The tails of the distribution must extend beyond the USL and LSL regardless of how much or how little variation the response possesses.
2. The area extending beyond the USL and LSL is equal to the probability that our product will produce a response value outside the specification limits.
3. We can calculate this probability.

For example, assuming a response whose mean is centered between the specification limits, if there are 6 standard deviations between the center and each of the specification limits, then there is approximately a 2 in a billion chance of exceeding the limits. This is a 6 sigma design. Six standard deviations were determined by Motorola in the 1980s to be acceptable for their business. This is a very small probability indeed! However, if applied to a digital system operating at nanosecond rates, there is an opportunity for two errors every second. Probabilities must be put in the context of the problem at hand.

If there are three standard deviations between each of the specification limits, then there is approximately a 0.27 percent chance of exceeding the limits. These probabilities are easily determined using a standard normal table or a software package such as Minitab. Readers are encouraged to determine these values for themselves.

Readers are also encouraged to remember that these probabilities apply only when the distribution is normal and centered between its limits. A centered response is a rare condition, and probabilities increase dramatically as the response drifts off center. Each time the response is sampled, it will have a different average and will only approximate a normal distribution. It is common to assume that a 6 sigma design will have a 3.4 in a million chance of exceeding the specification limits, not 2 in a billion. This assumes that, on average, over the long term, the mean of the response will vary by 1.5 times its standard deviation. We will discuss this in greater detail later in this chapter. The message here is that normality should be tested and the mean compared to the target so that correct probabilities are calculated.

In practice, DFSS does not demand that the limits span $+/-6$ standard deviations. This is a matter of economics, technical feasibility, and the legal, health, and societal consequences of operating beyond the specification limits. If death is a consequence, such as in the commercial aviation, medical, or pharmaceutical industries, then 6 sigma performance may not be adequate.

A more realistic approach is Design For N-Sigma (DFNS). It is the responsibility of a company's design, marketing, financial, and legal organizations to collectively determine N for each product and market.

DFSS uses metrics that summarize performance without directly using probabilities. A common metric for this is the capability index, C_p, shown in this equation:

$$Cp = \frac{USL - LSL}{6\sigma}$$

Its use implies that the mean of the response is centered between the specification limits. We will discuss an index that deals with an off-center condition later. Knowledge of the distribution is required only when probabilities are to be estimated. A normal distribution is commonly assumed.

This equation is deceptively simple, so let's look at its pieces. The six in the denominator expresses our desire to include plus and minus three standard deviations of the response. These are common tolerancing and control chart limits. If we assume that the response is normally distributed, then 99.73 percent of all values produced will lie within $+/-3$ standard deviations from the mean.

Sigma, σ, is the population standard deviation of the response. Sigma is estimated by the sample standard deviation. The numerator of the C_p equation captures the specification limits placed on the response by the customer or by the product development team on behalf of the customer. Assuming a normal distribution, this index quantifies how wide the specification limit is compared to 99.73 percent of the possible response values.

Let's assume that the standard deviation of the response is unity. If the specification limits span $+/-6$ standard deviations, the system has a C_p of 2, shown in the following equation. This is the literal goal of DFSS. In this situation the system is said "to possess Six Sigma performance," "to possess six sigma capability," or "to be six sigma capable." A three sigma system has a C_p of 1.

$$C_p = \frac{USL - LSL}{6\sigma} = \frac{12\sigma}{6\sigma} = 2$$

The critical quantity in the calculation of capability is sigma. There are many conditions under which to collect the observations from which sigma will be estimated. It is important to know the conditions under which the observations were made before the reported capability is accepted. If the observations are made with the system in a relatively noise-free environment, the standard deviation will be relatively small and the system's capability may be close to 2. However, we are interested in capability in the presence of noise, not in the pristine, benign, and friendly world of the development lab.

We use the capability index during product development to track the design's growing maturity. This is done in the presence of the noises that are expected to assail the product's performance in its intended-use environment. All the noises may not be known or able to be simulated during development, but three common noises are ambient temperature, altitude, and humidity. If the functions in the product are sensitive to these general noises, an environmental chamber must be available to assess the design's decreasing sensitivity to these noises as the design matures. The development team can then assess the readiness of the subsystem to be integrated into the evolving engineering model or of the readiness of the product to be transferred to manufacturing.

So far, our discussion of capability has been simplistic. A response does not remain centered and on target over time. If several samples of a response are taken, the mean of each sample will be different. Some of this is due to measurement error but, assuming that the measurement system is capable, most of the variation will be due to the system's inherent sensitivity to noise. This is the VOD that, as design engineers, we work to control. A study done by Evans (1975) showed that, in general, process means vary by about 1.5 times their standard deviation. Figure 24–2 shows this situation. Note that, in what follows, the shift could have been drawn to the left of the target with the same conclusions.

This situation illustrates the need for a capability index that accounts for an off-target condition. If we calculate C_p, we would get a value of 2. Yet the probability of being beyond the upper specification limit is far greater than a C_p of 2 implies. In this situation, the probability of being to the left of the LSL is negligible and can be assigned a value of 0. The probability of being to the right of the USL is approximately 3.4×10^{-6} or 3.4 per million. So, a new capability index is needed. The index chosen is C_{pk}. It is calculated as shown in this equation:

$$C_{pk} = \min\left(\frac{USL - \overline{X}}{3\sigma}, \frac{\overline{X} - LSL}{3\sigma} \right)$$

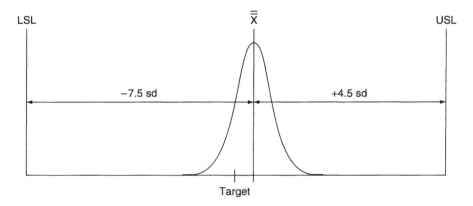

Figure 24–2 Capability with a 1.5 σ Shift

For our example, using a 1.5 standard deviation shift in the mean, $C_{pk} = 1.5$, as shown in the following equation.

$$C_{pk} = \min\left(\frac{USL - \overline{X}}{3\sigma}, \frac{\overline{X} - LSL}{3\sigma}\right) = \min\left(\frac{6\sigma - 1.5\sigma}{3\sigma}, \frac{1.5\sigma - (-6\sigma)}{3\sigma}\right) = 1.5$$

Thus, a design that yields 6 sigma performance ($C_p = 2$) in the short term can be expected to produce a C_{pk} of 1.5 in the long term when in production. As Reigle Stewart, a founder of Six Sigma toleracing, would say, "If Design can provide a C_p of 2 then Manufacturing must provide a C_{pk} of 1.5." Our focus changes from "Are we meeting the specification" to "What is our capability?"

C_p and C_{pk} allow concise communication regarding the performance of our design's critical responses. It is assumed that the capability obtained is a conscious and deliberate compromise between the required quality, the cost of achieving that quality, and the consequences of operating a nonconforming product.

If the mean of the response is centered between its limits, we use C_p. If it is not, we use C_{pk}. The value of C_p or C_{pk} is then compared to the desired value. Typically the values are $C_p = 2$ and $C_{pk} = 1.5$. If these or the agreed-upon values have not been attained, the design team must take action. The design is not mature enough to proceed to the next phase of development. The action to be taken is to center the mean and/or reduce the variation.

The techniques of robust design should be used to determine what region of the overall operating space results in system operation that is least sensitive to noise. Synergistic or ordinal interactions between control factors and noise are sought and exploited to reduce the system's sensitivity to the noise. These techniques fall within the domain of Taguchi methods and classical methods of robust design. They are covered in subsequent chapters in this book.

Key Detailed Inputs

```
The system to be assessed.
```

Specific Steps and Deliverables

```
Select the environment in which data will be collected.
```

A definition of the noises that are expected in the intended-use environment is crucial so that capability can be assessed under those conditions. Capability may be initially assessed with some noises held constant. As the design's maturity evolves, however, its readiness for the next step must be determined using capability in a stressful, noisy environment.

> **Select the responses to be assessed.**

Capability studies are not done on all of a system's responses, only the critical ones. The first step is to determine which system responses are critical. This must have been determined in the concept and early design phases. The application of critical parameter management beginning with VOC, then QFD, then Pugh concept selection defines and determines which responses are critical.

> **Select and qualify the measurement systems used to acquire the data.**

A capability study, as in any situation where data will be used to make decisions, requires a capable measurement system. The techniques covered in Chapter 23 on MSA should have been applied to the measurement system and its Gage R&R, P/T ratio, and the number of distinct categories determined.

> **Run the system in the prescribed environment and acquire the data.**

> **Calculate the sample standard deviation and mean.**

The response data are acquired over a period of time that allows variation in the response to be captured. This may take seconds, or it may take hours. It depends on the particular system and the rates at which the noises are varied. Once the data have been collected, the sample average and sample standard deviation serve as estimates of μ and σ. The data should be plotted in a trend chart or a control chart to look for shifts and drifts in the response over the acquisition time. Shifts and drifts in the response result in a poor capability index. Charting the response can help determine the root cause of such variation.

> **Calculate the capability using the spec limits and the appropriate capability index.**

The capability indices C_p and C_{pk} are calculated from the sample and the specification limits.

Compare the current capability to the desired capability.

Final Detailed Output

**Decision to transfer the system to the next
development phase or to continue robustness efforts.**

A C_p less than 2 indicates that the design team needs to identify the sources of the variation and continue to apply the techniques of robust design and/or empirical tolerancing. A C_p of 2 indicates that the design is sufficiently mature and can be safely transferred to Manufacturing.

An Example

An example of determining capability using Minitab will now be given. One hundred observations were taken on the temperature fall time in seconds of a blood analysis system running in an environmental chamber in which altitude, temperature, and humidity were varied. The observations were taken every minute for 100 minutes.

The response is well centered between the specification limits, as shown graphically in Figure 24–3. It is also evident by the fact that C_p is very close in value to C_{pk}. The response appears to be normally distributed. A normality test can be used to determine this.

Process capability C_p measures the best that the system is capable of. It is associated with short-term variation. The estimation of short-term variation is determined by the subgroup size of the data. In this case the data has not been subgrouped so the subgroup size is 1.

Minitab uses a default moving range of size 2 to determine the short-term standard deviation when the subgroup size is 1. The estimate of short-term variation is called σ_{within}.

The average moving range is a biased estimate of sigma. Minitab uses the bias adjuster d_2 to adjust the moving range to obtain an estimate of short-term variation. The equations used to calculate C_p are shown in the next equation block with a moving range size of 2.

$$d_2 = 1.128$$

$$\sigma_{within} = \frac{AvgMR}{d_2} = \frac{0.5044}{1.128} = 0.447134$$

$$C_p = \frac{USL - LSL}{6\sigma_{within}} = \frac{12 - 8}{6(0.44173)} = 1.49$$

Process performance P_p measures the performance of the system in the long term. It is associated with long-term variation. Long-term capability is calculated in the same way as C_p except that sigma is the long-term sigma instead of the short-term sigma. The estimation of

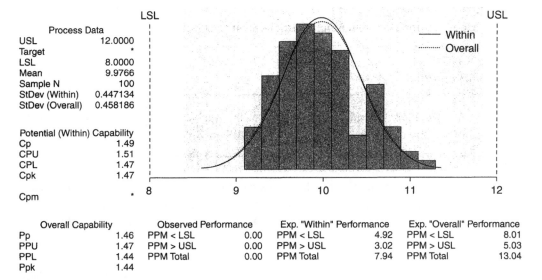

Figure 24–3 Process Capability Analysis for C2

long-term variation is based on the pooled sample standard deviation. The pooled sample stan-dard deviation is also a biased estimate of sigma. Minitab uses the bias adjuster c_4 to adjust the pooled sample standard deviation to obtain an estimate of the overall variation $\sigma_{overall}$. The equa-tions used to calculate P_p are shown in the following equation block with $n = 100$.

$$s = \sqrt{\frac{\sum_{i=1}^{n}(x_1 - \bar{x})^2}{n - 1}} = 0.4570$$

$$c_4 = \frac{4(n - 1)}{(4n) - 3} = 0.99083$$

$$\sigma_{overall} = \frac{s}{c_4} = 0.4581$$

$$P_p = \frac{USL - LSL}{6\sigma_{overall}} = \frac{12 - 8}{6(0.4581)} = 1.455$$

At best the specification limits span 9 sigma, as in:

$$USL - LSL = 4 \approx 9(\sigma_{within})$$

There are approximately 4.5 sigma on either side of the mean. The probability, assuming a normal distribution of producing a response value outside the specification limits, is therefore approximately 7 ppm, or seven out-of-spec values for every million opportunities. This may not sound too bad. However, if the response mean shifts by an average of 1.5 sigma over time, the out-of-spec condition can occur at a rate of 2,700 ppm.

System performance must be improved. The capability must be raised to 2. The techniques of robust design described in this book should be applied until the capability indices meet the required values.

Capability Study Checklist and Scorecard

Checklist for a Capability Study

Actions:	Was this Action Deployed? YES or NO	Excellence Rating: 1 = Poor 10 = Excellent
Select the environment in which data will be collected.	_____	_____
Select the responses to be assessed.	_____	_____
Select and qualify the measurement systems used to acquire the data.	_____	_____
Run the system in the prescribed environment and acquire the data.	_____	_____
Calculate the sample standard deviation and mean.	_____	_____
Calculate the capability using the spec limits and the appropriate capability index.	_____	_____
Compare the current capability to the desired capability.	_____	_____
Decision to transfer the system to the next development phase or to continue robustness efforts.	_____	_____

Scorecard for a Capability Study

The definition of a good scorecard is that it should illustrate, in simple but quantitative terms, what gains were attained from the work.

This suggested format for a scorecard can be used by a product planning team to help document the completeness of its work.

Critical to Function Response	Unit of Measure:	Target Capability:	Actual Capability:	Actions to Resolve Capability Gap:
1.	_____	_____	_____	_____
2.	_____	_____	_____	_____
3.	_____	_____	_____	_____

We repeat the final deliverables:

- Assessment of the stability of the critical-to-function responses
- Estimates of the mean and standard deviation of the critical-to-function responses
- Estimate of the capability index C_p and C_{pk} used to help decide if the system is ready for the next phase of development

References

Evans, D. H. (1975). Statistical tolerancing: The state of the art. *Journal of Quality Technology,* *7*(1), 1–12.

Harry, M., & Stewart, R. (1998). Six Sigma mechanical design tolerancing. Shaumburg, IL: Motorola University Press.

Multi-Vari Studies

Where Am I in the Process? Where Do I Apply Multi-Vari Studies Within the Roadmaps?

Linkage to the I²DOV Technology Development Roadmap:

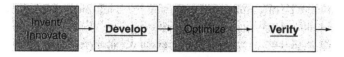

Multi-vari studies are conducted during the Develop and Verify phases of I²DOV to identify the factors that affect the system's critical-to-function responses.

Linkage to the CDOV Product Development Roadmap:

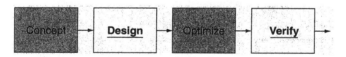

Multi-vari studies are conducted during the Design phase of CDOV to identify the factors that affect the product's critical-to-function responses. It is conducted during the Verify phase to identify factors that might have been missed in earlier phases.

What Am I Doing in the Process? What Do Multi-Vari Studies Do at this Point in the Roadmaps?

The Purpose of Multi-Vari Studies in the I²DOV Technology Development Roadmap:

Multi-vari studies are conducted during I²DOV technology development and certification to identify critical and noncritical factors and response of the technology.

The Purpose of Multi-Vari Studies in the CDOV Product Development Roadmap:

Multi-vari studies are conducted during CDOV product design and capability certification to identify critical and noncritical factors and response of the product.

What Output Do I Get at the Conclusion of this Phase of the Process? What Are Key Deliverables from Multi-Vari Studies at this Point in the Roadmaps?

- The coefficients of correlation between the many inputs and outputs of the system
- A list of the inputs and outputs that have been identified as potential critical parameters and responses
- A list of the inputs and outputs that have been eliminated as potential critical parameters and responses

The key deliverables of a multi-vari study are the coefficients of correlation between the many inputs and outputs of the system. The coefficients are used to eliminate some inputs and outputs from the list of potential critical parameters and responses. This will reduce the size and number of subsequent designed experiments that need to be run, thereby saving time and experimental effort.

Multi-Vari Study Process Flow Diagram

This visual reference identifies the detailed steps of multi-vari studies.

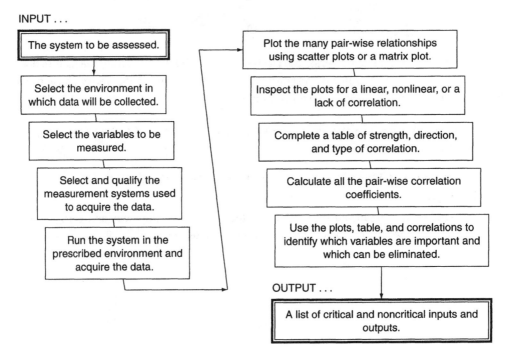

INPUT . . .

| The system to be assessed. |

| Select the environment in which data will be collected. |

| Select the variables to be measured. |

| Select and qualify the measurement systems used to acquire the data. |

| Run the system in the prescribed environment and acquire the data. |

| Plot the many pair-wise relationships using scatter plots or a matrix plot. |

| Inspect the plots for a linear, nonlinear, or a lack of correlation. |

| Complete a table of strength, direction, and type of correlation. |

| Calculate all the pair-wise correlation coefficients. |

| Use the plots, table, and correlations to identify which variables are important and which can be eliminated. |

OUTPUT . . .

| A list of critical and noncritical inputs and outputs. |

Verbal Descriptions for the Application of Each Block Diagram

This section explains the specific details of each step of a multi-vari study:

- *Key detailed input(s)* from preceding tools and best practices
- *Specific steps and deliverables* from each step to the next within the process flow
- *Final detailed outputs* that flow into follow-on tools and best practices

Multi-vari, a term coined by A. B. Seder in the 1950s, refers to a method used to detect correlation among many variables in a system. It is used in the Design and Verify Phases of CDOV to determine which factors and responses should and should not be studied in subsequent designed experiments. We cannot study all the important parameters and responses in a DOE. There are usually too many of them. We only study the few critical parameters and responses in a DOE.

Multi-vari is a passive experimental technique in that the behavior of the system is measured without purposefully varying the inputs to the system, as we would do in a designed experiment. The system's inputs and outputs are measured in the presence of everyday variation. The variation comes from changes in operators, temperature, voltage, pressure, raw materials, and so on.

A multi-vari analysis makes use of both graphical and analytic tools. The key graphical tools are scatter and matrix plots. The primary analytic tool is the coefficient of correlation.

Key Detailed Inputs

> **The system to be assessed.**

Specific Steps and Deliverables

> **Select the environment in which data will be collected.**

Multi-vari experiments are generally run without altering or adjusting the environment. Data are taken as the system experiences its "normal, "every-day" sources of variation.

> **Select the variables to be measured.**

Multi-vari studies are not done on all system inputs and outputs, only the suspected critical ones. The first step is to determine which are critical. This must have been determined in the Concept and early Design phases. The application of Critical Parameter Management beginning with VOC, then QFD, then Pugh concept selection defines and determines which inputs and outputs are critical.

> **Select and qualify the measurement systems used to acquire the data.**

A multi-vari study, as in any situation where data will be used to make decisions, requires a capable measurement system. The techniques covered in Chapter 23 on MSA should have been applied to the measurement system and its Gage R&R, P/T ratio, and the number of distinct categories determined.

> **Run the system in the prescribed environment and acquire the data.**

> **Plot the many pair-wise relationships using scatter plots or a matrix plot.**

Let's consider an example in which six system variables, A, B, C, D, E, and F, were simultaneously measured over a period of 6 hours. A is an output of the system, and B through F are inputs. A scatter plot can be used to visually indicate relationships between the variables. The scatter plot is a common graphical tool used to visually determine if two variables do or do not depend on each other. A scatter plot between A and B is shown in Figure 25–1. The plot suggests that a relationship does exist. We create scatter plots for all the pair-wise comparisons among all the measured variables. The scatter plots also indicate if the relationship is linear or nonlinear. This is important since the analytic measure of correlation we will use later, the Pearson coefficient of correlation, can only detect linear correlation.

A matrix plot is a convenient way to view all the pair-wise scatter plots. Minitab provides this utility in its graph menu pull-down (see example in Figure 25–2).

The matrix plot indicates that

A appears to be related to B, D, E, and F.

The A and B relationship appears to be nonlinear.

B appears to be related to E and F.

The B and F relationship appears to be nonlinear.

C appears to be related to D.

D appears to be related to E and F.

E appears to be related to F.

> **Inspect the plots for a linear, nonlinear, or a lack of correlation.**

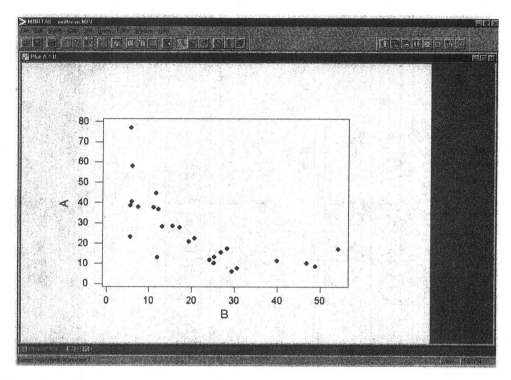

Figure 25–1 Scatter Plot of A and B Variables

> **Complete a table of strength, direction, and type of correlation.**

These relationships are summarized in Table 25–1. The first entry in a cell is the strength of the relationship, *S*trong, *M*edium, or *W*eak. The next entry is the direction of the relationship, (+) for positive and (−) for negative. A positive relationship means that both variables increase or decrease together. A negative relationship means that one variable increases as the other variable decreases. The last entry is the type of relationship, *L*inear or *N*onlinear. An empty cell indicates that no obvious relationship is seen in the plot.

Table 25–1 suggests that the output, A, is driven by the inputs B, D, E, and F. It is likely that B will be a critical input since it has a strong influence on A. It also indicates multi-colinearity—that is to say that some of the inputs are related to each other. Perhaps it is not possible to vary one input, say B, without causing or requiring F to vary. This knowledge can be used to our benefit. We can study only one of the input variables if we know the equation describing the relationship. We would choose to study the one that is easiest to vary and/or measure.

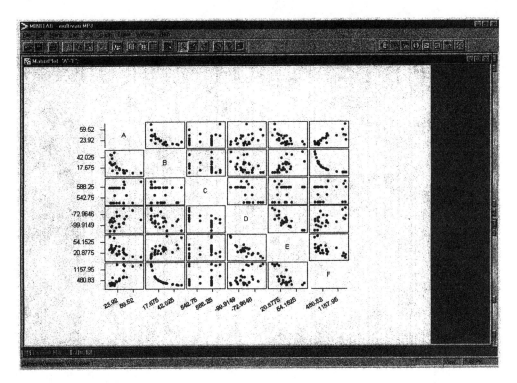

Figure 25–2 Matrix Plot of Variables A–F

Table 25–1 Sample Table of Strength, Direction, and Type of Correlation

Factor	A	B	C	D	E
B	S, −, NL				
C					
D	M, +, L	W, −, L	S, −, L		
E	M, −, L	M, +, L		S, −, L	
F	W, +, L	S, −, NL		W, +, L	M, −, L

> **Calculate all the pair-wise correlation coefficients.**

This is very useful information, but we don't want to make these decisions by visually interpreting a chart. We want the decision to be based on a numeric value and some level of statistical significance. The metric we use is the correlation coefficient, r.

The Pearson correlation coefficient, r_{x1x2}, is given in the following equation for paired observations of the variables x_1 and x_2. S_{x1} and S_{x2} are the standard deviations of x_1 and x_2, respectively.

$$r_{x1x2} = \frac{\sum_{i=1}^{n}[(x_{1i} - \bar{x}_1)(x_{2i} - \bar{x}_2)]}{(n-1)S_{x1}S_{x2}}$$

The correlation coefficient ranges from -1 to $+1$. Zero indicates no correlation. Values less than -0.8 or greater than $+0.8$ are considered a strong correlation.

The correlation coefficient only tests for a linear relationship. Two variables, like B and F in our example, can be strongly nonlinearly related and yet have a small magnitude for r. This is one reason why we use the matrix plot to visually identify correlation that the coefficient cannot detect.

Minitab can calculate the correlation coefficient for us for all the pair-wise combinations. It also reports a p-value for each coefficient. The null hypothesis here is that the correlation coefficient is zero. The alternative hypothesis is that the correlation coefficient is not zero. A p-value of 0.05 or less allows us to reject the null hypothesis and accept that the correlation is not zero.

The correlation coefficients for the 15 pair-wise combinations follows. The correlation of a factor with itself is not shown since it is always unity.

Correlations: A, B, C, D, E, F

	A	B	C	D	E
B	-0.690				
	0.000				
C	0.156	-0.212			
	0.448	0.299			
D	0.401	-0.465	-0.631		
	0.042	0.017	0.001		
E	-0.587	0.659	0.165	-0.805	
	0.002	0.000	0.420	0.000	
F	0.778	-0.810	0.164	0.533	-0.650
	0.000	0.000	0.424	0.005	0.000

Cell Contents: Pearson correlation P-Value

> **Use the plots, table, and correlations to identify which variables are important and which can be eliminated.**

What conclusions can be drawn from the data in the table? First, the correlation coefficients for AC, BC, CE, and CF cannot be distinguished from zero, at an alpha risk of 5 percent, given the noise in the data. Their p-values are greater than 0.05 so we do not reject the null hypothesis. We conclude that these variables are not correlated.

The output A is correlated with the four variables F, B, E, and D, listed in order of strength of the correlation. These are candidate factors for a DOE. The DOE will provide a more detailed understanding of the relationship. A DOE can provide an equation, also known as an *empirical model* or *transfer function, A = f(B,D,E,F)*, that analytically relates the inputs to the output. We use the transfer function to determine the relative strength with which variation in a factor causes variation in the response.

Final Detailed Output

> **A list of critical and noncritical inputs and outputs.**

We see from the table that B is highly correlated with F and that D is highly correlated with E. We could choose to study B or E instead of F or E. We would do this if B or E were easier to measure and adjust compared to F or E.

In some situations correlation is all we need to know to direct the work to improve the system's performance. A detailed transfer function from a DOE may be unnecessary. It may be economically possible to tightly tolerance an input that has a high correlation with the output. Sometimes all that is needed is to apply statistical process control to the inputs. If these methods do not result in the required performance improvements, more detailed system modeling and robust design techniques may be necessary. These techniques are covered in other sections of this book.

Multi-Vari Study Checklist and Scorecard

Checklist for Multi-Vari

Actions:	Was this Action Deployed? YES or NO	Excellence Rating: 1 = Poor 10 = Excellent
Select the environment in which data will be collected.	_____	_____
Select the variables to be measured.	_____	_____
Select and qualify the measurement systems used to acquire the data.	_____	_____
Run the system in the prescribed environment and acquire the data.	_____	_____

(*continued*)

Plot the many pair-wise relationships using scatter plots or a matrix plot.	_____	_____
Inspect the plots for a linear, nonlinear, or a lack of correlation.	_____	_____
Complete a table of strength, direction, and type of correlation.	_____	_____
Calculate all the pair-wise correlation coefficients.	_____	_____
Use the plots, table, and correlations, identify which variables are important and which can be eliminated.	_____	_____
A list of critical and noncritical inputs and outputs.	_____	_____

Scorecard for Multi-Vari

The definition of a good scorecard is that it should illustrate, in simple but quantitative terms, what gains were attained from the work. This suggested format for a scorecard can be used by a product planning team to help document the completeness of its work

System Variable Name:	Unit of Measure:	Input or Output:	Critical or Noncritical?
1.	_____	_____	_____
2.	_____	_____	_____
3.	_____	_____	_____

We repeat the final deliverables:

- The coefficients of correlation between the many inputs and outputs of the system
- A list of the inputs and outputs that have been identified as potential critical parameters and responses
- A list of the inputs and outputs that have been eliminated as potential critical parameters and responses

The key deliverables of a multi-vari study are the coefficients of correlation between the many inputs and outputs of the system. The coefficients are used to eliminate some inputs and outputs from the list of potential critical parameters and responses. This will reduce the size and number of subsequent designed experiments that need to be run, thereby saving time and experimental effort.

Reference

Breyfogle, F. W. (1999). *Implementing six sigma.* New York: Wiley Interscience.

Regression

Where Am I in the Process? Where Do I Apply Regression Within the Roadmaps?

Linkage to the I²DOV Technology Development Roadmap:

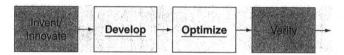

Regression is primarily used during the Develop and Optimize phases of I²DOV. It is used much less during the Invent/Innovate and Verify phases.

Linkage to the CDOV Product Development Roadmap:

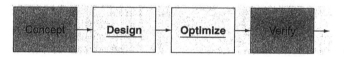

Regression is primarily used during the Design and Optimize phases of CDOV. It is used much less during the Concept and Verify phases.

What Am I Doing in the Process? What Does Regression Do at this Point in the Roadmaps?

The Purpose of Regression in the I²DOV Technology Development and the CDOV Product Development Roadmaps:

Regression is used to create an equation or transfer function from the measurements of the system's inputs and outputs acquired during a passive or active experiment. The transfer function is then used for sensitivity analysis, optimization of system performance, and tolerancing the system's components.

What Output Do I Get at the Conclusion of this Phase of the Process? What Are Key Deliverables from Regression at this Point in the Roadmaps?

- An estimation of the relative strength of the effect of each factor on the response
- An equation that analytically relates the critical parameters to the critical responses
- An estimate of how much of the total variation seen in the data is explained by the equation

Regression Process Flow Diagram

Visual reference of the detailed steps of regression illustrates the flow of tasks and responsibilities.

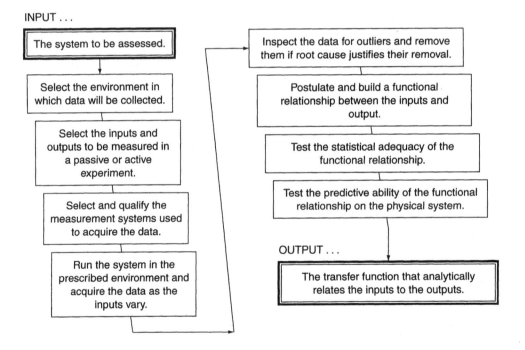

INPUT . . .

The system to be assessed.

Select the environment in which data will be collected.

Select the inputs and outputs to be measured in a passive or active experiment.

Select and qualify the measurement systems used to acquire the data.

Run the system in the prescribed environment and acquire the data as the inputs vary.

Inspect the data for outliers and remove them if root cause justifies their removal.

Postulate and build a functional relationship between the inputs and output.

Test the statistical adequacy of the functional relationship.

Test the predictive ability of the functional relationship on the physical system.

OUTPUT . . .

The transfer function that analytically relates the inputs to the outputs.

Verbal Descriptions for the Application of Each Block Diagram

This section explains the specific details of the regression process.

- *Key detailed input(s)* from preceding tools and best practices
- *Specific steps and deliverables* from each step to the next within the process flow
- *Final detailed outputs* that flow into follow-on tools and best practices

The behavior of the systems we design is governed by the laws of physics. An engineer can be thought of as an applied physicist. Our task is to purposefully manipulate the laws of physics to perform useful work for our customers. The laws we manipulate have been identified and written down by many famous men and women throughout history. These are the "famous dead guys and gals" like Newton, Kirchoff, Curie, Faraday, and Fourier. They have given us the ideal functions that mathematically describe the physics that underlie the technology we employ in our products.

For example, Ohm told us that the voltage through a resistor is a function of current and resistance, $v = ir$. This ideal function is exactly that, ideal. If there were no variation in nature, we would always obtain the same measured value for each value of current.

However, variation is everywhere. Variation causes repeated measurements of the voltage across a resistor to be different each time we make a measurement. This is shown in Figure 26–1. The open circles are what Ohm predicted we would measure, and the filled circles are our actual measurements.

So, Herr Ohm missed or purposefully left out a term. The missing term accounts for all the variation that causes the measured voltage to deviate from the voltage predicted by the ideal function. Some sources of variation are measurement system error, temperature variation, humidity, current variation, and self-heating. These sources of variation cannot be eliminated. They can only be reduced and controlled.

Ohm's law can now be extended to include this missing term. The resulting equation describes the behavior that we actually observe in addition to the physics embedded in that behavior. The "behavioral" version of Ohm's law is $v = ri + \varepsilon$. Voltage is the sum of Ohm's law and an error term ε. The error term accounts for the difference between what we measure and what Ohm predicted.

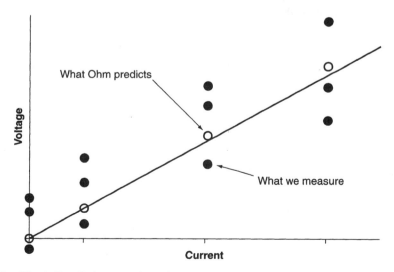

Figure 26–1   Ohm's Predictions vs. Actual Measurements

The situation can be generalized for any linearly related set of independent and dependent variables x and y as $y = \beta_0 + \beta_1 x + \varepsilon$. The terms y, β_0, β_1, and the error ε are random variables. The dependent variable x is also a random variable, but we treat it as fixed. We assume that x can be set and reset with great precision compared to its range. In our example, we assume we can set the current to its various values with great precision.

Our goal is to model or fit an equation to the observed behavior. This is called *empirical modeling,* and the resulting equation is called an empirical model, fitted equation, or transfer function.

As engineers, we have insight regarding the underlying physics at work in the system we are modeling. This allows us to make assumptions about the structure of the model we seek. In our example, given that Ohm's law is at work, a sensible fitted equation is $\hat{y}_i = b_0 + b_1 x_i$. The "hat" above the y indicates that it is an estimate of y. It is called a *fitted value,* or a prediction.

Notice that in the fitted model the βs have been replaced by bs. The bs serve as estimates of the true but unknown values of the coefficients, the βs. Also notice that there is no error term in the fitted equation. Uncertainty in the fitted equation is accounted for by uncertainty in b_0 and b_1.

The bs are calculated so that there is minimum total error between the values predicted by the equation and the actual observations. An observed value at the ith level of x is y_i. A calculated value using the fitted equation at the ith level of x is $\hat{y}_i$. The error between the two is $\varepsilon_i = y_i - \hat{y}_i$. This is called a *residual.* A large residual indicates that the predicted value does not match the observed value closely. We want our model to have small residuals at all values of x. The residual will vary between a positive and a negative value for each observation at each level of x. However, what we want is an error that increases and decreases monotonically. We can achieve this if we square the residuals and then sum them over all values of x. Our fitted model should minimize the sum of the squared errors. This is curve fitting by "least squares."

So, we substitute for the fitted value, square the error, and sum the overall levels of x. We'll assume, for now, that there is only one observation at each level of x. This is shown in this equation block:

$$e_i = y_i - \hat{y}_i = y_i - (\beta_0 + \beta_1 x_i)$$
$$e_i^2 = (y_i - \beta_0 - \beta_1 x_i)^2$$
$$S = \sum_{i=1}^{n} e_i^2 = \sum_{i=1}^{n} (y_i - \beta_0 - \beta_1 x_i)^2$$

Our goal is to minimize the sum of the squared errors. We do this by taking the partial derivative with respect to both β_0 and β_1 and setting them equal to zero, as in the following equation block.

$$\frac{\partial \sum_{i=1}^{n} e_i^2}{\partial \beta_0} = 0 = \sum_{i=1}^{n} (2\beta_0 + 2\beta_1 x_i - 2y_i)$$

$$\frac{\partial \sum_{i=1}^{n} e_i^2}{\partial \beta_1} = 0 = \sum_{i=1}^{n} (2\beta_0 x + 2\beta_1 x_i^2 - 2y_i x_i)$$

We can solve these equations for β. After some algebra, we obtain the normal equations, which have nothing to do with a normal distribution. Solution of the normal equations provides the least squares estimators of β—$\hat{\beta}_0 = b_0$ and $\hat{\beta}_1 = b_1$:

$$\hat{\beta}_1 = b_1 \frac{\sum_{i=1}^{n} [(x_i - \bar{x})(y_i - \bar{y})]}{\sum_{i=1}^{n} (x_i - \bar{x})^2} = \frac{SS_{xy}}{SS_{xx}}$$

$$\hat{\beta}_0 = b_0 = \bar{y} - b_1 \bar{x}$$

These are our best estimates of β_0 and β_1, and they result in a fitted line with a minimum sum of squared error between the predicted and the observed data.

An alternative form of the regression equation is obtained if we substitute for b_0 to obtain $\hat{y}_i = \bar{y} - b_1 \bar{x} + b_1 x_i = \bar{y} + b_1(x_1 - \bar{x})$. The alternate form illustrates an interesting property of straight-line regression. If we let $x_i = \bar{x}$, we see that the regression line goes through the point $(\bar{x}, \bar{y})$. We will use this to develop the ANOVA for regression.

We have our estimates of β_0 and β_1. We now need a way to test the statistical significance of these estimates. We use ANOVA to accomplish this. See Figure 26–2 for the following discussion. The distance of an observation from the overall mean of all the data can be broken into two parts. One part is the distance of the predicted value from the mean; this distance is *due* to the regression. The other part is the distance of an observation from the predicted value; this distance is *about* the regression.

This is written in the following equation for the ith observation.

$$(y_i - \bar{y}) = (\hat{y}_i - \bar{y}) + (y_i - \hat{y}_i)$$

But we want terms that relate to variance, and we want this for all the observations. So we square and then sum both sides over all the observations. If you do the math, you'll find that, in the summation, the cross-product term equals 0. The result is the next equation along with a "sum of squares" interpretation of the equation.

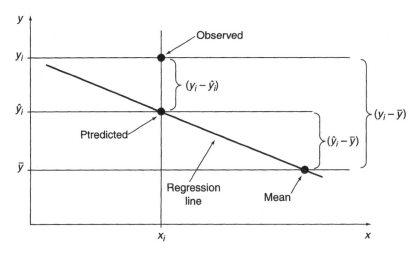

Figure 26–2 Decomposition of Variation

$$\sum_{i=1}^{n} (y_i - \bar{y})^2 = \sum_{i=1}^{n} (\hat{y}_i - \bar{y})^2 + \sum_{i=1}^{n} (y_i - \hat{y})^2$$

$$SS_{total} = SS_{due \; to \; the \; regression} + SS_{about \; the \; regression}$$

The total variation in the data is composed of the variation due to the regression and the variation about the regression. Variation about the regression is a measure of the error between the observations and the fitted line. This term is called a *residual* or *residual error*. If the regression were to fit all the observation perfectly, this term would be 0. We want variation due to the regression to be larger than the variation about the regression.

We can compare the variation due to the regression to the total amount of variation in all the data. This ratio is called R^2 and is given in the following equation. It is the square of the coefficient of multiple correlation and is typically given in percent.

$$R^2 = \frac{SS_{due \; to \; regression}}{SS_{total}} 100\% = \frac{\displaystyle\sum_{i=1}^{n} (\hat{y}_i - \bar{y})^2}{\displaystyle\sum_{i=1}^{n} (y_i - \bar{y})^2} 100\%$$

The regression line should explain a large amount of the total variation, so we want R^2 to be large. As R^2 decreases, the error becomes a larger part of the total variation. This is not desired.

In biological, psychological, and economic systems, R^2 in the range of 50 percent to 60 percent is considered very good. We, however, are engineers manipulating physics, mainly Newtonian. Our systems are better behaved and, in a sense, simpler than the ones previously described. We expect R^2 in the range of 80 percent to 98 percent. Choosing a poor model will reduce R^2. Sloppy experimentation, including a measurement system that is not capable, also reduces R^2. These situations cause the error term to be large and result in a small R^2. Knowledge of the underlying physics as well as excellence in experimentation and metrology is imperative.

R^2 provides a measure of how much variation our regression equation explains, but we must also determine if that variation is statistically significant when compared to the variation due to error. We can perform a statistically rigorous test if we constrain the error term ε that we spoke about earlier.

Remember we said the behavior we observe is due to the sum of physics and variation. The error term accounts for the deviation of each observation from what the ideal functions of physics would predict. If we require that $\varepsilon \sim N(0,\sigma^2)$, that is, ε is normally distributed with a mean of zero and a variance of σ^2, then we can devise a statistically rigorous test to determine if the variation due to regression is merely experimental error or if it models a real relationship.

This is similar to what we did in ANOVA. If ε is normally distributed, y is also normally distributed. If y is normally distributed, its square is chi-square distributed. The ratio of two chi-square variables is F-distributed. So we can calculate an F-ratio given by this equation:

$$F = \frac{MS_{due\ to\ regression}}{MS_{error}} = \frac{\dfrac{\sum\limits_{i=1}^{n} (\hat{y}_i - \bar{y})^2}{dof_{regression}}}{\dfrac{\sum\limits_{i=1}^{n} (y_i - \hat{y}_i)^2}{dof_{error}}}$$

The dof for regression is equal to the number of terms in the equation, not including b_0. The dof_{error} is easiest to find as $dof_{total} - dof_{regression}$ where dof_{total} is one less than the number of total observations.

We know the distribution of the F-ratio. It is F-distributed, so we can therefore determine the probability of getting F values if the null hypothesis were true. The null suggests that the variation explained by the regression is no different from error. This is equivalent to saying that the coefficients of the equation are 0. So, if the p-value associated with a given F-ratio is greater than 0.05, we fail to reject the null and conclude that the regression is no different from noise. If the p-value is less than 0.05, we reject the null and conclude that the regression equation explains variation that is greater than noise.

We can also perform a statistical test on the individual coefficients of the regression. We can test whether the intercept b_0 is equal to 0. We can test whether the slope b_1 is equal to 0. We use a t-test for b_0. If the p-value is greater than 0.05, we fail to reject the null that the coefficient is 0 and conclude that the intercept is 0. If the p-value is less than 0.05, we reject the null and conclude that the intercept is not 0.

We also use a t-test for b_1. If the p-value is greater than 0.05, we fail to reject the null that the coefficient is 0 and conclude that the slope is 0. If the p-value is less than 0.05, we reject the null and conclude that the slope is not 0. We are overdue for an example.

Key Detailed Inputs

> **The system to be assessed.**

The system is a simple electrical network consisting of a resistor and voltage supply.

Specific Steps and Deliverables

> **Select the environment in which data will be collected.**

In this example the data was acquired on a test bench located in a manufacturing area.

> **Select the inputs and outputs to be measured in a passive or active experiment.**

Let's say we measured the voltage across a 15Ω resistor at current levels from 1 to 5 amps. We measured the voltage three times at each level of current and randomly changed the current level each time. This is a simple, single-factor, single-response experiment. The data are shown in the scatter plot in Figure 26–3. We can "see" Ohm's law and we can see variation in the plot. No outliers are evident in the data.

> **Select and qualify the measurement systems used to acquire the data.**

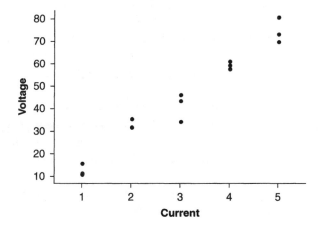

Figure 26–3   Scatter Plot on Data from Voltage Experiment

Regression, as in any situation where data will be used to make decisions, requires a capable measurement system. The techniques covered in Chapter 23 on MSA should have been applied to the measurement system and its Gage R&R, P/T ratio, and the number of distinct categories determined.

Run the system in the prescribed environment and acquire the data as the inputs vary.

Inspect the data for outliers and remove them if root cause justifies their removal.

A scatter plot and individual statistical process control chart can be used to identify potential outliers. If outliers are found, root cause should be sought. They can be removed if it is justifiable. In addition, the analysis can be conducted with and without the outliers and the differences compared.

Postulate and build a functional relationship between the inputs and output.

Our knowledge of the physics, the appropriate ideal function, and the form of the data in the plot all suggest that a linear, first order equation is a likely model with which to begin. Let's use Minitab to fit a line through the data. Choose Stat > Regression > Regression. We get the following window. Our response is Voltage and our single predictor is Current.

The session window shows the result of the regression analysis.

Regression Analysis: Voltage versus Current

```
The regression equation is
Voltage = -0.907187 + 15.0162 Current

S = 4.35971       R-Sq = 96.5 %        R-Sq(adj) = 96.2 %

Analysis of Variance

Source            DF          SS          MS          F        P
Regression         1     6764.57     6764.57    355.897    0.000
Error             13      247.09       19.01
Total             14     7011.67
```

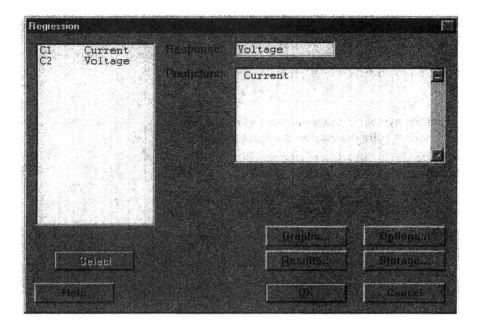

Let's calculate the coefficients of the equation, using the normal equations presented previously. Try doing this yourself.

$$b_1 = \frac{\sum_{i=1}^{n} [(x_i - \bar{x})(y_i - \bar{y})]}{\sum_{i=1}^{n} (x_i - \bar{x})^2} = \frac{SS_{xy}}{SS_{xx}} = \frac{450.486}{30} = 15.0162$$

$$b_0 = \bar{y} - b_1\bar{x} = 44.1414 - 15.0162*3 = -0.907187$$

Now let's calculate the various sums of squares. The equations are a bit different since we have three measurements at each level of current. A particular observation at a particular value of current is denoted $y_{i,j}$.

$$SS_{total} = \sum_{i=1}^{5} \sum_{j=1}^{3} (y_{i,j} - \bar{y})^2 = 7{,}011.67$$

$$SS_{reg} = \sum_{i=1}^{5} \sum_{j=1}^{3} (y_{i,j} - \hat{y}_i)^2 = 6{,}764.57$$

$$SS_{error} = SS_{total} - SS_{reg} = 247.09$$

Now let's account for the degrees of freedom. The total degrees of freedom is given by

$$dof_{total} = (\#\text{observations} - 1) = (15 - 1) = 14$$

The regression degrees of freedom is given by

$$dof_{reg} = \#\text{terms in the equation} = 1 \text{ (Don't include } b_0.)$$

The error degrees of freedom is given by

$$dof_{error} = dof_{total} - dof_{reg} = 14 - 1 = 13$$

Now calculate the mean sums of squares for regression and error.

$$MS_{reg} = \frac{SS_{reg}}{dof_{reg}} = \frac{6{,}764.57}{1} = 6{,}764.57$$

$$MS_{error} = \frac{SS_{error}}{dof_{error}} = \frac{247.09}{13} = 19.01$$

Test the statistical adequacy of the functional relationship.

We can estimate the amount of experimental error from the mean sums of square for error. In this example, experimental error is

$$\sqrt{MS_{error}} = 4.36 \text{ volts}$$

This tells us that we cannot tell whether an observed change of 4 volts is due to a change in actual resistor voltage or is just experimental error.

The F value for the regression is given by

$$F = \frac{MS_{reg}}{MS_{error}} = \frac{6{,}764.57}{19.01} = 355.897$$

This is a large F value, and it occurs with a very small probability. It is reported by Minitab as $p = 0.000$, but it is not zero, just very small. The likelihood of getting this F value is very small if the null hypothesis were true.

The null hypothesis is that the variation due to regression is no different from the variation due to error. We reject the null since $p < 0.05$ and conclude that the regression equation explains variation that is different from error. This is good. Our equation is useful.

We can calculate R^2, the coefficient of determination.

$$R^2 = (SS_{reg}/SS_{total})100\% = (6,764.57 / 7,011.67)100\% = 96.5\%.$$

This is good. The variation explained by the regression equation accounts for 96.5 percent of the total variation seen in the experiment.

The quality of the regression can also be assessed from a plot of the residuals versus the fitted values. The amount of variation not explained by the regression is contained in the residual error term. If the regression is useful it should explain most of the variation and the residual error should be small. There should be no observable structure to the variation in the residuals. If we plot the residuals against the fitted values, we should not see any structure in that plot.

The residuals versus fits plot for this example follows.

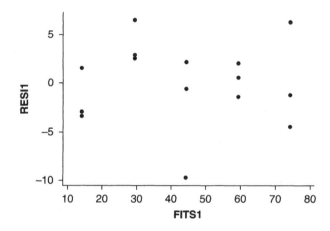

The plot shows no discernable structure. There's an unusual residual at a fitted value of approximately 43 volts. Our model appears to be a good one based on R^2 and the residuals versus fits plot. We do have a large amount of error though, and measurement system capability should be checked.

Minitab can produce a plot, along with the equation R^2, of the fitted line and the data. Use Stat > Regression > Fitted Line Plot. We get the plot window shown in Figure 26–4.

> **Test the predictive ability of the functional relationship on the physical system.**

We now have an equation that predicts voltage based on current. The equation includes all the sources of noise that caused the voltage to differ from what Ohm's law would predict. The equa-

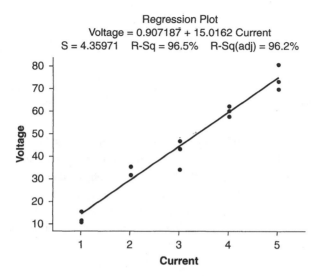

Figure 26–4   Fitted Line Plot for the Resistor Example

tion is *voltage* = −0.907187 + 15.0162 *current*. We can use the equation to predict the voltage to be obtained at current levels throughout the experimental range.

We can assess the predictive ability of the regression model using the PRESS metric. PRESS is the prediction sums of squares. It is found by removing the ith observation from the data and then creating a new regression equation. The new equation is used to predict the missing observation. The error for the ith case is $y_i - \hat{y}_i$. This is repeated for all the observations, and PRESS is calculated by summing overall the PRESS values, $PRESS = \Sigma \, (y_i - \hat{y}_i)^2$. A small PRESS value is desired, and the associated model is considered to have good predictive capability.

R-sq(pred) is determined by

$$R^2_{pred} = 1 - \frac{PRESS}{Total_{SS}}$$

In this example R-sq(pred) = 95.4 percent, which means that the model predicts new values as well as an observation that is already in the data set. The PRESS value and R-sq(pred) are shown in the following Minitab analysis.

Regression Analysis: Voltage versus Current

```
The regression equation is
Voltage = - 0.91 + 15.0 Current
```

```
Predictor            Coef        SE Coef             T         P
Constant           -0.907         2.640          -0.34     0.737
Current           15.0162         0.7960         18.87     0.000

S = 4.360                 R-Sq = 96.5%              R-Sq(adj) = 96.2%
PRESS = 323.656           R-Sq(pred) = 95.38%
```

The system should be run under a given condition and the actual and predicted values compared to assess the model's ability to predict actual behavior. This should be done anytime an equation has been developed to describe the behavior of a physical system.

We can also use the equation to determine the extent to which variation in current causes variation in voltage. We can determine how tightly we need our control or tolerance current and resistance so that the variation in voltage is kept to an acceptable limit. We will cover this in the section on optimization.

Final Detailed Outputs

> **The transfer function that analytically relates the inputs to the outputs.**

The transfer function that analytically relates the inputs to the outputs is restated below.

$$voltage = -0.907187 + 15.0162 \; current$$

Our discussion so far has been for an extremely simple case. We have looked at a single predictor, x, and one response, y. This is simple linear regression. It is instructive but of limited use.

Typically, we are interested in equations in which the predictors are nonlinear, (for example, squared, cubic, square root, or transcendental). Linear regression can handle this. In fact, linear regression only refers to the coefficients of the equation, not to the predictors. So $y = b_0 + b_1 (\ln(x))$ is linear regression while $y = b_0 + \ln(b_1)x$ and $y = b_0 + b_1 x^{b_1}$ are not.

We are also interested in equations that relate many predictors to many responses. This is multivariate statistics and is a bit beyond what we will discuss here. We will, however, discuss equations that relate many predictors to one response. This is called *multiple regression*. We must now switch from scalar notation to matrix notation if we are to explain even a simple multiple linear regression situation.

Imagine we make many measurements of our single response, $y_1, y_2, \ldots y_n$. Also assume our model has many terms in it, $x_1, x_2, \ldots x_r$. The xs can be linear or nonlinear but the coefficients must be linear. We can represent our behavioral equations as a series of equations shown here. Our response is a function of p predictors, and we have n observations of our response. The zs can be functions of the xs such as $z_1 = x_1 x_2^2$.

$$y_1 = \beta_0 + \beta_1 z_{1,1} + \beta_2 z_{1,2} + \ldots \beta_p z_{1,p} + \varepsilon_1$$
$$y_2 = \beta_0 + \beta_1 z_{2,1} + \beta_2 z_{2,2} + \ldots \beta_p z_{2,p} + \varepsilon_2$$
$$\vdots$$
$$y_n = \beta_0 + \beta_1 z_{n,1} + \beta_2 z_{n,2} + \ldots \beta_p z_{n,p} + \varepsilon_n$$

We can write this set of equations in a compact matrix form as $Y = Z\beta + \varepsilon$. In this representation, Y, β, Z, and ε are vectors and matrices as defined in the following equation. The matrix Z consists of linear or nonlinear functions of our x variables. For example, $z_1 = x_1^2 x_3^3$

$$Y = \begin{bmatrix} y_1 \\ \vdots \\ y_n \end{bmatrix} = Z \begin{bmatrix} 1 & z_{1,1} & \cdots & z_{1,p} \\ \vdots & \vdots & \ddots & \vdots \\ 1 & z_{n,1} & \cdots & z_{n,p} \end{bmatrix} \beta \begin{bmatrix} \beta_0 \\ \vdots \\ \beta_p \end{bmatrix} + \varepsilon \begin{bmatrix} \varepsilon_1 \\ \vdots \\ \varepsilon_n \end{bmatrix}$$

We can fit an equation to the observations. The form of the fitted equation, in matrix notation, is $\hat{Y} = ZB$, where B is a vector of coefficients that best estimates β in a least squares sense. We can solve this equation with a little matrix algebra shown below in the following equation block. Note that, in what follows, we assume $Z = X$ with no loss of generality.

$Y = XB$

Premultiply both sides by the transpose of X to get a square matrix:

$X'Y = X'XB$

Now, premultiply both sides by $(X'X)^{-1}$ which is the inverse of $X'X$ to isolate B:

$(X'X)^{-1} X'Y = (X'X)^{-1} X'XB = B$

So, we have solved for B:

$B = (X'X)^{-1} X'Y$

We can use matrix math to calculate the sums of squares for the ANOVA table as in the simple regression case. The F-ratio, R^2, and p will be the same as in simple regression.

Let's look at an example of multiple regression. Let's say we measured the time it took to cool $100\mu l$ of a fluid contained in a plastic pouch from 90°C to 30°C. The fluid was heated resistively and cooled by forced air. The fan speed (RPM) was varied as well as the volume of fluid (μl), ambient temperature (degrees C), altitude (m ASL), and humidity (%RH). The experiment was conducted in an environmental chamber so that altitude, temperature, and humidity could be varied.

Thermodynamics tells us that the relationship between the response and the predictors is nonlinear. We therefore expect to have some squared terms in the model. We must run an experiment that allows us to estimate linear and squared terms. A central composite design was chosen. We'll discuss how the experiment is related to the form of the equation we can build in the next chapter on designed experiments.

Analysis of the data using Minitab tells us which effects are statistically significant. These are shown in bold in the analysis table that follows. This table was obtained from Minitab's Stat > DOE > Response Surface > Analyze Response Surface Design.

Response Surface Regression: Cool Time (sec) versus Speed, Vol, . . .

The analysis was done using coded units.
Estimated Regression Coefficients for Cool Tim

Term	Coef	SE Coef	T	P
Constant	**5.0513**	**0.12991**	**38.883**	**0.000**
Speed	**-0.8413**	**0.06648**	**-12.654**	**0.000**
Vol	**0.7034**	**0.06648**	**10.580**	**0.000**
Temp	**0.4479**	**0.06648**	**6.737**	**0.000**
Alt	-0.0795	0.06648	-1.195	0.257
RH	0.0223	0.06648	0.335	0.744
Speed*Speed	**-0.6478**	**0.06014**	**-10.771**	**0.000**
Vol*Vol	0.0398	0.06014	0.661	0.522
Temp*Temp	0.0365	0.06014	0.607	0.556
Alt*Alt	-0.0296	0.06014	-0.492	0.632
RH*RH	-0.0899	0.06014	-1.495	0.163
Speed*Vol	**0.5853**	**0.08143**	**7.188**	**0.000**
Speed*Temp	0.0755	0.08143	0.927	0.374
Speed*Alt	-0.0100	0.08143	-0.122	0.905
Speed*RH	-0.0213	0.08143	-0.262	0.798
Vol*Temp	0.0534	0.08143	0.656	0.526
Vol*Alt	-0.1024	0.08143	-1.258	0.234
Vol*RH	-0.0410	0.08143	-0.503	0.625
Temp*Alt	0.0285	0.08143	0.350	0.733
Temp*RH	0.0840	0.08143	1.032	0.324
Alt*RH	0.0882	0.08143	1.083	0.302

S = 0.3257   R-Sq = 97.8%   R-Sq(adj) = 93.9%

The table tells us that our equation should contain the following terms: fan speed, volume, temperature, fan speed^2, and an interaction between speed $\times$ volume. We can now use Minitab's regression utility to "build" the equation.

Regression Analysis: Cool Time (sec) versus Speed, Vol, . . .

The regression equation is
Cool Time (sec) = 5.02 - 0.841 Speed + 0.703 Vol + 0.448 Temp
                  - 0.645 Speed2 + 0.585 SpeedXvol

Predictor	Coef	SE Coef	T	P
Constant	5.01675	0.06646	75.48	0.000
Speed	-0.84127	0.06067	-13.87	0.000
Vol	0.70342	0.06067	11.59	0.000
Temp	0.44793	0.06067	7.38	0.000
Speed2	-0.64488	0.05427	-11.88	0.000
SpeedXvo	0.58530	0.07431	7.88	0.000

$S = 0.2972$ R-Sq = 95.7%   R-Sq(adj) = 94.9%

The R^2 value is large. Approximately 96 percent of the variation in the data is explained by the regression equation. The experimental error is approximately 0.3 sec. We check the residuals vs. fits plot to further evaluate the regression equation. The plot is shown in Figure 26–5. It does not show any obvious patterns. Our regression equation is acceptable.

Let's talk about the interaction between speed and volume that appears in the equation. This is shown graphically in Figure 26–6. The diagram indicates that the change in the response as volume varies from low to high is smaller when speed is high than when speed is low. There is a change in the response that depends on speed. The slopes of the two lines are not parallel; this is how you recognize an interaction graphically. Now, the lines can be nonparallel, within the range studied, in two ways. They can cross or not cross. Figure 26–6 shows a synergistic or beneficial interaction since the lines do not cross. Figure 26–7 shows a generalized antisynergistic interaction.

Let's see how a synergistic interaction can be beneficial. Fluid volume is difficult to control in this example. Fluid volume is a noise factor. Fan speed, however, can be controlled and is a control factor. The design engineer has the ability to specify and control the speed of the fan tightly. Robustness is a measure of how much the mean and variance of the response changes in the presence of noise. Clearly, cool time changes less when the fan speed is high compared to

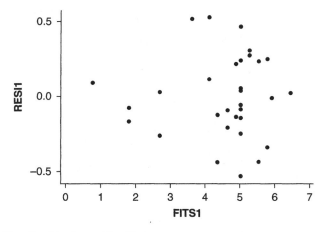

Figure 26–5 Plotting Residuals vs. Fits Plot

when it is low, as volume changes. The system is more robust and less sensitive to variation in fluid volume if we operate the system at a high fan speed. The synergistic interaction between a noise factor and a control factor enables robustness. It is precisely interactions of this form that we make use of in robust design.

Now we have an equation that models the behavior of our system, but what do we do with it? The transfer function or empirical model is extremely valuable. It tells us quantitatively how the critical inputs affect the critical output. If an input has a large coefficient, it strongly affects the output. Small variation in this input causes large variation in the output. This is a design component we must control. We will need to make our design less sensitive to it through the parameter design step of robust design. We may also need to tightly tolerance this critical input. An input with a small coefficient means that the response is not as sensitive to this design component. We

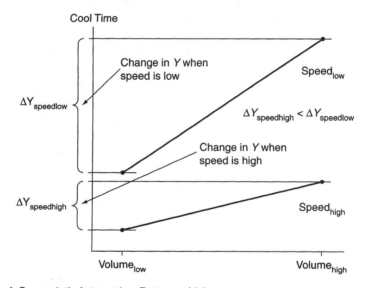

Figure 26–6   A Synergistic Interaction Between Volume

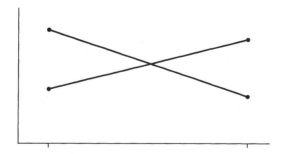

Figure 26–7   A General Antisynergistic Interaction

may be able to widen its tolerance, purchase it at a lower price, and use the money to help pay for components that require tight tolerances.

Another important use of the transfer function is Monte Carlo simulation. Monte Carlo simulation allows us to determine the statistics of the output from knowledge of the statistics of the inputs. We can use the mean, variance, and distribution of the inputs to estimate the mean, variance, and distribution of the output.

Here's how a Monte Carlo simulation works. Assume we have the following transfer function: $\hat{y} = b_0 + b_1x_1 + b_3x_2 + b_2x_3$. Further assume that each x has a known or estimated mean, variance, and distribution. We use a computer to "randomly sample" from each of the known distributions for each of the xs. This provides a value for each x. We substitute these values into the equation and calculate a value of y. Once again we randomly sample from each of the distributions for each of the xs and get new values. We substitute these values into the equation and calculate a new value of y. We do this many thousands of times. The number of simulations is only limited by how long each simulation takes and the computer's memory. We can then determine the mean, variance, and distribution of the outputs from the many simulated "observations" of y.

We can now predict the performance of a population of products in terms of mean and variance. We can also calculate the capability of our critical response if we know the specification limits. We can evaluate the effects of changes in the statistics of the design components as a result of changing suppliers or when changes in a supplier's raw materials or processes change a characteristic of a component supplied to us.

We have reviewed the basics of simple and multiple linear regression. We have built the regression equation from observed data. But we have not talked about how the data are produced—that is, the method by which we vary the xs to cause the changes in y. The way in which the xs are varied is tightly coupled to the analysis of the data and the form of the equations we can build. It is much more efficient to obtain and analyze data from a designed experiment than to just pick the levels of our predictors haphazardly. The creation of designed experiments is yet another crucial skill that the DFSS practitioner must master. This is the subject of Chapter 27.

Regression Checklist and Scorecard

Checklist for Regression

Actions:	Was this Action Deployed? YES or NO	Excellence Rating: 1 = Poor 10 = Excellent
Select the environment in which data will be collected.	_____	_____
Select the inputs and outputs to be measured in a passive or active experiment.	_____	_____
Select and qualify the measurement systems used to acquire the data.	_____	_____
Run the system in the prescribed environment and acquire the data as the inputs vary.	_____	_____

(continued)

Inspect the data for outliers and remove them if root cause
  justifies their removal.   _____   _____

Postulate and build a functional relationship between the inputs
  and output.   _____   _____

Test the statistical adequacy of the functional relationship.   _____   _____

Test the predictive ability of the functional relationship
  on the physical system.   _____   _____

Scorecard for Regression

The definition of a good scorecard is that it should illustrate in simple but quantitative terms what gains were attained from the work. This suggested format for a scorecard can be used by a product planning team to help document the completeness of its work.

Transfer Function Critical Response:	Transfer Function Critical Parameters:	R^2:	VOC Need That this Transfer Function Addresses:	Response Unit of Measure:
1.	_____	_____	_____	_____
2.	_____	_____	_____	_____
3.	_____	_____	_____	_____

We repeat the final deliverables:

- An estimation of the relative strength of the effect of each factor on the response
- An equation that analytically relates the critical parameters to the critical responses
- An estimate of how much of the total variation seen in the data is explained by the equation

References

Draper, N. R., & Smith, H. (1998). *Applied regression analysis.* New York: Wiley-Interscience.

Neter, J., Kutner, M. H., Nachtsheim, C. J., & Wasserman, W. (1996). *Applied linear statistical models.* Boston: McGraw-Hill.

Design of Experiments

Where Am I in the Process? Where Do I Apply DOE Within the Roadmaps?

Linkage to the I^2DOV Technology Development Roadmap:

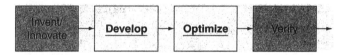

Design of experiments is primarily used during the Develop and Optimize phases of I^2DOV. It is used much less during the Invent/Innovate and Verify phases.

Linkage to the CDOV Product Development Roadmap:

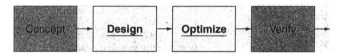

Design of experiments is primarily used during the Design and Optimize phases of CDOV. It is used much less during the Concept and Verify phases.

What Am I Doing in the Process? What Does DOE Do at this Point in the Roadmaps?

The Purpose of DOE in the I^2DOV Technology Development and the CDOV Product Development Roadmaps:

The purpose of a DOE in I^2DOV or CDOV is to obtain data from which to quantitatively estimate the effects of the factors on the system's responses without having to measure the system's response at every point in its operating space.

What Output Do I Get at the Conclusion of this Phase of the Process? What Are Key Deliverables from DOE at this Point in the Roadmaps?

- A quantitative assessment of the criticality of the parameters studied. The assessment is in terms of sums of squares and epsilon.
- The data from which a transfer function can be created
- An estimate of experimental error

DOE Process Flow Diagram

This visual reference of the detailed steps of DOE illustrates the flow of tasks in this process.

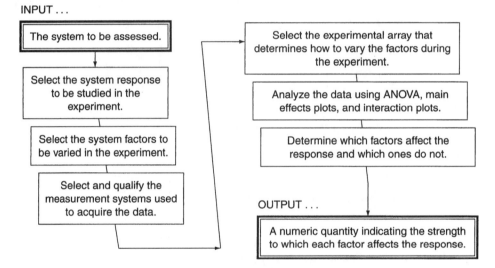

Verbal Descriptions for the Application of Each Block Diagram

This section explains the specific details of the DOE process:

- *Key detailed input(s)* from preceding tools and best practices
- *Specific steps and deliverables* from each step to the next within the process flow
- *Final detailed outputs* that flow into follow-on tools and best practices

Key Detailed Inputs

> **The system to be assessed.**

A DOE allows us to estimate the performance of a system within its operating space without having to measure the system's response at every point in that space. Imagine a system that is driven by three factors, each of which vary between some limits. The operating space is a three-dimensional volume. If the limits all have the same range, the volume is a cube, as shown in Figure 27–1.

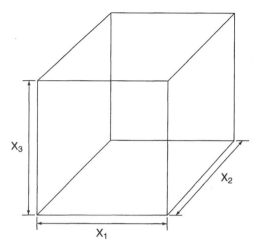

Figure 27–1 Experimental Space for Three Factors

This cube defines the space over which we will experiment. We could try to measure the system's performance at every point in the cube. This would take forever since there are an infinite number of operating points. Since we can only measure some of the points, the question becomes, which points and how many? We could test at, say, 10,000 randomly chosen points. This Monte Carlo experiment has the drawback of requiring many tests, and it may not cover the space evenly, since it is traversed randomly.

Another approach is to divide the range of each factor into some number of increments, say 10, and then test at each combination of the increments. Three factors, each at 10 levels, requires 10*10*10 = 1,000 tests. This is better than the Monte Carlo experiment since it may have less tests and covers the space uniformly. But this is still a large number of tests. We want to find the minimum number of tests to run and still adequately estimate the system's behavior. How do we determine this?

As noted in the previous chapter, the form of the equation we can build is tightly coupled to the way in which the factors are varied in the experiment. What does this mean? Let's stay with our example of a system that is affected by three factors, x_1, x_2, and x_3, and for now is assumed to be linear. This equation describes all the effects of all the factors on the response:

$$\hat{y} = b_0 + b_1x_1 + b_2x_2 + b_2x_3 + b_{12}x_1x_2 + b_{13}x_1x_3 + b_{23}x_2x_3 + b_{123}x_1x_2x_3.$$

This is a big equation. It contains the effects of the factors by themselves, x_1, x_2, and x_3, called *main effects;* the 3 two-way interactions, x_1x_2, x_1x_3, and x_2x_3; and the 1 three-way interaction, $x_1x_2x_3$.

Let's do an information analysis on this equation. Assuming two levels to each factor, each main effect accounts for one dof. Each two-way interaction accounts for 1*1 = 1 dof. The three-way accounts for 1*1*1 = 1 dof. We can keep track of this in Table 27–1.

Table 27-1   Information Analysis of Equation

Effect	DOF
3 main effects at 1 dof each	3
3 two-way interactions at 1 dof each	3
1 three-way interactions at 1 dof	1
Total	7

There are seven dof in this equation. If we include the intercept b_0 as an additional piece of information, we have a total of eight pieces of information. So, we must estimate eight coefficients, b_0 through b_7, to completely fill in or parameterize the equation. Where do the estimates come from? They come from the data generated by an experiment. We get a piece of information each time we experiment on a unique combination of the factors. To estimate the eight coefficients, we must test eight unique treatment combinations, no less. We should, of course, take replicate measurements to estimate error.

What are the eight unique combinations? We can list them in standard or Yates order if we assign a -1 to the factor at its low level and a $+1$ to the factor when at its high level. Only two levels are needed for our factors because we are assuming, for now, a linear system (only two points are needed to draw a straight line).

We have replaced the physical values of our factors by their experimental units. For example, if we investigate temperature from 100°C to 200°C, then -1 corresponds to 100°C and $+1$ to 200°C. The experimental design is shown in Table 27-2 where A, B, and C have been substituted for x_1, x_2, and x_3, respectively. The Yates naming scheme is also shown. When a factor appears at its high level, its letter is placed in the Yates list. A 1 is assigned when all the factors are at their low levels.

This experimental design is called a *full factorial* because it contains all the possible combinations of the three factors. There are $2*2*2 = 2^3 = 8$ unique treatment combinations. We only

Table 27-2   A Three-Factor, Two-Level Experiment in Yates Order

tc	Yates	A	B	C
1	(1)	-1	-1	-1
2	a	$+1$	-1	-1
3	b	-1	$+1$	-1
4	ab	$+1$	$+1$	-1
5	c	-1	-1	$+1$
6	ac	$+1$	-1	$+1$
7	bc	-1	$+1$	$+1$
8	abc	$+1$	$+1$	$+1$

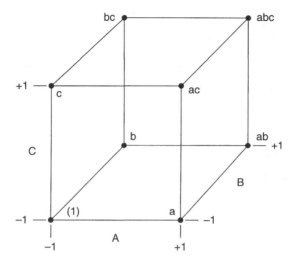

Figure 27–2   Identifying the Eight Observation Points

need eight observations in the experiment space to adequately estimate the system's behavior, if it is linear. These eight points lie at the eight corners of the cube in Figure 27–2.

We can do the same information analysis for a nonlinear system with three factors. A quadratic curve is a simple type of nonlinearity. We need a minimum of three points to estimate this type of curve. Each factor must now be varied over three levels. The equation now contains 27 terms, $3*3*3 = 3^3 = 27$. The equation is very large so we'll only look at the dof table (Table 27–3).

There are 26 effects, plus the b_0 intercept, for a total of 27 pieces of information. We need a 27-run, full factorial experiment to estimate all 27 coefficients. This is getting big. Imagine four factors at three levels. That's $3^4 = 81$ unique combinations, requiring at least 81 runs in the experiment—more, if we want replicates to estimate error.

We can now see how the structure of the transfer function determines the size of the experiment we must run. Every treatment combination provides the ability to estimate a coefficient. We must be careful about how many effects we want to estimate. The cost of estimating effects

Table 27–3   DOFs for Nonlinear System with Three Factors

Effects	DOF
3 linear main effects x_1, x_2, x_3	3
3 quadratic main effects x_1^2, x_2^2, x_3^2	3
3 linear two-way interactions	3
9 quadratic two-way interactions	9
8 quadratic three-way interactions	8
Total	26

is experimental runs. There's another property of factorial designs we should mention. They are orthogonal. That is to say, the correlation between any two columns in the matrix is zero.

Another way to say this is that there is no multi-colinearity among the factors. This property allows us to estimate the effects of each factor independently from any other effect. There's another benefit to an orthogonal design. If you recall from our discussion of regression (Chapter 26), the solution of the coefficients was given by $B = (x'x)^{-1} x'y$.

In order to solve this, we must be able to invert $x'x$. This matrix can only be inverted if its determinant is nonzero. If the experiment is orthogonal, and of sufficient rank, then $x'x$ is guaranteed to be invertible. Let's try an example in which we know what the outcome should be.

> **The system to be assessed.**

Say we know that our system is driven by three factors, *A, B,* and *C,* and that it is adequately described by the following model: $\hat{y} = 20 + 2A + 5B + 10C + 1.5AB + 2.5AC$.

Specific Steps and Deliverables

> **Select the system response to be studied in the experiment.**

In this example $\hat{y}$ is the response. This is a simulation example so the physical meaning of the response is arbitrary.

> **Select the system factors to be varied in the experiment.**

In this example *A, B,* and *C* are the factors. This is a simulation example so the physical meaning of these factors is arbitrary.

> **Select and qualify the measurement systems used to acquire the data.**

A DOE, as in any situation where data will be used to make decisions, requires a capable measurement system. The techniques covered in Chapter 23 on MSA should have been applied to the measurement system and its Gage R&R, P/T ratio, and the number of distinct categories determined. This example is a simulation so there is no error in the measurement system. In fact, noise must be added to avoid getting identical results each time the simulation is repeated.

> **Select the experimental array that determines how to vary the factors during the experiment.**

Let's use the equation to determine what type of experiment we should run. The model requires the estimation of three main effects and 2 two-way interactions. It is assumed that a linear model will adequately describe the system's behavior. So we can run a two-level experiment. The desired effects account for six degrees of freedom. A three-factor full factorial experiment requires eight treatment combinations. We choose to run the full factorial and "waste" two dofs.

We will vary each of the four factors from -1 to $+1$. The experimental design with the data follows. Each value of Y is found by substituting the appropriate values of A, B, and C into the transfer function identified previously.

A	B	C	Y
-1	-1	-1	7
1	-1	-1	3
-1	1	-1	14
1	1	-1	16
-1	-1	1	22
1	-1	1	28
-1	1	1	29
1	1	1	41

> **Analyze the data using ANOVA, main effects plots, and interaction plots.**

We can use Minitab to analyze the results of this simulated experiment. The session window output follows.

Fractional Factorial Fit: Y versus A, B, C

```
Estimated Effects and Coefficients for Y (coded units)

Term              Effect           Coef
Constant                        20.0000
A                 4.0000          2.0000
B                10.0000          5.0000
C                20.0000         10.0000
A*B               3.0000          1.5000
A*C               5.0000          2.5000
B*C              -0.0000         -0.0000
A*B*C             0.0000          0.0000
```

As expected, Minitab has no trouble estimating the coefficients of the equation. They are shown under the "Coef" column. There is no noise in the data, so the coefficients are estimated exactly.

The transfer function given by Minitab is $\hat{y} = 20 + 2A + 5B + 10C + 1.5AB + 2.5AC$. The BC and ABC effects are 0. Minitab cannot report any statistics because we do not have repeated or replicated data. We cannot simply rerun the simulation because we'll get exactly the same data. We must artificially add noise to the calculated value of Y. The added noise can be used to simulate measurement system noise.

So let's add noise to each value of Y. The noise we'll add is normally distributed with a mean of 0 and a variance of 0.5, written as noise $\sim N(0, 0.5)$. Now we can have replicate values of Y. The new experiment, repeated twice, is as follows:

A	B	C	y	Noise N(0,0.5)	Y plus Noise
−1	−1	−1	7	0.30138	7.3014
1	−1	−1	3	0.16978	3.1698
−1	1	−1	14	−0.11259	13.8874
1	1	−1	16	−0.08244	15.9176
−1	−1	1	22	−0.34998	21.6500
1	−1	1	28	1.07401	29.0740
−1	1	1	29	0.16081	29.1608
1	1	1	41	0.88053	41.8805
−1	−1	−1	7	−0.74774	6.2523
1	−1	−1	3	−0.03694	2.9631
−1	1	−1	14	0.43611	14.4361
1	1	−1	16	−0.51814	15.4819
−1	−1	1	22	−0.36958	21.6304
1	−1	1	28	−0.20907	27.7909
−1	1	1	29	−0.11390	28.8861
1	1	1	41	−0.64902	40.3510

We can use Minitab once again to analyze the results of this simulated experiment. The session window output is:

Fractional Factorial Fit: Y plus Noise versus A, B, C

```
Estimated Effects and Coefficients for Y (coded units)

Term          Effect        Coef      SE Coef          T          P
Constant                  19.9896      0.1492     134.01      0.000
A             4.1780       2.0890      0.1492      14.00      0.000
B            10.0212       5.0106      0.1492      33.59      0.000
```

C	20.1268	10.0634	0.1492	67.47	0.000
A*B	2.6371	1.3186	0.1492	8.84	0.000
A*C	5.2643	2.6321	0.1492	17.65	0.000
B*C	0.0121	0.0060	0.1492	0.04	0.969
A*B*C	0.0129	0.0065	0.1492	0.04	0.967

The coefficients differ from the values of the known transfer function. The difference is due to the noise we added. We also have p-values since we have replicates.

> **Determine which factors affect the response and which ones do not.**

Here, the null hypothesis is that a coefficient is zero. We reject this hypothesis for all but the BC and ABC effects. We can also use Minitab's balanced ANOVA utility to determine if an effect is statistically significant:

ANOVA: Y plus Noise versus A, B, C

Factor	Type	Levels	Values	
A	fixed	2	−1	1
B	fixed	2	−1	1
C	fixed	2	−1	1

Analysis of Variance for Y plus N

Source	DF	SS	MS	F	P
A	1	69.82	69.82	196.14	0.000
B	1	401.70	401.70	1128.38	0.000
C	1	1620.35	1620.35	4551.60	0.000
A*B	1	27.82	27.82	78.14	0.000
A*C	1	110.85	110.85	311.38	0.000
B*C	1	0.00	0.00	0.00	0.969
A*B*C	1	0.00	0.00	0.00	0.967
Error	8	2.85	0.36		
Total	15	2233.39			

Here the null hypothesis is that a term in the model has no effect on the mean of the response. We reject this hypothesis for all but the BC and ABC terms since their p-values are greater than 0.05. The transfer function reported by Minitab is

$$\hat{y} = 19.99 + 2.09A + 5.01B + 10.06C + 1.32AB + 2.63AC$$

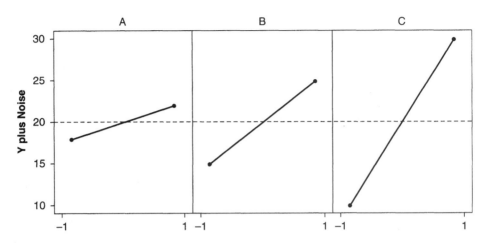

Figure 27-3 Main Effects Plot—Data Means for Y plus Noise

We can use Minitab to look at the factor effects graphically in the main effects plot in Figure 27–3.

We can see from this plot that *C* has the greatest effect on the response due to its relatively steep slope. *B* is next, then *A*. Keep in mind that we cannot determine statistical significance from this plot, only relative strengths. We must refer to the ANOVA table to determine statistical significance. Also keep in mind that *A*, *B*, and *C* are involved in statistically significant two-way interactions. We must therefore inspect the interactions plot to see how the *AB* and *AC* interactions affect the response before we consider the main effects.

The interaction plot is shown in Figure 27–4.

We see that the lines in the interaction plots for *AB* and *AC* are not parallel. The *BC* interaction lines are parallel. This indicates that *AB* and *AC* interact and that *BC* do not interact. We can use these plots to select the levels of the significant factors to maximize or minimize the response. Like the main effects plot, we cannot determine statistical significance of these effects from the plot, only relative strengths. We must refer to the ANOVA table to determine statistical significance.

Final Detailed Outputs

> **A numeric quantity indicating the strength with which each factor affects the response.**

We can see that based on the analysis, *A*, *B*, *C*, *AB*, and *AC* are statistically significant effects. Their p-values are less than 0.05. This means that the variation caused by each effect is sufficiently larger than the experimental noise to be distinguished from noise. But what of their rela-

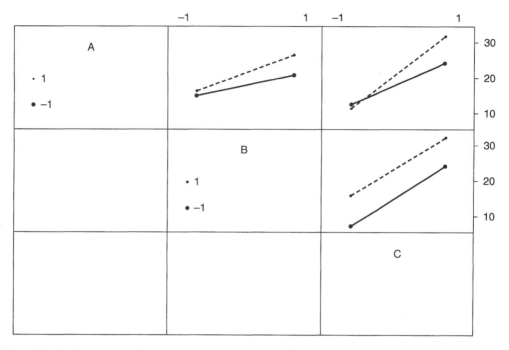

Figure 27–4 Interaction Plot (data means) for *Y* plus Noise

tive strength to cause a change in the response? We need a measure of practical significance. We can use the epsilon^2 metric to compare the relative strength of each effect on the response. We use the sums of squares from ANOVA to calculate epsilon^2, as shown in the following table.

Effect	Sum of Squares	epsilon^2	%epsilon^2
A	69.82	69.82/2,230.54=0.031	3.1%
B	401.70	401.70/2,230.54=0.18	18%
C	1620.35	1620.35/2,230.54=0.72	72%
AC	27.82	27.82/2,230.54=0.012	1.2%
AB	110.85	110.85/2,230.54=0.05	5%
Total	2,230.54		

The table indicates that while all the effects are statistically significant, 90 percent of the variation in the response is caused by *B* and *C* alone. *B* and *C*, and perhaps *A* due to its involvement in the relatively strong (5 percent) interaction, are the system's critical to function parameters. These parameters must be managed through the method of Critical Parameter Management.

Thus, Minitab performs as expected. Simulation is a helpful way of learning DOE and learning the limitations of estimating coefficient values as noise increases. The reader should practice this with other transfer functions.

Fractional Factorial Designs

As the number of effects to be estimated and their polynomial order increases (i.e., x_1, x_1^2, x_1^3, x_1x_2, $x_1^2x_1$), the number of treatment combinations in the experiment must increase. It is rare, though, that all the terms in the equation are required to adequately model the behavior of the system. We are wasting time, money, and materials on experimental runs that estimate trivial effects. If we have evidence to suggest which effects are not needed, we can save some experimentation.

Three-way interactions are not common in optomechatronic systems. Chemical, biological, and economic systems are exceptions.

If the effect of the three-way interaction is small, we can substitute a new factor for the interaction. Consider three factors at two levels each. The entire, Full-Factorial experimental array in Yates order with the identity column I is shown in Table 27–4. The patterns of -1 and $+1$ for the interactions are found by multiplying the appropriate columns. For example, the first three rows of the AB column are found as follows: $(-1)*(-1) = +1$, $(+1)*(-1) = -1$, and $(-1)*(+1) = -1$. The columns of -1 and $+1$ are called contrasts and tell us how to calculate the effects associated with the respective columns.

If the effect of the ABC interaction is small, which three-way interactions typically are, then we can "use" the ABC column to estimate some other effect. One use of this column is for blocking. **Blocking** is a technique used to determine if a factor that is not really part of the experiment affects the response. For example, if an experiment requires two batches of a material to complete and we suspect that the batch will have an effect on the response, we must keep track of which runs were done using which batch.

Treatment combinations 1, 4, 6, and 7 may pertain to batch one and treatment combinations 2, 3, 5, and 8 to batch two.

We block on "batch" and use the ABC column for this purpose. We can then test if "batch" has an effect on the response.

We can also use the ABC column to estimate the effect of a fourth factor, D. The effect of D is now algebraically added to the effect of ABC. The effect we calculate is the sum of the two

Table 27–4   Three-Factor Experiment Showing Interaction Contrasts

tc	Yates	I	A	B	C	AB	AC	BC	ABC
1	(1)	1	-1	-1	-1	$+1$	$+1$	$+1$	-1
2	a	1	$+1$	-1	-1	-1	-1	-1	$+1$
3	b	1	-1	$+1$	-1	-1	$+1$	$+1$	$+1$
4	ab	1	$+1$	$+1$	-1	$+1$	-1	-1	-1
5	c	1	-1	-1	$+1$	$+1$	-1	-1	$+1$
6	ac	1	$+1$	-1	$+1$	-1	$+1$	$+1$	-1
7	bc	1	-1	$+1$	$+1$	-1	-1	-1	-1
8	abc	1	$+1$	$+1$	$+1$	$+1$	$+1$	$+1$	$+1$

effects. This is OK if the effect of ABC is small compared to that of D. The factor D is said to be **aliased** or **confounded** with ABC.

We write this as $D = $ ABC. ABC is called the *design generator.* We cannot tell if the effect we measure is due mostly to D or ABC. In fact, if the effects of D and ABC are equal and opposite, their sum would be zero and we would conclude that D has no effect, even though both D and ABC have profound effects on the response. So, we must be careful when using this technique.

What do we gain by confounding D with ABC? We gain the ability to estimate four factors in the same number of runs we were using to estimate three factors. The new experiment is shown in Figure 27–5.

As expected, there's no free lunch. We must give something up. Actually, we give up two things. We give up the ability to estimate the effect of ABC. This is OK if the ABC effect is small. We also give up the ability to estimate some or all of the two-way interactions. Some or all of the two-way interactions are now aliased with each other. We can determine the aliasing scheme by using the design generator ABC.

If we multiply both sides of the generator by D, we get $I = $ ABCD, which is called the defining contrast. I is the identity column. The term on the right is called the *word,* and the word is good. It tells us about the aliasing structure of our design and the resolution of our design. The resolution is the length of the word. It is four in this case, so this is a resolution IV design. The aliasing of an effect is determined by multiplying both sides of the defining contrast by the effect. For example, to determine what A is aliased with, we multiply both sides by A. The product of any factor times itself is I, that is, AA $= I$. Also the product of any factor times I is the factor, that is, AI $= A$.

So, AI $=$ AABCD $= A = $ BCD. This tells us that A is confounded with the three-way interaction BCD. In a similar manner, ABI $=$ ABABCD $=$ AB $=$ CD. This tells us that the two-way interaction AB is confounded with the two-way interaction CD. Luckily, we don't have to calculate this. Minitab reports the aliasing structure for us. Let's try it.

Choose from the menu, Stat > DOE > Factorial > Create Factorial Design. The following window will appear.

tc	A	B	C	D
1	-1	-1	-1	-1
2	$+1$	-1	-1	$+1$
3	-1	$+1$	-1	$+1$
4	$+1$	$+1$	-1	-1
5	-1	-1	$+1$	$+1$
6	$+1$	-1	$+1$	-1
7	-1	$+1$	$+1$	-1
8	$+1$	$+1$	$+1$	$+1$

Figure 27–5 A Four-Factor Fractional Factorial Experiment

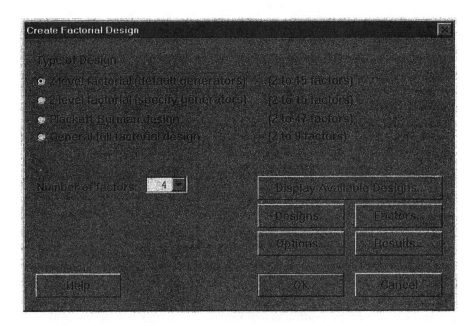

Choose four factors and then select Designs. The next window will appear.

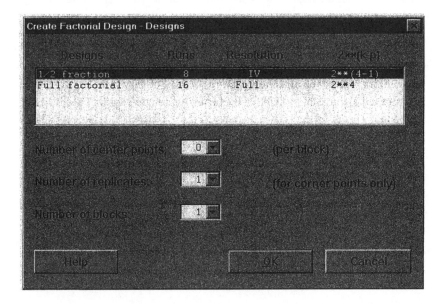

Choose the ½ fraction, which has eight runs. Click OK twice and we get the following in the session window along with a design in the worksheet.

Factorial Design

```
Fractional Factorial Design

Factors:       4      Base Design:       4, 8     Resolution:   IV
Runs:          8      Replicates:        1        Fraction:     1/2
Blocks:        none   Center pts (total): 0

Design Generators: D = ABC

Alias Structure

I + ABCD

A + BCD
B + ACD
C + ABD
D + ABC
AB + CD
AC + BD
AD + BC
```

Here we see that Minitab has created a Fractional Factorial design. There are four factors with eight runs. It is resolution IV and it is a half fraction. It is half of a Full Factorial because it has only half the number of runs required to estimate four factors in a Full Factorial.

The design generator is $D = ABC$. The aliasing structure is shown in the session window. We see that all the main effects are confounded with three-way interactions. This is good. Three-way effects are rare. We also see that some of the two-way interactions are confounded with each other. This is bad if those two-way effects are large. A resolution IV design is not a good design to run if you need to estimate effects of all the two-way interactions. In the case of four factors, we must either run the Full Factorial or fold the design (Montgomery, 1997).

It is not until we have a resolution V design that we are safe with regard to two-way interactions. Let's create a design with five factors. Choose from the menu, Stat > DOE > Factorial > Create Factorial Design. The window shown on the top of page 564 will appear.

Choose five factors and then select Display Available Designs. The next window will appear.

We can see that for five factors we can choose a resolution III with 8 runs, a resolution V with 16 runs, or a Full Factorial with $2^5 = 32$ runs.

There's no selection option in this window. Click OK and choose Designs. The following window shown on the bottom of page 564 appears.

Choose the 16-run resolution V option. Click OK twice, and the information appears in the session (see top of page 565).

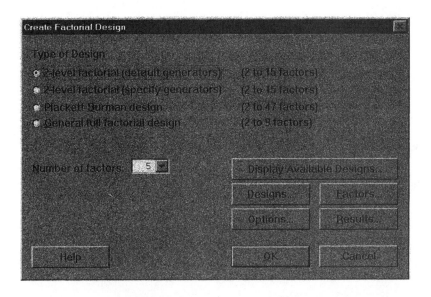

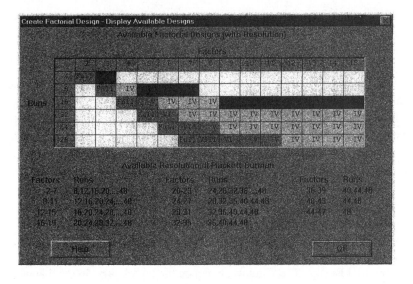

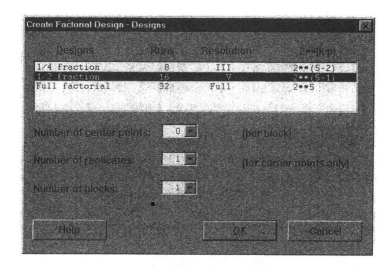

Factorial Design

Fractional Factorial Design

Factors:	5	Base Design:	5, 16	Resolution:	V
Runs:	16	Replicates:	1	Fraction:	1/2
Blocks:	none	Center pts (total):	0		

Design Generators: E = ABCD

Alias Structure

I + ABCDE

A + BCDE
B + ACDE
C + ABDE
D + ABCE
E + ABCD
AB + CDE
AC + BDE
AD + BCE
AE + BCD
BC + ADE
BD + ACE
BE + ACD

```
CD + ABE
CE + ABD
DE + ABC
```

We see that all main effects are aliased with four-way interactions, and all two-way interactions are aliased with three-ways. This is very good. Three-ways are rare, and four-ways are almost unheard-of.

We see that Fractional Factorial designs allow us to estimate the effects of more factors in fewer runs. A recommended approach to experimentation is to begin with a two-level Fractional Factorial. Don't expend all your time, money, and material on one big experiment. An experiment answers many questions; that is why we run it. However, an experiment also uncovers many new questions. You'll probably need to run subsequent experiments to answer the new questions. The recommendation is to begin with a two-level Fractional Factorial. But what if the system is non-linear? A two-level design can only estimate linear effects. We can, however, run a two-level design and determine if the system is linear. We do this by including a center point in the experiment. A center point is at the center of the experiment space. It is the treatment combination with all factors set at their zero level.

Let's take the example of two factors at two levels. The experiment space is the square shown in Figure 27–6.

In this case, we have two factors at two levels requiring four runs. These are the black dots in Figure 27–6. We also measure the response at the center point.

The center point is the point in the experiment space where the factors are set to zero. We run the experiment and analyze the results. We can create the following linear model from the data: $\hat{y} = b_0 + b_1A + b_2B$. This is the dotted line shown in Figure 27–6.

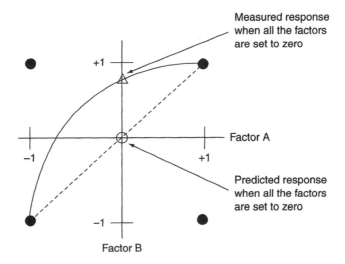

Figure 27–6 Center Point of Experiment Space

However, we have observations actually measured at the center point. This is shown as the open triangle. We can use the linear equation to calculate what the response would be at the center point if the system were linear. Obviously, the calculated value is b_0 since A and B are set to 0 in the equation.

If the difference between the measured and calculated center point values are statistically equal, we say that the system does not have curvature. It is linear and a linear equation is adequate. This is the dotted line shown in Figure 27–6. If they are statistically different, we say the system has curvature. It is nonlinear. A linear equation is not adequate. We don't know the system's trajectory. We don't know the form of the nonlinearity. It may be the solid, curved line shown in the figure or some other curve. The simplest nonlinearity is a squared or quadratic relationship. We need to run at least a three-level experiment to estimate a quadratic equation.

There's another benefit to running center points. If we take replicate measurements at the center point, we can get an estimate of pure error. This estimate allows us to obtain p-values in the ANOVA table from which we can determine the statistical significance of each effect.

If the system is determined to be nonlinear and greater detail is needed about the functional relationship of the factors to the response, we move on to the response surface methods of experimentation. This is the subject of the next section.

DOE Checklist and Scorecard

Checklist for DOE

Actions:	Was this Action Deployed? YES or NO	Excellence Rating: 1 = Poor 10 = Excellent
Select the system response to be studied in the experiment.	_____	_____
Select the system factors to be varied in the experiment.	_____	_____
Select and qualify the measurement systems used to acquire the data.	_____	_____
Select the experimental array that determines how to vary the factors during the experiment.	_____	_____
Analyze the data using ANOVA, main effects plots, and interaction plots.	_____	_____
Determine which factors affect the response and which ones do not.	_____	_____
A numeric quantity indicating the strength to which each factor affects the response.	_____	_____

Scorecard for DOE

The definition of a good scorecard is that it should illustrate in simple but quantitative terms what gains were attained from the work. This suggested format for a scorecard can be used by a product planning team to help document the completeness of its work.

Factor Name:	Statistical Significance, p-value:	Practical Significance, ϵ^2:	Critical Parameter:	Unit of Measure:
1.	_____	_____	_____	_____
2.	_____	_____	_____	_____
3.	_____	_____	_____	_____

We repeat the final deliverables:

- A quantitative assessment of the criticality of the parameters studied. The assessment is in terms of sums of squares and epsilon.
- The data from which a transfer function can be created.
- An estimate of experimental error.

Reference

Montgomery, D. C. (1997). *Design and analysis of experiments*. New York: Wiley & Sons.

PART VI

Tools and Best Practices for Optimization

The tools and best practices discussed in this section focus on deliverables required at gate reviews for the third phases in the I^2DOV and CDOV processes.

The tools and best practices chapters are structured so that the reader can get a strong sense of the process steps for applying the tools and best practices. The structure also facilitates the creation of PERT charts for project management and critical path management purposes. A checklist, scorecard, and list of management questions concludes each chapter.

Taguchi Methods
for Robust Design

Where Am I in the Process? Where Do I Apply Robust Design Within the Roadmaps?

Linkage to the I²DOV Technology Development Roadmap:

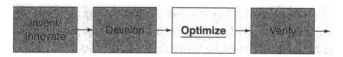

Robust design is conducted within the Optimize phase during technology development. It is applied to new technology subsystems that are being developed for integration into a new platform or for integration into existing system designs.

Linkage to the CDOV Product Development Roadmap:

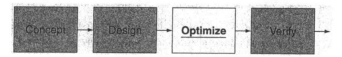

Robust design is also conducted within the Optimize phase during product development. Here we apply robust design to a specific product with specific noises. This focused application of robust design assures that the subsystems are insensitive to the specific product delivery and use noises and specific system integration noise factors that were not clear at technology optimization and verification during the I²DOV process.

What Am I Doing in the Process? What Does Robust Design Do at this Point in the Roadmaps?

The Purpose of Robust Design in the I²DOV Technology Development Roadmap:

In I²DOV we use the ideal function from the Develop phase for a subsystem technology to define a generic set of product line noise factors, signal factors (critical adjustment parameters), control factors (critical functional parameters), and critical functional responses. We conduct analytical (Monte Carlo simulation) and experimental procedures (designed experiments) to develop critical functional parameter set points that provide insensitivity to the generic set of product line noise factors. We then use a small number of signal factors (CAPs) and one or more other specialized control factors that are capable of changing the slope of the ideal function (beta shifting) to adjust the mean response on to the desired target, known as the *Dynamic Taguchi Method.*

The Purpose of Robust Design in the CDOV Product Development Roadmap:

In CDOV we once again use the ideal function for a subsystem to define a specific product focused set of noise factors, signal factors (CAPs), control factors (critical functional parameters), and critical functional responses. We conduct Monte Carlo simulation and experimental procedures (designed experiments) to leave the subsystems in a documented state of insensitivity to the set of specific product noise factors. We use a small number of signal factors (CAPs) and one or more other specialized control factors that are capable of changing the slope of the ideal function (beta shifting) to adjust the mean response on to the desired target.

What Output Do I Get at the Conclusion of Robust Design? What Are Key Deliverables from Robust Design at this Point in the Roadmaps?

- *Optimized signal-to-noise metrics* that characterize the useful interactions between control factor and noise factor interactions for the subsystem critical functional response(s)
- *Optimized critical functional parameter set points* that control the subsystem's robustness
- *Optimized critical adjustment parameters* that tune the mean value of the subsystem's critical functional response(s) on to its desired targets
- *Subsystem Critical Parameter Management data* that is documented in the CPM relational database:
  CFR mean
  CFR standard deviation
  CFR coefficient of variation [(s/mean) $\times$ 100]
  CFR S/N ratio
  CFR Cp index (early quantification of the design's ability to meet specifications)

The Robust Design Process Flow Diagram (found on page 573)

A visual reference of the detailed steps of robust design illustrating the flow of the process follows.

INPUT . . .

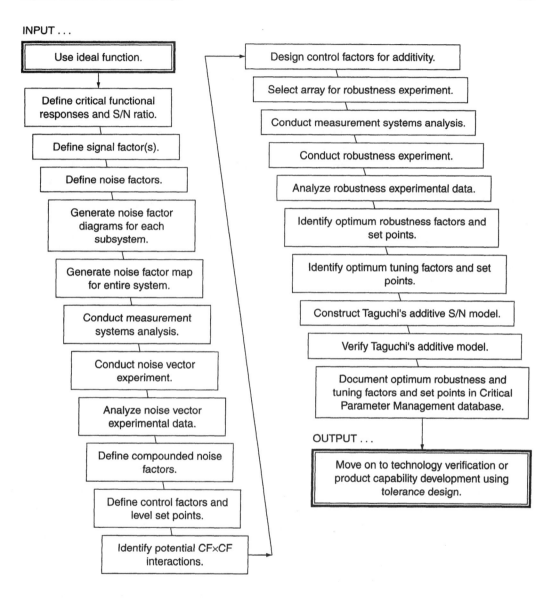

Verbal Descriptions for the Application of Each Block Diagram

This section explains the specific details of what to do in the robustness optimization process:

- *Key detailed input(s)* from preceding tools and best practices
- *Specific steps and deliverables* from each step to the next within the process
- *Final detailed outputs* that flow into follow-on tools and best practices

Key Detailed Inputs

> **Use ideal function.**

The primary input into the robust design process from the preceding phase is the set of mathematical models that describe the ideal/transfer functions of the subsystems and subassemblies. An ideal/transfer function expresses physical laws and engineering design principles in the form of a mathematical equation that defines an idealized functional relationship between a dependent variable (Y) and independent design variables (xs). The ideal/transfer function is a deterministic math model that typically does not contain the effects of noise, only the linear and nonlinear main effects and interactive effects of the control factors on the critical functional response.

Ideal/transfer functions are often simplified to the point that they can be expressed as plots of a general, linearized input-output relationship, as in the following diagram.

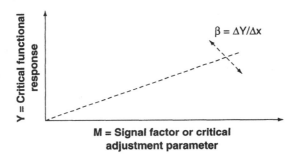

Additional input from prior tools and best practices:

- *Requirement targets and tolerances* from subsystem Houses of Quality within the QFD process (concept engineering phase)
- *Input-output constraint diagrams, functional flow diagrams, process maps, cause and effect diagrams, and FAST diagrams* from the concept generation process (concept engineering phase)
- *Failure modes and effects tables* for use in subsystem noise diagrams and system noise map from design FMEA process (design engineering phase)
- *Superior subsystem and CFR data acquisition system design concept* from Pugh concept selection process (concept engineering)
- *Main effects and interaction plots* from sequential design of experiments process during concept generation process (concept engineering phase)
- *Refined design concept* from design for manufacturing and assembly process and modular design (design engineering phase)
- *Capable data acquisition system* from measurement system analysis process (design engineering phase)

Specific Steps and Deliverables

Define critical functional responses and S/N ratio.

The ideal function relates a variable that must be defined as a measurable critical functional response to a signal factor, also called a *critical adjustment parameter.* The critical functional response must be a continuous, measurable engineering scalar or vector variable that is directly related to the fundamental functionality of the subsystem. It must be directly tied to the efficiency of the flow and transformation of mass and/or energy of the subsystem. If the mass is not flowing or changing phase, its current physical state, under the influence of some form of energy input, must be considered to be the CFR, such as deflection or expansion of a solid material. It is important that you measure functions and avoid measuring an attribute of quality such as reliability, defects, or yield (% acceptable). We avoid these "quality" metrics for three main reasons:

1. They are only indirectly linked to the physical principles that underwrite the fundamental functionality of the design and its sensitivity to realistic sources of noise.
2. They require larger samples of data to calculate and reach valid statistical conclusions about the design's performance.
3. It is hard to draw concrete engineering conclusions during design using metrics that were developed as retrospective, after-the-fact measures of a design's quality.

We must use metrics that provide immediate feedback about our progress in design performance characterization and improvement, while we intentionally stress test the design well before finalizing nominal set points.

Most cases of robust design will be evaluated using the dynamic S/N ratio. The dynamic S/N ratio measures the strength of the power of a proportional, engineered relationship between a signal factor and the CFR with respect to the power of the variation around that proportional relationship. In a word, the "signal" of the ideal function is relative to the noise (assignable cause variation) that is active during a mass/energy conversion process within a design.

Dynamic S/N = $10\log[\beta^2/\sigma^2]$

where $[\beta^2]$ is the slope of the best fit line through a set of data relating a CFR to changes of a critical adjustment parameter.

The effect of Taguchi noise factors (assignable causes of variation) on β^2 is expressed in the term σ^2. We will explore signal and noise factors presently.

Define signal factor(s).

The signal factor is an "engineered" control factor that has the capacity to adjust the mean of the critical functional response. That is why it is also referred to as a *critical adjustment parameter.* It also has a moderate-to-low level effect on the standard deviation of the critical functional response. If the critical adjustment parameter (CAP) does not have a linear relationship with the critical functional response (CFR), the CFR data can be transformed using some form of log or exponential function to "linearize" the relationship.

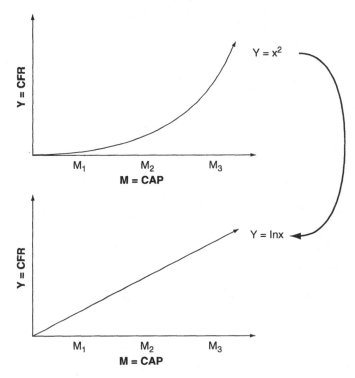

The nonlinear effect in the ideal function has been transformed into a linear effect by using a natural log transformation on the CFR data.

> **Define noise factor(s).**

The subsystem ideal function is used to define and measure inefficiencies and disruptions in the flow, transformation, and state of energy, mass, and information (logic and control signals) for each subsystem. These disruptive elements can come from three sources.

1. **External noise factors:** Factors entering the subsystem from outside sources
2. **Unit-to-unit noise factors:** Factors such as material or part-to-part variation
3. **Deterioration noise factors:** Factors causing dynamic wear or temporal degradation

These are referred to as *Taguchi noise factors* and lie at the root of the physics of variation for a subsystem. They are special or assignable cause sources of variation. They cause statistically significant variation in critical functional responses, well beyond random or common cause sources of variation.

Generate noise factor diagrams for each subsystem.

Subsystem noise diagrams are schematic representations of the three sources of variation that assail and disrupt each subsystem's critical functional response. They also contain the output noises that can assail and disrupt other subsystems or the user's environment:

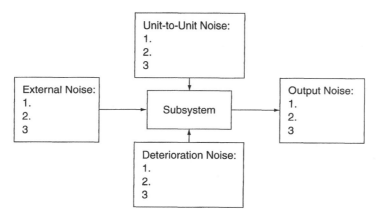

Generate noise factor map for entire system.

A system noise map is a series and/or parallel network of all the subsystem noise diagrams. These maps illustrate all the key noise flow paths from one subsystem to another throughout the integrated system. They document how variation is transmitted across the system (see figure on top of next page).

Conduct measurement systems analysis.

Although measurement systems analysis has been done previously, prior to the screening design of experiments during concept engineering, it is wise to certify the capability (P/T ratio and Gage R&R) of the data acquisition system prior to measuring the subsystem critical functional responses during the noise experiment (see Chapter 23).

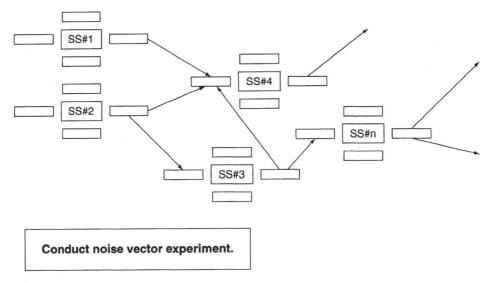

Conduct noise vector experiment.

This step requires the use of a two-level Full or Fractional Factorial orthogonal array (designed experiment) to study the main effects of the candidate noise factors. Interactions between noise factors can be evaluated if deemed necessary. The usual approach is to only study the main effects of up to 11 noise factors ($N_A - N_K$) by using the L_{12} Plackett-Burman Array (Figure 28–1). Less than 11 noise factors may be studied in this array without compromising its balance, or orthogonality. This array distributes any specific aliasing effects from any of the noise factors more or less uniformly between all columns across the array. Thus, the interactive effects of noise factors are spread out to all columns promoting the ability to see their main effects above and beyond random noise and interactivity in the data.

It is highly recommended that replicate data ($Y_1 - Y_n$) be taken during the noise experiment to help quantify the random variability in the sample data. This is called the *error variance,*

L_{12}	N_A	N_B	N_C	N_D	N_E	N_F	N_G	N_H	N_I	N_J	N_K	Y_1	Y_2	...	Y_n
1	1	1	1	1	1	1	1	1	1	1	1				
2	1	1	1	1	1	2	2	2	2	2	2				
3	1	2	2	2	2	1	1	1	2	2	2				
4	1	2	1	2	2	1	2	2	1	1	2				
5	1	1	2	1	2	2	1	2	1	2	1				
6	1	1	2	2	1	2	2	1	2	1	1				
7	2	2	2	2	1	1	2	2	1	2	1				
8	2	2	2	1	2	2	2	1	1	1	2				
9	2	1	1	2	2	2	1	2	2	1	1				
10	2	1	2	1	1	1	1	2	2	1	2				
11	2	2	1	2	1	2	1	1	1	2	2				
12	2	2	1	1	2	1	2	1	2	2	1				

Figure 28–1 The L_{12} Array

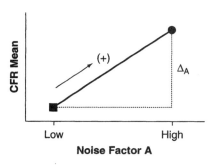 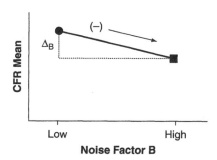

Figure 28–2 Analysis of Means Calculations for Noise Factors A and B

$(\sigma^2)_{error}$. A minimum of three to five replicates per experimental run is recommended. This is used to calculate the F-ratio (from ANOVA) for each noise factor to assess its statistical significance relative to its effect on the mean of the CFR.

> ## Analyze noise vector experimental data.

The main effects plots from a two-level noise experiment will define the magnitude and directional effect of each noise factor on the CFR. In Figure 28–2, we see the results of the analysis of means calculations for the noise factors A and B from a designed experiment.
    Note the following details:

- Noise factor A has a positive directional effect (+) on the mean of the CFR.
- Noise factor B has a negative directional effect (−) on the mean of the CFR.
- Noise factor A has a large magnitude effect (Δ_A) on the mean of the CFR.
- Noise factor B has a small magnitude effect (Δ_B) on the mean of the CFR.

Any measurable engineering response that has both magnitude and direction is defined as a *vector.* Here we see noise factors A and B have plots that define their vector effect.

> ## Define compounded noise factors.

With vectors defined for each noise factor, we are now ready to perform a grouping function that greatly improves the efficiency of how we induce Taguchi noise during a robust design experiment.

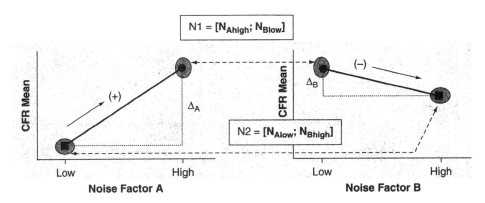

Figure 28–3   Noise Factor A and B Interactions

As we look at the main effects plots, we see that noise factor A and noise factor B have the following common traits:

- The CFR mean is driven up when noise factor A is at its high level and noise factor B is at its low level.
- We "compound" or group them together because they both have the same directional effect on the mean value of the CFR: N1 = $[N_{Ahigh}; N_{Blow}]$ drives the mean up (see Figure 28–3).
- The CFR mean is driven down when noise factor A is at its low level and noise factor B is at its high level.
- Similarly, we "compound" or group the same two noise factors together at their opposite levels because they both have the same directional effect on the mean value of the CFR: N2 = $[N_{Alow}; N_{Bhigh}]$ drives the mean down.

We can now adequately introduce stress to interact with the control factors within a designed experiment by applying the compounded noise factors to each Fractional Factorial experimental treatment combination in a "Full Factorial" manner:

		A	B	C	N1	N2	
	1.	1	1	1			
	2.	1	2	2			
	3.	2	1	2			
	4.	2	2	1			

Control Factors ⟶

Treatment Combinations

Compounded Noises ⟵

CFR Data

Taguchi recommends we attempt to expose every run in the robustness experiment to a full and consistent set of stressful, compounded noise factors. In this approach we allow the control factors to have a Full Factorial interaction with a consistent set of noise factors. Doing a partial study of control factor-to-noise factor interactivity sub-optimizes the intent of robust design experimentation. Our goal is to identify control factor-to-noise factor interactions (usually measured as changes in S/N values) that can be used to identify set points that leave the CFR minimally sensitive to noise. S/N ratios are designed to draw our attention to CF×NF interactivity. The bigger the difference between two S/N values within an experiment, the more interaction is occurring between a control factor and a compounded noise factor. This is the major measurement objective within the Taguchi method of designed experimentation for the development of a robust design. Notice that most of the "design" that goes into a Taguchi experiment is not related to the structure of a complex array but in the nature of the control factors that are loaded into a simple array and their relationship to compounded noise factors as they influence a critical functional response.

Define control factors and level set points.

Control factors are the "designed" engineering parameters that, in conjunction with the signal factor(s), control the ideal function of the design. Recall that a signal factor is a special class of control factor that has been designed to have a large effect on the mean of a CFR and a relatively small effect on the standard deviation of a CFR. These factors are developed and designed into a functioning subsystem. We want to make purposeful changes in the set points of these controllable factors to see how they interact with the Taguchi noise factors. The purpose of robust design is largely focused on provoking opportunities for control factors to interact with noise factors. When this is allowed to happen we can see the following trend in the interaction plots between CFs and NFs:

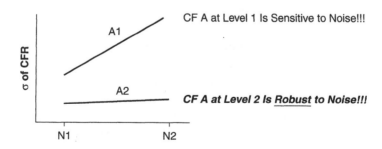

Control factors can break down into two categories.

    1. *Candidate critical functional parameters (CFP):* A critical functional parameter is a functional output from an engineered group of component specifications that are not considered to be a complete subsystem. They are measurable as scalar or vector quantities that include examples such as force, pressure, stiffness, displacement, velocity, acceleration, temperature, or current. We can have a CFP such as the stiffness of a spring.

The length of the coil, the diameter of the spring wire, and the spring material comprise some of the component specifications that can be designed to create the critical functional parameter of stiffness. We would then assess the main effect of stiffness at various levels on the robustness of a critical functional response. Sometimes we refer to CFPs as low-level CFRs. They are really CFRs relative to the component-level specifications used to determine their value, but they are CFPs when they are integrated with other CFPs and component specifications to produce a subsystem level critical functional response.

2. *Candidate critical-to-function specifications (CTF):* A critical-to-function specification is a controllable element or attribute of a component or material (solid, liquid, or gas). CTF specifications are scalar values such as a diameter, length, width, viscosity, density, area, resistance, volume, weight, and so on. A CTF specification is the most basic of control factors that might be loaded into a robust design experiment. Sometimes CTF specifications are integrated to create a critical functional parameter, as demonstrated in the previous example.

Identify potential CF×CF interactions.

Control factor-to-control factor interactions should have been documented during the Develop phase from I^2DOV or the Design phase from CDOV when the superior design concepts were modeled for basic functionality. CF×CF interaction or codependency is evaluated using an appropriately structured array from classical design of experiments during concept engineering. Once a statistically significant CF×CF interaction is defined, it can be assessed for its directionality and magnitude in rigorous mathematical terms. This modeled relationship between two codependent control factors and the CFR will form the basis for identifying interaction suppression schemes such as additivity grouping or sliding levels. We attempt to control or suppress strong, bidirectionally correlated control factor interactions during robust design experimentation.

During robust design we can identify and include CF×CF interactions as part of the parameter optimization process, but typically we just focus on the main effects of the candidate critical functional parameters or critical-to-function specifications. The real focus of robust design is on the interaction between noise factors and control factors:

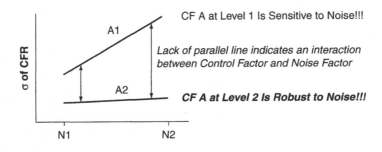

> **Design control factors for additivity.**

To suppress strong, negative interactions during robust design evaluations we must consider using design for additivity techniques to force the control factor main effects to stand out well above the effects of interactions and random noise (σ^2_{error}). In this way we can identify and sum up their additive effect on maximizing the robustness of the CFR to the Taguchi noises.

Design for additivity creates a set of control factors that have been intentionally engineered for low control factor-to-control factor interactivity. This enhances the likelihood of the successful verification of Taguchi's additive S/N model during the final step of robust design.

When we understand which control factors are strongly interactive, to the point of possessing a lack of unidirectional (monotonic) effect on the CFR (called an *antisynergistic interaction*), we can develop two strategies to suppress this unwanted effect.

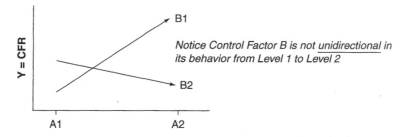

This is an example of two control factors that possess a "troublesome" antisynergistic interaction we want to suppress.

Two main methods are used to turn interactive control factors into additive control factors with respect to their unidirectional effect on the CFR: additivity grouping and sliding levels.

Additivity Grouping

Additivity grouping uses engineering knowledge of the functional contributions of two codependent control factors to logically align their level set points to derive a unidirectional response from the CFR:

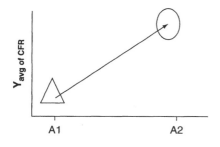
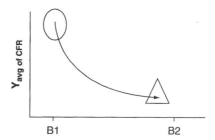

After grouping the control factors that cause the average of the CFR to go high together and the control factors that drive the CFR average low, we see the following result.

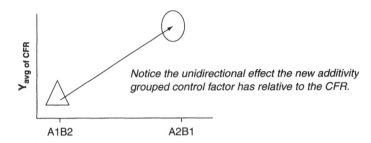

Notice the unidirectional effect the new additivity grouped control factor has relative to the CFR.

Now we can study the new control factor for its contribution to robustness (S/N gain) in the design without fear of the interaction affecting the robustness confirmation experiment.

Sliding Levels

It is possible to numerically quantify the magnitude and directionality of an interaction between two control factors. This is typically done in the Develop or Design phases. When we possess this information, we can adjust one control factor's set points by **sliding the level** of the set points in a designed experiment so that the effect of the interaction is eliminated. When this is done, the combined effect of the two previously interactive control factors is made to be unidirectional (monotonic) with respect to the CFR. The sliding level adjustment is applied to the set points of one of the interactive control factors by developing a correction factor:

- When factor A is at 10 units and factor B is at 10 units we see the CFR $= 110$.
- When factor A is at 20 units and factor B is at 10 units we see the CFR $= 210$.
- The rate of change of A (Δ_A) is $20-10 = 10$ units in the positive ($+$) direction.
- The rate of change of the CFR (Δ_{CFR}), related to Δ_A, is 100 units in the positive ($+$) direction:

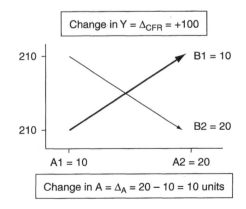

- When factor A is at 10 units and factor B is at 20 units we see the CFR = 210.
- When factor A is at 20 units and factor B is at 20 units we see the CFR = 110.
- B1 and B2 are not causing a unidirectional change in the CFR.
- The rate of change of A (Δ_A) is 10 units in the positive (+) direction.
- The rate of change of the CFR (Δ_{CFR}) is 100 units in the negative (−) direction.

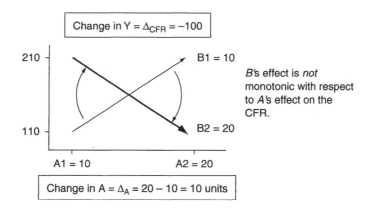

We can calculate a correction factor to be applied to develop a sliding level for factor B if we want to compensate for the negative (antisynergistic) interaction. For every positive change in the level of A, a scaled, negative change in the level of factor B is needed to eliminate the interaction (nonmonotonic plot of B with respect to A):

- Correction (sliding) factor = [B's level − 10] for every (+)100 units of the CFR you want to change in a unidirectional relationship with positive changes in A's level.

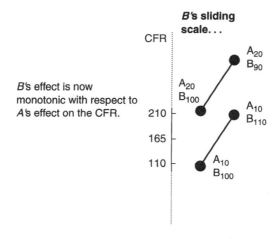

- When A=1 and B changes +10 units, the CFR increases by 100 units.

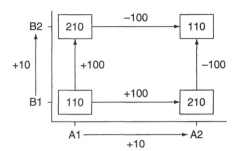

- When A=2 and B changes +10 units, the CFR decreases by 100 units; to compensate we can change B's level to smaller values to cause the CFR to consistently increase.

Monotonicity is attained when we make changes in both A and B levels and the CFR always changes in the same unidirectional way. Sliding scales are often required when we have time-energy input relationships. Baking is such an example. If we lengthen baking time, we may have to lower the energy input or vice versa in some scaled relationship or we may burn up the materials inadvertently within a designed experiment.

It is important to understand which control factors are strongly antisynergistic as we have seen for A and B in this example. If we use a sliding scale relationship to assign levels for interacting CFs in a robustness experiment, the likelihood of accidentally aliasing or confounding of the resulting main effect data will be low. This will help assure that we choose the right control factors and level set points for inclusion into Taguchi's additive model. This process will aid in the confirmation of our predicted S/N improvements from the individual control factors we have chosen to study.

> ### Select array for robustness experiment.

Once the control factors have been selected and appropriate level set points are assigned, a designed experiment can be selected. The **orthogonal array** (Figure 28–4) is essentially a matrix or table of columns that hold control factors and rows that contain experimental runs of control factor combinations to be evaluated in the presence of the compounded noise factors. The CFR is measured for each row under the N1 and N2 noise conditions for each level of the signal factor (typically three to five levels). So, it is quite typical for a robustness experimental run to just require two data points per signal factor (CAP) set point. We can take as many replicate data points for each noise value as we deem necessary to quantify experimental error in the data, typically due to meter error, inaccuracies in resetting the CF levels, or other unknown random or special cause sources of variation.

Robust design uses orthogonal or balanced arrays, like the L4 array in Figure 28–4, to provide a balanced opportunity for each control factor to exhibit its effect on the critical functional

L4 Orthogonal Array: 3 Degrees of Freedom	Control Factors			Signal Factor Level 1 to *n*	
	Control Factor A: 1 Degree of Freedom	Control Factor B: 1 Degree of Freedom	Control Factor C or A×B Interaction: 1 Degree of Freedom	Compounded Noise Factor: N1 (Take replicates as necessary)	Compounded Noise Factor: N2 (Take replicates as necessary)
Run 1:	Level 1	Level 1	Level 1	CFR Data	CFR Data
Run 2:	Level 1	Level 2	Level 2	CFR Data	CFR Data
Run 3:	Level 2	Level 1	Level 2	CFR Data	CFR Data
Run 4:	Level 2	Level 2	Level 1	CFR Data	CFR Data

Figure 28–4 L4 Orthogonal Array

response under the stressful influence of the Taguchi noise factors. Notice how each of the control factors gets to express its effect twice at level 1 and twice at level 2. This is the balancing property known as **orthogonality.** There is a simple way to calculate how large an orthogonal array is required to contain all the control factors (and CF×CF interactions, if necessary) we want to study for their effect on the CFR's robustness.

The number of control factor levels defines how many degrees of freedom (dof) the control factor is exploring in the design space. Design space is the entire range covered by all control factors and the ranges of their level set points. The number of control factor levels minus 1 (#CF Levels − 1 = dof) will equal the number of degrees of freedom required to quantify the main effect of the CF. For CFs with two levels (typical in noise experiments, where noises are the CFs) we require 1 dof and for CFs with three levels, typical in robustness experiments, we require two dof.

For an entire designed experiment consisting of an orthogonal array, the dof required are always one less than the number of runs in the array. So for our L4 example in Figure 28–4 the dof equal three (4 runs − 1). Each CF requires one dof for a total of three and that is exactly how many dof the L4 in its factorial layout has to offer.

The calculation rule for sizing an array for our selected CF at their respective levels is

$$\text{dof required} = \Sigma[(\# \text{ of CFs}) \times (\# \text{ of levels} -1)]$$

If we have some control factors at two levels and others at three levels, we must sum the individual CFs at two levels and then add their required dof to the sum of the CFs at three levels to obtain the total required dof.

Once we know the required dof, we can look up an array that possesses a number of dof equal to or greater than the required number of dof. It is not necessary to load a CF in every column in an array. It is not unusual to see designed experiments with empty columns. This does not compromise the array's balance. If we want to compromise the array's balance, we simply alter the level set points away from their prescribed values or fail to run one or more of the rows. That will sufficiently ruin the orthogonality of the experiment and the integrity of our CFR data (obviously we do not recommend this approach)!

We can choose from many orthogonal arrays within the menus of Minitab data analysis software. Figure 28–5 offers an example of the architecture of a typical dynamic experimental layout.

> **Conduct measurement systems analysis.**

Again, although measurement systems analysis has been done prior to the noise experiments, it is wise to certify the capability (P/T ratio and Gage R&R) of the data acquisition system prior to measuring the subsystem critical functional responses during the robustness experiment.

A	B	C	D	E	Signal Factor Level 1		Signal Factor Level 2		Signal Factor Level 3	
					N_1	N_2	N_1	N_2	N_1	N_2
						Outer array, which contains CFR data as gathered under "*Full Factorial*" conditions of signal and noise factors				
	Inner array, which contains control factor set points									

Figure 28–5   Architecture of Dynamic Experimental Layout

Conduct robustness experiment.

Once the experiment has been designed and the data acquisition system has been certified as capable to measure the CFRs, we are ready to physically conduct the robustness experiment.

It is important that great care and discipline be exercised as we set up each signal, control, and noise factor set point for each run within the orthogonal array. If we are sloppy in the way each set point is adjusted to its specified value from the array, variability will build up in the CFR data and inflate the error variance within the data. This experimental source of noise is to be suppressed and kept to a minimum.

One way to suppress errors in the data from systematic sources of variability outside of the purposeful changes we are inducing from the control factor levels within the orthogonal array is to randomize the order of the experimental runs. We simply write down the number of each run on a separate piece of paper, place them all in a container, mix the paper slips thoroughly, and then blindly draw them from the container. This will randomize the run order. Usually this is not necessary in a Taguchi experiment because the effects of the compounded noise factors, N1 and N2, far outweigh the much smaller effects of systematic or random sources of experimental error.

In a robustness experiment it is common to load the array and run it in an order that makes the changes in control factors easy and economical. Notice that in all the Taguchi style arrays (Latin square format as opposed to Yates order format) the first columns have orderly patterns of changes in the level set points. The deeper we go across the columns of the array, the more random and frequent the level set point changes become. We should place any difficult or expensive-to-change control factors in the first few columns in the array. This will save time and money during our robustness optimization experiments.

L$_{18}$	A	B	C	D	E	F	G	H	SF 1		SF 2		SF 3	
									N$_1$	N$_2$	N$_1$	N$_2$	N$_1$	N$_2$
1	1	1	1	1	1	1	1	1						
2	1	1												
3	1	1												
4	1	2												
5	1	2												
6	1	2												
7	1	3												
8	1	3												
9	1	3												
10	2	1												
11	2	1												
12	2	1												
13	2	2												
14	2	2												
15	2	2												
16	2	3												
17	2	3												
18	2	3												

Figure 28–6   L18 Dynamic Experiment Layout

In Figure 28–6, we see the most common of all Taguchi experimental layouts, the L18 dynamic experiment.

Note that the SF (signal factor) values are also called critical adjustment parameter set points.

> **Analyze robustness experimental data.**

Analysis of data from a robustness optimization experiment is easy to conduct. Numerous software products will perform the required calculations in a few milliseconds. This book features graphical and numerical results from Minitab, which is strong in its ability to support extensive analysis of robust design data.

The three main statistical measures of critical functional response data, as it represents sample data developed in the presence of stressful sources of Taguchi noise factors, are as follows:

1. The signal–to–noise ratio of the CFR
2. The standard deviation of the CFR
3. The slope (β) of the CFR (the family of mean values developed as we adjust the signal factor or CAP)

Competent analysis of robustness data requires the assessment of numerical values in the form of main effects plots for all three of these important metrics.

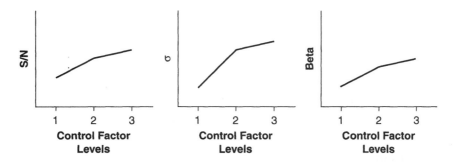

With data like this for all control factors, it is hard to miss the important features of a CF's impact on the CFR. CFs tend to either control the slope (**beta**) of the function, the standard deviation (σ), or both. Your task is to prove that you know how each CF can improve minimization of variability or adjust the slope. Adjustment of the slope means that you are changing the sensitivity of the functional relationship between the CFR and the signal factor (critical adjustment factor).

> **Identify optimum robustness factors and set points.**

Some control factors are good for making the CFR insensitive to the effects of noise. The steepness of the slope of the S/N plot in conjunction with the standard deviation plot will help you see which factors are best for making the design robust.

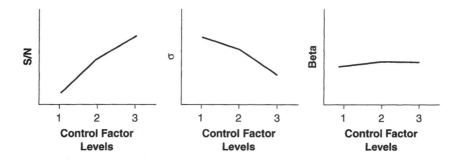

In many instances a CF will affect both the slope and standard deviation, and we must use our judgment as to how best to use a CF to optimize the design's performance. The choices are fairly clear: minimize sensitivity to noise or tune the mean by adjusting beta.

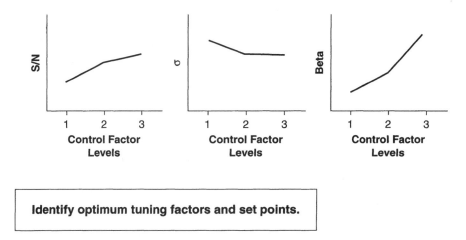

Identify optimum tuning factors and set points.

We use the signal factor to adjust the mean in relatively "gross" terms. The best we can get from this kind of adjustment is a local optimum. We use response surface methods to locate a global optimum when adjusting the mean.

We use the control factor that has the strongest linear effect on beta to fine-tune the mean value of the CFR relative to a specific target. The target may change as the product's specifications are changed during the development of the product. A design is considered *dynamic* when it has this built-in dual capability to have the CFR moved at will by slope adjustment (fine-tuning through beta sensitivity adjustment) as well as signal factor adjustment (coarse adjustment). This works well for local optimization.

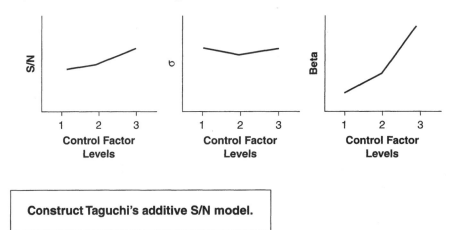

Construct Taguchi's additive S/N model.

Once we have used the main effects plots to help identify which control factors are going to have a significant effect on the optimization of robustness, we can isolate and sum their contributions to the overall CFR S/N gain. We need to use the overall mean signal-to-noise (S/N_{avg}) values calculated by averaging the S/N values from each of the rows of the robustness experiment (in this case rows 1 through 4):

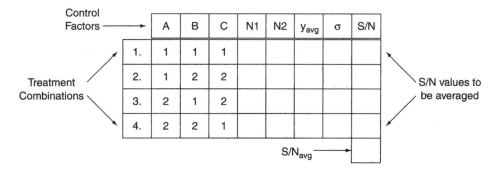

The additive model calculates a predicted optimum S/N built on the sum of all the significant control factor S/N contributions included in the prediction. CFs that have weak S/N gain in the main effects plots are usually excluded from the additive model. In the following equation, we have included all the factors in the additive model:

$$\textbf{S/N}_{\textbf{Pred opt}} = [(S/N_{A \text{ opt}}) - (S/N_{avg})] + [(S/N_{B \text{ opt}}) - (S/N_{avg})] + [(S/N_{C \text{ opt}}) - (S/N_{avg})]$$

The predicted optimum S/N value is now in need of confirmation through verification experiments at the predicted optimum set points for each control factor.

Verify Taguchi's additive model.

The predicted optimum S/N value must be verified by two actions. We begin by setting up the design at the optimum set points that have been identified as helpful in improving robustness. There will usually be one or two other control factors (beta shifters) that are found to have a significant effect on shifting beta (β), the slope of the "best fit line" that symbolizes the ideal function of the design. These fine-adjustment factors allow us to shift the slope to either a higher or lower level, thus tuning the sensitivity of the relationship between the critical adjustment parameter (CAP, or signal factor) and the critical functional response (CFR).

Once the design has been set up at its optimum levels for robustness and on-target performance, we turn our attention once again to properly stressing the design with exactly the same compounded noise factors used in the original robustness experiment. The new, robust design is now evaluated under the noise conditions over and over as many times as is deemed necessary to assure the data is truly repeatable. It is common to see the verification runs repeated a minimum of five times.

To confirm the robustness improvements are sufficiently matching the predicted S/N values, we need to calculate a range of acceptability, a tolerance beyond which we say the results are not confirmed. The verification tolerance is calculated using the following process:

1. Identify the largest and smallest S/N values from the original robust design experiment (S/N_{max} and S/N_{min}).

2. Calculate the difference between these values: $(S/N_{max} - S/N_{min}) =$ full scale range of S/N across the original experiment.

3. Divide the full scale range of S/N by either 3dB or 4dB, depending on how conservative we want the S/N tolerance limit to be. Note we use 3dB because it represents the amount of change required to cut the variance of the CFR in half.

4. Use the S/N tolerance limit $[(S/N_{max} - S/N_{min})/3]$ to compare with the average results from the verification experiments. If the average of the verification S/N is within the range established by the S/N tolerance limit, especially on the low end, we accept the fact that the predicted S/N performance has indeed been confirmed. If the average from the verification data surpasses the S/N tolerance limit, especially on the low side, we conclude that we have not confirmed and that there is either some form of control factor interaction or data acquisition or analysis error that must be discovered and corrected.

Document optimum robustness and tuning factors and set points in Critical Parameter Management database.

Now that the design's critical functional response has been made robust against the compounded set of noise factors and its critical adjustment parameter used to adjust the mean of the CFR on to the desired target, it is time to document our results. The following items are documented within the Critical Parameter Management database:

1. Optimum S/N value for the CFR after verifying the additive model
2. Baseline (benchmark) S/N value of the CFR prior to conducting robust design
3. Optimum nominal set points for each control factor that had a significant effect on the gain in robustness of the CFR
4. Optimum set points for the control factors that were used to shift Beta (β) to place the mean value of the CFR on to its desired target
5. The Cp of the CFR after robustness optimization, which can be quantified with and without the noise factor effects being present if we desire

The critical parameter database can contain any level of detail beyond this that we feel is important to track. The key thing to remember is that we want to be able to define and track relationships between all the critical parameters within the entire system. Robust design data typically is associated with system and subsystem level CFRs. System CFRs are measured for robustness, and their S/N data is stored at the system level within the relational database architecture. Subsystem CFR data related to robustness is stored in the subsystem level within the CPM database. Critical adjustment parameter information relative to a subsystem CFR mean performance is stored in the subsystem CAP level within the CPM database.

The components that are significant control factors for robustness and beta shifting are documented as critical-to-function (CTF specs) specifications within the component level critical parameter database.

The following CPM database scorecards help illustrate how to document data.

Documenting System Level CFR Robustness Performance:

DFSS-Critical Parameter Management Scorecard
System Requirement Document Line Items-to-System Critical Functional Response Map

(Initiated in System Concept Design Phase–Phase 1)

SRD Line Item:	System CFR:	Target	USL	LSL	Mean	σ	COV	S/N	Cp	Cpk

Documenting Subsystem Level CFR Robustness Performance:

DFSS-Critical Parameter Management Scorecard
Subsystem-to-System Critical Functional Response Relationship Map

(Initiated in Subsystem Concept Design and Parameter Optimization Phase–Phase 1)

SRD #	System CFR:	Subsystem CFR:	Target	USL	LSL	Mean	σ	COV	S/N	Cp	Cpk

Documenting Subsystem-CAP Level CFR Mean Shifting Performance:

DFSS-Critical Parameter Management Scorecard
Critical Adjustment Parameter-to-Subsystem CFR Relationship Map

(Initiated in System Integration and Optimization Phase–Phase 2)

SRD #	Subsys. CFR:	CAP Spec:	CAP Target	CAP USL	CAP LSL	CFR Mean	CFRs	CFR Cp	CFR Cpk	δCFR/ δCAP

Documenting Component Level CFR Performance:

DFSS-Critical Parameter Management Scorecard
Component-to-Subsystem Critical Functional Response Relationship Map

(Initiated in Product Design Phase–Phase 3)

SRD #	Subsystem CFR:	Component CTF Spec:	Target	USL	LSL	Mean	σ	Cp	Cpk	δCFR/ δCTF

Once we have updated the Critical Parameter Management database, it is now time to move on to our next set of tasks.

Final Detailed Outputs

Move on to technology verification or product capability development using tolerance design.

Once robust design has been conducted on subsystems and the integrated system, the team can move on to the verification of capability phases.

Robust Design Checklist and Scorecard

This suggested format checklist can be used by an engineering team to help plan its work and aid in the detailed construction of a PERT or Gantt chart for project management of the robustness optimization process within the Optimize phase of the CDOV or I²DOV development process. These items also serve as a guide to the key items for evaluation within a formal gate review.

Checklist for Robust Design

Actions:	Was This Action Deployed? YES or NO	Excellence Rating: 1 = Poor 10 = Excellent
1. Define the ideal function.	_____	_____
2. Define critical functional responses and S/N ratio.	_____	_____
3. Define the transducers, instrumentation, and your complete data acquisition system.	_____	_____
4. Identify the candidate noise factors.	_____	_____

5. Conduct a noise experiment. _____ _____

6. Identify and plot the magnitude and directionality of noise factors using analysis of means (ANOM). _____ _____

7. Identify statistically significant noise factors using ANOVA. _____ _____

8. Compound noise factors for use in main robustness experiment (N_1 and N_2). _____ _____

9. Identify control factors for the design. _____ _____

10. Select a properly sized orthogonal array to evaluate the robustness contribution and mean adjustment capability from each control factor. _____ _____

11. Run the designed experiment in the presence of the uniformly applied, compounded noise factors. _____ _____

12. Perform analysis of means (ANOM) and plot the main effects and selected interactions of the mean, standard deviation, and S/N metric for each control factor. _____ _____

13. Select the optimum set point for the control factors that have a strong effect on the robustness of the critical functional response (these become critical-to-function parameters). _____ _____

14. Identify critical adjustment parameters, the control factors that have a strong effect on the mean of the CFR and a reasonably weak effect on the standard deviation. _____ _____

15. Construct Taguchi's additive model from the significant control factor signal-to-noise values (units are dBs). _____ _____

16. Conduct verification runs at the optimum S/N set points for the robustness control factors after using the CAPs to put the mean onto the target. _____ _____

17. If the new S/N values are reasonably close to the predicted S/N from the additive model, the robustness set points are confirmed—you are ready for tolerance design. _____ _____

18. If the new S/N values are well below the predicted S/N from the additive model, the robustness set points are not confirmed—you must go back and find the reason why. _____ _____

Scorecard for Robust Design

The definition of a good robustness scorecard is that it should illustrate in simple but quantitative terms what gains were attained from the work. This suggested format for a scorecard can be used by an engineering team to help document the completeness of its work.

Subsystem name: _____

Subsystem critical functional response: _____

Specific engineering units of subsystem critical functional response: _____

Number of statistically significant noise factors: _____

Number of control factors evaluated for robustness: _____

Number of control factors selected to include in additive model: _____

Number of control factors selected to adjust beta: _____

Baseline or benchmarked initial subsystem S/N value: _____

Verified optimal subsystem S/N value: _____

Prerobustness design capability (Cp) index for the subsystem CFR: _____

Postrobustness design capability (Cp) index for the subsystem CFR: _____

Table of Critical Parameter Set Points:

Critical Parameter Name:	Optimal Nominal Set Point:	Estimated Tolerance: USL	Estimated Tolerance: LSL	Used for Robustness or Beta Shifting:
1.	_____	_____	_____	_____
2.	_____	_____	_____	_____
3.	_____	_____	_____	_____

Table of Critical Noise Factor Set Points:

Critical Noise Factor Name:	Magnitude of Effect:	Direction of Effect: (+ or −)	F Ratio/$F_{crit.}$:	% Contribution or ε^2 Value:
_____	_____	_____	_____	_____
_____	_____	_____	_____	_____
_____	_____	_____	_____	_____

We repeat the final deliverables from robust design:

- Optimized signal-to-noise metrics for the critical functional response(s)
- Optimized critical-to-function parameter set points that control the subsystem's robustness
- Optimized critical adjustment factors that tune the mean value of the subsystem's critical functional response(s)

- Subsystem critical parameter management data that is documented in the CPM relational database:
  CFR mean
  CFR standard deviation
  CFR coefficient of variation [(s/mean) $\times$ 100]
  CFR S/N ratio
  CFR Cp index

References

Fowlkes, W. Y., & Creveling, C. M. (1995). *Engineering methods for robust product design.* Reading, MA.: Addison-Wesley.

Phadke, M. S. (1989). *Quality engineering using robust design.* Englewood Cliffs, NJ.: Prentice Hall.

Taguchi, G., Chowdhury, S., & Taguchi, S. (2000). *Robust engineering.* New York: McGraw-Hill.

Response Surface Methods

Where Am I in the Process? Where Do I Apply Response Surface Methods Within the Roadmaps?

Linkage to the I²DOV Technology Development Roadmap:

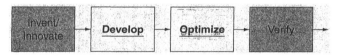

RSM is primarily used during the Develop and Optimize phases of I²DOV. It is used much less during the Invent/Innovate and Verify phases.

Linkage to the CDOV Product Development Roadmap:

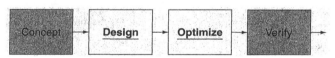

RSM is primarily used during the Design and Optimize phases of CDOV. It is used much less during the Concept and Verify phases.

What Am I Doing in the Process? What Does RSM Do at this Point in the Roadmaps?

The Purpose of RSM in the I²DOV Technology Development and CDOV Product Development Roadmaps:

Response surface methods allow the construction of a detailed mathematical model and a visual map of system performance. The purpose of the model and maps are to finely tune and optimize the system's performance around a desired operating point.

What Output Do I Get at the Conclusion of RSM? What Are Key Deliverables at this Point in the Roadmaps?

- An equation, typically a second order polynomial, that analytically describes the behavior of the system around a desired operating point
- A series of two-factor-at-a-time response maps that graphically describe the behavior of the system around a desired operating point

RSM Process Flow Diagram

This visual reference illustrates the flow of the detailed steps of RSM:

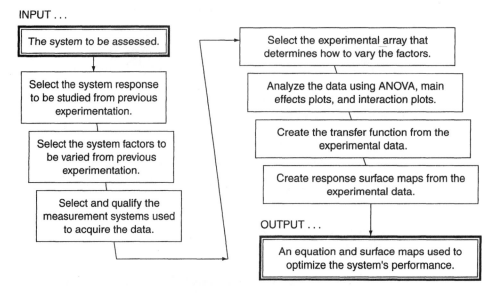

Verbal Descriptions of the Application of Each Block Diagram

This section explains the specific details of the RSM process:

- *Key detailed input(s)* from preceding tools and best practices
- *Specific steps and deliverables* from each step to the next within the process flow
- *Final detailed outputs* that flow into follow-on tools and best practices

Key Detailed Inputs

> **The system to be assessed.**

Response surface methods allow the construction of a detailed mathematical model and a visual map of system performance. The maps are similar to the topographic maps used for hiking. A topographic map shows contour intervals that indicate changes in elevation as one moves in any magnetic heading along the surface of the earth. The topographic map provides a visualization of the peaks and valleys of the terrain.

Consider altitude to be a critical response and the X and Y directions of travel to be two critical parameters of the response. The topographic map allows us to determine the values of X and Y that place us on the top of a mountain along the Continental Divide or at the bottom of Death Valley. It also tells us where the flat plains of Kansas are located. The mountain peaks and valley bottoms are regions of high sensitivity and poor robustness. Small changes in the X or Y direction result in large changes in altitude. The plains of Kansas are a region of insensitivity and robustness of altitude to changes in X and Y. We can move a large distance in any direction without much change in altitude. This is robustness.

What are the experimental methods needed to obtain the detailed mathematical models and response surface maps? A response surface, in general, is curved and bumpy. In order to model this surface, we need a transfer function with quadratic or higher terms in it. A two-level design is not capable of estimating quadratic effects. We need an experiment with at least three levels. This can get very large. There is, however, an economical design that gives the needed information. It is called a central composite design (CCD). The CCD is a composite of a Full Factorial with a center point and additional points called star or axial points.

Let's look at a simple two-factor CCD, shown in Figure 29–1 for two factors X_1 and X_2. The Full Factorial portion of the CCD is shown by the points located at $(+1, +1), (+1, -1), (-1, +1)$,

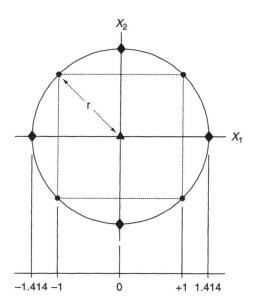

Figure 29–1   Two-Factor Central Composite Design

$(-1, -1)$. The center point of the design is shown by the triangle located at the point $(0,0)$. So far, the design can only estimate first order main effects, interactions, and test for curvature. A generalized transfer function of this form is shown in this equation.

$$y = b_0 + b_1x_1 + b_2x_2 + \ldots b_nx_n + b_{12}x_1x_2 \ldots + b_{ij}x_ix_j$$

We discussed this type of experiment earlier in Chapter 27 on DOE. There are several possible outcomes of such an experiment. It is possible that interaction terms are not significant but the transfer function does not adequately model system behavior. Another possibility is that interactions are significant but, even with the interactions in the model, the transfer function still does not adequately model system behavior. Analysis of the residuals, R^2, and lack of fit will indicate whether the transfer function adequately models system behavior.

These two possible outcomes indicate that the transfer function may be too simple. Additional terms may be required. If a residuals vs. fits plot indicates a curved pattern, a squared or quadratic term is probably needed. However, we need additional experimental data to estimate squared terms. How can we get the additional information without wasting the runs we have already made? We can accomplish this by adding additional points to the design, called *star* or *alpha points*. These are shown as diamonds in Figure 29–1. Figure 29–1 shows the geometry of the CCD. The alpha points are on the circumference of a circle. The radius of the circle is

$$r = \sqrt{(-1)^2 + (+1)^2} = \sqrt{2} = 1.4142$$

So the star points are at $(-1.4142, 0)$, $(+1.4142, 0)$, $(0, +1.4142)$, and $(0, -1.4142)$. Each factor now has five levels: 1.4142, $+1$, 0, -1, and -1.4142.

The additional axial points now make it possible to estimate squared main effects and interactions without aliasing. The general form of the obtainable transfer function is shown in this first equation and in an equivalent form in the second:

$$y = b_0 + b_1x_1 + \ldots b_nx_n + b_{11}x_1^2 + \ldots b_{nn}x_n^2 + b_{12}x_1x_2 + \ldots b_{n-1,n}x_{n-1}x_n$$

$$y = b_0 + \sum_{i=1}^{n} b_ix_i + \sum_{i=1}^{n} b_{ii}x_i + \sum_{i}^{n-1}\sum_{j}^{n} b_{ij}x_ix_j$$

The discussion for two factors can be extended to three factors and more. The geometry of a CCD is shown in Figure 29–2 for three factors A, B, and C. The circles are the factorial points, the triangle is the center point, and the diamonds are the axial points.

Minitab can be used to create and analyze response surface experiments. Let's work through an example.

Specific Steps and Deliverables

Select the system response to be studied from previous experimentation.

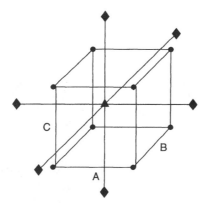

Figure 29–2 Three-Factor Central Composite Design

We'll revisit the cool time experiment from Chapters 26 and 27 on regression and DOE. The critical-to-function response is the time it takes to cool the fluid from an upper set point to a lower set point.

Select the system factors to be varied from previous experimentation.

Previous experimentation was conducted to determine that fan speed, fluid volume, ambient temperature, ambient altitude, and ambient humidity are critical system parameters.

Select and qualify the measurement systems used to acquire the data.

RSM, as with any process where data will be used to make decisions, requires a capable measurement system. The techniques covered in Chapter 23 on MSA should have been applied to the measurement system and its Gage R&R, P/T ratio, and the number of distinct categories determined.

Select the experimental array that determines how to vary the factors.

Prior experimentation and analysis of the physical principles of thermodynamics indicated that the system was nonlinear. This information led to a CCD as the chosen experiment. The experiment was conducted in an environmental chamber capable of altitude, temperature, and humidity control.

Analyze the data using ANOVA, main effects plots, and interaction plots.

The data were analyzed using Minitab's response surface analyzer utility. In Minitab, choose Stat > DOE > Response Surface > Analyze Response Surface Design. The following window appears. Select the response.

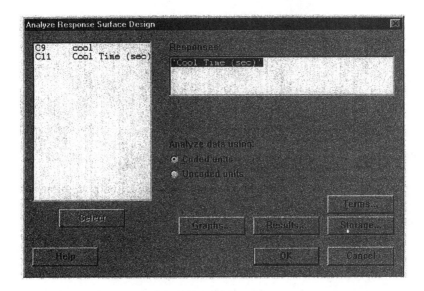

Then choose Terms, and the next window appears. The full quadratic model was requested. This allows estimation of a model of the form shown in the previous equation and its equivalent form.

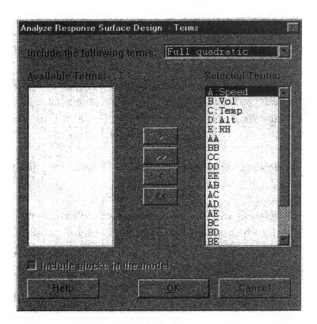

Select OK and analyze the data to obtain the following session window output. The significant effects are shown in bold.

Response Surface Regression: Cool Time (sec) versus Speed, Vol, Alt, RH

The analysis was done using coded units.

Estimated Regression Coefficients for Cool Tim

Term	Coef	SE Coef	T	P
Constant	**5.1358**	**0.19240**	**26.693**	**0.000**
Speed	**-0.8487**	**0.09846**	**-8.620**	**0.000**
Vol	**0.6878**	**0.09846**	**6.986**	**0.000**
Temp	**0.4476**	**0.09846**	**4.546**	**0.001**
Alt	0.0979	0.09846	0.994	0.342
RH	0.0234	0.09846	0.238	0.816
Speed*Speed	**-0.6243**	**0.08906**	**-7.010**	**0.000**
Vol*Vol	0.0202	0.08906	0.227	0.825
Temp*Temp	-0.0265	0.08906	-0.298	0.771
Alt*Alt	-0.0214	0.08906	-0.241	0.814
RH*RH	-0.1250	0.08906	-1.404	0.188
Speed*Vol	**0.6086**	**0.12059**	**5.046**	**0.000**
Speed*Temp	-0.0609	0.12059	-0.505	0.624
Speed*Alt	0.0740	0.12059	0.614	0.552
Speed*RH	0.0131	0.12059	0.108	0.916
Vol*Temp	0.0440	0.12059	0.365	0.722
Vol*Alt	-0.0183	0.12059	-0.151	0.882
Vol*RH	-0.0518	0.12059	-0.429	0.676
Temp*Alt	-0.0023	0.12059	-0.019	0.985
Temp*RH	0.1321	0.12059	1.095	0.297
Alt*RH	-0.0153	0.12059	-0.127	0.901

S = 0.4824   R-Sq = 95.3%   R-Sq(adj) = 86.8%

Analysis of Variance for Cool Tim

Source	DF	Seq SS	Adj SS	Adj MS	F	P
Regression	20	51.9469	51.9469	2.5973	11.16	0.000
Linear	5	33.6943	33.6943	6.7389	28.96	0.000
Square	5	11.8151	11.8151	2.3630	10.16	0.001
Interaction	10	6.4375	6.4375	0.6438	2.77	0.055
Residual Error	11	2.5595	2.5595	0.2327		
Lack-of-Fit	**6**	**1.8222**	**1.8222**	**0.3037**	**2.06**	**0.223**
Pure Error	5	0.7373	0.7373	0.1475		
Total	31	54.5064				

The analysis indicates that speed, speed^2, volume, temperature, and the speed temperature interactions are statistically significant. The remaining terms can be removed from the model and reanalyzed. In Minitab, choose Stat > DOE > Response Surface > Analyze Response Surface Design. The following window appears.

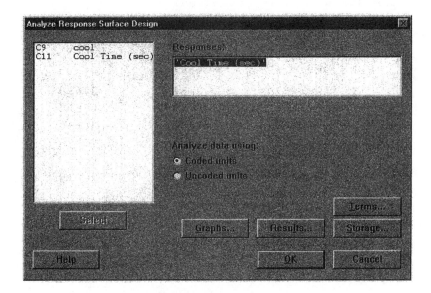

Select the response. Then choose Terms, and the next window appears.

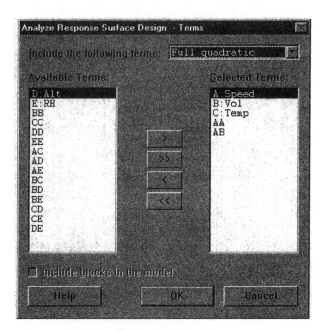

Choose the five significant terms and select OK to analyze the data. The session window output follows.

Response Surface Regression: Cool Time (sec) versus Speed, Vol, Temp

The analysis was done using coded units.

Estimated Regression Coefficients for Cool Tim

Term	Coef	SE Coef	T	P
Constant	5.0135	0.08565	58.532	0.000
Speed	-0.8487	0.07819	-10.854	0.000
Vol	0.6878	0.07819	8.797	0.000
Temp	0.4476	0.07819	5.725	0.000
Speed*Speed	-0.6141	0.06994	-8.781	0.000
Speed*Vol	0.6086	0.09576	6.355	0.000

S = 0.3831   R-Sq = 93.0%   R-Sq(adj) = 91.7%

Analysis of Variance for Cool Tim

Source	DF	Seq SS	Adj SS	Adj MS	F	P
Regression	5	50.6913	50.6913	10.1383	69.09	0.000
Linear	3	33.4513	33.4513	11.1504	75.99	0.000
Square	1	11.3144	11.3144	11.3144	77.11	0.000
Interaction	1	5.9256	5.9256	5.9256	40.38	0.000
Residual Error	26	3.8151	3.8151	0.1467		
Lack-of-Fit	9	0.8705	0.8705	0.0967	0.56	0.812
Pure Error	17	2.9446	2.9446	0.1732		
Total	31	54.5064				

The session window information indicates significant terms, relatively high R^2, and no significant lack of fit. This may be a good model. Let's look at the residuals. The residuals vs. fits plot (Figure 29–4) looks random and the residuals (Figure 29–3) are normal at an alpha of 0.5.

Create the transfer function from the experimental data.

Based on the analysis, we can create the transfer function shown in this equation:

Cool Time = 5.0135 − 0.8487 Speed + 0.6878 Vol + 0.4476 Temp

−0.6141 Speed^2 + 0.6086 SpeedXVol

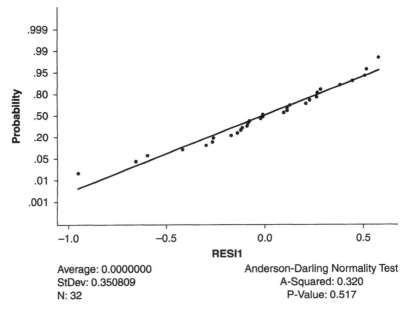

Average: 0.0000000
StDev: 0.350809
N: 32

Anderson-Darling Normality Test
A-Squared: 0.320
P-Value: 0.517

Figure 29–3 Normal Probability Plot

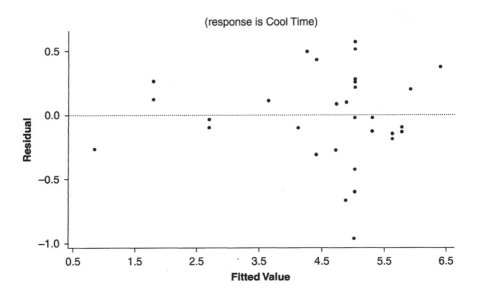

Figure 29–4 Residuals versus the Fitted Values

This equation is a mathematical description of the behavior of the system from which we can create surface plots and optimize the system's performance.

> **Create response surface maps from the experimental data.**

Surface plots allow us to visualize the system's behavior. Unfortunately, we can only plot two factors at a time in a surface plot. We have three significant factors so we need three response surface plots. We plot speed vs. temperature, speed vs. volume, and volume vs. temperature. The three plots are shown in Figures 29–5a, b, and c.

The speed vs. volume and speed vs. temperature plots are curved due to the quadratic effect of speed and the interaction between speed and volume. The volume vs. temperature plot is a plane because volume and temperature appear as linear terms and they do not interact.

Final Detailed Outputs

> **An equation and surface maps used to optimize the system's performance.**

We can now visually determine the levels of the factors that place the response at a maximum or minimum. We can also see the locations of robustness. For example, using Figures 29–5a and b, we can see that cool time is sensitive to changes in fan speed at high fan speeds.

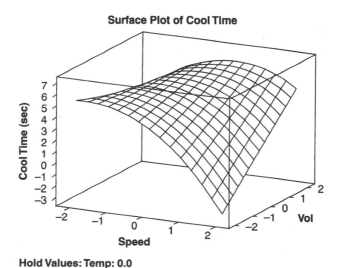

Surface Plot of Cool Time

Hold Values: Temp: 0.0

Figure 29–5a   Speed vs. Volume Surface Plot

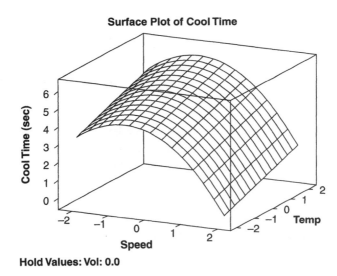

Figure 29–5b   Speed vs. Temperature Surface Plot

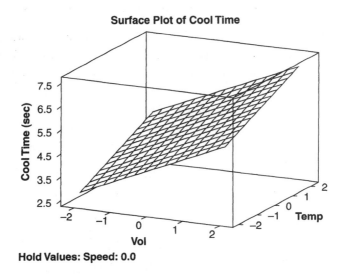

Figure 29–5c   Volume vs. Temperature Surface Plot

The slope of the surfaces are steep in this region. The flat, inclined plane shown in Figure 29–5c indicates that there is no region in which cool time is least sensitive to changes in volume or temperature. It does indicate, however, that cool time is shortest at low temperature and low volume.

The transfer function is extremely valuable. It is the prize of DFSS. The transfer function allows us to optimize the performance of the system, tolerance the system, and predict capability. We will use the transfer function in the next chapter to optimize the performance of the system and predict capability.

RSM Checklist and Scorecards

Checklist for RSM

Actions:	Was this Action Deployed? YES or NO	Excellence Rating: 1 = Poor 10 = Excellent
Select the system response to be studied from previous experimentation.	_____	_____
Select the system factors to be varied from previous experimentation.	_____	_____
Select and qualify the measurement systems used to acquire the data.	_____	_____
Select the experimental array that determines how to vary the factors.	_____	_____
Analyze the data using ANOVA, main effects plots, and interaction plots.	_____	_____
Create the transfer function from the experimental data.	_____	_____
Create the response surface maps from the experimental data.	_____	_____
Document an equation and surface maps used to optimize the system's performance	_____	_____

Scorecard for RSM

The definition of a good scorecard is that it should illustrate in simple but quantitative terms what gains were attained from the work. This suggested format for a scorecard can be used by a product planning team to help document the completeness of its work.

Response Surface Transfer Function Critical Response:	Response Surface Transfer Function Critical Parameters:	R^2:	VOC Need That this Response Surface Transfer Function Addresses:	Response Unit of Measure:
1.	_____	_____	_____	_____
2.	_____	_____	_____	_____
3.	_____	_____	_____	_____

We repeat the final deliverables:

- An equation, typically a second order polynomial, that analytically describes the behavior of the system around a desired operating point
- A series of two-factor-at-a-time response maps that graphically describe the behavior of the system around a desired operating point

Reference

Myers, R. H., & Montgomery, D. W. (1995). *Response surface methodology.* New York: Wiley Interscience.

Optimization Methods

Where Am I in the Process? Where Do I Apply Optimization Within the Roadmaps?

Linkage to the I²DOV Technology Development Roadmap:

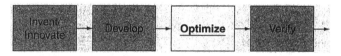

Optimization is primarily used during the Optimize phase of I²DOV. It is used much less during the Invent/Innovate and Verify phases.

Linkage to the CDOV Product Development Roadmap:

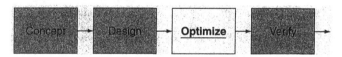

Optimization is primarily used during the Optimize phase of CDOV. It is used much less during the Concept, Design, and Verify phases.

What Am I Doing in the Process? What Does Optimization Do at this Point in the Roadmaps?

The Purpose of Optimization in the I²DOV Technology Development and the CDOV Product Development Roadmaps:

Optimization is an analytic, numeric, and graphical process for determining the system's operating point that simultaneously provides an on-target minimum varitation response. A compromise is typically required between the target response value and its variation due to one or more control factors that affect both mean value and variance in a conflicting manner.

What Output Do I Get at the Conclusion of Optimization? What Are Key Deliverables at this Point in the Roadmaps?

- A list of the factor nominals and tolerances that yield capable, on-target, minimum variance performance

Optimization Process Flow Diagram

This visual reference illustrates the flow of the detailed steps of optimization.

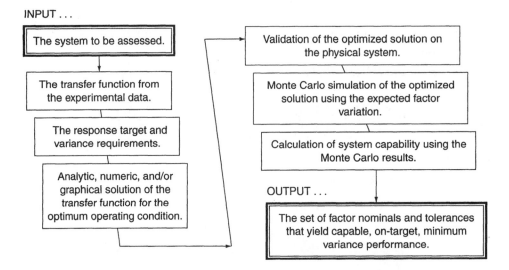

Verbal Descriptions of the Application of Each Block Diagram

This section explains the specific details of what to do during the optimization process:

- *Key detailed input(s)* from preceding tools and best practices
- *Specific steps and deliverables* from each step to the next within the process flow
- *Final detailed outputs* that flow into follow-on tools and best practices

The system to be assessed.

Let's review the steps we followed to obtain the fluid cooling transfer function.

The transfer function from the experimental data.

The experimental approach taken on the fluid cooling problem is called *sequential experimentation*. Sequential experimentation allows us to build a transfer function incrementally without wasting experimental resources estimating unneeded terms. We stop at each step in the sequence and decide if we have sufficient information to adequately model the behavior of our system. If not, we continue and run another experiment.

Sequential experimentation begins with a relatively small Fractional Factorial experiment with a center point that allows estimation of main effects and curvature. If curvature is not significant and if a "main-effects only" model adequately describes the system, we're done. The system is additive and an additive transfer function is sufficient. A general additive transfer function is shown here:

$$y = b_0 + b_1x_1 + b_2x_2 + \dots b_nx_n$$

Analysis of the fluid cooling data indicated that the effects of fan speed, fluid volume, and temperature were significant. However, the center point was also significant, indicating that the system was nonlinear. We therefore could not construct a simple additive transfer function. When curvature is significant, we must investigate the nature of the nonlinearity. The nonlinearity could be due to an interaction like A $\times$ B or due to a quadratic term like A^2 or from several quadratic terms and/or interactions acting together. The next step in the sequence of experimentation is to add some or all of the treatment combinations from the Full Factorial that are not included in the Fractional Factorial. This allows us to determine if any two-way interactions are significant. If any two-way interactions are found to be significant, we add them to our transfer function and determine if the new transfer function adequately describes the system's behavior. We use the residuals vs. fits plot to see if there is structure to the plot. A general transfer function with interactions is shown in this equation:

$$y = b_0 + b_1x_1 + b_2x_2 + \dots b_nx_n + b_{12}x_1x_2 \dots + b_{ij}x_ix_j$$

In our fluid cooling example, we ran the remaining eight treatment combinations, resulting in a Full Factorial design. Our transfer function was improved by the addition of an interaction term between fan speed and fluid volume, but the residuals vs. fits plot showed a pattern. The pattern indicated that there was still information in the data that was not being explained by the model, even with the two-way interaction in the model. The pattern suggested that a quadratic term may be needed in the transfer function. So, the next step was to run an experiment capable of estimating quadratic terms. We chose a central composite design (CCD).

The CCD allows us to estimate the linear and quadratic main effects and linear two-way interactions. The CCD is a Full Factorial with center points and additional star or alpha points. The Fractional Factorial with center point and the added Full Factorial runs provides the "central" part of the CCD. These treatments had already been run. We only need to run the star points to acquire the additional information needed to complete the CCD and estimate the quadratic terms. The final transfer function obtained is shown in this equation:

Cool Time = 5.0135 − 0.8487 Speed + 0.6878 Vol + 0.4476 Temp −
0.6141 Speed^2 + 0.6086 SpeedXVol

The method of sequential experimentation has provided us with a transfer function that adequately describes the behavior of our system over a small operating space. But what do we do with it? One use of the transfer function is to determine the levels at which to set fan speed, fluid volume, and temperature to optimize cool time. Another use of the transfer function is Monte Carlo simulation. Monte Carlo simulation allows us to determine the descriptive statistics of the response. The statistics of the response are a result of how the transfer function combines the statistics of the factors. We can estimate the capability of our design using the descriptive statistics and the specification limits.

> **The response target and variance requirements.**

> **Analytic, numeric, and/or graphical solution of the transfer function for the optimum operating condition.**

Let's use the transfer function to determine the factor levels that optimize cool time. The signs of the coefficients provide some information concerning how to minimize the response. In this case, speed should be set at its highest level, and volume and temperature should both be set at their lowest levels. However, a compromise between speed and volume might be required due to their interaction.

We can use the transfer function to find the partial derivative of cool time with respect to each of the five terms in the model, set them to zero, and solve. We would have five equations in five unknowns. We would need starting guesses to begin the optimization search. In general, optimization by this method can be a tedious task. It is easier to use a software package to do this than to solve by hand. We'll use Minitab.

In the cool time example, we ran a CCD. Minitab can use this type of design to optimize the response after we have analyzed the CCD design and obtained our transfer function. Using Stat > DOE > Response Surface > Response Optimizer, we obtain the window at the top of page 619.

Select the response to be optimized and then choose Setup to get the next window, shown on the bottom of page 619.

The minimum value for cool time is zero. Choose a target goal with a target value of zero seconds. We'll bound this by a region around 0, -1 to 1 seconds. Click OK and run the optimizer to get the next window at the top of page 620.

Minitab has identified the levels to set the factors at to give an average cool time of zero:

Speed	=	1.4383 units
Volume	=	-1.5823 units
Temperature	=	-0.4165 units

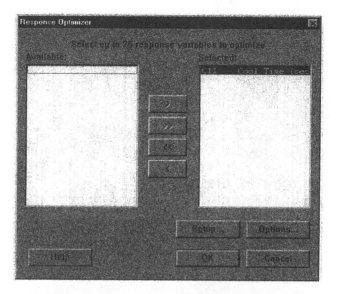

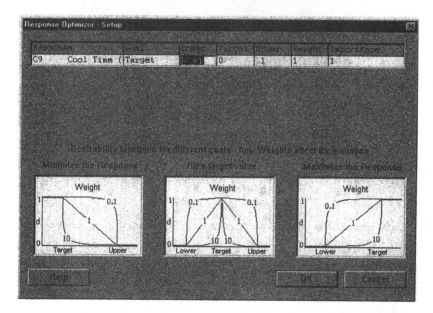

These values are in experimental or coded units. They must be converted to physical units. We can use Minitab to accomplish this. In Minitab we choose Stat > DOE > Display Designs. The window in the middle of page 620 appears.

Choose Uncoded Units and click OK. The worksheet is now in physical units as opposed to experimental units. The physical units must have been previously entered, of course. Now redo the optimization. The results that follow at the bottom of page 620 are in physical units.

Optimal D 1.0000	Hi Cur Lo	Speed 2.0 [1.4383] −2.0	Vol 2.0 [−1.5823] −2.0	Temp 2.0 [−0.4165] −2.0

Cool Tim
Targ; 0.0
y = 0.0
d = 1.0000

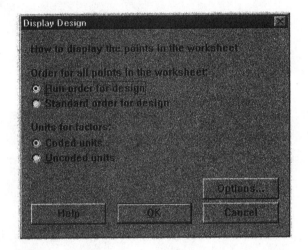

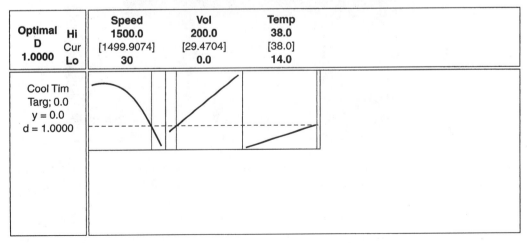

Optimal D 1.0000	Hi Cur Lo	Speed 1500.0 [1499.9074] 30	Vol 200.0 [29.4704] 0.0	Temp 38.0 [38.0] 14.0

Cool Tim
Targ; 0.0
y = 0.0
d = 1.0000

The physical values to obtain an average cool time of zero are:

$RPM_{optimal} = 1500$ RPM

$Volume_{optimal} = 29.5$ μl

$Temperature_{optimal} = 38°C$

Unfortunately, zero cool time is not physically realizable, the system will not function at a fluid volume of only 29.5 μl, and 38°C is too hot. We can find the factor levels that give a target cool time of 2 seconds. In Minitab we choose Stat > DOE > Display Designs and choose Setup. The following window appears.

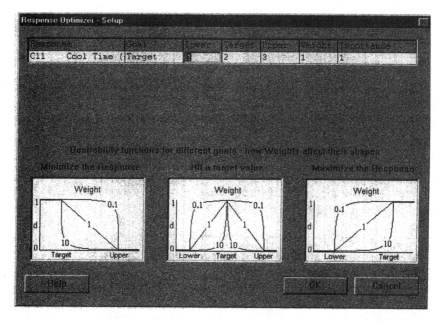

Choose a target goal with a target value of 2 seconds. We'll bound this by a region around 2, 1 to 3 seconds. Click OK and run the optimizer to get the next window, shown at the top of page 622. The physical values to obtain an average cool time of 2 seconds are:

$RPM_{optimal} = 1500$ RPM

$Volume_{optimal} = 172$ μl

$Temperature_{optimal} = 14°C$

Validation of the optimized solution on the physical system.

The physical system must now be set up and run at this set of factor levels to verify that the predicted time is obtained. The system will not produce the predicted time every time it is run. The

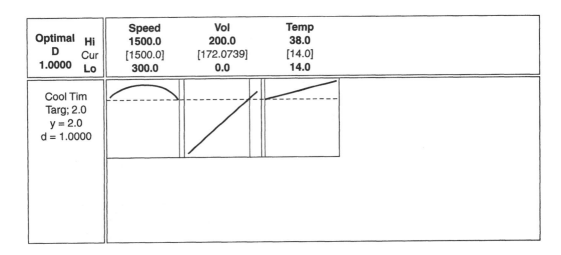

Optimal		Speed	Vol	Temp
D	Hi	1500.0	200.0	38.0
	Cur	[1500.0]	[172.0739]	[14.0]
1.0000	Lo	300.0	0.0	14.0

Cool Tim
Targ; 2.0
y = 2.0
d = 1.0000

predicted time is an average time. The system must be run several times at these factor levels. A confidence interval on the mean will then confirm if the system is performing as predicted. If it is not performing as predicted, either the prediction is wrong or the system has changed since it was last tested. A rerun of one or more of the treatment combinations from the original CCD can determine if the system has changed. The new observations won't agree with the original observations if the system has changed. If they do agree, the prediction is incorrect. A data entry or manipulation error may be the cause, or the predictive model may be inadequate. Check your analysis and check your residuals.

> **Monte Carlo simulation of the optimized solution using the expected factor variation.**

We have answered the question, "What are the factor values that produce an optimum and a target response?" A more practical question to ask is "What cool times can I expect from the system given the values of speed, volume, and temperature at which I expect to operate the system?"

We can use the transfer function in a Monte Carlo simulation to answer this question. We need two things to accomplish this. We must have a transfer function and we must have the descriptive statistics of the factors in the transfer function. We have the transfer function:

Cool Time = 5.0135 − 0.8487 Speed + 0.6878 Vol + 0.4476 Temp −
0.6141 Speed^2 + 0.6086 SpeedXVol

Now for the descriptive statistics of the factors in the transfer function. Fan speed is assumed to be normally distributed with a speed of 1200 +/− 300 RPM. Three sigma control on fan speed was obtained, so its standard deviation is 300/3 = 100 RPM. Fluid volume is assumed to be normally distributed with a volume of 100 +/− 20 μl. Three sigma control is assumed, so its standard deviation is 20/3 = 6.7μl.

Ambient temperature is assumed to be uniformly distributed with a temperature of 20 +/− 2°C. Temperature will range from 18 to 22 degrees. These descriptive statistics are summarized:

Fan Speed	∼ N(1200, 100)
Volume	∼ N(100, 6.7)
Temperature	∼ U(18, 21)

These physical units must be converted into experimental units. We performed a five-factor CCD. The alpha value for this is 2. We obtained this information from Minitab when we created the experimental design. The experimental value for the center point of the experiment is the point 0, 0, 0, 0, 0 for speed, volume, temperature, altitude, and RH. The analysis indicated that only speed, volume, and temperature were significant. The physical values for the center point for these three factors are given in Table 30–1 as well as the physical values at the +2 alpha position. So, for speed, the experimental units span 2 − 0 = 2 units, while the physical units span 1500 − 900 = 600 units. The slope is 2/600 = 0.003 experimental units/physical units. This is calculated for all three factors in Table 30–1.

Next, we need to convert the physical operating points and variation into experimental operating points and variation. This is shown in Table 30–2.

Table 30–3 summarizes the factor distributions in both physical and experimental units. We use the experimental units in the Monte Carlo simulation since the transfer function is also in experimental units.

Table 30–1 Physical and Experimental Relationship Among the Factors

Factor	Experimental Center	Physical Center	Physical +2 Alpha Point	Slope
Fan Speed (RPM)	0	900 RPM	1500	2/(1500 − 900) = 0.003
Volume (μl)	0	100	200	2/(200 − 100) = 0.02
Temperature (°C)	0	26	38	2/(38 − 26) = 0.167

Table 30–2 Relationship of Physical and Experimental Variation

Factor	Physical Operating Point and Variation	Experimental Operating Point	Experimental Variation
Fan Speed (RPM)	1200 +/− 100	0 + (1200 − 900)*0.003 = 1	100*0.003 = 0.33
Volume (μl)	100 +/− 6.7	0 + (100 − 100)*0.02 = 0	6.7*0.02 = 0.134
Temperature (°C)	20 +/− 2	0 + (20 − 26)*0.167 = −1	2*0.167 = 0.33

Table 30–3 Experimental and Physical Factor Distributions

Factor	Physical Distribution	Experimental Distribution
Fan Speed	N(1200, 100)	N(1, 0.333)
Volume	N(100, 6.7)	N(0, 0.134)
Temperature	U(18, 22)	U(−1.33, 1.33)

We now have all the information needed to run a Monte Carlo simulation. We'll use Crystal Ball,® which is a statistical package that runs under Microsoft® Excel. Crystal Ball® makes it easy to conduct Monte Carlo simulations and optimization.

Monte Carlo simulation uses a computer to simulate building and running a design thousands of times. The simulations are done under conditions that simulate the variation in the parts and conditions expected during actual use. It is an extremely efficient way to evaluate the effects of a design's critical parameters on the design's critical responses. It is a fantastic what-if tool.

The graphical and tabular output from a 100,000 run Crystal Ball ®Monte Carlo simulation is shown in the following chart using the factor descriptive statistics summarized previously.

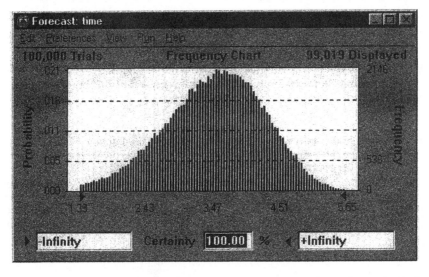

This simulation took about 8 minutes to perform, perhaps orders of magnitude faster than physically building and testing the system. The graphical output shows 100,000 values of cool time obtained from repeated random sampling from the assumed distributions for speed, volume, and temperature. These values are substituted into the transfer function, and the function is evaluated 100,000 times.

The results indicate that cool time is likely to be normally distributed with a mean of 3.5 seconds and a standard deviation of 0.8 seconds. This is the performance that can be expected from the system when it is run under the expected operating conditions.

Calculation of system capability using the Monte Carlo results.

We can use this information to calculate the capability of the system. All we need is the upper and lower specification limits. The upper limit is 6 seconds and the lower bound is 0 seconds. Capability is calculated in this equation:

$$C_p = \frac{USL - LSL}{6\sigma} = \frac{6 - 0}{6(0.8)} = 1.25$$

This is a little better than 3 sigma performance. Assuming a normal distribution, approximately $2{,}700/2 = 1{,}350$ per million (this is a one-sided limit) systems would perform outside the upper specification limit, without a mean shift. It would be worse with the expected long-term mean shifts and drifts. This is poor performance. We would like 6 sigma performance, and a capability index of 2. How can we achieve this? We must reduce variation. What must the variation be reduced to for 6 sigma performance? We solve the capability equation for sigma.

$$\sigma = \frac{USL - LSL}{6C_p} = \frac{6 - 0}{6*2} = 0.5$$

How can we achieve a standard deviation of 0.5? We must search for engineering methods to reduce the variation in the three factors that affect cool time. Reducing the variation in room temperature is not an option. This would require the customer to install expensive HVAC equipment just to use the product. Variation in fluid volume could be improved by using a fixture, but this would increase the time to fill the system and thereby reduce throughput. Speed control is the best option. Closed-loop speed control is used to reduce the variation from $+/- 300$ RPM to $+/- 100$ RPM. Three sigma control on fan speed is obtained so its standard deviation is $100/3 = 33$ RPM. We must convert this to experimental units. We use the formula given in Table 30–2. The new tolerance for fan speed is 0.11 in experimental units. Now, let's rerun the Monte Carlo simulation with the new standard deviation value for fan speed. The simulation results are as follows:

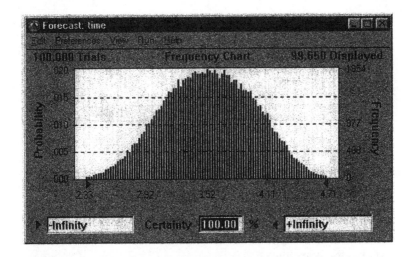

The system variation has been reduced to 0.45 seconds. We can calculate the new capability.

$$C_p = \frac{USL - LSL}{6\sigma} = \frac{6 - 0}{6(0.45)} = 2.2$$

We have achieved our goal of 6 sigma performance. However, the response is not centered between the limits. It is closer to the upper limit. We calculate C_{pk}.

$$C_{pk} = \frac{USL - \overline{X}}{3\sigma} = \frac{6 - 3.55}{3(0.45)} = 1.8$$

We now have to decide if it is worth increasing the fan speed to center the response to bring the capability up to 2. Once that decision has been made, we can transfer the design to manufacturing with confidence. If a control plan, including SPC, is placed on the components that are used to build the system, the system will perform as designed. Approximately 3.4 per million systems produced will have cool times that exceed the upper limit of 6 seconds.

Final Detailed Outputs

> **The set of factor nominals and tolerances that yield capable, on-target, minimum variance performance.**

The set of factor nominals and tolerances that yield acceptably capable performance with regard to the response target values and variance are summarized here. The tolerances are $+/-$ three standard deviations.

Fan Speed	1200 +/− 100	RPM
Volume	100 +/− 20	μl
Temperature	20 +/− 2	°C

Optimization Checklist and Scorecard

Checklist for Optimization

Actions:	Was this Action Deployed? YES or NO	Excellence Rating: 1 = Poor 10 = Excellent
The transfer function from the experimental data.	_____	_____
The response target and variance requirements.	_____	_____
Analytic, numeric, and/or graphical solution of the transfer function for the optimum operating condition.	_____	_____
Validation of the optimized solution on the physical system.	_____	_____
Monte Carlo simulation of the optimized solution using the expected factor variation.	_____	_____

(*continued*)

Calculation of system capability using the Monte Carlo
results.   _____   _____

The set of factor nominals and tolerances that yield
capable, on-target, minimum variance performance.   _____   _____

Scorecard for Optimization

The definition of a good scorecard is that it should illustrate in simple but quantitative terms what gains were attained from the work. This suggested format for a scorecard can be used by a product planning team to help document the completeness of its work.

Control Factor:	Control Factor Optimal Nominal Value:	Control Factor Optimal Tolerance:	Required Control Factor Capability:	Control Factor Unit of Measure:
1.	_____	_____	_____	_____
2.	_____	_____	_____	_____
3.	_____	_____	_____	_____

We repeat the final deliverable:

- A list of the factor nominals and tolerances that yield capable, on-target, minimum variance performance

References

Myers, R. H., & Montgomery, D. W. (1995). *Response surface methodology*. New York: Wiley Interscience.

Vanderplaats, G. N. (1984). *Numerical optimization techniques for engineering design*. New York: McGraw-Hill.

P A R T V I I

Tools and Best Practices for Verifying Capability

The tools and best practices discussed in this section focus on deliverables required at gate reviews for the fourth phase in the I^2DOV and CDOV processes.

These chapters are structured so that the reader can get a strong sense of the process steps for applying the tools and best practices. The structure also facilitates the creation of PERT charts for project management and critical path management purposes. A checklist, scorecard, and list of management questions end each chapter.

Analytical Tolerance Design

Where Am I in the Process? Where Do I Apply Analytical Tolerance Design Within the Roadmaps?

Linkage to the I²DOV Technology Development Roadmap:

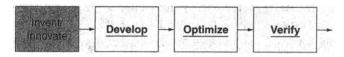

Analytical tolerance design is conducted across the Develop and Verify phases during technology development.

Linkage to the CDOV Product Development Roadmap:

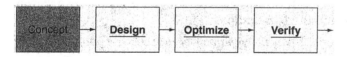

Analytical tolerance design is conducted within the Design, Optimize, and Verify phases during product development.

What Am I Doing in the Process? What Does Analytical Tolerance Design Do at this Point in the Roadmaps?

The Purpose of Analytical Tolerance Design in the I²DOV Technology Development Roadmap:

Analytical tolerance design is used in the Develop phase of the I²DOV process to establish the initial functional and assembly tolerances for the baseline technology concepts that have been selected from the Pugh concept selection process. It is used to understand the sensitivities that exist within and between components, subassembly, and subsystem interfaces.

Analytical tolerance design is used later in the Verify phase to assess integration sensitivities as the platform is integrated and evaluated for performance under nominal (random variation sources) and stressful (induced assignable cause variation sources, or Taguchi noises) noise conditions.

The Purpose of Analytical Tolerance Design in the CDOV Product Development Roadmap:

Analytical tolerance design is used in the Design phase of the CDOV process to establish the initial functional and assembly tolerances for the baseline product design concepts (subsystems, subassemblies, and components) that have been selected from the Pugh concept selection process. It is used to understand the sensitivities that exist between components, subassembly, and subsystem interfaces.

Analytical tolerance design is used later in the Optimize phase to analyze and balance system-to-subsystem interface sensitivities. It is used during the Verify Certification phase to assess remaining integration sensitivity issues as the final product design certification units are evaluated for performance under nominal (random variation sources) and stressful (induced assignable cause variation sources) noise conditions.

What Output Do I Get at the Conclusion of Analytical Tolerance Design? What Are Key Deliverables from Analytical Tolerance Design from its Application in the Roadmaps?

- *Optimized critical functional parameter tolerance set points* that control subsystem and subassembly form, fit, and function contributions to system level functional and assembly variation
- *Optimized critical adjustment parameter tolerances* used to tune the mean value of the system, subsystem, and subassembly critical functional response(s)
- *Optimized critical-to-function specification set points* that control component contributions to functional and assembly variation
- *Critical Parameter Management data* that is documented in the CPM relational database:
  System level variance models
  Subsystem level variance models
  Subassembly level variance models
  Component level variance models
  Manufacturing process level variance models
  CFP tolerances and their percentage contribution to CFR variation
  CAP tolerances and their percentage contribution to CFR variation
  CTF spec tolerances and their percentage contribution to CFR variation

The Analytical Tolerance Design Flow Diagram

This visual reference illustrates the flow of the detailed steps of the analytical tolerance design process.

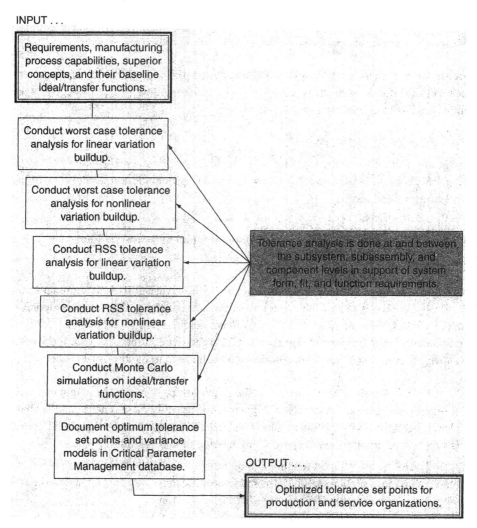

INPUT . . .

Requirements, manufacturing process capabilities, superior concepts, and their baseline ideal/transfer functions.

Conduct worst case tolerance analysis for linear variation buildup.

Conduct worst case tolerance analysis for nonlinear variation buildup.

Conduct RSS tolerance analysis for linear variation buildup.

Conduct RSS tolerance analysis for nonlinear variation buildup.

Conduct Monte Carlo simulations on ideal/transfer functions.

Document optimum tolerance set points and variance models in Critical Parameter Management database.

Tolerance analysis is done at and between the subsystem, subassembly, and component levels in support of system form, fit, and function requirements.

OUTPUT . . .

Optimized tolerance set points for production and service organizations.

Verbal Descriptions for the Application of Each Block Diagram

This section explains the specific details of what to do in analytical tolerance design:

- *Key detailed input(s)* from preceding tools and best practices
- *Specific steps and deliverables* from each step to the next within the process flow
- *Final detailed outputs* that flow into follow-on tools and best practices

Key Detailed Inputs

> **Requirements, manufacturing process capabilities, superior concepts, and their baseline ideal/transfer functions.**

The baseline form, fit, and function models, manufacturing process capabilities, and requirements they seek to fulfill define each subsystem, subassembly, and component concept. These are the major inputs to the analytical tolerancing process.

Specific Steps and Deliverables

The text in this section has been reproduced from the book *Tolerance Design* (1997) by C. M. Creveling. This book fully explains, illustrates, and provides examples on the details within each of the following subprocesses.

> **Conduct worst case tolerance analysis for linear variation buildup.**

The worst case method is an extreme representation of how variability will show up in the assembly processes. That is to say, there is a low statistical probability that every component will come in at its high or low tolerance limit in significant numbers. If the component tolerances are set at 6 standard deviations of the component CTF specification distributions, the worst case method analyzes component variation contributions at those extreme limits ($+/- 6\sigma$).

1. Analyze the geometric and spatial relationships between the components as they fit together to make an assembly. Then assess the assembly's geometric and spatial relationships relative to the product's functional capability. What is the functional purpose of the stacked-up components in specific engineering terms and units?

2. Define the specific functional requirement of the assembly. This may be done by quantifying the overall assembly form and fit specifications (where G = a Gap, line-to line contact, or an interference fit—Gnom., Gmax., and Gmin.)

3. Either estimate (from look-up tables, manufacturing process data, or recommendations) or allocate (from previously determined overall assembly specifications) an initial tolerance for each component in the assembly. Each component must be assessed on an individual basis for variance contribution.

4. Add or subtract each tolerance from the associated nominal set point so that each component dimension can then be added to form either a maximum possible assembly value (Gmax.) or a minimum assembly value (Gmin.). All the largest possible components are added to arrive at the maximum assembly size. All the smallest components are added to yield the smallest possible assembly size. Select a point on one side or the other of the unknown gap being evaluated as a starting point for the tolerance stack model.

The technique for generating a tolerance stack is called *vector chain diagramming*.

5. The space or assembly gap, contact point, or interference (G) can now be assessed because it is now clear what the extremes of the component assembly can be at the individual tolerance limits. The nominal gap, contact point, or interference fit dimension (Gnom.) can be calculated and its tolerances documented. Sometimes Gnom. and its tolerances will be specified as a requirement and the component parts have to be developed to sum up to meet Gnom. and its assigned tolerance values. In this process, the component tolerances and their nominals will have to be tuned to support the gap, contact point, or interference specifications. This is referred to as *tolerance allocation*.

Let's look at a case where linear worst case tolerance analysis can be applied. The case involves two mating parts that must fit together. The parts represent an example of pin and hole tolerance analysis.

Figure 31–1 illustrates the parts. Figure 31–2 shows two worst case scenarios for the two mating parts. Figure 31–3 illustrates one possible set of specifications for Part A and Part B.

We shall explore these tolerances to see if they work for a worst case scenario.

In Figure 31–4 the analysis begins at the right edge of the right pin. One must always select a logical starting point for a stack analysis. Typically, it will be at one side of an unknown critical gap dimension (distance from the start of *a* to the end of *g*) as we have selected in Figure 31–4.

We will step across the assembly from point *a* to point *g* to sum up all the stack up dimensions according to their sign.

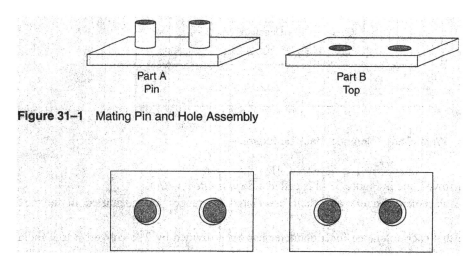

Figure 31–1 Mating Pin and Hole Assembly

Part A
Pin

Part B
Top

Figure 31–2 Two Worst Case Scenarios

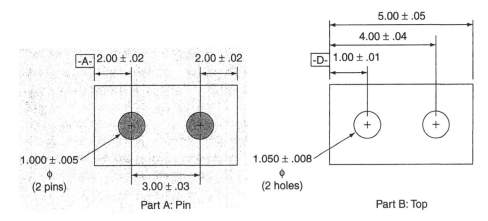

Figure 31–3   Specifications for Part A and Part B

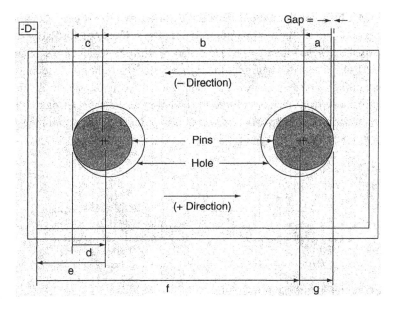

Figure 31–4   Worst Case Tolerance Stack-up Example

Each arrowed line in Figure 31–4 is called a *displacement vector.*

The displacement vectors and their associated tolerances are quantified in the next example.

Notice that each tolerance for a diameter has been divided by 2 to convert it to a radial tolerance.

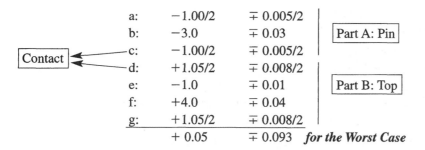

a:	$-1.00/2$	$\mp 0.005/2$	
b:	-3.0	∓ 0.03	Part A: Pin
c:	$-1.00/2$	$\mp 0.005/2$	
d:	$+1.05/2$	$\mp 0.008/2$	
e:	-1.0	∓ 0.01	Part B: Top
f:	$+4.0$	∓ 0.04	
g:	$+1.05/2$	$\mp 0.008/2$	
	$+0.05$	∓ 0.093	*for the Worst Case*

Analyzing the results we find the following conditions: $+0.05$ nominal gap and $+0.093$ tolerance buildup for the worst case in the *positive* direction.

This case tells us that the nominal gap is $+0.050$, and the additional growth in the gap will be $+0.093$, due to tolerance buildup, for a total gap of 0.143. Thus, a measurable gap will exist when all the tolerances are at their largest values.

For the worst case when all dimensions are at their smallest allowable dimensions, we see the following scenario: $+0.05$ nominal and -0.093 tolerance buildup for the worst case in the *negative* direction.

This case tells us that the nominal gap is $+0.050$, and the reduction in the gap will be -0.093, due to negative tolerance buildup, for a total of -0.043 (interference). In this worst case scenario, the parts could not mate.

Conduct worst case tolerance analysis for nonlinear variation buildup.

---

1. Construct a geometric layout of each component with its nominal set point and the initial recommended (bilateral) tolerances. Define the critical functional features, sizes, and surfaces. Some dimensions on the component will not contribute to the assembly stack-up. A common error in tolerance analysis is the inclusion of tolerance parameters that are not functionally involved in the assembly tolerance.

2. Construct an assembly drawing of the components. This will graphically illustrate the geometric relationships that exist between the components. Again, define the critical functional features, sizes, and surfaces. At this point, the engineering team should define the required assembly tolerance. This tolerance range will be the target range for the non-linear tolerance adjustment process to hit.

3. Employ plane geometry, trigonometry, algebra, and calculus as necessary to define the arithmetic (functional) relationship that exists between the component dimensions and the assembly dimension(s).

$$y = f(\chi_1, \chi_2, \chi_3, \dots \chi_n)$$

The ideal/transfer function math model (vector chain diagram converted to a math model) must directly express the critical path of the functional interfaces.

4. Use the ideal/transfer function math model to solve for the nominal assembly dimension by using the nominal component values (components with *no* unit-to-unit variation). Reaffirm the initial assembly tolerance required for proper functional product performance. This will typically involve some form of engineering functional analysis, including experimental prototype evaluations.

5. Employ differential calculus to solve for the partial derivatives of the math model terms that represent the component tolerance sensitivities with respect to the assembly tolerance.

$$Nom_y \approx \left|\frac{\partial f}{\partial \chi_1}\right|\chi_1 + \left|\frac{\partial f}{\partial \chi_2}\right|\left|\frac{\partial f}{\partial \chi_3}\right|\chi_3 + \ldots + \left|\frac{\partial f}{\partial \chi_n}\right|\chi_n$$

The partial derivatives are solved with the components all set at their nominal values (typically at the tolerance range midpoint). The tolerances can be unilateral or equal bilateral. The differentiation is typically done using manual analytical techniques or by using computer-aided numerical analysis techniques. Some problems can be difficult to differentiate. Using the definition of the derivative can simplify the process of calculating a sensitivity. Also available for help are computer-aided analytical differentiation programs like Mathematica, Macsyma, MathCAD, and Maple.

For linear tolerance stacks, the partial derivatives are all equal to 1.

6. With the values for each component sensitivity quantified, the worst case tolerance equation can be used to determine how a calculated assembly tolerance range is functionally related to the required assembly tolerance range. We do this by inserting the initial (bilateral) component tolerances and their sensitivities into the following equation. Note that in this process we use the absolute value of the sensitivities.

$$tol_y = \left|\frac{\partial f}{\partial \chi_1}\right|tol_1 + \left|\frac{\partial f}{\partial \chi_2}\right|tol_2 + \left|\frac{\partial f}{\partial \chi_3}\right|tol_3 + \ldots + \left|\frac{\partial f}{\partial \chi_n}\right|tol_n$$

The evaluation of the recommended initial component tolerances with respect to the required assembly tolerance, as constrained by the calculated sensitivities, will be an iterative process. The goal is to adjust the component tolerances to get the assembly tolerance range as close to the required range as possible. The ultimate outcome of the iterations will be a revised set of component tolerances, probably a little different from the original recommended tolerances, that provide an approximate match to the required assembly tolerance. The tolerance range for the assembly, as calculated by the adjusted component tolerances, will likely be located either a little above or below the required tolerance range. It is not unusual that the component nominal values must be shifted (called "centering" between the tolerance limits) to put the assembly tolerance range on top of the required assembly tolerance range (or as close as reasonably possible).

7. After shifting the component tolerances to get the assembly tolerance range near to the required range, the remaining offset between the required range and calculated range can be shifted by carefully shifting the nominal component set point values. The guiding principle in component nominal shifting is to do it in proportion to each component tolerance. Greenwood and Chase (1998) recommend each component be shifted by an equal fraction (f) of each respective tolerance.

The needed shift in the assembly tolerance range must be calculated.

$$\Delta y = \left[\frac{(y_{high} + y_{low})}{2} \right]_{Required} - \left[\frac{(y_{high} + y_{low})}{2} \right]_{Actual}$$

The following equation is then used to determine the amount the assembly will shift for given component nominal shifts:

$$\Delta y = \frac{\partial(f)}{\partial \chi_1} \Delta \chi_1 + \frac{\partial(f)}{\partial \chi_2} \Delta \chi_2 + \ldots \frac{\partial(f)}{\partial \chi_n} \Delta \chi_n$$

When we know how far the assembly tolerance range is off from the required range (Δy) we can then go back and solve for how much to shift each component nominal value (Δx). We establish the proper direction (sign) and amount of nominal shift by the following expression:

$$\Delta x_i = SGN(\Delta y) SGN \frac{\partial(f)}{\partial x_1} (f) tol_i$$

This expression simply states that the change needed in the nominal value of parameter (i) is going to depend on the sign of the shift change required in the assembly variable (y) and the sign of the sensitivity times the fraction of the tolerance. The equation that will be used to solve for (f) will be:

$$\Delta y = \frac{\partial(f)}{\partial x_1} (SGN)(SGN)(f)(tol_1) + \frac{\partial(f)}{\partial x_2} (SGN)(SGN)(f)(tol_2) + \ldots$$

$$(f) = \frac{\Delta y}{\frac{\partial(f)}{\partial x_1} (SGN)(SGN)(tol_1) + \frac{\partial(f)}{\partial x_2} (SGN)(SGN)(tol_2) + \ldots}$$

8. Once the common shift fraction (f) is determined, the component nominals can be recalculated to define the new midpoint (nominal) values. This is done by using the following expression:

$$\chi_{(i \, shifted)} = \chi_{(i \, nominal)} \pm (f)(tol_i)$$

9. The new shifted nominals and their tolerances can now be used to calculate the new assembly nominal value and the resulting tolerance range about the assembly nominal. Referring back to the nominal and tolerance equations:

$$Nom_y \approx \left|\frac{\partial f}{\partial \chi_1}\right|\chi_1 + \left|\frac{\partial f}{\partial \chi_2}\right|\chi_2 + \left|\frac{\partial f}{\partial \chi_3}\right|\chi_3 + \dots + \left|\frac{\partial f}{\partial \chi_n}\right|\chi_n$$

$$tol_y = \left|\frac{\partial f}{\partial \chi_1}\right|tol_1 + \left|\frac{\partial f}{\partial x_2}\right|tol_2 + \left|\frac{\partial f}{\partial \chi_3}\right|tol_3 + \dots + \left|\frac{\partial f}{\partial \chi_n}\right|tol_n$$

This new assembly tolerance range can then be compared to the required assembly tolerance range for suitability. The match should be reasonably close at this point. Another iteration can be performed if necessary to get the actual assembly tolerance range even closer to the required assembly tolerance range. This iteration can easily be done by repeating steps 7-9 after recalculating new sensitivity values based on the new shifted component nominal values (repeat of step 5).

See Creveling (1997) for a complete example of this fairly involved process.

Conduct RSS tolerance analysis for linear variation buildup.

The *root sum of squares* method is a more realistic representation of how variability will show up in your assembly processes. That is to say, there is a low statistical probability that every component will come in at its high or low tolerance limit in significant numbers. If the component tolerances are set at 6 standard deviations of the component CTF specification distributions, the worst case method analyzes component variation contributions at those extreme limits ($+/- 6\sigma$). The RSS method analyzes component variation contributions at less extreme, more likely limits of ($+/- 1\sigma$). RSS does not stand for *"realistic statistical study,"* but it could!

1. Analyze the geometric and spatial relationship between the components as they fit together to make an assembly. Then assess the assembly's geometric and spatial relationship relative to the product's functional capability. What is the functional purpose of the stacked-up components in specific engineering terms and units?

2. Define the specific functional requirement of the assembly. This may be done by quantifying the overall assembly form & fit specifications (. . . **where G = a Gap, line-to line contact, or an interference fit.**—Gnom., Gmax., and Gmin.).

3. Either estimate (from look up tables, manufacturing process data, or recommendations) or allocate (from previously determined overall assembly specifications) an initial tolerance for each component in the assembly. Each component must be assessed on an individual basis for variance contribution.

4. Each tolerance is initially assumed to be set at $+$ or $-$ 3 standard deviations from the nominal set point. Thus, the standard deviation of the production process output is assumed to be approximately Tol./3. It is highly recommended that one obtain manufacturing data to support this assumption and make adjustments accordingly. If the tolerances are set at some level greater than or less than 3 standard deviations of production process output distribution for the component CTF specifications, this must be used instead of just assuming 3 standard deviations equal the tolerance limits.

5. The assembly tolerance can be assessed because it is now assumed that the component part tolerances are based upon 3 sigma levels from reasonably normal distributions. This is done by employing the RSS equation to the vector chain diagram model:

$$T_{Assy} = \sqrt{T_1^2 + T_2^2 + T_3^2 + \ldots + T_n^2}$$

The nominal gap, contact, or interference dimension (Gnom.) can be calculated and its tolerances assigned. Sometimes Gnom. and its tolerances will be specified as a requirement (usually due to some functional requirement at the subsystem or subassembly level) and the component parts have to be developed to sum up to the meet Gnom. and its tolerance values. In the case when an allocation process must be used, the component tolerances and nominals will have to be tuned to support the Gap specifications. This is referred to as *tolerance allocation*. The same disciplined summation process as we defined in the worst case method, including using the sign convention of displacements to the right being positive, should be used.

6. From Six Sigma literature we have the option to use the relationship:

$$\sigma_{adjusted} = \frac{Tolerance}{3C_p}$$

to state that an adjustable format for assigning tolerances is available for our use based upon the short-term process capability (Cp). This is done when we have process data to prove that the 3 sigma paradigm is *not* in effect (Cp does not $=$ 1). The known Cp can be used to adjust each component tolerance so that the effect is aggregated up to the assembly tolerance. Then we can calculate an estimate for the gap standard deviation:

$$\sigma_{Gap} = \sqrt{\left(T_e/3Cp\right)^2 + \sum_{i=1}^{m}\left(T_{pi}/3Cp_i\right)^2}$$

From this expression we can then calculate the probability of being above or below a specified gap using the Z transform; see Chapters 3 and 5 in Creveling (1997):

$$Z_{G_{\min}} = \frac{G_{\min} - G_{nominal}}{\sqrt{\left(T_e/3C_p\right)^2 + \sum_{i=1}^{m}\left(T_{pi}/3C_{pi}\right)^2}}$$

and

$$Z_{G_{\max}} = \frac{G_{\max} - G_{nominal}}{\sqrt{\left(T_e/3C_p\right)^2 + \sum_{i=1}^{m}\left(T_{pi}/3C_{pi}\right)^2}}$$

In general form:

$$Z_{Gap} = \frac{G_{\max\,or\,\min} - G_{nominal}}{\sigma_{Gap}}$$

Let's look at the case where linear RSS case tolerance analysis is applied.[1]

The case involves two mating parts that must fit together. The parts represent a pin and hole tolerance analysis. Figure 31–5 illustrates the parts. Figure 31–6 shows two scenarios for the two mating parts.

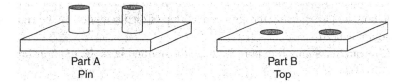

Figure 31–5 Mating Pin and Hole Assembly

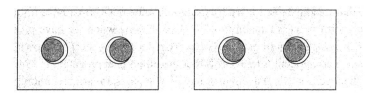

Figure 31–6 Two RSS Case Scenarios

[1] This case was developed at Eastman Kodak Company by Margaret Hackert-Kroot and Juliann Nelson.

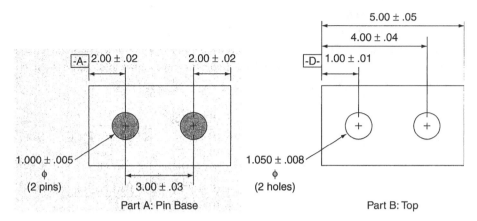

Part A: Pin Base   Part B: Top

Figure 31–7   Specifications for Part A and Part B

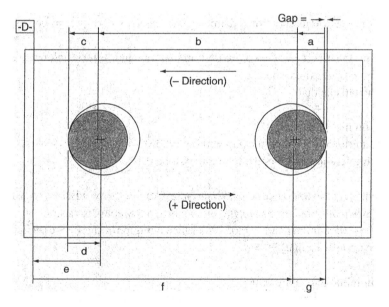

Figure 31–8   RSS Case Tolerance Stack-up Example

Figure 31–7 illustrates one possible set of specifications for Part A and Part B. We shall explore these tolerances to see if they work for an RSS scenario. In Figure 31–8 the analysis begins at the right edge of the right pin. One must always select a logical starting point for a stack analysis. Typically, it will be at one side of an unknown critical gap dimension—in this case, the distance from the start of *a* to the end of *g*.

We will step across the assembly from point *a* to point *g* to sum up all the stack-up dimensions according to their sign.

Each arrowed line in Figure 31–8 is called a *displacement vector.* The sign for each displacement vector is illustrated in the figure. The displacement vectors and their associated tolerances are quantified as follows:

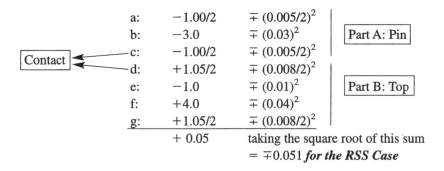

$$
\begin{array}{lll}
\text{a:} & -1.00/2 & \mp (0.005/2)^2 \\
\text{b:} & -3.0 & \mp (0.03)^2 \quad \boxed{\text{Part A: Pin}} \\
\text{c:} & -1.00/2 & \mp (0.005/2)^2 \\
\text{d:} & +1.05/2 & \mp (0.008/2)^2 \\
\text{e:} & -1.0 & \mp (0.01)^2 \quad \boxed{\text{Part B: Top}} \\
\text{f:} & +4.0 & \mp (0.04)^2 \\
\text{g:} & +1.05/2 & \mp (0.008/2)^2 \\
\hline
& +0.05 & \text{taking the square root of this sum} \\
& & = \mp 0.051 \textit{ for the RSS Case}
\end{array}
$$

Notice that each tolerance for a diameter has been divided by 2 to convert it to a radial tolerance.

Analyzing the results we find a +0.05 nominal gap and +0.051 tolerance buildup for the RSS case in the *positive* direction.

This case tells us that

- The nominal gap is +0.050.
- The additional growth in the gap will be +0.051.
- Due to tolerance buildup, the total gap will be 0.101.

Thus, when all the tolerances are at their largest values, there will be a measurable gap. For the RSS case when all dimensions are at their smallest allowable dimensions, we see the following scenario: + 0.05 nominal and − 0.051 tolerance buildup for the RSS case in the *negative* direction. This case tells us that

- The nominal gap is +0.050.
- The reduction in the gap will be −0.051, due to *negative* tolerance buildup.
- Total gap will be −0. 001 (interference).

In this scenario the parts could not mate.

If the worst case conditions are set up so that the pins are biased toward contacting the inner surfaces of the holes (the second scenario in Figure 31–6), the gap results are the same as the scenario we just worked out, because of the symmetry of the tolerancing scheme in Figure 31–7 (this need not be true of other pin–hole situations).

> **Conduct RSS tolerance analysis for nonlinear variation buildup.**

In some cases the geometry of tolerance buildup is nonlinear. This is often due to trigonometric functions existing within the vector chain diagram. With a few changes, the nonlinear approach to tolerance analysis can be used in the root sum of squares format. We just have to modify step 6 from the worst case method from nonlinear stack-up modeling. With this modification in place, the rest of the tolerancing process can be followed as described in the worst case nonlinear process flow diagrams. See Creveling (1997) for an example of this process.

7. Worst case process modified for the RSS case: With the values for each component sensitivity quantified, the RSS tolerance equation can be used to determine how a calculated assembly tolerance range functionally and statistically relates to the required assembly tolerance range. We do this by inserting the square of the initial (bilateral) component tolerances and the square of their sensitivities into the following equation. We then take the square root of the sum of squares (the *root sum of squares*) to define the assembly or dependent tolerance.

$$tol_y = \sqrt{\left(\frac{\partial f}{\partial x_1}\right)^2 tol_1^2 + \left(\frac{\partial f}{\partial x_2}\right)^2 tol_2^2 + \left(\frac{\partial f}{\partial x_3}\right)^2 tol_3^2 + \ldots + \left(\frac{\partial f}{\partial x_n}\right)^2 tol_n^2}$$

The evaluation of the recommended initial component tolerances with respect to the required assembly tolerance, as constrained by the calculated sensitivities, will be an iterative process. In this step the goal is to adjust the component tolerances to get the assembly tolerance range as close to the required range as possible. The ultimate outcome of the iterations will be a revised set of component tolerances, probably a little different from the original recommended tolerances, that provide an approximate match to the required assembly tolerance. The tolerance range for the assembly, as calculated by the adjusted component tolerances, will likely be located either a little above or below the required tolerance range. It is not unusual that the component nominal values must be shifted to put the assembly tolerance range on top of the required assembly tolerance range (or as close as reasonably possible). Remember the additional option of including Cp or Cpk values in defining the tolerance values in this process:

$$Tolerance_i = 3C_{pk} \times \sigma_{adjusted}$$

> **Conduct Monte Carlo simulations on ideal/transfer functions.**

Monte Carlo simulations on personal computers are fast, cost-effective ways to explore tolerance sensitivity relationships between the xs and their percent contribution to the Ys in the flow-up of variational effects across and within the boundaries of the levels of the system architecture.

Monte Carlo simulations follow a fairly simple process:

1. Generate an ideal/transfer function for the function and assembly requirement under evaluation (in assembly cases these are created using the vector chain diagram of the tolerance stack-up).

$$Y = f(x)$$

This is the model the Monte Carlo simulation exercises as it runs through the assembly or function generation process trials or iteration calculations.

2. Gather data to construct a histogram of the distribution of the individual x input variables. This provides the team with the Δx values as represented by the mean and standard deviation of the distribution for the x variables. These are either subsystem or subassembly CFPs or component CTF specifications in CPM terms. The normality of the distributions should also be quantified. The Monte Carlo simulation randomly samples from these x variable distributions and enter their values into the ideal/transfer function model:

$$\Delta Y = f(\Delta x)$$

The model is loaded with many hundreds or thousands of trial or sample values of the x variables as randomly generated and placed in the ideal/transfer function model by the simulation software.

3. Build, load, and run the simulation in the computer. Crystal Ball® Monte Carlo simulation software is a common tool that can be effectively used to conduct the simulation process. A complete tutorial description and example of how Monte Carlo simulation is done is documented in Chapter 8 of Creveling's (1997) *Tolerance Design.*

4. Analyze the results from the Monte Carlo simulation. The typical output from a Monte Carlo simulation is a forecasted sample distribution of the Y variable. The output distribution has a sample mean and sample standard deviation that is representative of the hundreds or thousands of trials that were run in the simulation. Capability (Cp) calculations for the Y variable can be made to see if the ideal/transfer function is capable with the induced noise from the variation of the input x variables.

5. Refine the ideal/transfer function nominal (mean) x input values or standard deviation input values. Rerun the simulation to compare the Y outputs against the requirements.

How Crystal Ball® uses Monte Carlo Simulation

Crystal Ball® implements Monte Carlo simulation in a repetitive three step process:

1. For every assumption cell, Crystal Ball® generates a number according to the *probability distribution* you defined and placed into the spreadsheet.
2. Crystal Ball® commands the spreadsheet to calculate a new response value from the math model you have entered into the Excel *forecast* cell.
3. Crystal Ball® then retrieves each newly calculated value from the forecast cell and adds it to the graph in the forecast windows.

The iterative nature of the Monte Carlo process can be simply illustrated by the flow chart in Figure 31–9.

It would not be unusual to specify that the Monte Carlo routine should be run through 1,000 or more iterations. This provides an ample set of data

- Used to help estimate a useful portion of the real-world effects that can occur from random unit-to-unit variation
- Based on an identifiable set of probability distributions that are obtained from production samples

A probability distribution can be determined from two sources:

- Sample data that is truly representative of the nature of variability present in the output components
- The output of a mathematical function (equation) that is just an estimate that expresses the output as a probability of occurrence value between 0 and 1 or a percentage between 0 percent and 100 percent.

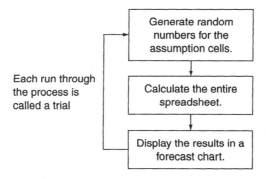

Figure 31–9   Monte Carlo Process

The probability is based on independent parameters we typically refer to as *sample* or *population statistics*.

Typical input parameters for probability distributions are based on whether the distributions are discrete (binary events) or continuous (analog events). For discrete distributions we often see input parameters stated as:

- Probability of an event (P)
- Total number of trials (n) available for the event to occur or not occur

There are other parameters for discrete distributions. Consult the Crystal Ball® manual or a text on probability and statistics to obtain additional information concerning the discrete distribution statistics.

The continuous distributions typically use the well-known parameters μ (the population mean) and σ (the population standard deviation):

- Sample statistics for the mean and standard deviation are used to estimate these population parameters.
- You will almost always be working with sample values for the mean and standard deviation.

Again, you need to consult the Crystal Ball® manual or a text on probability and statistics to obtain additional information concerning other types of continuous distribution statistics.

In conclusion, the shape of each distribution provides a graphic illustration of how likely a particular value or range of values is to occur. Probability distributions show us the pattern that nature has assumed for certain phenomena as they disperse themselves when allowed to occur in a random or unbiased condition.

The following types of probability distributions are available in Crystal Ball®:

Uniform	Exponential
Normal	Weibull (Rayleigh)
Triangular	Beta
Binomial	Gamma
Poisson	Logistic
Hypergeometric	Pareto
Lognormal	Extreme Value

Crystal Ball® also has the ability to permit custom (user-specified) distributions. The software package has a unique portfolio of user-adjustable distributions, which allows one to provoke a diverse set of variations in the model trial runs through the program's random number generator. Each one of the probability distributions has a specific geometric nature and a set of defining parameters that give it its distinctive form to represent the range of probabilities for an event or response value.

Returning to the two types of probability distributions:

- The *discrete* form accounts for integer values such as any one value appearing on a die being thrown.
- The *continuous* probability distribution can be described by any rational value that might occur, such as in reading an analog thermometer which outputs a continuum of measured responses.

Discrete distributions do not have decimals associated with a response value. The events they measure are individual occurrences such as five flights leaving an airport per hour or three deaths per year. Continuous distributions use decimals since they work with continuous phenomena that can have a fine range of values that can be infinitely spread between one value to another. For example, we might find that a car's velocity may be 72.454 miles per hour at one point and 66.375 miles per hour at another point of measurement.

We will briefly discuss and display the main probability distributions that are used in Crystal Ball's® Monte Carlo simulator. We want you to be familiar with the general shape of the distributions and the reason why these probability distributions are used to represent actual phenomena.

Figure 31–10 is the first of two galleries displaying 12 of the 16 probability distributions resident in Crystal Ball.®

Although Crystal Ball® has 16 distributions available for use, we will focus on those displayed in the window shown in Figure 31–10.

Certain distributions are more commonly used in Monte Carlo analysis. The first three distributions we will discuss are referred to in the Crystal Ball® manual as the common probability

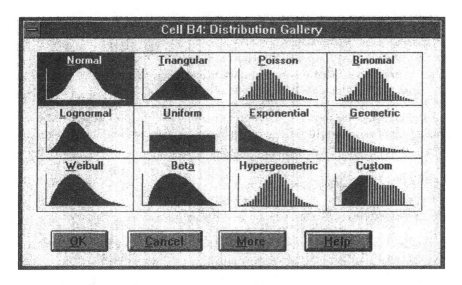

Figure 31–10 Crystal Ball's® Main Distribution Gallery

distributions, due to the fact that it is not always easy to assign a distribution to a particular situation without first knowing something about the nature of variability resident in the production process of the component being toleranced. Thus, the common distributions are the default selections to employ until one has a better alternative.

Characterizing Probability Distributions for Tolerance Analysis Applications

The usual method of identifying the distribution type that matches or approximates the behavior of an input variable, typically a component dimension, being used in a Monte Carlo simulation is to gather a large, statistically valid sample of data (30–100 samples) from a similar component that is currently being produced. This data is used to construct a histogram. The histogram can then be used in comparison with one or more of the distributions displayed in the distribution Gallery. The histogram, along with engineering judgment, will generally suffice to make a reasonable selection for the tolerance analysis input assumptions. Access to a professional statistician with manufacturing experience can be of significant help in the process of identifying the appropriate probability distribution to use in many different situations.

The Common Probability Distributions Available in Crystal Ball®

The Normal Distribution: The normal or Gaussian distribution is a common method of expressing the variation and central tendency of the output from many types of manufacturing processes. The normal distribution has three distinctive characteristics about how it expresses uncertainty:

1. Some value of the uncertain (input) variable is the most likely (the mean of the distribution).
2. The uncertain (input) variable could as likely be above the mean as it could be below the mean (symmetrical about the mean).
3. The uncertain variable is more likely to be in the vicinity of the mean than far away (accounts for the bell shape of the distribution).

The Uniform Distribution: The uniform distribution is helpful in cases where it is just as likely that the input variable will assume one value in the range of possible values as any other. The uniform distribution has three distinctive characteristics about how it expresses uncertainty:

1. The minimum value (in the range) is fixed.
2. The maximum value (in the range) is fixed.
3. All values between the minimum and the maximum occur with equal likelihood.

The Triangular Distribution: The triangular distribution accounts for the case where there is clear knowledge of a range between which likely values will occur with a single, *most likely* value located somewhere within the range of possible values. The triangular distribution has three distinctive characteristics about how it expresses uncertainty:

1. The minimum value (in the range) is fixed.
2. The maximum value (in the range) is fixed.
3. The most likely value falls between the minimum and maximum values, forming a triangular-shaped distribution, which shows that values near the minimum or the maximum are less likely to occur than those near the most likely value.

Step-by-Step Process Diagrams for a Crystal Ball® Monte Carlo Analysis

Step 1: Open/create a spreadsheet in Excel under the Crystal Ball® function window.

Step 2: Set up the problem by defining the input assumptions.

    2.1. Define your assumptions:
- An assumption is an estimated value that will come from a specified probability distribution that you select as an input to the spreadsheet math model.
- Crystal Ball® has 16 different distributions from which you can select.

    2.2. Select and enter the distribution for your assumption. The selected probability distribution will represent the nature of the production process output distribution for the dimension or characteristic to which you are applying the tolerance. You can select the distribution by using the Cell Menu and the Define Assumption Command.

    2.3. Repeat Step 2.2 for each component or element in your tolerance design case.

Step 3: Set up the problem by defining the *output forecasts*.

    3.1. Establish the functional relationships (math models) that exist between the various input assumption terms to produce the output response. A forecast is the value taken on by the dependent variable, y, as each independent variable, x_i (input assumption value), changes as the Monte Carlo simulation runs through each trial.

The forecast formula is represented in general form as:

$$y = f_1(x_1, x_2, x_3 \ldots x_n)$$

More than one forecast can be generated by defining several forecast formulas within the same spreadsheet:

$$z = f_2(x_1, x_2, x_3 \ldots x_n)$$

    3.2. Select Define Forecast from the Cell Menu. Specify the engineering units being calculated in the forecast formula.

    3.3. Repeat Step 3.2 for each forecast formula in your tolerance analysis.

Step 4: Run a simulation. Running a Monte Carlo simulation means that you will exercise the spreadsheet math model(s) at least 200 times by using the software's random number generator in conjunction with the distributions that represent your input assumptions. At least 200 trials will be calculated to represent the uncertainty or likelihood of what can happen in your tolerance case.

  4.1. Choose Run Preferences from the Crystal Ball® Run Menu.

  4.2. Type in the number of trials you would like to run (200 minimum). In our cases you will not need to provide a seed number.

Step 5: Interpret the results. Determine the level of certainty in the forecasted results desired. Crystal Ball® reports the forecasted outputs from the math models in a graphic display that contains both probability and frequency distributions of the results. The *probability axis* (left side vertical axis) of the chart states the likelihood of each output value or group of values in the range of 0 probability to 100 percent probability (0.000 −1.000). The *frequency axis* (right side vertical axis) of the chart states the number of times a particular value or group of values occurs relative to the total number of runs in the simulation.

This process involves how to change the Display Range of the forecast, which enables you to look at specific portions of the output response distribution and to identify the probability or frequency of certain responses of interest occurring.

In tolerance analysis these responses happen to be the values that are occurring within the tolerance range. Crystal Ball® initially shows all the distribution of trials; then it is up to you to modify the range of values Crystal Ball® includes in the forecast certainty window.

Crystal Ball® allows the user to alter the minimum and maximum values (tolerance range) of the range of output to be selected and evaluated. The Certainty Box responds by displaying the percent certainty of occurrence of the values within the specified range. In this way we can quantify the percentage of values falling within the required range of tolerance for an assembly or system. This concludes the basic capability of Crystal Ball® to analyze a tolerance problem.

Crystal Ball® has additional capabilities to take the user deeper into the analysis of tolerances. The user can receive the following statistical information concerning the output response data and distribution:

mean

median

mode

standard deviation

variance

skew

kurtosis

coefficient of variability

range minimum

range maximum

range width

standard error of the mean

All of these quantitative statistical measures are defined in Chapter 3. Crystal Ball® can also construct a table of the distributed results by percentile increments of 10 percent (or by other increments one chooses to set). Crystal Ball® allows the user to compare multiple tolerance analyses within a large system as one alters the assumptions or tolerance ranges. Crystal Ball® uses trend charts to facilitate this graphical analysis process.

Sensitivity Analysis Using Crystal Ball®

Crystal Ball® allows the user to quantify the effect of each of the input assumptions on the overall output response. This means the software can calculate the sensitivity coefficients, including the proper sign associated with the directionality of their effect. In Crystal Ball® terms, this means we can assess the sensitivity of the forecast (output response) relative to each assumption (input variable). Some practitioners of tolerance sensitivity analysis may be uncomfortable with the process of the derivation and solution of partial derivatives.

Crystal Ball® is a friendly addition to the user's tolerance analysis tool kit because it does all the sensitivity analysis within the software algorithms. Crystal Ball® *ranks* each assumption cell (tolerance distribution input) according to its importance to each forecast (output response distribution) cell.

The Sensitivity Analysis Chart displays these rankings as a bar chart, indicating which assumptions are the most important and which are the least important in affecting the output response. From the Crystal Ball® manual, we find that the Sensitivity Analysis Chart feature provides three key benefits:

1. You can find out which assumptions are influencing your forecasts the most, reducing the amount of time needed to refine estimates.
2. You can find out which assumptions are influencing your forecast the least, so that they may be ignored (or at least strategically identified for alternative action) or discarded altogether.
3. As a result, you can construct more realistic spreadsheet models and greatly increase the accuracy of your results because you will know how all of your assumptions affect your model.

Crystal Ball® uses a technique that calculates *Spearman rank correlation coefficients* between every assumption and every forecast cell while the simulation is running. In this way the sensitivities are capable of being quantified within Crystal Ball® for any case being considered.

Again, quoting from the Crystal Ball® manual regarding how sensitivities are calculated:

Correlation coefficients provide a meaningful measure of the degree to which assumptions and forecasts **change together.** If an assumption and a forecast have a high correlation coefficient,. . . it means
that the assumption has a significant impact on the forecast (both through its uncertainty and its model
sensitivity (from the **physical *and* statistical** parameters embodied in the system being analyzed).

The physical parameters are embedded in the math model placed in the forecast cell. The
statistical parameters are embedded in the distributions selected as assumptions.

To quantify sensitivity magnitude and directionality:

($+$) Positive coefficients indicate that an increase in the assumption is associated with an
increase in the forecast.

($-$) Negative coefficients imply the reverse situation.

The larger the absolute value of the correlation coefficient, the stronger the relationship.

Referring back to the worst case tolerancing material, in which we discussed defining signs
for the directionality of the sensitivity coefficients in nonlinear evaluations, we found that this
could be a little confusing. With Crystal Ball® the sign is automatically calculated along with the
sensitivity value. Again, the concern for the tolerance analyst is

properly setting up the mathematical relationship (forecast) between the assumptions so that Crystal
Ball correctly calculates the *magnitude and direction* of the sensitivity coefficient.

Crystal Ball® can be of help if one is assessing tolerance issues between components, subassemblies, and major assemblies as one integrates an entire product or system:

Crystal Ball® also computes the correlation coefficients for all pairs of forecasts while the simulation
is running. You may find this sensitivity information useful if your model contains several interme-
diate forecasts (assemblies or subsystems) that feed into a final forecast (system).

Crystal Ball® has an option that helps clarify the output for sensitivity analysis.

. . . The Sensitivity Preference dialog box lets you display the sensitivities as a percentage of the contribution to the variance of the target forecast.
    This option, called *Contribution to Variance,*

        • doesn't change the order of the items listed in the Sensitivity Chart
        • makes it easier to answer questions such as,. . .
        . . . . . . "what percentage of the variance or uncertainty in the target forecast is due to
        assumption x?"

It is important to note that this method is only an approximation!
It is *not precisely* a variance decomposition (such as a formal ANOVA would provide).

Crystal Ball® calculates Contribution to Variance by squaring the rank correlation coefficients and normalizing them to 100%.

It is important to recognize some scenarios that can occur in the application of Crystal Ball's® sensitivity analysis option.

> The sensitivity calculation may be inaccurate for certain correlated assumptions. For example,. . . if an important assumption were highly correlated with an unimportant one, the unimportant assumption would likely have a high sensitivity with respect to the target forecast (output). Assumptions that are correlated are flagged as such on the Sensitivity Chart. In some circumstances, turning off correlations in the Run Preference dialog box may help you to gain more accurate sensitivity information.

For tolerance applications this warning will be valid when a design assembly is made up from components that are antisynergistically interactive. This means that one component has a dependency on another component for its effect on the response. "In practice, this problematic type of interaction should be identified and corrected long before tolerancing is undertaken," the manual notes.

Some designs are inherently interactive. In these cases one must be vigilant as to how these interdependencies affect the response variable. Sometimes the interactions are necessary and helpful. A word to the wise is sufficient: Know where the design interactions exist and at what strength they act to influence the output response. Fortunately, Crystal Ball® has a mechanism to decouple the assumptions from their correlations so you can assess each one in relative independence. In general, it is considered good design practice to create and optimize components, subsystems, and subassemblies, and systems that are reasonably free from elements that have strong antisynergistic interactions.

When design elements act with interdependence that is synergistic and monotonic with respect to one another, then performance generally becomes more stable and predictable (Figure 31–11), especially when strong noises are present. (See Chapter 28 on robust design for help in developing products and processes that do not possess strong interactions.)

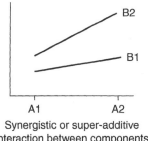

Synergistic or super-additive
interaction between components

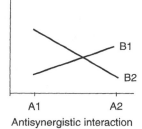

Antisynergistic interaction
between components A and B

Figure 31–11   Assumptions with Interactions

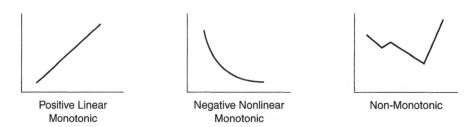

| Positive Linear | Negative Nonlinear | Non-Monotonic |
| Monotonic | Monotonic | |

Figure 31–12   Monotonicity

With super-additivity or synergistic interactivity, the correlation is still present, but one can see how the elements work together to drive the response in a useful direction. Antisynergistic interactivity between two component contributions to the response is typically counteractive. The directionality of the response is reversed as one factor changes. This situation often forces the design team to apply tight tolerances as a countermeasure to this problematic form of sensitivity. Additionally, many designers often end up tightening tolerances on unimportant parameters because they are correlated to important ones. This leads to unnecessary expense during manufacturing.

The sensitivity calculation may be inaccurate for assumptions whose relationships with the target forecast are not monotonic. A *monotonic relationship* (Figure 31–12) means that . . . an increase in the assumption (input) tends to be accompanied by a strict increase in the forecast (output), or an increase in the assumption tends to be accompanied by a strict decrease in the forecast.

If you are worried about being dependent on rusty calculus skills and knowledge of parameter interactivity in tolerance analysis, you can relax and focus your mental energy on the issues of:

- Properly setting up the assumption cells with the appropriate probability distributions
- Defining the right mathematical relationship between each assumption cell (component tolerance) and the forecast cell (assembly or system tolerance)

This is where the tolerance analysis is either going to be set up right or wrong. You need to look no further in basic skill requirements and development for competent tolerance analysis than in these two critical areas. For advanced tolerance analysis, all the math, computer, and experimental skills discussed in the text will need to be in the practitioner's full command.

An Example of How to Use Crystal Ball®   Let's run a Monte Carlo simulation on a linear example:

The simple linear stack problem is shown at the top of page 657.

We begin by opening Crystal Ball®. This is done by double clicking on its icon in the applications window. Crystal Ball® will open in an Excel spreadsheet (Figure 31–13).

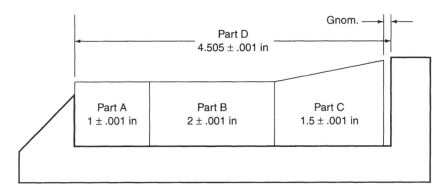

We save this file by opening the File Menu and select the Save As function. We name this file *Tolex1.xls*.

The next step is to setup the Crystal Ball® spreadsheet to help us in working with a tolerance problem.

1. In cell A1 type "Tolerance Study for a Linear Case".
2. In cell A3 type "Part A:"
3. In cell A4 type "Part B:"
4. In cell A5 type "Part C:"
5. In cell A6 type "Part D:"

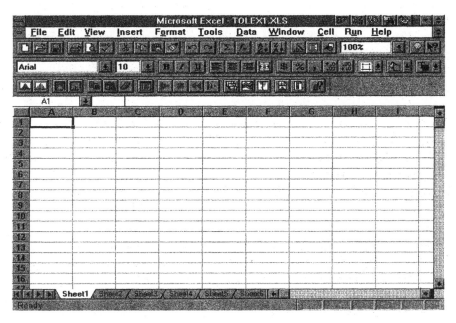

Figure 31–13 Opening Crystal Ball®

These are the names of the four components. Next we enter six specific pieces of data that Crystal Ball® will need as we put it to work doing the tolerance analysis. The six pieces of data that we enter are as follows:

1. Nominal dimension for each part (Nom.)
2. Tolerance on each nominal dimension (Tol.)
3. The type of distribution that describes the uncertainty in the nominal dimension (Distr.)
4. The tolerance divided by 3 standard deviations (Tol./3)
5. The nominal dimension plus the tolerance (Nom. + Tol.)
6. The nominal dimension minus the tolerance (Nom. − Tol.)

To enter and align these six pieces of data with the four components, follow these simple steps:

1. In cell B2, type "Nom."
2. In cell C2, type "Tol."
3. In cell D2, type "Distr."
4. In cell E2, type "Tol./3"
5. In cell F2, type "Nom. + Tol."
6. In cell G2, type "Nom. − Tol."

Cells B3 through B6 hold the numbers we call *Assumption Values*. We can now enter the nominal values 1.0000, 2.0000, 1.5000, and 4.5050 in cells B3 through B6. The appropriate tolerances can then be entered into the cells C3 through C6. It is assumed that the user understands the basics of spreadsheet use and can adjust the cell widths to accommodate the cell entries so that the Crystal Ball® spreadsheet has an orderly and useful structure. It is also assumed one has a basic knowledge of how to program a spreadsheet to do calculations. Even so, we will illustrate how this is done.

The next step is to enter the simple linear math model that sums the individual component contributions to form a residual assembly gap dimension. We call this output response the *forecast*. It is the uncertainty associated with this gap dimension that we would like to study. The gap cannot be allowed to fall below 0.001 in.

To enter the math model we perform the following process and enter the model information in the appropriate cell:

1. Select cell B8 and type "=B6-B3-B4-B5" (this is the assembly gap math model).
2. Click on the *check mark* in the box to the left of the bar that you used to enter the equation (note it is made up from the Assumption Value cells).

Cell B8 now holds the tolerance equation that employs values representing probabilistic values that will be determined by the random number generator in conjunction with the specific distribution you assign to Parts A, B, C, and D. This value, the assembly gap dimension in units of inches, is called the *forecast value*.

Our final tasks are to enter the appropriate distribution for each component and program the rest of the spreadsheet cells to hold their proper respective values based on the nominals and tolerances assigned to each component part.

Parts A and C are known to be represented by uniform distributions. Parts B and D are represented by normal distributions.

This is entered into cells D3 through D6. Later, we will specifically ask Crystal Ball® to assign these distributions to actively control values that are generated in the assumption cells. Cells in columns E, F, and G are calculated from cells located in columns C and D. Column E holds the calculated standard deviations based on the three sigma tolerance assumption. Each cell for column E is calculated by dividing the respective tolerance from column C by 3. Thus, column E will have the following entries:

E3 will hold "=C3/3"
E4 will hold "=C4/3"
E5 will hold "=C5/3"
E6 will hold "=C6/3"

Cells F3 through F6 and G3 through G6 hold the tolerance limits as they are either added to or subtracted from the nominal dimensions for each component:

F3 will hold "=B3+C3"   G3 will hold "=B3−C3"
F4 will hold "=B4+C4"   G4 will hold "=B4−C4"
F5 will hold "=B5+C5"   G5 will hold "=B5−C5"
F6 will hold "=B6+C6"   G6 will hold "=B6−C6"

The values in these cells will be used to modify the output from various forecasts. The modifications will help in the analysis of the output distribution from the simulation runs.

This concludes the basic spreadsheet construction process for preparing to perform a Monte Carlo tolerance analysis. It matters little whether the tolerance analysis is performed on linear or nonlinear cases as far as the spreadsheet setup is concerned.

The key issues involve the model the user derives and enters into the forecast cell and the distributions assigned to the assumption cells. If the functional relationship is correctly expressing the nature of the component stack-up or interface geometry and the appropriate distributions are properly assigned as assumptions, the rest of the analytical procedure is handled by asking Crystal Ball® for the appropriate output charts and graphs. The spreadsheet shown in Figure 31–14 is representative of how most tolerance cases will be set up.

We are now ready to assign the specific distributions to each component assumption cell. To do this, we select the assumption cell by clicking on it and then Selecting the Cell Menu. Under the Cell Menu, we see the command Define Assumption. Select this command by clicking on it.

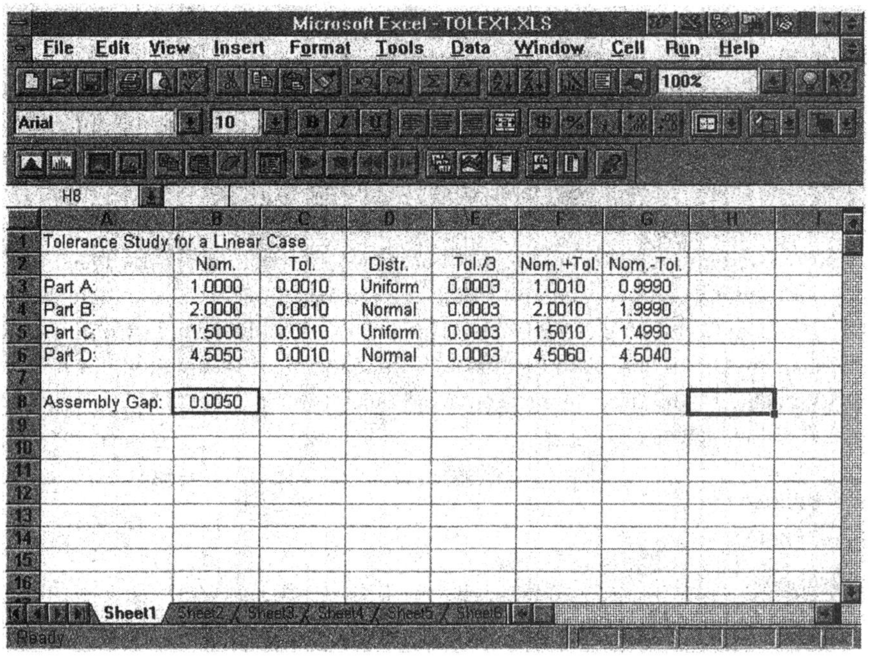

Figure 31–14   Example of Tolerance Case Setup

A window appears with the Distribution Gallery. Click on the distribution you want associated with the component represented in the cell. This window is displayed in Figure 31–15 with the uniform distribution selected (darkened).

Click OK and the Distribution Gallery disappears. We have selected the uniform distribution.

A new window appears as shown in Figure 31–16. This is the window for entering specific parameter information for the uniform distribution. Enter the values for the upper and lower tolerances into the boxes labeled Min and Max. Click Enter and then click OK.

We have successfully entered the uniform distribution for the assumption cell for Part A. Now we return to the Crystal Ball® spreadsheet window. We can proceed with defining the assumption cell for Part B. Simply repeat the process of opening the Cell Menu and selecting Define Assumption.

The window in Figure 31–17 is open. Click on the normal distribution since it represents how Part B will express its uncertainty in the assembly.

Click OK to tell Crystal Ball® to select the normal distribution. The normal distribution window appears as in Figure 31–18.

In this window we need to enter the Mean = 2.0000 and Std. Dev = 0.0003 values for Part B. Click Enter and then click OK. This returns us to the main Crystal Ball® spreadsheet where we can repeat this process for Parts C and D, respectively.

We are now ready to use the Define Forecast command located in the Cell Menu just below the Define Assumption command. First, click on cell B8 to activate the Forecast cell we are

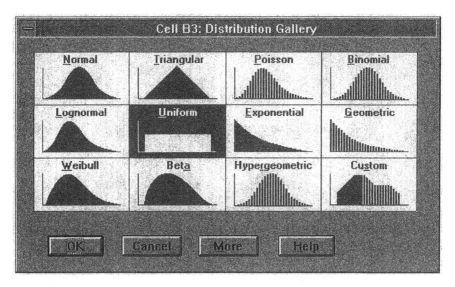

Figure 31–15   Crystal Ball® Distribution Gallery

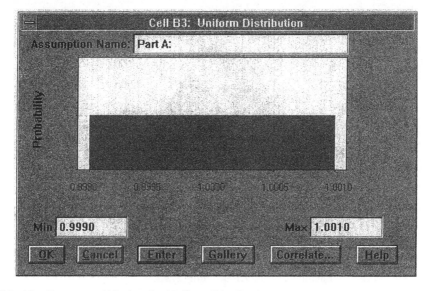

Figure 31–16   Parameter Window for Uniform Distribution

defining. Then click on the Cell Menu and select the Define Forecast command. The window in Figure 31–19 appears.

The Units box needs to be filled out by typing in *inches*. Click OK. This returns us to the main Crystal Ball® spreadsheet and signals the successful definition of the Forecast cell.

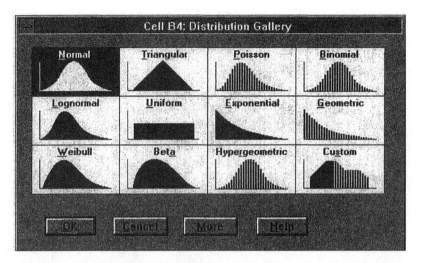

Figure 31–17   Setting Distribution for Part B

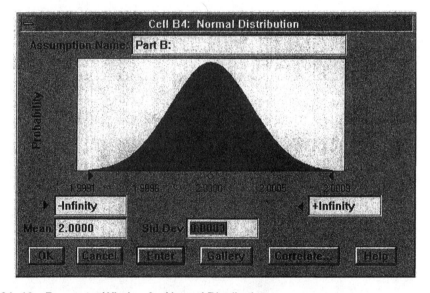

Figure 31–18   Parameter Window for Normal Distribution

Running the Monte Carlo Simulation

The first thing required before running the simulation is the specification of a few preferences. These user-defined items are necessary before Crystal Ball® can properly iterate through the calculations.

Select the Run Menu and then select the Run Preferences option. Figure 31–20 displays the Run Preferences window. Note that a number of items can be changed in this window.

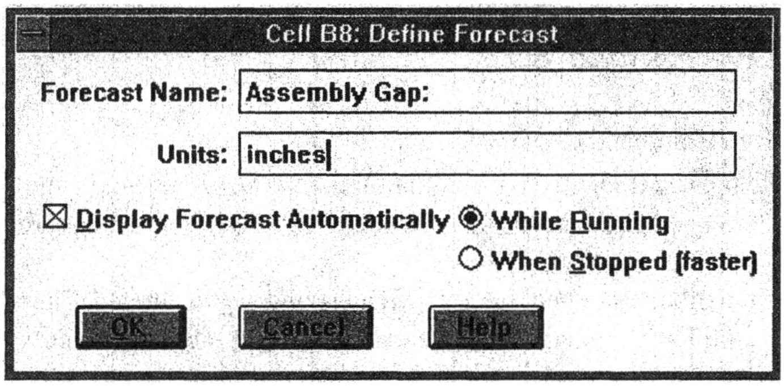

Figure 31–19 Define Forecast Window

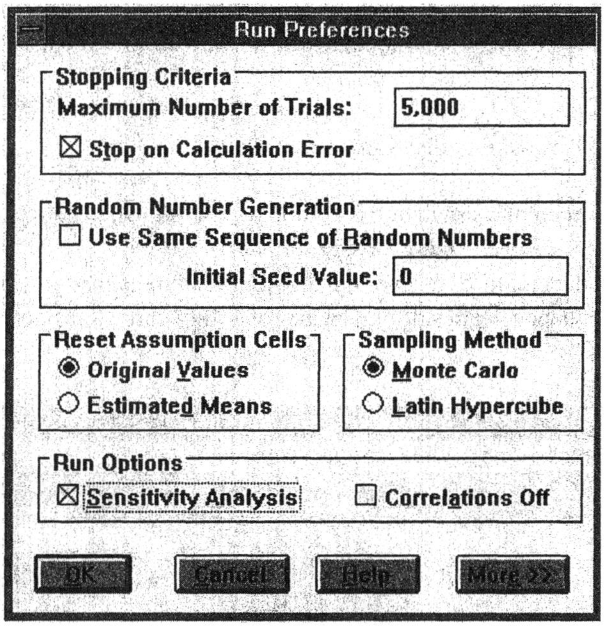

Figure 31–20 Run Preferences Window

For our case at hand, the only Preferences we need to set are:

Maximum Number of Trials: 5,000
Sampling Method: Monte Carlo
Sensitivity Analysis: Turned on (x in box)

Now that these settings are enabled, click OK.

To run the simulation, follow these steps:

Select the Run Menu.

Select the Run Command to start the simulation.

The Run Menu is altered once the simulation is running. The Run Command is replaced by a Stop Command that enables the simulation to be temporarily interrupted by the user.

Once the simulation is stopped, the Run Menu is once again altered. The Stop Command is replaced by the Continue Command. This reactivates the simulation exactly where it left off. Once the simulation is running again, the Stop Command is back in place under the Run Menu.

We can stop and start the simulation at will until the specified number of trials is completed. We can view one of two different windows to see how the simulation is progressing *during* the process, which can be an interesting and fun thing to observe!

- Under the View Menu, select the Statistics window to see how the Forecast statistics are evolving.
- Under the View Menu, select the Frequency Chart window to see how the Forecast frequency and probability distribution is evolving.

The completed run of the simulation for 5,000 trials is shown in both Frequency Chart (Figure 31–21) and Statistics Table (Figure 31–22) formats.

As we can see, the graphical form of the distribution of response values is generated by the simulation both in frequency (right side axis) and probability units (left side axis). Notice that two

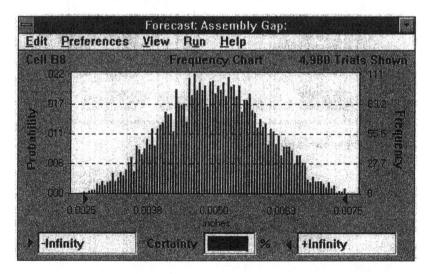

Figure 31–21   Simulation Frequency Chart

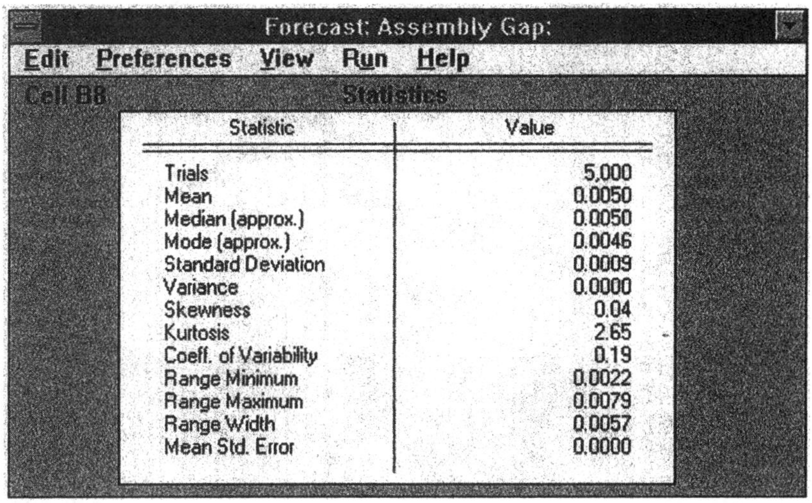

Statistic	Value
Trials	5,000
Mean	0.0050
Median (approx.)	0.0050
Mode (approx.)	0.0046
Standard Deviation	0.0009
Variance	0.0000
Skewness	0.04
Kurtosis	2.65
Coeff. of Variability	0.19
Range Minimum	0.0022
Range Maximum	0.0079
Range Width	0.0057
Mean Std. Error	0.0000

Figure 31–22 Simulation Statistics Table

arrow heads are located on the horizontal axis of the chart. These markers are adjustable and are linked to the certainty box located between the arrow heads. We can enter the upper and lower tolerance limits being evaluated for the assembly gap. In our case, we want all values to remain $> 0.001''$. The certainty box will display the percent certainty of occurrence for the range of values contained within the arrow heads.

In the example in Figure 31–21, the values displayed between 0.0025 and 0.0075 contain 100 percent of the results from the simulation. Thus, we are certain, based on our assumptions, that we will not have any significant occurrence of assembly interference. This is how the software helps us determine the number of assemblies that will be outside the tolerance limits.

The statistical table in Figure 31–22 displays the results of the simulation in various quantitative measures that provide the descriptive statistics we can use to help understand the nature of variability associated with the assembly.

The sensitivity analysis is displayed in the Sensitivity Chart (Figure 31–23).

The sensitivity chart displays just how much, graphically and quantitatively, each component in the assembly affects the output response (the assembly gap).

Notice that Crystal Ball® automatically applies the right sign to each component sensitivity coefficient. Also note the effect each assumed distribution had on the sensitivities. [See Chapter 7 in *Tolerance Design* (Creveling, 1997) for a discussion on this distributional form of sensitivity.] The uniform distribution has twice the sensitivity effect in comparison to the normal distribution for equal tolerance ranges of $0.0006'$.

Sensitivity Charts can also display the percent contribution to variation (epsilon squared) for each tolerance parameter (Figure 31–24).

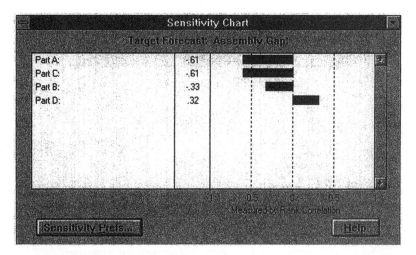

Figure 31–23   Simulation Sensitivity Chart

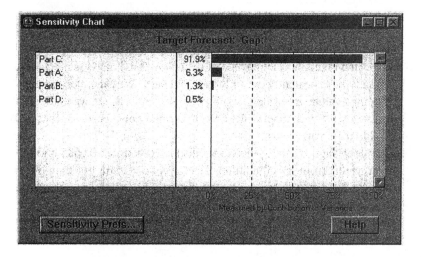

Figure 31–24   Percent Contribution to Variation

Epsilon squared is a term used to describe the proportion of the total variance contributed by each of the variables that can significantly affect the measured response in a design. In Figure 31–24, one can see that Part C contributes almost 92 percent of all the variation in the gap analysis. This is clearly the most significant tolerance term in the model.

> **Document optimum tolerance set points and variance models in Critical Parameter Management database.**

The optimized tolerance set points and the ideal/transfer function models and variance models used to optimize them are documented in the Critical Parameter Management database. These models are essential for product and process Critical Parameter Management during production and service support activities once the design is certified and transferred over to the production and service organizations. When these teams have to change the design, they have a clear set of information to guide them.

Final Detailed Outputs

> **Optimized tolerance set points for production and service organizations.**

The outputs from analytical tolerance design are product, production process, and service process tolerance set points and variance models that have been built from the ideal/transfer functions.

Analytical Tolerance Design Checklist and Scorecard

This suggested format for a checklist can be used by an engineering team to help plan its work and aid in the detailed construction of a PERT or Gantt chart for project management of the analytical tolerance design process within the Design, Optimize, and Capability Verify phase of the CDOV or I^2DOV development process. These items also serve as a guide to the key items for evaluation within a formal gate review.

Checklist for Analytical Tolerance Design

Actions:	Was this Action Deployed? YES or NO	Excellence Rating: 1 = Poor 10 = Excellent
1. Conduct worst case tolerance analysis for linear variation buildup.	_____	_____
2. Conduct worst case tolerance analysis for nonlinear variation buildup.	_____	_____
3. Conduct RSS tolerance analysis for linear variation buildup.	_____	_____
4. Conduct RSS tolerance analysis for nonlinear variation buildup.	_____	_____
5. Conduct Monte Carlo simulations on ideal/transfer functions.	_____	_____
6. Document optimum tolerance set points and variance models in Critical Parameter Management database.	_____	_____

Scorecard for Analytical Tolerance Design

The definition of a good tolerance scorecard is that it should illustrate in simple but quantitative terms what gains were attained from the work. This suggested format for a scorecard can be used by an engineering team to help document the completeness of its work.

Subsystem, subassembly, or component name: _____

Subsystem, subassembly, or component critical functional response or CTF specification: _____

Specific engineering units of subsystem and subassembly CFRs or component CTF spec: _____

Number of CFPs and CTF specs evaluated for sensitivity: _____

Number of CFPs and CTF specs selected to include in variance model: _____

Number of CFPs and CTF specs selected to adjust CFR capability: _____

Baseline or benchmarked initial subsystem, subassembly, or component CFR or CTF spec variance value: _____

Verified optimal subsystem, subassembly, or component CFR or CTF spec variance value: _____

Pretolerance design capability (Cp) index for the subsystem, subassembly, or component CFR or CTF spec: _____

Posttolerance design capability (Cp) index for the subsystem, subassembly, or component CFR or CTF spec: _____

Table of Critical Tolerance Parameter Set Points:

Critical Tolerance Parameter Name:	Optimal Nominal Set Point:	Final Tolerance: USL	Final Tolerance: LSL	Used for Capability Improvement:
1.	_____	_____	_____	_____
2.	_____	_____	_____	_____
3.	_____	_____	_____	_____

We repeat the final deliverables from analytical tolerance design:

- Optimized critical functional parameter set points that control subsystem and subassembly contributions to system level functional and assembly variation
- Optimized critical adjustment parameter tolerances used to tune the mean value of the system, subsystem, and subassembly critical functional response(s)
- Optimized critical-to-function specification set points that control component contributions to functional and assembly variation

- Critical Parameter Management data that is documented in the CPM relational database:

  System level variance models

  Subsystem level variance models

  Subassembly level variance models

  Component level variance models

  Manufacturing process level variance models

  CFP tolerances and their percentage contribution to CFR variation

  CAP tolerances and their percentage contribution to CFR variation

  CTF spec tolerances and their percentage contribution to CFR variation

References

Creveling, C. M. (1997). *Tolerance design.* Reading, MA.: Addison-Wesley.

Drake, Jr., P. (1999). *Dimensioning and tolerancing handbook.* New York: McGraw-Hill.

Greenwood, W. H., & Chase, K. W. (1998). *Worst case tolerance analysis with non-linear problems.* Manufacturing Review, Vol. 1. No. 1. ASME.

Harry, M., & Stewart, R. (1988). *Six Sigma mechanical design tolerancing.* Schaumberg, IL.: Motorola University Press.

Empirical Tolerance Design

Where Am I in the Process? Where Do I Apply Empirical Tolerance Design Within the Roadmaps?

Linkage to the I²DOV Technology Development Roadmap:

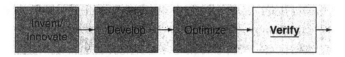

Empirical tolerance design is conducted within the Verify phase during technology development.

Linkage to the CDOV Product Development Roadmap:

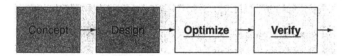

Empirical tolerance design is conducted within the Optimize and Verify phases during product development.

What Am I Doing in the Process? What Does Empirical Tolerance Design Do at this Point in the Roadmaps?

The Purpose of Empirical Tolerance Design in the I²DOV Technology Development Roadmap:

Empirical tolerance design is used in the Verify phase to assess integration sensitivities as the platform is integrated and evaluated for performance under nominal (random variation sources) and stressful (induced, assignable cause variation sources) noise conditions. It is used to understand the sensitivities that exist between component, subassembly, and subsystem interfaces.

The Purpose of Empirical Tolerance Design in the CDOV Product Development Roadmap:

Empirical tolerance design is used in the Optimize phase to analyze and balance system-to-subsystem interface sensitivities. It is also used to understand the sensitivities that exist between component, subassembly, and subsystem interfaces.

It is used during the Capability Verify phase to assess remaining integration sensitivity issues as the final product design certification units are evaluated for performance under nominal (random variation sources) and stressful (induced, assignable cause variation sources) noise conditions.

What Output Do I Get at the Conclusion of Empirical Tolerance Design? What Are Key Deliverables from Empirical Tolerance Design from its Application in the Roadmaps?

- *Optimized critical functional parameter set points* that control subsystem and subassembly contributions to system level functional and assembly variation
- *Optimized critical adjustment parameter tolerances* used to tune the mean value of the system, subsystem, and subassembly critical functional response(s)
- *Optimized critical-to-function specification set points* that control component contributions to functional and assembly variation
- *Critical Parameter Management data* that is documented in the CPM relational database:
  System level variance models
  Subsystem level variance models
  Subassembly level variance models
  Component level variance models
  Manufacturing process level variance models
  CFP tolerances and their percentage contribution to CFR variation
  CAP tolerances and their percentage contribution to CFR variation
  CTF spec tolerances and their percentage contribution to CFR variation

The Empirical Tolerance Design Flow Diagram (found on page 673)

The visual reference illustrates the flow of the detailed steps of the empirical tolerance design process.

Verbal Descriptions for the Application of Each Block Diagram

This section explains the specific details of each step in the empirical tolerance design process:

- *Key detailed input(s)* from preceding tools and best practices
- *Specific steps and deliverables* from each step to the next within the process flow
- *Final detailed outputs* that flow into follow-on tools and best practices

INPUT . . .

```
┌─────────────────────────────┐
│  S/N and mean plots from    │
│  robust design process,     │
│  initial tolerance set      │
│  point estimates from       │
│  analytical tolerance       │
│  design, and distribution   │
│  data from the supply       │
│  chain.                     │
└─────────────────────────────┘
              │
              ▼
┌─────────────────────────────┐
│ Define the engineering      │
│ design or manufacturing     │
│ process parameters to be    │
│ evaluated.                  │
└─────────────────────────────┘
```

Define the engineering design or manufacturing process parameters to be evaluated.

Develop the quality loss function for the output response.

Define the manufacturing or assembly standard deviation (1s) above and below the mean set point for each critical parameter under evaluation.

Select an appropriate orthogonal array with enough degrees of freedom to facilitate the proper evaluation of the effect of the 1 standard deviation set points for each parameter on the measured output response.

Align the appropriate compounded noise factors with the orthogonal array experiment.

Set up and run the designed experiment.

Conduct a formal analysis of variance on the data using Minitab.

Calculate the total quality loss for the baseline design.

Quantify the quality loss due to each parameter in the baseline design.

Calculate the percentage each parameter contributes to the overall variability in the output response.

Calculate the process capability (Cp) for the design.

Construct the variance equation.

Use the new, lower parameter variance values in the loss function to calculate the new, lower loss values.

Meet with the manufacturing engineers responsible for the production processes that make the components to define the increase in component costs associated with the reduction in each parameter variance (s^2).

Obtain the new, upgraded components and rerun the parameters in the orthogonal array experiment at the new standard deviation limits as defined by the reduction in parameter variances.

Calculate the new Cp from the customer tolerance limits and the new standard deviation coming from the upgraded design data (using the new ANOVA values) from the verification experiment.

OUTPUT . . .

Optimized tolerances based on reduced standard deviations for the components.

Key Detailed Inputs

> **S/N and mean plots from robust design process, initial
> tolerance set point estimates from analytical tolerance
> design, and distribution data from the supply chain.**

Results from robust design, analytical tolerancing, and manufacturing process capability studies on component variability provide the necessary inputs for empirical tolerance design.

Measurement system analysis must be applied to the data acquisition system to assure the data is low in measurement system noise.

Specific Steps and Deliverables

> **Define the engineering design or manufacturing
> process parameters to be evaluated.**

The team must review the signal-to-noise, standard deviation, and mean plots from robust design. These will indicate which CFPs and CTF specifications have high sensitivity (large slopes) in terms of S/N and mean responses. Parameters that have high sensitivity to both mean and S/N are good candidates for tolerance tightening. This is so because when small changes in the CFP or CTF specifications are induced, the mean and standard deviation of the CFR are strongly affected.

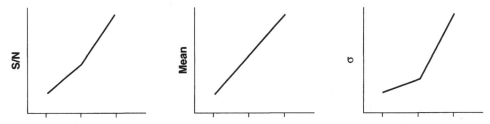

Parameters that are known to have a small effect on the CFR mean, S/N, and standard deviation will not deliver useful results when attempting to reduce CFR variability through tolerance design methods.

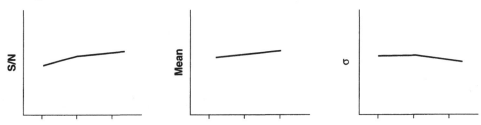

Parameters that exhibit this mild behavior are not good candidates for tolerance design.

Develop the quality loss function for the output response.

To properly account for balancing cost and economical loss associated with performance variation, the team needs to develop the quality loss function for the CFR.

$$\$ \text{Loss}(y) = k[\sigma^2 + (y_{avg} - \text{Target})^2]$$

where $\$ \text{Loss}(y)$ is the dollars lost due to the CFR (y) being off target due to two reasons:

1. Variation (σ^2) about a local mean (y_{avg}) that reduces short-term capability (Cp)
2. Variation of the local mean (y_{avg}) away from the desired target that reduces long-term capability (Cpk)

The economic side of the loss function is embodied in the economic coefficient (k).

$$k = A_0/\Delta^2_0$$

This coefficient is developed by two parameters:

1. (A_0) is the term for dollars lost due to the variance and mean growing to a level where a fail-to-satisfy level of CFR performance is reached.
2. Δ^2_0 is the fail-to-satisfy tolerance range (VOC tolerance limits: USL and LSL values for a reasonable percentage of customers).

The current state of affairs in loss function development is that most financial managers are suspicious and unsure that it should be trusted to make significant business decisions. This is because the term (A_0) contains "soft" numbers mixed with "hard" numbers. Hard numbers include scrap, rework, cost of poor quality, and warranty dollars. The soft numbers include loss of business due to lost market share, customer dollar loss due to off-target performance, and repair or replacement costs. These soft numbers are recognized by financial professionals, but their accuracy is a matter of concern. It is not easy or simple to build accurate soft numbers for estimating (A_0). The best we can advise is to invest the time to quantify (A_0) for critical functional responses that are real economic drivers in the product.

Define the manufacturing or assembly standard deviation (1s) above and below the mean set point for each critical parameter under evaluation.

With the candidate subsystem and subassembly CFPs and component CTF specifications identified, the team now must determine the variability associated with these x input variables. The goal
of the empirical tolerance study is to realistically "wiggle" the critical-to-function x variables and
measure their percent contribution to the "wiggle" they induce in the Y output variable.

The question to answer is, "How much should each critical input be wiggled?" The answer
lies in how the manufacturing process actually puts wiggle in the components. It does so in rec-
ognizable patterns over time called *distributions*. The team must go out and quantify the nature
of variation likely to be seen in components by measuring the mean and standard deviation from
distributions of components that are like the ones being developed. If the components in development are not in production, the team finds components that have very similar materials, form,
and features to stand in as reasonable representations. In this component level case, the standard
amount of wiggle we want to induce in an empirical tolerance study is $+/-$ 1 standard deviation.

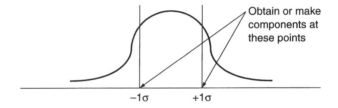

If part of our tolerance study includes the adjustment variability associated with a CAP-
CFR relationship or critical functional parameter variation from a subassembly or subsystem,
then we must estimate what a reasonable amount of expected variation from those input variables
will be. The overriding priority in setting up the wiggle is that it ought to be about one-third of
your expected worst-case "wiggle" condition. If your UCL and LCL values are equal to 3 standard deviations, then it stands to reason that a reasonable or likely amount of expected variation
is 1 standard deviation. This is the basis for root sum of square tolerancing in the previous chapter. We are essentially getting ready to run a designed experiment that emulates an RSS tolerance
analysis.

> **Select an appropriate orthogonal array with enough degrees of freedom
> to facilitate the proper evaluation of the effect of the 1 standard deviation
> set points for each parameter on the measured output response.**

For the purposes of demonstrating how empirical tolerancing is applied to components, we will
focus on the inducement of $+/-$ 1σ variation about the "robust nominals" for these evolving
critical-to-function specifications. This means we typically need to select a two-level Fractional
Factorial experiment. Often we use a Placket-Burman array to screen through the CTF specs for
their main effects and (ε^2) values.

The candidate component CTF specs need to be properly loaded into an appropriately sized
orthogonal array (designed experiment). Two issues must be addressed at this point in the process:

1. Selection of an experiment that has enough degrees of freedom to evaluate the main effects of the candidate parameters
2. Sizing the experiment to include major interactions between certain CTF specs that will play a key role in variance reduction and constraint

Mild interactions that are statistically significant but relatively weak in their contribution (ε^2) to overall variability in the CFR are usually not included in the experiment. This reduces the size and cost of the experiment without jeopardizing the validity of the results.

> **Align the appropriate compounded noise factors with the orthogonal array experiment.**

The component CTF specs that are being set at their $+/-$ 1 standard deviation values must be evaluated under the same compounded noise factors that were used during robust design. This helps ensure that the unit-to-unit variation noise represented in the component CTF specs being set at their $+/-$ 1 standard deviation values are also stressed by external and deteriorative sources of variation as well. This helps further the robustness of the design by making sure the tolerances that will be derived by the final ANOVA results from the designed experiments are robust.

> **Set up and run the designed experiment.**

Once the candidate tolerance control factors are loaded inside the designed experiment and the noise factors are arranged outside the array in their compounded format, it is time to run the experiment and measure the CFR.

Measurement system capability should have been previously confirmed so that noise in the data due to measuring the CFR is at least 10 times smaller than the magnitude of the CFR differences you are trying to detect.

Figure 32–1 shows what an L12 Placket-Burman Tolerance Experiment might look like when saturated (totally loaded) with CTF specs (designated as C):

Please note this is strictly a screening experiment designed specifically to look at main effects only. This is an excellent array to use if you do not have a need to evaluate interactions in a tolerance context.

> **Conduct a formal analysis of variance on the data using Minitab.**

The data from the compounded noise columns from the designed experiment contains the wiggle information from the array's 1s (components at (-1σ) levels) and 2s (components at ($+1\sigma$) levels) plus the effects of the assignable cause variation embedded in the compounded noise factors.

Candidate Tolerance Control Factors...												Noise Factors	
L_{12}	C_A	C_B	C_C	C_D	C_E	C_F	C_G	C_H	C_I	C_J	C_K	N_1	N_2
1	1	1	1	1	1	1	1	1	1	1	1		
2	1	1	1	1	1	2	2	2	2	2	2		
3	1	2	2	2	2	1	1	1	2	2	2		
4	1	2	1	2	2	1	2	2	1	1	2		
5	1	1	2	1	2	2	1	2	1	2	1		
6	1	1	2	2	1	2	2	1	2	1	1		
7	2	2	2	2	1	1	2	2	1	2	1		
8	2	2	2	1	2	2	2	1	1	1	2		
9	2	1	1	2	2	2	1	2	2	1	1		
10	2	1	2	1	1	1	1	2	2	1	2		
11	2	2	1	2	1	2	1	1	1	2	2		
12	2	2	1	1	2	1	2	1	2	2	1		

Figure 32–1 The L12 Array with Compounded Noise Factors (N1 and N2)

The compounded noise variables data needs to be averaged prior to application of the ANOVA process.

The ANOVA table will provide a wealth of information that can be used to determine the following information:

1. Each $[SS_{factor}/ SS_{Total}] \times 100 = \varepsilon^2_{factor}$ (percent contribution of the tolerance control factor to total variation of the CFR, or *epsilon squared*)
2. Each $[SS_{factor}/ dof_{Total}] = \sigma^2_{factor}$ (variance contribution of the tolerance control factor as part of the total variation of the CFR for 2 level experiments only!)
3. The total variance value for the CFR under evaluation $[SS_{Total}/dof_{Total}] = \sigma^2_{Total}$ (the sum of all the tolerance control factor variances will equal this total variance of the CFR for 2 level experiments only!)

Note: This technique applies to the output from an ANOVA table that is the result of a 2 level factorial designed experiment with the replicated data averaged (such as the data for each compounded noise factor). It will provide information about differences between the main effects. Weak main effects are pooled to represent the error variance.

Calculate the total quality loss for the baseline design.

The quality loss function can easily be calculated for the CFR by using the total variance (σ^2_{Total}) from the ANOVA data.

Assuming we can adjust the mean onto the target using a critical adjustment parameter, the loss function for a nominal best case is:

$$\$\, Loss(y) = k[\sigma^2 + (y_{avg} - Target)^2]$$
$$\$\, Loss(y) = k[\sigma^2 + 0]$$
$$\$\, Loss(y) = k[\sigma^2_{Total}]$$

This is the baseline loss of the design prior to adjusting the loss by reducing variability in the tolerance control factors.

> **Quantify the quality loss due to each parameter in the baseline design.**

The quality loss contribution due to each tolerance control parameter is calculated as follows:

$$\$\, Loss(y) = k[\sigma^2_{factor}]$$

The total quality loss is therefore

$$\$\, Loss(y) = k[\sigma^2_{factor\,A} + \sigma^2_{factor\,B} + \ldots + \sigma^2_{factor\,n}]$$

Each parameter contributes its own specific dollar loss due to the magnitude of its variance contribution. This dollar amount will be used to balance the cost to reduce the baseline variance vs. the cost to lower the variance to a new level.

> **Calculate the percentage each parameter contributes to the overall variability in the output response.**

Each parameter can have its specific percent contribution to total variation in the CFR calculated by using the following information from the ANOVA table.

$$[SS_{factor}/SS_{Total}] \times 100 = \varepsilon^2_{factor}$$

This equation represents the percent contribution of the tolerance control factor to total variation of the CFR. This helps the team see which parameters are causing the most variation in the CFR. This further identifies which parameters are critical to function. Parameters that have weak epsilon squared values are good candidates for tolerance loosening. When a tolerance can be relaxed, often its cost can be lowered. These cost savings can be used to help offset the cost of the tolerance tightening that will soon occur on the parameters with high epsilon squared values. This is part of the cost balancing that is done in the Optimize and Verify phases of the CDOV process.

> ## Calculate the process capability (Cp) for the design.

The ANOVA data provides the SS_{Total} and the dof_{Total}, which are needed to calculate the total variance for the CFR.

$$[SS_{Total}/dof_{Total}] = \sigma^2_{Total}$$

Taking the square root of the variance provides the standard deviation (σ^2_{Total}) for the CFR. This can be used, in conjunction with the USL and LSL, to calculate the design performance capability.

$$Cp = [(USL - LSL)/6\sigma^2_{Total}]$$

This is not the capability under random conditions but rather the capability when unit-to-unit, external, and deterioration noises (assignable cause variation) are intentionally being induced at realistic levels. It is wise to measure design CFRs for capability with and without assignable cause variation. The goals are for CFRs to attain a $Cp = 2$ without noise (under nominal, random conditions) and a minimum $Cp = 1$ with assignable cause noise "turned on." It is up to each program management team to decide what the acceptable value of Cp with noise is going to be. Typical values range between $Cp = 1 - 1.33$.

> ## Construct the variance equation.

At this point the Cp index for the CFR, under stressful noise conditions, has been calculated and compared to a desired goal ($Cp > 1$). If the Cp value is less than 1, then corrective action must be taken.

The metric used to guide corrective action is the individual component variance. A variance equation can be constructed from the individual component variances and can be solved to calculate a new, lower overall variance for the CFR. The variance equation is used to balance variance reductions across the candidate tolerance control factors.

$$\sigma^2_{Total} = [\sigma^2_{factor\ A} + \sigma^2_{factor\ B} + \ldots + \sigma^2_{factor\ n}]$$

We must add two factors to the variance equation to help us in the balancing of the overall CFR variation (σ^2_{Total}). The first factor that must be included in the equation is the epsilon squared (ε^2_{factor}) terms for each significant tolerance control factor.

$$\sigma^2_{\text{Total}} = \left[\sigma^2_{\text{factor A}} \times \left(\varepsilon^2_{\text{factor A}}\right) + \ldots \sigma^2_{\text{factor } n} \times \left(\varepsilon^2_{\text{factor } n}\right)\right]$$

We need to include just one more term that is new to the process. The *variance adjustment factor (VAF)* is the decimal value of reduction for each tolerance control factor. It is always a decimal value less than 1 and greater than 0.

$$\sigma^2_{\text{Total}} = \left[\left(\text{VAF}^2\right) \times \sigma^2_{\text{factor A}} \times \left(\varepsilon^2_{\text{factor A}}\right)\right] + \ldots + \left[\left(\text{VAF}^2\right) \times \sigma^2_{\text{factor } n} \times \left(\varepsilon^2_{\text{factor } n}\right)\right]$$

Now we have the current variance, its percent contribution to overall variation, and the amount of variance reduction required to meet the overall variance reduction requirement.

The team can set a Cp goal and then back-calculate the variance adjustment factors through an iterative process. For a Cp goal of 1, we would set up the equations as follows:

$$\sigma_{\text{Total}} = \text{USL} - \text{LSL}/6. \ldots$$

Squaring (σ_{Total}), we establish the new variance target (σ^2_{Total}) for the CFR. Now each individual variance can be adjusted according to whatever is economically and physically reasonable. This goal must be carefully attained through iterative balancing across all of the significant tolerance control factors.

Use the new, lower parameter variance values in the loss function to calculate the new, lower loss values.

Each tolerance control factor has a new variance calculated for it based on discussions with representatives from the supply chain. The team can try to allocate the reduced variance values, but it is wise to seek the input of the people who are actually going to be responsible for consistently delivering the components with the reduced variances.

Return to the quality loss function equation and recalculate the new, reduced component loss values using the reduced variance values. These new loss values can be added to establish the new total loss value due to the new 1σ values that have been reduced from the original 1σ values. The reduction in loss will be proportional to the aggregated reduction in the variances to reach the desired Cp value of 1.

Meet with the manufacturing engineers responsible for the production processes that make the components to define the increase in component costs associated with the reduction in each parameter variance (s^2).

Now it is time to "get real" and find out if our suppliers can provide the reduced variances and at what cost. This is where negotiations and trade-offs will cause several iterations to balance the variance equation. Each component cost increase must be compared to the quality loss value for the component to see if it is worth the expense. It is not uncommon for a ratio of 50–100:1 of loss reduction to cost increase. It all depends on the specific costs and losses that confront us in this case.

For component costs that are reasonably low in comparison to the quality loss reduction, a decision is made to proceed with the expense. For cases where the cost to reduce variation is high or even exceeds the quality loss reduction, we do not proceed with that variance reduction. We iterate through all the statistically significant tolerance control factors until we reduce variance of the CFR while controlling our costs for the reductions. This can become quite a challenging balancing act!

For factors that are found to be statistically insignificant and that have low epsilon squared values, we consider loosening their current tolerances and use those cost savings to help pay for the tighter tolerances on the CTF specifications.

> **Obtain the new, upgraded components and rerun the parameters in the orthogonal array experiment at the new standard deviation limits as defined by the reduction in parameter variances.**

Once the balancing process is completed and the new costs are in alignment with our cost vs. quality loss reduction goals, it is time to obtain samples of the new component $+/- 1\sigma$ values. These are new components at the reduced variance levels after the organization has agreed to and paid for their new, lower variances (as reflected in the $+/- 1\sigma$ values).

If the components (at $+/- 1\sigma$) cannot be obtained by part sorting from a production process, then the parts must be made "off-process" as close to the production materials, surface finishes, and geometric dimensions as possible.

The original designed experiment can then be rerun at the new, reduced, $+/- 1\sigma$ set points with the same compounded noises.

> **Calculate the new Cp from the customer tolerance limits and the new standard deviation coming from the upgraded design data (using the new ANOVA values) from the verification experiment.**

The ANOVA data allow us to recalculate the quality loss values, the component variances, the epsilon squared values, and, of most importance, *the new Cp index*. This verification experiment should validate that the new, retoleranced components are constraining variation and producing the desired Cp performance even in the presence of stressful sources of "assignable cause variation." Now, the team has balanced cost and functional performance in what we hope will be a win-win situation for the organization and our customers. A complete example of this process is found in Chapter 17 of *Tolerance Design* (Creveling, 1997).

Final Detailed Outputs

> **Optimized tolerances based on reduced standard deviations for the components.**

The output is an ability to calculate tolerances based on reduced component variances that enable the design to meet its Cp goals under both nominal (random variation) conditions and under stressful (assignable cause variation) conditions.

Empirical Tolerance Design Checklist and Scorecard

This suggested format for a checklist can be used by an engineering team to help plan its work and aid in the detailed construction of a PERT or Gantt chart for project management of the empirical tolerance design process within the Optimize phase of the CDOV or I^2DOV development process. These items also serve as a guide to the key items for evaluation within a formal gate review.

Checklist for Empirical Tolerance Design

Actions:	Was this Action Deployed? YES or NO	Excellence Rating: 1 = Poor 10 = Excellent
1. Define the engineering design or manufacturing process parameters to be evaluated.	_____	_____
2. Develop the quality loss function for the output response.	_____	_____
3. Define the manufacturing or assembly standard deviation (1s) above and below the mean set point for each critical parameter under evaluation.	_____	_____
4. Select an appropriate orthogonal array with enough degrees of freedom to facilitate the proper evaluation of the effect of the 1 standard deviation set points for each parameter on the measured output response.	_____	_____
5. Align the appropriate compounded noise factors with the orthogonal array experiment.	_____	_____
6. Set up and run the designed experiment.	_____	_____
7. Conduct a formal analysis of variance on the data using Minitab.	_____	_____
8. Calculate the total quality loss for the baseline design.	_____	_____
9. Quantify the quality loss due to each parameter in the baseline design.	_____	_____
10. Calculate the percentage each parameter contributes to the overall variability in the output response.	_____	_____

(continued)

11. Calculate the process capability (Cp) for the design.   _____   _____
12. Construct the variance equation.   _____   _____
13. Use the new, lower parameter variance values in the
    loss function to calculate the new, lower loss values.   _____   _____
14. Meet with the manufacturing engineers responsible
    for the production processes that make the
    components to define the increase in component costs
    associated with the reduction in each parameter
    Variance (s^2).   _____   _____
15. Obtain the new, upgraded components and rerun the
    parameters in the orthogonal array experiment at the
    new standard deviation limits as defined by the
    reduction in parameter variances.   _____   _____
16. Calculate the new Cp from the customer tolerance
    limits and the new standard deviation coming from
    the upgraded design data (using the new ANOVA
    values) from the verification experiment.   _____   _____

Scorecard for Empirical Tolerance Design

The definition of a good empirical tolerance design scorecard is that it should illustrate in simple but quantitative terms what gains were attained from the work. This suggested format for a scorecard can be used by an engineering team to help document the completeness of its work.

Subsystem, subassembly, or component name: _____

Subsystem, subassembly, or component critical functional response or CTF specification:

_____

Specific engineering units of subsystem and subassembly CFRs or component CTF spec:

_____

Number of CFPs and CTF specs evaluated for sensitivity: _____

Number of CFPs and CTF specs selected to include in variance model: _____

Number of CFPs and CTF specs selected to adjust CFR capability: _____

Baseline or benchmarked initial subsystem, subassembly, or component CFR or CTF spec variance value: _____

Verified optimal subsystem, subassembly, or component CFR or CTF spec variance value:

_____

Pretolerance design capability (Cp) index for the subsystem, subassembly, or component CFR or CTF spec: _____

Posttolerance design capability (Cp) index for the subsystem, subassembly, or component CFR or CTF spec: _____

Table of Critical Tolerance Parameter Set Points:

Critical Tolerance Parameter Name:	Optimal Nominal Set Point:	Final Tolerance: USL	Final Tolerance: LSL	Used for Capability Improvement:
1.	_____	_____	_____	_____
2.	_____	_____	_____	_____
3.	_____	_____	_____	_____

Table of Critical CFR Effects:

Affected CFR Name:	Magnitude of Effect:	Direction of Effect: (+ or −)	F Ratio/$F_{crit.}$:	% Contribution or ε^2 Value:
_____	_____	_____	_____	_____
_____	_____	_____	_____	_____
_____	_____	_____	_____	_____

We repeat the final deliverables from empirical tolerance design:

- Optimized critical functional parameter set points that control subsystem and subassembly contributions to system level functional and assembly variation
- Optimized critical adjustment parameter tolerances used to tune the mean value of the system, subsystem, and subassembly critical functional response(s)
- Optimized critical-to-function specification set points that control component contributions to functional and assembly variation
- Critical Parameter Management data that is documented in the CPM relational database:
  System level variance models
  Subsystem level variance models
  Subassembly level variance models
  Component level variance models
  Manufacturing process level variance models
  CFP tolerances and their percentage contribution to CFR variation
  CAP tolerances and their percentage contribution to CFR variation
  CTF spec tolerances and their percentage contribution to CFR variation

Reference

Creveling, C. M. (1997). *Tolerance design.* Reading, MA: Addison Wesley.

Reliability Evaluation

This chapter was written by Edward Hume. Ted is as seasoned a veteran in Six Sigma methods as one can get. He practiced Six Sigma in design and management positions over much of his career. Ted is an electrical engineer specializing in the design and reliability sciences.

Where Am I in the Process? Where Do I Apply Reliability Evaluation Within the Roadmaps?

Linkage to the I²DOV Technology Development Roadmap:

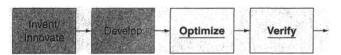

Reliability evaluations are conducted within the Optimize and Verify phases during technology development.

Linkage to the CDOV Product Development Roadmap:

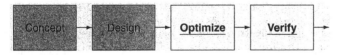

Reliability evaluations are conducted within the Optimize and Verify phases during product development.

What Am I Doing in the Process? What Does Reliability Evaluation Do at This Point in the Roadmaps?

The Purpose of Reliability Evaluation in the I^2DOV Technology Development Roadmap:

Reliability evaluation is applied to qualitatively define and quantitatively assess the baseline reliability of new technology platforms and subsystems being developed for integration into a new platform or for integration into existing product designs. Preventative application of numerous tools and best practices can then be used to improve reliability, as required, prior to the technology being transferred into a commercialization program.

The Purpose of Reliability Evaluation in the CDOV Product Development Roadmap:

Reliability evaluations are first applied to quantify the reliability of subsystems, subassemblies, and components and then applied *to* the integrated system after robustness optimization of the subsystems and subassemblies. Preventative application of numerous tools and best practices can also be used to improve the reliability as required prior to design certification.

What Output Do I Get at the Conclusion of Reliability Evaluations? What Are Key Deliverables from Reliability Evaluations at This Point in the Roadmaps?

- Phase 3B: Measured reliability performance of robust subsystems, subassemblies, and components before and after they are integrated into the system
- Phase 4A: Measured reliability performance of cost balanced product design certification systems
- Phase 4B: Measured reliability performance of production product systems
- Component, subassembly, and subsystem reliability performance
  - Predicted reliability vs. measured reliability
- Component, subassembly, and subsystem reliability performance improvement activities
  - Required deliverables from specific tools and best practices that improve reliability by reducing variability and sensitivity to deterioration and external noise factors

The Reliability Evaluation Flow Diagram (found on page 689)

The visual description illustrates the flow of the detailed steps of the reliability evaluation process.

Detailed Descriptions for the Application of Each Block Diagram

This section explains the specific details of what to do during the reliability evaluation process:

- *Key detailed input(s)* from preceding tools and best practices
- *Specific steps and deliverables* from each step to the next within the process flow
- *Final detailed outputs* that flow into follow-on tools and best practices

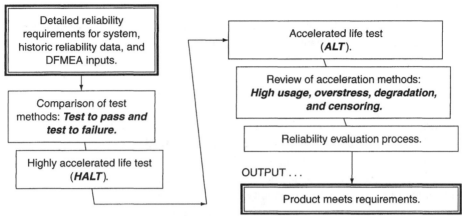

Key Detailed Inputs

Detailed reliability requirements for system, historic reliability data, and DFMEA inputs.

These detailed reliability requirements become inputs to defining test requirements:

- Measurable definitions of reliability requirements
- Environmental conditions
- Governmental agency requirements

These requirements are used to define the type of test, test parameters, sample size, and length of testing. A detailed test plan is developed and updated throughout the testing process.

Specific Steps and Deliverables

Comparison of test methods: *Test to pass and test to failure.*

Test to Pass

Customers, regulatory agencies, and company internal documents often specify requirements in terms of test to pass. Examples of this are as follows:

- Three relays are to be tested for 100,000 cycles.
- Five valves must withstand a 500 pounds per square inch internal pressure.
- Ten motors are to be tested for 100,000 hours of operation.

In all cases the products are tested under specified test conditions and must complete the test without failure.

When are these tests developed and to what purpose? Some of the tests are part of the initial test process developed to ensure product reliability:

- The tests represent an expected design life plus margin.
- The tests represent expected stress levels plus margin.

When high failure rates occur in the field, application tests are often added. These new tests have been shown to differentiate between products with high field failures and those that have been improved to eliminate the failures. Completing the tests successfully provides a level of confidence that the product will succeed in its application. As a manufacturer of the product, however, you have little prior information on the probability of the product passing the test. In addition, you have no information about the relative performance of two products that pass the test. Relative reliability provides a credible method of evaluating product even when the correlation to field reliability has not been established. The following are common reasons for product evaluation:

- To evaluate product, material, and process changes
- To compare new product to current and competitive products

An exercise that we use in our training is to bend paper clips until they fail. The process used is to open the paper clip up with a 180° bend and then close it with another 180° bend. Each 180° bend is counted. The bending process is continued until the clip breaks. This is intended to be an illustrative process. Suppose our test to pass criteria is that four paper clips must complete 11 bends without fail. The following is a set of pass-fail data representing six tests:

Test 1	P	P	P	P
Test 2	P	P	P	P
Test 3	P	P	P	P
Test 4	P	P	P	*F*
Test 5	*F*	P	P	P
Test 6	P	P	P	P

As you can see, 92 percent of all the paper clips pass at 11 cycles. However, only 66 percent of the six tests meet our test to pass criterion. After six tests we can see that the paper clips don't meet the test to pass requirement, but we ran four tests before finding that out. Suppose the first three tests were part of the development and verification process and the fourth was a part of the final customer qualification. The most common response to the failure of the fourth test is to launch an effort to find out what has changed in the product or manufacturing process. Little thought is given to the fact that perhaps nothing has changed and that the failures represent normal variation for the product. The bottom line is that testing to pass only provides pass-fail results. No statistics on the test sample or projections of the population performance are available from this method of testing.

Many part qualification, design verification, and production validation tests combine capability and reliability/durability tests where the entire set of tests is specified in terms of a test to pass. This type of qualification testing is also devoid of statistical thinking. As the complexity increases, so does the need for statistical thinking.

Test to Pass Tool Summary

Limitations

- Probability of success is not known for individual tests or for combinations of tests.
- Relative performance of products that pass the test is not known.

Benefits

- Passing the test provides a level of confidence that the product will succeed in its application.

Recommendations

- If tests are driven by external test requirements, continue testing beyond pass requirements until enough failures have occurred to establish the probability of success.
- If tests are internal, over time find new statistically based tests to accomplish the intended results.

Test to Failure

Customers, regulatory agencies, and company internal documents also specify reliability in statistical terms. Examples of this are as follows:

- Demonstrate that the relay will have less than 1 percent failures at 20,000 cycles and less than 10 percent failures at 100,000 cycles with 90 percent confidence.
- Demonstrate that the valve will have less than 1 percent failures, due to damage from the overpressure, at 500 pounds per square inch pressure with 90 percent confidence.
- Demonstrate that the motor will have less than 1 percent failures at 30,000 cycles and less than 5 percent failures at 100,000 cycles with 95 percent confidence.

We can use the algorithms in Minitab to fit our data with a distribution and predict reliability performance just as we did in Chapter 20, Reliability Prediction. The three-step process is repeated as follows:

1. Fit a known distribution to the data.
2. Review the shape of statistical functions $f(t)$, $R(t)$, and $h(t)$, and observe the distribution parameters.
3. Project B1, 10,. . .(1 percent, 10 percent failures), life (months, years, miles), and project reliability/survival probabilities at times such as 10 years, 100,000 miles, and so on.

We will use the data from our paper clip experiment again, this time in parametric form, to demonstrate the process. Our specification will be less than 1 percent failures at 9 bends and less than 10 percent failures at 11 bends with 90 percent confidence.

Step 1.   Fit a known distribution to the data. We will use the first three tests, including 12 paper clips, for our initial data set. Open Minitab and enter time to failure for the first 12 paper clips into one column labeled "Paper Clip-I". The sample data follow:

Test 1	15	26	14	12
Test 2	13	16	13	21
Test 3	16	12	16	14
Test 4	15	16	15	*10*
Test 5	*10*	16	15	14
Test 6	18	17	16	14

Using pull-down menus, follow this path: Stat, Reliability/Survival, Distribution ID Plot-Right Cens. Load "Paper Clip-I" into the dialog box and run the program. Graphical results showing the best fit are shown in Figure 33–1.

We pick the lognormal distribution because it has the lowest Anderson Darling number and the best visual fit.

Step 2.   Review the shape of statistical functions $f(t)$, $R(t)$, and $h(t)$. Follow this path: Stat, Reliability/Survival, DistributionOverview Plot-Right Cens. Load "Paper Clip-I" into the dialog box and select the lognormal distribution. Graphical results are shown in Figure 33–2.

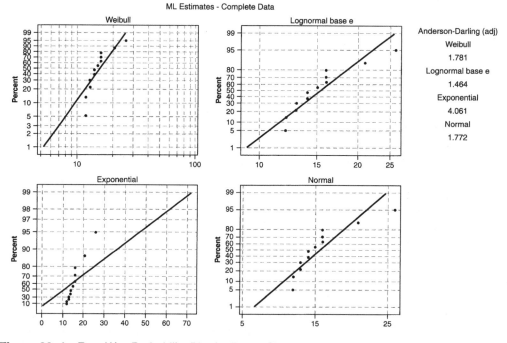

Figure 33–1   Four-Way Probability Plot for Paper Clip-I

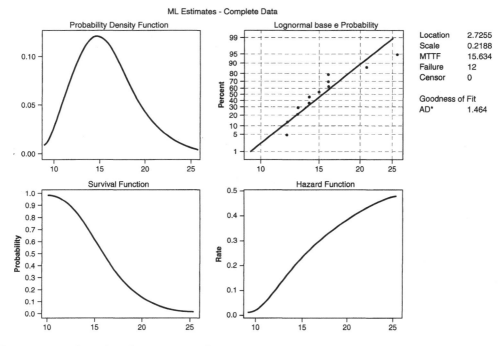

Figure 33–2 Overview Plot for Paper Clip-I

The DistributionOverview Plot-Right Cens. has a Hazard Function, $h(t)$, that represents a wear-out condition.

Let's predict reliability performance and compare it to requirements.

Step 3. Using pull-down menus, follow this path: Stat, Reliability/Survival, Parametric Dist Analysis Plot-Right Cens. Load "Paper Clip-I" into the dialog box and select the lognormal distribution. Select the Estimate button and enter into:

- Estimate survival probabilities for these times (value): − 11 bends
- Confidence Level: − 90
- Confidence Intervals: − Lower

Tabular data from the session window follow:

Table of Percentiles

Percent	Percentile	Standard Error	90.0% Normal Bound Lower
1	9.1746	1.1157	7.8506
10	11.5314	.9831	10.3379

A summary of performance vs. specification for 12 paper clips:

	% Failures	Nominal Number of Bends	Lower 90% Confidence Limit Number of Bends
Specification	≤ 1%		9.0
Performance	1%	9.2	7.9
Specification	≤ 10%		11.0
Performance	10%	11.5	10.3

Return to Minitab and enter time to failure for all 24 paper clips into one column labeled "Paper Clip". Lognormal is still the best distribution, and the following is the output from Parametric Dist Analysis Plot-Right Cens.

Table of Percentiles

Percent	Percentile	Standard Error	90.0% Normal Bound Lower
1	9.2912	0.7362	8.3941
10	11.4700	0.6371	10.6819

A summary of performance vs. specification for 24 paper clips:

	% Failures	Nominal Number of Bends	Lower 90% Confidence Limit Number of Bends
Specification	≤ 1%		9.0
Performance	1%	9.3	8.4
Specification	≤ 10%		11.0
Performance	10%	11.5	10.7

What Have We Learned?

- Our testing required only 12 paper clips to get a good estimate of population reliability.
- Testing with 24 clips only serves to reduce the standard error and thus lower the confidence limit spread.

Since the nominal performance is above specification, it is possible to test additional clips until the lower confidence limit equals the specification. Using the nominal bends, 90 percent lower confidence requirement, and the standard error, you can estimate the sample size required to bring the lower confidence limit into specification. For the 1 percent specification it would require a total sample size on the order of 300, and for the 10 percent specification it would require a total sample size on the order of 100. Given that the margin is low and any change in material or process could easily move the performance below specification, it would make sense not to continue testing but rather to improve the paper clip's margin to specification. We have learned a lot more about margin to specifications and can draw conclusions on how to proceed based on testing 12 clips to failure.

If the specification that four paper clips must pass 11 cycles with no failures, defined previously in "Test to Pass," is an external requirement, you can use Minitab to determine the probability of survival for the test. Tabular survival probabilities from the session window follow:

Table of Survival Probabilities

		95.0% Normal CI	
Time	Probability	Lower	Upper
11.0000	0.9236	0.6716	0.9921

Nominal survival probability of four paper clips tested to pass is $(0.9236)^4 = .73$. The probability of passing the test is 73 percent. Stated another way, one of four tests is expected to fail. Again we draw the same conclusion, that the product or process needs to be improved. With this analysis there is no hunt for what changed in the product or manufacturing process to cause the failures. The cause is inherent in the nominal design.

This analysis can be expanded to predict the probabilities of part qualification, design verification, and production validation tests that consist of a number of different capability and reliability/durability tests, all requiring zero failures. Treat all of the tests as series elements in a reliability block diagram as described in Chapter 20. For the capability tests, collect parametric data and calculate the probability of passing at specification limits. For reliability/durability tests, test to failure and calculate reliability/survival probabilities at the test limits. Multiply all the passing and survival probabilities to get the probability of passing the qualification, verification, or validation test. This process will also highlight the weaknesses in product performance.

Test to Failure Tool Summary

Limitations

- The analysis is intended for one failure mode per data set.

Benefits

- Test to failure moves reliability from the black and white world of test to pass into the real world with uncertainty managed by statistics.
- Probability of success is known for individual tests or for combinations of tests.
- Relative performance of products that are tested can be used for benchmarking purposes.

Recommendations

- Use test to failure and statistical analysis whenever possible.

Highly accelerated life test (*HALT*).

You have heard the phrase that a chain is only as strong as its weakest link. HALT is intended to find the weak links in the design and manufacturing process in a short period of time. Developed by Dr. Greg Hobbs (2000), it provides a unique approach to reliability growth. To find the weak links HALT uses every stimulus and combination of stimuli of potential value. Some examples of stimuli are as follows:

Vibration	Temperature	Temperature Cycling
Voltage	Voltage Cycling	Humidity

Stress levels are stepped well beyond expected environmental conditions until the fundamental limits of the product technology or environmental equipment are reached. An example of fundamental limits of technology would be the temperature at which a plastic structure starts to lose its strength. During the process we find operating and destruct limits for temperature and vibration. In addition, a number of failures occur that represent the weak links in the product. The goal is to find and fix all the weak links. They represent weaknesses in the product that are likely to become failures in its application. The following is an example of a HALT process and summary of HALT test data. The information is from *Summary of HALT and HASS Results at an Accelerated Reliability Test Center* and QualMark Corporation training. QualMark specializes in HALT.

Halt Process

1. Temperature stress
   a. Cold step stress—Start at 20°C and step in 10°C increments to determine operating and destruct limits.
   b. Hot step stress—Start at 20°C and follow same process.

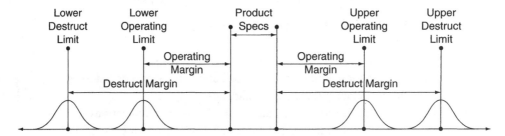

2. Vibration step stress—Start at 3–5 Grms and increase in 3–5 Grm steps until operating and destruct limits are determined. When possible, modifications are made to the product to extend the limits.

3. Rapid temperature transitions—Continuously ramp between hot and cold temperature limits as fast as the chamber and product will allow. Temperature limits set just below operating limits.

4. Combined environment—Test 30-minute temperature profiles with 10-minute dwell times at temperatures close to operating limits. Vibration starts at 3–5 Grms during the first profile and increases 5–10 Grms in each subsequent profile.

Results[2]: 47 different products, used in environments from office to airplane, have operating and destruct limits that are summarized in the following table.

	HALT Limits					
Attribute	**Thermal Data, °C**				**Vibration Data, Grms**	
	LOL	**LDL**	**UOL**	**UDL**	**VOL**	**VDL**
Average	−55	−73	+93	+107	61	65
Most Robust	−100	−100	+200	+200	215	215
Least Robust	+15	−20	+40	+40	5	20
Median	−55	−80	+90	+110	50	52

As you can see, the range in operating and destruct limits is large. Much of the difference is due to the products' different environmental requirements. The differences in operating and destruct limits, however, provide credible evidence that HALT can differentiate between different products' expected reliability.

A Pareto analysis of failure modes caused by HALT testing is listed in the following table:

Failure Mode	**Qty**
Troubleshooting in progress	85
Broken lead	53
Failed component, cause unknown	24
Component fell off (nonsoldered)	9
Screws backed out	9
Circuit design issue	8
Connector backed out	5
Tolerance issue	4
Broken component	4

The types of failures found reflect typical field failures. There may be a bias in favor of those failures that occur in harsher environments, but not always. Results should prove valid for office environments as well.

HALT Tool Summary

Limitations

- Results do not correlate directly with field reliability specifications.
- Most experience with HALT is with electronic products.

Benefits

- Short cycle of learning
- Low effort to perform the test
- Failure modes identified that are caused by interactions between temperature and vibration stresses

Recommendations

- Use to identify weaknesses in new products, especially when the technology is new to the company.
- Use to provide a quick reliability estimate by comparing HALT results to those of known internal and/or competitor products.

Accelerated life test (*ALT*).

ALT is a test process using acceleration factors to cause product failures in a shorter time than seen in actual field conditions. It is intended to be an accelerated simulation of field performance. The investment in time and resources to set up and run an ALT program is significant, and a good system to collect field failure data is required. Be prepared to stick with the program for a period of years before it becomes successful. ALT entails a nine-step process:

1. Review current field and test failure data on similar products.
2. Identify all potential failure modes and causes in the products that you are to test.
3. Select a comprehensive set of stresses that are significant in causing failures.
4. Select high usage and overstress acceleration factors.
5. Develop a test that incorporates all significant stresses and planned acceleration factors.
6. Run the test with similar products with current field failure information.
7. Monitor the test for failures on a real-time basis if possible.
8. Do a root cause analysis of each failure.
9. Correlate test and field failures to see if they fit the same distribution and have the same slope on their probability plot (Minitab's parametric distribution analysis) and the same distribution of failure modes.
   a. If the reliability functions and distribution of failure modes are similar, calculate the acceleration factor.
   b. If the reliability functions and/or distribution of failure modes are different, then return to step 5 to modify the test parameters to resolve the differences.

An accelerated life test process provides timely feedback on reliability for newly developed products and modified products. Timely feedback allows problems to be identified and fixed during the Optimize phase and to verify performance during the Verify phase. The ALT process needs to be updated continuously to take into account new failure modes, new environmental conditions, new technology, and improving reliability.

ALT Tool Summary

Limitations

• ALT requires significant resources over a number of years to be able to use internal testing to predict field reliability.

Benefits

• Relative performance of products that are tested can be used for benchmarking purposes.
• Timely feedback on projected field reliability provides the feedback necessary to make significant improvements in reliability.

Recommendations

• Make the investment to develop an ALT program if reliability is an important customer need.
• Use ALT to do internal and industry comparisons to establish relative reliability.
• Use ALT as a tool to drive reliability improvement.
• Update the ALT process to take into account changes in technology and environmental requirements.
• Update the ALT process to increase acceleration as product performance improves.

Review of acceleration methods: *High usage, overstress, degradation, and censoring.*

Acceleration factors are used to shorten the time required to predict field reliability. Each of these factors is distinctly different in approach, but all accomplish the same result: shortened test time.

High Usage

High usage is the easiest and most predictable method to reduce test time. It involves either increased speed or reduced off-time without actually changing other operating conditions that would result in a stress increase.

Increased speed examples

• Run bearings faster.
• Increase the rate of relay closures.
• Increase frequency of the AC voltage applied when testing insulation electrical breakdown.

Reduced off-time example

• Appliances that typically only run a few hours a day are run 24 hours a day.

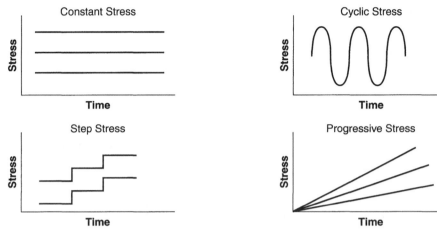

Figure 33–3 Application of Stress Loading

In both cases the acceleration factor = new speed/original speed, or new on-time/original on-time. Since high usage reduces test time without changing stress levels it should be used to the extent possible in all cases of accelerated life testing.

Overstress

The product is run at higher-than-normal levels of accelerating stress(es) to shorten product life. Typical accelerating stresses are temperature, voltage, mechanical load, temperature cycling, humidity, and vibration. We can apply stresses in a number of ways, as shown in Figure 33–3. Constant and cyclic stresses are the most common methods used with accelerated life testing. Step stress or progressive stress methods are used with HALT testing.

Our intent with accelerated life testing is to simulate field reliability in the shortest period of time. Increases in stress levels must be done with care to maintain the same failure modes and the same slope of the probability plot. As an example we return to our paper clip, which is exposed to +/− 30° bends in field operation. We want to accelerate failures by testing at increased angles of bend. A fixture is used and the clips are rotated through +/− 30°, +/−50°, and +/−90°, counting the number of rotations it takes for the clip to break. The resulting failure data is shown on the following page.

We will use Minitab's parametric distribution analysis right-censored algorithm to generate probability plots for all three angles (Figure 33–4). We continue to use lognormal as the distribution to fit the data. Use Minitab's pull-down menus: Stat, Reliability/Survival, Parametric Dist Analysis Plot-Right Cens. Load "Angle Failure Data" and run the algorithm.

Note that the slopes of the lines for 30 and 50 degrees are quite close while the slope for the 90 degree line is quite different. This is apparent visually when looking at the curves and also nu-

Angle = 30	Angle = 50	Angle = 90
173.5	47.5	18
129.5	69.5	18
142.5	69.5	16
173.5	34.0	18
112.5	44.5	16
165.5	40.0	17
130.5	72.5	18
207.5	62.5	13
157.5	49.5	18
90.5	53.5	19
111.5	53.5	17
101.5	55.5	18
102.5	58.5	20
119.5	36.5	17
117.5	44.5	19

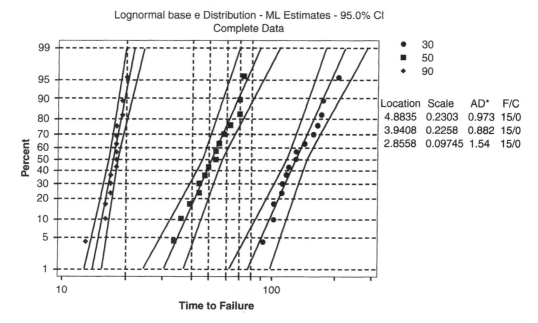

Figure 33–4 Probability Plot for 30–90

merically when looking at the scale factor. Slope is represented by scale factor with lognormal distributions and shape factor with Weibull distributions. Output from the session window allows us to look at 10 percent and 50 percent failure at 30 and 50 degrees to calculate acceleration. In this case the acceleration is 2.6 times. With many products it is hard to get large acceleration factors from one overstress condition. Data is presented in the following table.

Table of Percentiles

| | | (Bends) | Standard | 95.0% Normal CI | |
| | | | | Lower | Upper |
Angle	Percent	Percentile	Error		
30	10	38.5331	3.0311	33.0276	44.9562
50	10	98.3287	7.8913	84.0171	115.0781

Acceleration = 2.6

| | | (Bends) | Standard | 95.0% Normal CI | |
| | | | | Lower | Upper |
Angle	Percent	Percentile	Error		
30	50	51.4612	2.9996	45.9055	57.6893
50	50	132.0903	7.8553	117.5576	148.4195

Acceleration = 2.6

Degradation

In some products, failure is defined as a point when performance degrades to the minimum specification level. In these cases you should try to establish a model of the degradation, using regression, at appropriate overstress levels. When a stable model has been verified, it is possible to use extrapolation carefully to shorten the time required to make a decision on a product's reliability. Examples of product specifications where degradation can be used are tire tread thickness and insulation breakdown voltage. Models of tread thickness and insulation breakdown can be developed for a standard test process and used to extrapolate expected life.

Censoring

Censoring is a process of dealing with products when we are not sure exactly when they failed or when they will fail. The most common use of censoring is to analyze life test results when units do not complete the test because:

1. They were pulled off to open up test slots for other products or to do some other form of testing or analysis.
2. After completing the allotted test time, the functioning units are pulled off test.
3. After reaching a certain percentage failure, the test is terminated.

The second and third reasons directly shorten the time to make a decision on a product's reliability. What do we lose when not all the units have failed? Let's use the paper clip data and artificially create different levels of censored data. We will determine which number of cycles to start the censoring. For example, when we tell Minitab to censor at value 17, all values of 17 and beyond will be censored, resulting in 20 uncensored values and 4 censored values. Use Minitab's pull-down menus: Stat, Reliability/Survival, Parametric Dist Analysis Plot-Right Cens. Load "Paper Clip Data." Select Censor and load the censor value (17, 16) into Censor Value. Run the algorithm.

The results tables and the table of values follow. The tables have been sorted in ascending order, and Cs have been added to indicate censoring in this table only. They are not used by Minitab's algorithm.

	Paper Clip Data	
Cycles to Failure	**Censor at 17 Uncensored 20 Right Censored 4**	**Censor at 16 Uncensored 14 Right Censored 10**
10	10	10
10	10	10
12	12	12
12	12	12
13	13	13
13	13	13
14	14	14
14	14	14
14	14	14
14	14	14
15	15	15
15	15	15
15	15	15
15	15	15
16	16	C
16	16	C
16	16	C
16	16	C
16	16	C
16	16	C
17	C	C
18	C	C
21	C	C
26	C	C

	Table of Percentiles					
Uncensored	**Right Censored**			**Standard**	**95.0% Normal CI**	
Values	**Values**	**Percent**	**Percentile**	**Error**	**Lower**	**Upper**
24	0	10	11.4700	0.6371	10.2869	12.7891
20	4	10	11.2097	0.7453	9.8401	12.7699
14	10	10	11.0462	0.9685	9.3022	13.1173

As you can see, for 10 percent failures the predicted nominal value of bends goes from 10.5 to 10 as the number of clips censored goes from 0 to 10. In addition, the 95 percent confidence limits increase with the number censored. From a practical point of view, results can be analyzed as additional failures occur and the test can be stopped once an acceptable lower confidence band has been reached.

Table of Survival Probabilities

Uncensored Values	Right Censored Values	Time	Probability	95.0% Normal CI	
				Lower	Upper
24	0	11.0000	0.9318	0.8181	0.9808
20	4	11.0000	0.9134	0.7850	0.9735
14	10	11.0000	0.9024	0.7660	0.9689

Survival probabilities vary in the same way.

Reliability evaluation process.

---

1. Develop a test plan based on detailed inputs:
   a. Select one or more of the tests that have been described.
   b. Determine time requirements.
   c. Develop an initial plan for samples size.
2. Use a control plan and build test units:
   a. Prototype units should be built with a prototype control plan.
   b. Pilot units should be built with a start-up control plan.
3. Execute the test plan:
   a. Complete each of the planned tests.
   b. Complete a root cause analysis on all failures.
   c. Project reliability when possible.
4. Evaluate performance to requirements:
   a. Meets requirements.
      i. Document all corrective actions taken.
      ii. Update reliability models.
      iii. Update FMEAs.
      iv. Document lessons learned.
      v. Update design rules.
      vi. Update scorecards.
   b. Does not meet requirements.
      i. Take corrective actions on the product design or build process.
      ii. Start the test process over again.

Final Detailed Outputs

> **Product meets requirements.**

- Document all corrective actions taken.
- Update reliability models.
- Update FMEAs.
- Document lessons learned.
- Update design rules.
- Update scorecards.

Reliability Evaluation Checklist and Scorecard

This suggested format for a checklist can be used by an engineering team to help plan its work. The checklist will also aid in the detailed construction of a PERT or Gantt chart for project management of the reliability evaluation process within the Optimize and Verify Phases of the CDOV or I^2DOV development process. These items will also serve as a guide to the key items for evaluation within a formal gate review.

Checklist for Reliability Evaluations

Actions:	Was this Action Deployed? YES or NO	Excellence Rating: 1 = Poor 10 = Excellent
1. Comparison of test methods: *Test to pass and test to failure.*		
2. Highly accelerated life test (HALT).		
3. Accelerated life test (ALT).		
4. Review of acceleration methods: *high usage, overstress, degradation, and censoring.*		
5. Reliability evaluation process.		
6. Document lessons learned		
7. Update design rules		
8. FMEAs updated		

Scorecard for Reliability Evaluations

The definition of a good reliability evaluation scorecard is that it should illustrate in simple but quantitative terms what gains were attained from the work. This suggested format for a scorecard can be used by an engineering team to help document the completeness of its work:

Table of Documented Reliability Failures and Corrective Actions:

Item/Function Name:	Failure Mode:	Corrective Action:
1.	———	———
2.	———	———
3.	———	———

We will repeat the final deliverables from reliability evaluations:

- Phase 3B: Measured reliability performance of robust subsystems, subassemblies, and components before and after they are integrated into the system
- Phase 4A: Measured reliability performance of cost balanced product design certification systems
- Phase 4B: Measured reliability performance of production product systems
- Component, subassembly, and subsystem reliability performance
  - Predicted reliability vs. measured reliability
- Component, subassembly, and subsystem reliability performance improvement activities
  - Required deliverables from specific tools and best practices that improve reliability by reducing variability and sensitivity to deterioration and external noise factors

References

Crowe, D., & Feinberg, A. (2001). *Design for reliability.* Boca Raton, FL: CRC Press.

Hobbs, G. K. (2000). *Accelerated reliability engineering.* New York: Wiley.

Nelson, W. (1983). *Vol. 6: How to analyze reliability data.* New York: ASQC Quality Press.

Nelson, W. (1990). *Accelerated testing: Statistical models, test plans and data analyses.* New York: Wiley.

Statistical Process Control

Where Am I in the Process? Where Do I Apply QFD Within the Roadmaps?

Linkage to the I²DOV Technology Development Roadmap:

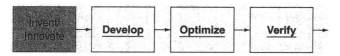

SPC is used in increasing amounts during the Develop, Optimize, and Verify phases of I²DOV. It is seldom used during the Innovate phase.

Linkage to the CDOV Product Development Roadmap:

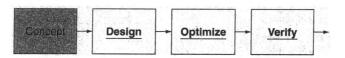

SPC is used in increasing amounts during the Design, Optimize, and Verify phases of CDOV. It is seldom used during the Concept phase.

What Am I Doing in the Process? What Does SPC Do at this Point in the Roadmaps?

The Purpose of SPC in the I²DOV Technology Development Roadmap:

SPC is used to monitor the critical responses of the technology subsystem to assess stability and predictability. This is done in an environment that exposes the system to the generic noises (i.e., temperature, altitude, and humidity) it will experience when used in a product.

The Purpose of SPC in the CDOV Product Development Roadmap:

SPC is used to monitor the critical responses of the product to assess stability and predictability. This is done in an environment that exposes the product to the generic and specific noises it will experience in production, shipment, and the intended-use environment.

What Output Do I Get at the Conclusion of SPC? What Are Key Deliverables from SPC at This Point in the Roadmaps?

- Graphical representation of the stability and predictability of the critical-to-function responses
- Estimates of sigma of the critical-to-function responses
- Estimates of system capability

SPC Process Flow Diagram

This visual reference illustrates the flow of the detailed steps of statistical process control.

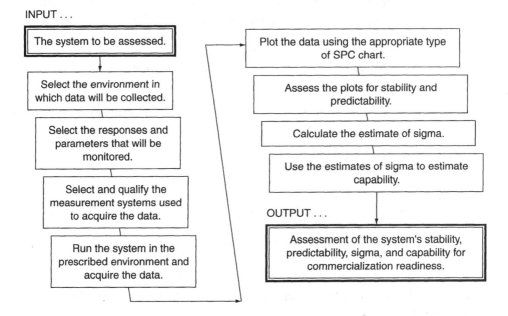

Verbal Descriptions of the Application of Each Block Diagram

This section explains the specific details of what to do during the SPC process:

- *Key detailed inputs* from preceding tools and best practices
- *Specific steps and deliverables* from each step to the next within the process flow
- *Final detailed outputs*

During the Verify phase of technology development and the Verify phase of product development, the readiness of the technology for product development and the readiness of the product for manufacture is assessed. The assessment consists of many checklist items. Stability, predictability, and the capability of the system's critical-to-function responses and parameters are key quantitative checklist items.

Key Detailed Inputs

> **The system to be assessed.**

A system is said to be "in statistical control" when the observed variation is random. This type of random variation is called *common cause* variation since it is the common, everyday variation that all systems experience. Common cause variation cannot be eliminated. It can only be reduced and controlled. A system that is not in statistical control exhibits shifts, drifts, and patterns in its responses. The means and standard deviations vary over time. This type of variation is called *assignable cause* variation since the observed shifts and drifts can be attributed to specific causes such as changes in input voltage or parameter changes due to changes in temperature.

Stability refers to the statistics of the measured variable over time. The mean, standard deviation, and distribution remain relatively unchanged when the response is stable.

The product is not safe for product development or manufacture if the mean and standard deviation of its CTF responses change appreciably in the presence of noise. This is a nonrobust condition. Shifts and drifts in the mean and standard deviation can easily be seen on an SPC chart. An abrupt upward shift and a gradual upward drift, as shown in the "individual" SPC charts in Figures 34–1 and 34–2, are obvious types of assignable cause variation.

Many types of SPC charts are in use. The two major classes are charts for variables or continuous data and charts for attribute data. In DFSS, we are primarily interested in the charts for

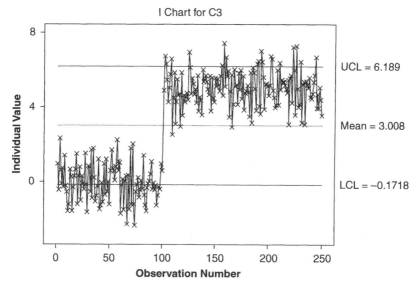

Figure 34–1 Abrupt Shift in Data: Example of Assignable Cause Variation

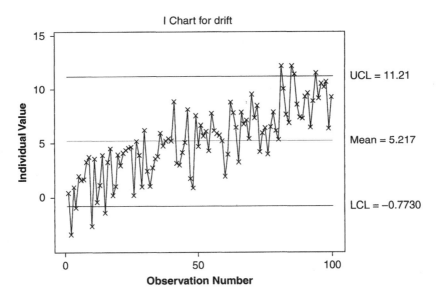

Figure 34–2   Upward Drift in Data: Example of Assignable Cause Variation

continuous data since these are the engineering quantities we manipulate and control. Attribute charts track metrics such as the number of failures and percent pass or reject rates. These lagging indicators of quality measure quantities that are distant from the actual physical functions being performed in the system.

This is not to say that attribute charts have no value. They are of tremendous value on the manufacturing floor. They clearly indicate where to allocate a company's limited resources for process improvement activities.

Attribute charts can also be correlated to sources of variation that have affected the product's quality and thereby help in the search for assignable causes. An improvement to attribute charts is the use of a continuous type of SPC chart to track the system's continuous CTF responses that were identified earlier through CPM. This is still a lagging indicator in that the response is produced after a critical function combines its inputs to form the response. This is an improvement, however, in that these responses are directly tied to the functions being performed in the product.

An even better situation is to use a continuous type of SPC chart to track the continuous CTF parameters that were identified in CPM. These are leading indicators of quality that can be used to predict the response before the function takes place. The transfer functions obtained from a DOE provide the link between the measured quality of the function's inputs and the unmeasured quality of the function's outputs.

We will look at two types of statistical control charts used to chart variables or continuous data, the *Individuals and Moving Range* (I&MR) chart and the *Xbar&R* chart.

Dr. Walter S. Shewhart is credited with proposing the theory of control charts in the 1930s. All SPC charts share a common structure. They display the observations or a function of the observations, they display a measure of central tendency of the observations using a centerline (CL), and they display a measure of variability of the data using upper control limits (UCL) and lower control limits (LCL). The centerline is the mean of the data, and the control limits are drawn some number of standard deviations above and below the centerline, typically three. The general form of the control chart for the variable X is shown in the following equation, where L is typically three.

$$UCL = \mu_X + L\sigma_X$$
$$CL = \mu_X$$
$$LCL = \mu_X - L\sigma_X$$

Many types of control charts fall within each of the two major classes mentioned earlier. These are described in the texts listed at the end of this section. We will focus on two commonly used charts for variables data, the I&MR chart and the Xbar&R chart. We'll begin with the I&MR chart.

An I&MR chart is shown in Figure 34–3 for 100 observations of a CTF response. This chart was created within Minitab's quality control utility. The upper chart in Figure 34–3 displays the individual observations, their mean, and the UCL and LCL. The data must be recorded and plotted in the order in which they were taken to preserve their relationship to time. The estimate of μ is the sample mean of the data, $\hat{\mu} = \bar{x} = -0.03816$, in this case.

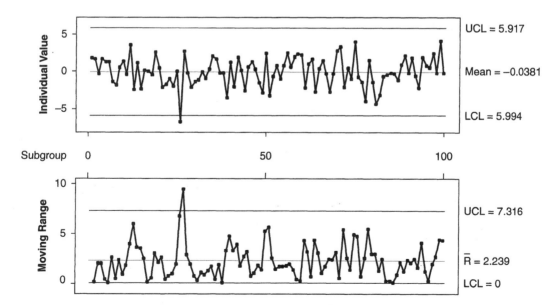

Figure 34–3   I&MR Chart for C2

The estimate of sigma is obtained from the average moving range. A moving range is created by calculating the difference between the current observation and an observation some distance away from the current observation.

The next observation then becomes current and the difference is calculated again. This is repeated until the end of the data is reached. Minitab uses a default distance of two for the moving range, but the user can enter a different value. The ith moving range for two successive observations is given in the following equation:

$$MR_i = |x_i - x_{i-1}|$$

This quantity is calculated for all m moving ranges and the average moving range then is calculated using this equation:

$$\overline{MR} = \frac{\sum_{i=1}^{m} MR_i}{m}$$

In the example given, the average moving range is 2.239. This is shown as $\overline{R}$ in the lower chart in Figure 34–3. The average moving range is a biased estimate of the variation of the response variable. It must be corrected based on the number of observations used in the moving range. The correction factor is called d_2, and is a lookup value. In this example, there are two observations in each moving range, so $d_2 = 1.128$. Our estimate of sigma is

$$\hat{\sigma}_x = \frac{\overline{MR}}{d_2} = \frac{2.239}{1.128} = 1.985$$

The centerline and the 3 sigma UCL and LCL for the individuals chart in Figure 34–3 are given in this equation block:

$$UCL = \hat{\mu}_x + 3\hat{\sigma}_x = -0.03816 + 3(1.985) = 5.917$$
$$CL = \hat{\mu}_x = -0.03816$$
$$LCL = \hat{\mu}_x - 3\hat{\sigma}_x = -0.03816 - 3(1.985) = -5.994$$

The limits for the moving range chart are found in a manner similar to the individual charts, as shown in this equation:

$$UCL = \overline{MR} + 3\hat{\sigma}_R$$
$$CL = \overline{MR}$$
$$LCL = \overline{MR} - 3\hat{\sigma}_R$$

The average moving range is a biased estimate of the variation of the range of the response variable. It must be corrected using two factors, d_2 and d_3, to obtain an unbiased estimate of the variation in the range. These factors are also found in a lookup table. In this example, there are two observations in each moving range, so $d_2 = 1.128$ and $d_3 = 0.853$.

Our estimate of sigma is

$$\hat{\sigma}_R = d_3 \frac{\overline{MR}}{d_2} = \frac{0.853}{1.128}\overline{MR}$$

We can substitute this in to the previous equation block to get the following equation block.

$$UCL = \overline{MR} + 3\hat{\sigma}_R = \overline{MR} + 3d_3\frac{\overline{MR}}{d_2} = \left(1 + 3\frac{d_3}{d_2}\right)\overline{MR}$$

$$CL = \overline{MR}$$

$$LCL = \overline{MR} - 3\hat{\sigma}_R = \overline{MR} - 3d_3\frac{\overline{MR}}{d_2} = \left(1 - 3\frac{d_3}{d_2}\right)\overline{MR}$$

This equation can be simplified if we let

$$D_4 = \left(1 + 3\frac{d_3}{d_2}\right) \text{ and } D_3 = \left(1 - 3\frac{d_3}{d_2}\right)$$

D_3 is set to zero if it is less than zero. For a 3 sigma limit chart with a moving range size of two, $D_3 = 0$ and $D_4 = 3.267$. The range chart centerline and control limits are given in the following equation and are shown in Figure 34–3.

$$UCL = D_4\overline{MR} = 3.267(2.239) = 7.315$$

$$CL = \overline{MR} = 2.239$$

$$LCL = D_3\overline{MR} = 0$$

The discussion so far has assumed that the response is normally distributed. When this is true, the probability of observing a response value above, within, or below the control limits can be determined. The control limits are three standard deviations away from the mean. If the response is normally distributed, there is a 99.73% chance that an observation will be within the control limits. Therefore, there is a $100\% - 99.73\% = 0.27\% = \sim 0.3\%$ chance that an observation will be outside the control limits. This is a small probability indeed. If observations are seen outside the limits, it indicates that the statistics of the response are not stable. The mean and/or the standard deviation are likely to have changed. Therefore, the system is not in control.

However, one observation outside the control limit may not be sufficient to conclude that the system is out of control. The probability of an observation being outside the limits is 0.0027, not zero. We expect an observation to be outside the limits for every

$$\frac{1}{0.0027} = 370.37 \text{ observations}$$

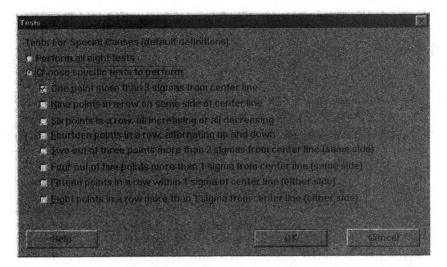

Figure 34–4   Minitab Tests Window for SPC

So, if an individual control chart contains more than 370 observations, we should expect to see one observation outside the limits, even though the system is in control. Many tests, in addition to being outside the control limits, can be applied to detect out-of-control conditions. Minitab provides a list of tests that can be selected, shown in Figure 34–4. These tests help determine when a system is out of control and when action should be taken to bring the system back into control.

Using the control chart we can graphically observe trends and shifts in the response. We can determine if the observed variation is just random, common cause variation. If so, no action should be taken, assuming the observed mean is on target. We can also determine if the system has changed as a result of special cause variation. We can then take action to restore control.

We have assumed normality. If the response is not normally distributed, we can proceed in several ways. One approach is to mathematically transform the response so that the transformed data is normally distributed. The Box-Cox power transformation is a useful transformation when the raw data is unimodal (one humped). Another approach is to make use of the central limit theorem.

Recall an implication of the central limit theorem whereby the sum of many nonnormal random variables tends to be normal. This approach enables the averaging of subsets or groups of data that logically belong together. The determination of a logical group is based on the manner in which the data are produced and the manner in which they are acquired by the measurement system. For example, if five observations are made every hour for 20 hours, a grouping of the 100 observations into 20 groups of five observations is a logical grouping of the data.

We create an Xbar&R chart by logically grouping the observations, averaging the grouped observations, determining an estimate of variation of the grouped observations, and then chart-

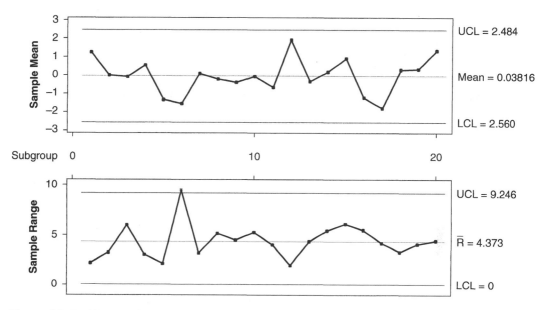

Figure 34–5 Xbar&R Chart for C2

ing the centerline and control limits. An Xbar&R chart is shown in Figure 34–5 using the same data used in the I&MR chart and, in this case, an arbitrarily chosen group size of five.

The Xbar chart plots the average of each group. The centerline is the average of the averages and is $\overline{X} = -0.03816$. The control limits are determined using the average range $\overline{R}$. The average range is the average of all the group ranges. A group range is the difference between the maximum observation and the minimum observation within each group. In this example $\overline{R} = 4.373$.

The variation must first be corrected, using the standard error of the mean, to account for the fact that we are using means and not individual observations. The range is a biased estimate of the variation of the response variable.

It must be further corrected using d_2 to obtain an unbiased estimate of the variation in the average response. In this example, there are five observations in each range, so $d_2 = 2.326$. Our estimate of sigma of the mean response is therefore

$$\hat{\sigma}_x = \frac{\overline{R}}{d_2\sqrt{n}} = \frac{4.373}{2.326\sqrt{5}} = 0.841$$

The centerline and control limits for the Xbar chart are given in this equation:

$$UCL = \overline{\overline{x}} + 3\hat{\sigma}_x = \overline{\overline{x}} + 3\frac{\overline{R}}{d_2\sqrt{n}} = -0.03186 + 3(0.841) = 2.484$$

$$CL = \bar{\bar{x}} = -0.03186$$

$$LCL = \bar{\bar{x}} - 3\hat{\sigma}_{\bar{x}} = \bar{\bar{x}} - 3\frac{\bar{R}}{d_2\sqrt{n}} = -0.03186 - 3(0.841) = -2.56$$

The calculation of the centerline and control limits for the R chart are

$$UCL = \bar{R} + 3\hat{\sigma}_R = \bar{R} + 3d_3\frac{\bar{R}}{d_2} = \left(1 + 3\frac{d_3}{d_2}\right)\bar{R}$$

$$CL = \bar{R}$$

$$LCL = \bar{R} - 3\hat{\sigma}_R = \bar{R} - 3d_3\frac{\bar{R}}{d_2} = \left(1 - 3\frac{d_3}{d_2}\right)\bar{R}$$

The previous equation can be simplified if we once again let

$$D_4 = \left(1 + 3\frac{d_3}{d_2}\right) \text{ and } D_3 = \left(1 - 3\frac{d_3}{d_2}\right)$$

In this example, each range is calculated using five observations so $d_2 = 2.326$ and $d_3 = 0.864$. Our estimate of sigma is

$$\hat{\sigma}_R = \frac{d_3\bar{R}}{d_2} = \frac{(0.864)4.373}{2.326} = 1.624$$

For a 3 sigma limit chart with a range size of five, $D_3 = 0$ and $D_4 = 2.115$. The range chart centerline and control limits are given in this equation:

$$UCL = D_4\bar{R} = 2.115(4.373) = 9.246$$

$$CL = \bar{R} = 4.373$$

$$LCL = D_3\bar{R} = 0$$

These limits are shown in Figure 34–5.

If we assume or confirm that the means are normally distributed, there is a 0.27% chance of the average of five observations being outside the limits. Notice, of course, that the limits are tighter for the Xbar chart than for the individuals. This must be the case since the variation of the average is always less than that of the individual observations. Also realize that there is a much greater chance that the means are normally distributed than the individual observations due to the central limit theorem.

In all cases, we must not confuse or mix specification limits with control limits. We never plot specification limits on a control chart. A system may be in control and at the same time out of specification. This is a bad situation, at best a 3 sigma system with a capability close to unity. We can, at best, expect a short-term defect rate of 2,700 ppm.

The system provides 6 sigma performance when the specification limits are three standard deviations beyond the control limits. This system has a capability close to two. We can, at best, expect a long-term defect rate of 3.4 ppm. This assumes a long-term 1.5 sigma variation in the mean.

We have seen how to use an SPC chart to determine if a system's critical-to-function response is in control in the presence of noise. If this is the case and its capability index is two or better, the system has earned the right to progress to the next step in the product development process.

We are now ready to consider an example.

Specific Steps and Deliverables

> **Select the environment in which the data will be collected.**

We want to assess the level of control and stability of the system in an environment that closely simulates the intended-use environment. This may require an environmental chamber and/or other methods of simulating the intended-use environment.

> **Select the responses and parameters that will be monitored.**

The primary quality characteristic of interest in this example is the leak rate measured in units of bar/sec. Leak rate is a nonnegative, continuous random variable. Normality is not expected. The USL = 12 and LSL = 0.

> **Select and qualify the measurement systems used to acquire the data.**

A measurement system analysis was performed on the test fixture. Gage R&R and P/T were acceptable, but a slow drift was observed over a week's time. It was found to be due to a plastic hose in the test fixture. The plastic tube would expand over time and cause a steady downward drift in measured leak rate. A metal tube was installed and the fixture determined to be capable.

> **Run the system in the prescribed environment and acquire the data.**

The leak rates of 250 units were measured as a trial sample. They are shown in the plot in Figure 34–6. The circled observations were due to assembly errors and were removed from the data.

The sample histogram and normality test are shown in Figures 34–7 and 34–8, respectively.

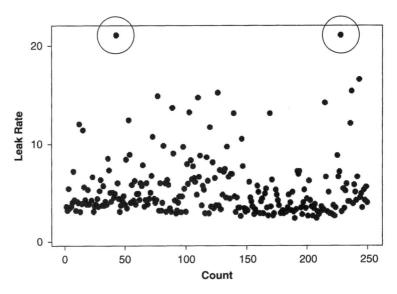

Figure 34–6 Leak Rate Trial Sample

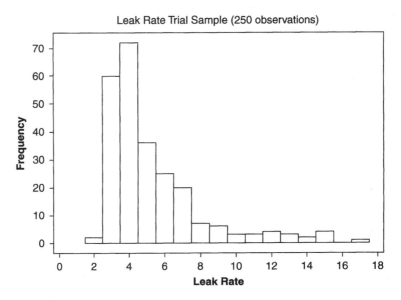

Figure 34–7 Sample Histogram

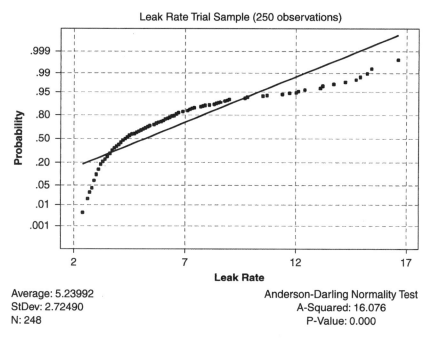

Figure 34–8   Normal Probability Plot

The data are nonnormal, as expected. Several approaches were tried to normalize the data. A Box-Cox transformation was found (lambda = −0.899), but the transformed data failed the Anderson-Darling test for normality. A different approach was tried. Observations were grouped in sizes of 5 and 10, and then averaged. Normality was tested for each size. A size of 10 produced normally distributed means (Figure 34–9).

> **Plot the data using the appropriate type of SPC chart.**

An Xbar&R chart with a subgroup size of 10 is an appropriate control chart for this situation. This was produced using subsequent observations and is shown in Figure 34–10.

> **Assess the plots for stability and predictability.**

Obviously the response is not in control. The lack of control indicates its mean and/or standard deviation are changing. The system's performance is not predictable. The reason for the lack of control must be found. Is the system responding to noise or to changes in its critical-to-function

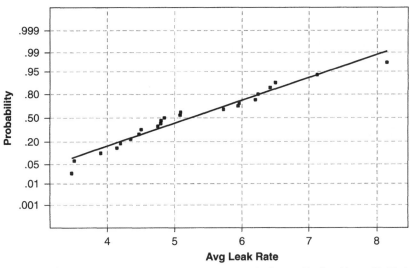

Average: 5.2348
StDev: 1.6721
N: 25

Anderson-Darling Normality Test
A-Squared: 0.476
P-Value: 0.219

Figure 34–9   Average Leak Rate for $n = 10$

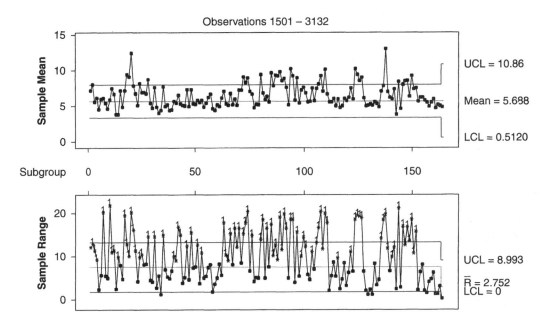

Figure 34–10   Xbar&R Chart for Leak Rate

parameters? The system's critical parameters must be measured along with the response. This will indicate if tighter control must be applied to the input parameters. External noises should also be measured. Perhaps additional robustness efforts must be applied to reduce the system's sensitivity to these sources of variation.

In either case, there is much work to be done here. Hopefully, robustness and tolerancing efforts will bring the system into control. However, this situation may be the result of poor concept selection and indicate a fundamental flaw in the design. The gate reviews of a disciplined product development process are designed to minimize this as a reason for poor control.

> **Calculate the estimate of sigma.**

Sigma is estimated to be $\hat{\sigma} = \dfrac{\overline{R}}{d_2} = \dfrac{2.752}{3.078} \approx 0.9$ for $n = 10$.

> **Use the estimates of sigma to estimate capability.**

Capability is estimated to be $C_p = \dfrac{\text{USL} - \text{LSL}}{6 \hat{\sigma}} = \dfrac{12 - 0}{6(0.9)} = 2.2$.

This sounds good. However, remember that the system is not in control. This is an overly optimistic estimate of system performance. The capability index cannot be used as the sole factor in determining system performance. Good capability is a necessary but insufficient requirement.

Final Detailed Outputs

> **Assessment of the system's stability, predictability, sigma, and capability for commercialization readiness.**

The system has good capability. However, it is not stable, so its performance is not predictable. This system is not ready for the next step in commercialization. The reason for the lack of control must be found.

SPC Checklist and Scorecard
Checklist for SPC

Actions:	Was this Action Deployed? YES or NO	Excellence Rating: 1 = Poor 10 = Excellent
Select the environment in which data will be collected.	_____	_____
Select the responses and parameters that will be monitored.	_____	_____
Select and qualify the measurement systems used to acquire the data.	_____	_____
Run the system in the prescribed environment and acquire the data.	_____	_____
Plot the data using the appropriate type of SPC chart.	_____	_____
Assess the plots for stability and predictability.	_____	_____
Calculate the estimate of sigma.	_____	_____
Use the estimates of sigma to estimate capability.	_____	_____
Assessment of the system's stability, predictability, sigma, and capability for commercialization readiness.	_____	_____

Scorecard for SPC

The definition of a good scorecard is that it should illustrate in simple but quantitative terms what gains were attained from the work. This suggested format for a scorecard can be used by a product planning team to help document the completeness of its work.

Critical Response:	Is it Stable and in Control?	Capability Index:	Ready for the Next Step?	Unit of Measure:
1.	_____	_____	_____	_____
2.	_____	_____	_____	_____
3.	_____	_____	_____	_____

We repeat the final deliverables:

- Graphical representation of the stability and predictability of the critical-to-function responses
- Estimates of sigma of the critical-to-function responses
- Estimates of system capability

References

Montgomery, D. (1997). *Introduction to statistical quality control.* New York: Wiley & Sons.
Wheeler, D. J. (1995). *Advanced topics in statistical process control.* Knoxville, TN: SPC Press.

Linking Design to Operations

This Epilogue was written by Stephen A. Zinkgraf, Ph.D., CEO and President of Sigma Breakthrough Technologies, Inc.

The penultimate goal of Design For Six Sigma is to create new products, technologies, and processes that lead measurably and directly to company growth by meeting or beating the business plan. The best global metric I've seen to describe the success of a commercialization program is a simple ratio:

Commercialization success = $ spent on R&D / $ in revenue from new products

I love this metric because it reminds us that R&D is a business process with the accountability to generate new revenue. If the ratio is one or greater, your company probably does not have an effective commercialization program. I see large, Fortune 500 companies—even those noted for product development—produce a very poor ratio for commercialization success. One company I've worked with that is known for product innovation saw their commercialization success ratio to be about one, which was not a very good return-on-investment for the millions they'd spent on R&D.

Why is this ratio so hard to improve? Anyone who has been involved with operations management in a manufacturing company has had the unique experience of supporting the launch of a new product or new process. A typical launch of a new product in most companies inflicts enormous pain on all parties involved. The product usually has huge production problems with low yields at all process steps, and with repair loops and repair activities popping up everywhere. Line operators and supervisors run amuck screaming for technical assistance. The operations managers claim that the product was thrown over the wall by design. The design engineers traditionally spend a large percentage of their time on the plant floor "troubleshooting," and engineering change notices flood the system.

When a company implements a new Six Sigma program, the company will usually start the process in their operations functions and focus on cost cutting. About 50 percent to 75 percent of all the operations Six Sigma projects will have a design change on a product or process as a solution. In fact, companies that have not implemented Design For Six Sigma (DFSS) will see their design engineers spend 25 percent to 40 percent (or more) working out issues on products already launched into the manufacturing operations.

Design engineers essentially become part-time operations process engineers. In fact, a lot of factories have a core of engineers that exist to work primarily on product issues. This is a huge cost-of-poor-quality (COPQ) to those companies since these resources are not focused on growth and new products or services but on maintaining current products and processes. Since growth is the primary success factor for companies today, poor upfront design practices directly impact the company's long-term success.

An example from the Honeywell (formerly AlliedSignal) Six Sigma deployment launched in late 1994 illustrates the point. In 1995, the Engineered Materials Sector of the former AlliedSignal earned about $100 million from the first year of its Six Sigma deployment into operations. They estimated that about 75 percent of that result could be tied directly to engineering changes in product or process specifications—in other words, from a major product or process redesign.

Had AlliedSignal had a viable DFSS program, their operations program might have only produced $25 million, while the other $75 million would have been earned later from the initial launch of the new products. This is also a COPQ that is not tracked by design groups. If it was, companies would move much more rapidly into Design For Six Sigma than they normally do. We see companies usually initiating their DFSS programs in 1 to 2 years after they start their operations programs. I'll say that Six Sigma has become an industry standard in managing companies primarily because of poor design processes. These poor design processes allow operations process improvement projects to yield huge benefits because proper upfront preventative work was not done.

As Robert Copper, author of *Winning at New Products,* once quoted, "Beating a new product into submission in operations for a year will save the company six weeks of upfront preventative work." Therein lies the operations goal and focus of Design For Six Sigma: to seamlessly introduce new products, technologies, processes, and services into the company. By following the roadmaps and using the tools summarized in this book, companies will be able to move a new product into operations quickly and with far less pain.

This methodology sets the stage for the company to meet the business plan goals of the new product within a relatively short period of time. I define product launch cycle-time as the time between the initial concept and the product finally meeting the business plan as to cost, quality, and delivery. The usual definition of product launch cycle-time is the time from developing the initial concept to the first manufacturing run. This latter definition does not take into account the business impact of the new product. DFSS is definitely a major part of the product development business process and should be treated accordingly.

Design For Six Sigma is the antithesis of the basic notion of Six Sigma. Six Sigma exists primarily to remove variability from all systems. Design For Six Sigma actually addresses proficient imposition of variability on the system. You see, not all variability is bad and, in fact, if a company doesn't aggressively impose variability on itself, the company will cease to exist.

Ford Motor Company nearly went out of business in its early years because Ford insisted on making cars only one color—black. General Motors, in contrast, offered a variety of colors and captured a large percentage of the automotive market. So, by General Motors foisting the variability of multiple colors on its manufacturing system, GM was able to go head-to-head with a major competitor. To generalize, if a company is not always introducing the variability associated with new products, technologies, or services, that company will ultimately lose in the marketplace.

However, all variability, whether good (new products or services) or bad (scrap and repair costs), has an adverse effect on manufacturing operations. If the reader has had any experience with paint lines on the manufacturing floor (usually one of the biggest sources of poor yield and high COPQ), he or she will readily recognize that GM's introduction of multiple colors must have had a huge impact on their manufacturing processes—either forcing them to deploy multiple paint lines or to go through the pain of complex color changeovers. Variability from multiple colors negatively impacted quality, cost, and speed.

With the two concepts in mind: (1) Not all variability is bad and (2) variability—good or bad—negatively impacts manufacturing operations, we can put DFSS into perspective. DFSS identifies the most predominant sources of variation in the product system. Through Critical Parameter Management (CPM), the important process inputs (X's) and outputs (Y's) are identified early in the commercialization process along with a recommended control plan. The CPM process gives the manufacturing operations a head start on planning for new product launches. CPM ensures operations that the probabilities of a smooth new product launch are high.

While all failure modes cannot be anticipated, by doing the upfront analysis the operations folks can deploy the new manufacturing system and iron out any unanticipated problems very quickly. In a company good at DFSS, Six Sigma in operations will focus on smaller issues. Black Belts and Green Belts will move away from the traditional project mix of 80 percent reactive projects and 20 percent proactive projects to 20 percent reactive projects and 80 percent proactive projects. In fact, companies that are good at DFSS will have more resources in operations available to work with customer and key suppliers than companies that are not good at DFSS.

With a good DFSS program, Six Sigma projects designed to improve operations performance will take less time and resources to complete. Let's look at the following table. I have listed the major steps and tools in a typical Six Sigma improvement project along with a comparison of the activities with a poor DFSS system and with a good DFSS system. You will notice that every step in the Process Improvement Methodology™* is positively affected by good DFSS practices.

---

*Process Improvement Methodology™ is a federally trademarked term of Sigma Breakthrough Technologies, Inc. Term used with permission of Sigma Breakthrough Technologies, Inc.

Comparison of Process Improvement between Poor and Good DFSS Processes

Process Improvement Step	Poor DFSS System	Good DFSS System
I. Measure Phase		
Process Map	Develop from scratch—critical parameters are unclear	Modify and Update—critical parameters identified from DFSS
Cause and Effects Matrix	Develop from scratch—critical parameters unclear	Modify and Update—initial analysis done by DFSS team
Gage Studies	Perform time-consuming gage studies—project is stalled until completed	All major measurement systems analyzed during DFSS—only secondary systems need analysis
Capability Studies	Develop from scratch—lack of clarity of customer requirements	Capability data available from DFSS—customer requirements are clear and well documented
II. Analysis Phase		
Failure Modes and Effects Analysis	Develop from scratch—time-consuming and necessarily constrained	Update and modify FMEA from DFSS—critical parameters clearly defined—FMEA is reviewed rather than created
Multi-Vari Study	Study has wide scope entailing more data because of general lack of technical knowledge of the process	Study is focused—critical parameters are clear—data collection efforts are less intensive
III. Improve Phase		
Experimental Design	More experiments—more input variables studied because of lack of technical knowledge of the process	Few experiments—more focused because critical parameters are already known
Lean Manufacturing Tools	Lean tools likely to be successful because of an initially poorly designed process	Lean tools are unlikely to be necessary because of optimal process design during DFSS
IV. Control		
Develop Process Control Plan	Develop from scratch—time-consuming to implement systems	Updated and modified—control systems in place at product launch because of DFSS
Put Performance Metrics in Place	Metrics developed from scratch and systems to collect data developed	Metrics reviewed and modified but data collection systems already in place from DFSS
Establish Long-Term Capability	Longer study required with more variables involved	Short studies with the right variables from CPM

The Process Improvement Methodology™ roadmap depicted in this table is federally copyrighted by and used with permission from Sigma Breakthrough Technologies, Inc.

Process improvement efforts in operations will be more focused and much faster in companies that define the critical parameters and control systems for new products during the DFSS process. Ultimately, operations Black Belts and Green Belts will be focused on preventative activities rather than reactive activities. Because of minimizing defects early in the new product launch cycle, fewer customer complaints and field problems will occur. Operations resources will extend their influence into the customer and supply base, establishing better relationships and strategic partnerships where everyone is much more successful.

Ultimately, product development engineers will work only a small percentage of the time to troubleshoot new products in operations. Because new product launches are smoother and business plan objectives are more quickly met, the company with a good DFSS process is now set to launch more new products each year than ever before.

As the saying goes, "In research and development (R&D), research creates surprises and development prevents surprises." That is exactly what DFSS has to offer. By clearly understanding customer needs using Concept Engineering and identifying critical parameters early in the commercialization process, creating innovative concepts to meet those needs and providing for seamless new product launches in operations, a DFSS-driven company will have more successful products and hit their business plan goals at levels much greater than their competitors.

Enjoy your journey into DFSS. Since product commercialization is the most complex of all business processes, your journey will not be easy. However, if you visualize your company being a leader in developing successful and innovative new products and being able to commercialize at light speed through seamless new product launches into operations, you'll be getting close to the real benefit of DFSS. See you on the road to Six Sigma!

Design For Six Sigma Abbreviations

ANOM analysis of means

ANOVA analysis of variance

CAD computer-aided design

CAPs critical adjustment parameters

CCD central composite design

CDOV concept, design, optimize, and verify

CF control factor

CFPs critical functional parameters

CFRs critical functional responses

CGI capability growth index

CL centerline (in SPC charts)

CNFs compounded noise factors

COPQ cost of poor quality

COV coefficient of variation

Cp capability index

Cp$_d$ design capability

Cpk measures long-term performance capability

CPM Critical Parameter Management

Cp$_m$ manufacturing process capability

CTF critical-to-function

DFMEA design failure modes and effects analysis

DFNS Design For N-Sigma

DFSS Design For Six Sigma

DFx Design For "x"

DOE design of experiments

dof degree of freedom

ESS environmental stress screening

FMEA failure modes and effects analysis

FTA fault tree analysis

Gage R&R repeatability and reproducibility

Gnom. nominal gap, contact, or interference dimension

HALT highly accelerated life testing

HAST highly accelerated stress testing

HOQ House of Quality

I&MR chart Individuals and Moving Range chart

I^{2}DOV invent/innovate, develop, optimize, and verify

KJ analysis (for Jiro Kawakita)

LCL lower control limit

LSL lower specification limit

LTB larger-the-better

MAIC measure, analyze, improve, and control

MSA measurement systems analysis

MSD mean square deviation

MTBF mean time between failure

MTTF mean time to failure
NTB nominal-the-best
NUD new, unique, and difficult
pdf probability density function
PLS product line strategy
P/T ratio precision-to-tolerance
PERT chart (program evaluation review technique)
PRESS prediction sum of squares
QFD quality function deployment
R&D community (research and development)
R&R study (repeatability and reproducibility)

RBD reliability block diagram
RPN risk priority numbers
RSM response surface methods
S/N signal-to-noise ratio
SPC statistical process control
SRD system requirements document
STB smaller-the-better
UCL upper control limit
UMC unit manufacturing cost
USL upper specification limit
VOC voice of the customer
VOP voice of the product/process

Glossary

A

absolute zero   A property of a quality characteristic that allows all measurements to begin from the number zero and go up the rational number line from that point. Many designs have an absolute zero scale of measurement and should have their data measured relative to that scale.

additive model   A mathematical function that expresses a dependent variable, y, in terms of the sum (simple addition) of functions of *relatively* independent variables represented by the control factors in an optimization experiment. The additive model assumes mild-to-moderate interactivity between the control factors and assigns the numerical contribution to the response for this interactivity to the error term in the model: $y = f(A) + g(B) + h(C) + \ldots + error$. This model is approximate in nature and is useful for engineering optimization but not for rigorous statistical math modeling to define fundamental functional relationships (ideal/transfer functions).

additivity   The natural or intentionally induced independence of each individual control factor's effect on a measured response.

adjustment factor   Any control factor that has a strong effect on the mean performance of the response. There are two special types of adjustment factors: scaling factors and tuning factors. Adjustment factors are used in static and dynamic applications of the two-step optimization process. In Critical Parameter Management, they are referred to as *critical adjustment parameters* (CAPs).

aliasing   See *confounding.*

ANOM (*AN*alysis *Of* the *M*ean)   An analytical process that quantifies the mean response for each individual control factor level. ANOM can be performed on data that is in regular engineering units or data that has been transformed into some form of signal-to-noise ratio or other data transform. Main effects and interaction plots are created from ANOM data.

ANOVA (*AN*alysis *Of* the *VA*riance)   An analytical process that decomposes the contribution each individual control factor has on the overall experimental response. The ANOVA process is also capable of accounting for the contribution of interactive effects between control factors and experimental error in the response, provided enough degrees of freedom are established in the experimental array. The value of epsilon squared (percentage contribution to overall CFR variation) is calculated using data from ANOVA.

antisynergistic interaction   An interaction between two or more control factors that have a negative, nonmonotonic effect on a measured critical functional response. These interactions can either be due to physical law (real, physical, fundamental interaction) or from mismatches between continuous, engineering x variable inputs and attribute or "quality metric" output Y variables (virtual interaction).

array   An arithmetically derived matrix or table of rows and columns that is used to impose an order for efficient experimentation. The rows contain the individual experiments. The columns contain the experimental factors and their individual levels or set points.

asymmetric loss function   A version of the nominal-the-best loss function that accounts for the possibility of different economic consequences as quantified by the economic coefficients k_{lower} and k_{upper} on either side of the target value.

average quality loss   The application of the quadratic loss function to numerous samples of a product or process resulting in the calculation of a term called the *mean square deviation* (MSD). The average qual-

ity loss approximates the nature of loss over a variety of customer economic scenarios, thus providing a generalized evaluation of the average or aggregated loss.

B

balanced experiment   An experiment where each experimental factor is provided with an equal opportunity to express its effect on the measured response. The term balance is synonymous with the *orthogonality* of a matrix experiment. This property precludes the ability of multi-colinearity from becoming a problem in data analysis.

benchmarking   The process of comparative analysis between two or more concepts, components, subassemblies, subsystems, products, or processes. The goal of benchmarking is to qualitatively and quantitatively identify a superior subject within the competing choices. Often the benchmark is used as a standard to meet or surpass. Benchmarks are used in building Houses of Quality, concept generation, and the Pugh concept selection process.

best practice   A preferred and repeatable action or set of actions completed to fulfill a specific requirement or set of requirements during the phases within a product development process.

beta (β)   The Greek letter, β, is used to represent the slope of a best fit line. It indicates the linear relationship between the signal factor(s) (critical adjustment parameters) and the measured critical functional response in a dynamic robustness optimization experiment.

blocking   A technique used in classical DOE to remove the effects of unwanted, assignable cause noise or variability from the experimental response so that only the effects from the control factors are present in the response data. Blocking is a data purification process used to help assure the integrity of the experimental data used in constructing a statistically significant math model.

build-test-fix methods   A traditional, undisciplined engineering process used to iterate product development and design activities until a "successful" product is ultimately produced. These processes can be lengthy and costly to employ. The main difficulty with this approach is inefficiency and production of suboptimal results requiring even more redesign effort into the production phase of the product life cycle. Build-test-fix produces design waste and rework.

C

calibration   Setting a meter or a transducer (measurement instrumentation) to a known standard reference that assures the quantification of reliable measures of physical phenomena. Calibration makes the accuracy and repeatability of measurements as good as possible given the inherent capability of the instrument.

capability growth index   The calculated percentage between 0 percent and 100 percent that a group of system, subsystem, or subassembly CFRs have attained in getting their Cp indices to equal a value of 2, indicating how well their critical functional responses have attained 6 sigma performance during product development.

capability index   Cp and Cpk indices that calculate the ratio of the voice of the customer vs. the voice of the product or process. Cp is a measure of capability based on short-term or small samples of data—usually what is available during product development. Cpk is a measure of

long term or large samples of data that include not only variation about the mean but also the shifting of the mean itself—usually available during steady state production.

central composite design (CCD)   An experimental design technique used in response surface methods to build nonlinear math models of interactive and nonlinear phenomena. A major tool used in the CDOV sequential design of experiments process for design and process optimization after robust design to place the mean of a CFR on target with high accuracy. The main inputs to a CCD are CAPs.

classical design of experiments (DOE)   Experimental methods employed to construct math models relating a dependent variable (the measured critical functional response) to the set points of any number of independent variables (the experimental control factors). DOE is used sequentially to build knowledge of fundamental functional relationships (ideal/transfer functions) between various factors and a response variable.

classification factor   A parameter that uses integers, simple binary states, or nonnumerical indicators to define the levels that are evaluated in the experimental process. Classes can be stages, groups, or conditions that define the level of the classification factor. This type of factor is used in cases where continuous variables cannot adequately define the state of the factor set point. We usually try to avoid using these in DFSS.

coefficient of variation (COV)   The ratio obtained when the standard deviation is divided by the mean. The COV is often expressed as a percentage by multiplying the ratio by 100. It is an indication of how much variability is occurring around the mean response. Often it is of interest to detect a quantitative relationship between the changes in the mean response and the corresponding changes in the standard deviation. This will be used in the cooptimization of certain designs where the variance scales along with the mean. One form of Taguchi's signal-to-noise ratios is based on the square of this ratio.

column   The part of an experimental array that contains experimental factors and their various levels or set points.

commercialization   Business process that harnesses the resources of a company in the endeavor of conceiving, developing, designing, optimizing, verifying design and process capability, producing, selling, distributing, and servicing a product.

compensation   The use of feed-forward or feedback control mechanisms to intervene when certain noise effects are present in a product or process. The use of compensation is only done when insensitivity to noise cannot be attained through robustness optimization.

complete   The property of a single critical functional response that promotes the quantification of all the dimensions of the ideal function of a design element.

component   A single part in a subassembly, subsystem, or system. An example would be a stamped metal part prior to having anything assembled to it.

component requirements document   The document that contains all the requirements for a given component. They are often converted into a quality plan that is given to the production supply chain to set the targets and constrain the variation allowed in the incoming components.

compound control factors   Control factors that have been strategically grouped (addi-

tivity grouping) to help minimize the possibility of strong, disruptive interactivity between the individual control factors. Grouped control factors are comprised of two or more factors that have their individual effects on the response understood and are grouped together so that they cocontribute in the same way to the measured response. Control factors can also be grouped when necessary to allow the use of a smaller orthogonal array. This technique is primarily used in robust design after the ideal function is identified.

compound noise factors   Noise factors that have been strategically grouped by the nature of their strength and effect on the directionality of the measured response. They are used to increase the efficiency of inducing statistically significant noises in robustness experiments.

concept engineering   The initial phase of the DFSS approach to product development. This part of the commercialization process is used to link the voice of the customer with the inventive and technical capabilities of the business (voice of the product and process) to help develop and design superior product concepts.

concurrent engineering   The process of simultaneously carrying out various product design, manufacturing, and service development activities in parallel rather than in series. The outcome is reduced cycle time, higher quality, and lower product commercialization costs. Concurrent engineering is noted for its focus on multifunctional teams that work closely to improve communication of seemingly conflicting requirements and priorities. Their goal is to formulate effective synergy around the customers' needs to bring the project to a speedy and optimal conclusion. It prevents problems and is a baseline competency in DFSS.

confounding The situation created when one control factor main effect or interaction is numerically blended with one or more other control factor main effects and/or interactive effects by intentionally or unintentionally allocating the same degrees of freedom in a matrix experiment to multiple factors. Often the term *aliasing* is used interchangeably with confounding. Confounding can be helpful or harmful depending on the amount of interactivity between the confounded factors.

conservation of energy The physical law that states that energy is neither created or destroyed. It is used as a basis for designing experiments to build and understand ideal functions and additive robustness models in DFSS. Energy models are additive models due to this "conservative" law (e. g., *Bernouli's equation*).

conservation of mass The physical law that states that mass is neither created or destroyed. It is used as a basis for designing experiments to build and understand ideal functions and additive robustness models in DFSS. Mass models are additive models due to this "conservative" law (e. g., *Lavosier's equation*).

continuous factor A parameter that can take on any value within a range of rational numbers to define the levels that are evaluated in the sequential experimental design process. We prefer to use these as x and Y variables (scalars and vectors) for modeling in DFSS.

contour mapping A two-parameter graphical output from a computer analysis of the math model created from any number of classically designed experiments (usually response surface experiments). It is used to predict points or areas of local or global optimum performance based on empirical data. It does not typically account for the effects of noise on the predicted response.

control factor The factors or parameters (CFP or CTF spec) in a design or process that the engineer can control and specify to define the optimum combination of set points for satisfying the voice of the customer.

critical adjustment parameter (CAP) A specific type of CFP that controls the mean of a CFR. CAPs are identified using sequential DOE and engineering analysis. They are the input parameters for response surface methods for the optimization of mean performance optimization after robust design is completed. They enable Cpk to be set equal to Cp, thus enabling entitlement to be approached if not attained.

critical functional parameter (CFP) An input variable (usually an engineered additivity grouping) at the subassembly or subsystem level that controls the mean or variation of a CFR.

critical functional response (CFR) A measured scalar or vector (complete, fundamental, continuous engineering variable) output variable that is critical to the fulfillment of a critical customer requirement. Some refer to these critical customer requirements as CTQs.

Critical Parameter Management (CPM) The process within DFSS that develops critical requirements and measures critical functional responses to design, optimize, and verify the capability of a product and its supporting network of manufacturing and service processes.

critical-to-function specification (CTF) A dimension, surface, or bulk characteristic (typically a scalar) that is critical to a component's contribution to a subassembly, subsystem, or system level CFR.

criticality A measurable requirement or functional response that is highly important to a customer. All requirements are important but only a few are truly critical.

crossed array experiment The combination of inner and outer orthogonal arrays to intentionally introduce noise into the experiment in a disciplined and strategic fashion. This is done to produce data that can be transformed into control factor × noise factor interaction plots to isolate robust set points for the control factors. S/N ratios replace such plots when compounded noises replace the outer arrays in a robustness experiment.

D

degrees of freedom (dof) The capacity of an experimental factor or an entire matrix experiment to produce information. The dof_f for an experimental factor is always one less than the number of levels assigned to the factor. The dof_{exp} for an entire matrix experiment is always one less than the number of experimental runs in the experiment. Whether accounting for experimental factor dof or matrix experiment dof, one dof is always employed to calculate the average response. This is why one dof is always subtracted from the total.

design capability (Cp_d) The Cp index for a design's critical functional response in ratio to its upper and lower specification limits (VOC-based tolerance limits).

design of experiments (DOE) A process for generating data that uses a mathematically derived matrix to methodically gather and evaluate the effect of numerous param-

eters on a response variable. Designed experiments, when properly used, efficiently produce useful data for model building or engineering optimization activities.

deterioration noise factor A source of variability that results in some form of physical deterioration or degradation of a product or process. This is also referred to as an *inner noise* because it refers to variation inside the controllable factor levels.

double dynamic signal-to-noise case A variation of the dynamic S/N case that employs two signal factors. The two signal factors (CAPs) are the functional signal factor, which is a factor that directly and actively controls some form of input energy into the design, and the process signal factor, which possesses a linear relationship to the output response and indirectly controls (attenuates) the energy transformation occurring in the design.

drift The tendency of a product or process output or set point to move off target.

dynamic signal-to-noise case A method of studying the linearity, sensitivity, and variability of a design by relating the performance of a signal factor (CAP), numerous control factors, and noise factors to the measured CFR. These cases relate the power of the proportionality of the signal factor (CAP) to the measured CFR with the power of the variance in the CFR due to noise.

E

economic coefficient The economic coefficient is used in the quality loss function. It represents the proportionality constant in the loss function of the average dollars lost (A_0) due to a customer reaction to off-target performance and the square of the deviation from the target response (Δ_0^2).

This is typically, but not exclusively, calculated when approximately 50 percent of the customers are motivated to take some course of economic action due to poor performance (but not necessarily outright functional failure). This is often referred to as the *LD 50 point* in the literature.

energy flow map A representation of an engineering system that shows the paths of energy divided into productive and nonproductive work. Analogous to a free body diagram from an energy perspective. They account for the law of conservation of energy and are used in preparation for math modeling and design of experiments.

energy transformation The physical process a design or product system uses to convert some form of input energy into various other forms of energy that ultimately produce a measurable response. The measurable response may itself be a form of energy or the consequence of energy transformations, which have taken place within the design.

engineering metrics A scalar or vector that is usually called a CTF spec, CFP, CAP, or CFR. They are greatly preferred as opposed to quality metrics (yield, defects, and so on) in DFSS.

Engineering Process A set of disciplined, planned, and interrelated activities that are employed by engineers to conceive, develop, design, optimize, and verify the capability of a new product or process design.

environmental noise factors Sources of variability that are due to effects that are external to the design or product, also referred to as *outer noise*. They can also be sources of variability that one neighboring subsystem imposes on another neighboring subsystem or component. Examples include vibration, heat, contamination, misuse, or overloading.

epsilon squared Percentage contribution to overall CFR variation calculated from ANOVA data: $(SS_{factor}/SS_{total}) \times 100 = \varepsilon^2$

experiment An evaluation or series of evaluations that explore, define, quantify, and build data that can be used to model or predict functional performance in a component, subassembly, subsystem, or product. Experiments can be used to build fundamental knowledge for scientific research, or they can be used to design and optimize product or process performance in the engineering context of a specific commercialization process.

experimental efficiency This process-related activity is facilitated by intelligent application of engineering knowledge and the proper use of designed experimental techniques. Examples include the use of Fractional Factorial arrays, control factors that are engineered for additivity, and compounded noise factors.

experimental error The variability present in experimental data that is caused by meter error and drift, human inconsistency in taking data, random variability taking place in numerous noise factors not included in the noise array, and control factors that have not been included in the inner array. In the Taguchi approach, variability in the data due to interactive effects is often but not always included as experimental error.

experimental factors Independent parameters that are studied in an orthogonal array experiment. Robust design classifies experimental factors as either control factors or noise factors.

experimental space The combination of all the control factor, noise factor, and signal factor (CAP) levels that produce the range of measured response values in an experiment.

F

F-ratio The ratio formed in the ANOVA process by dividing the mean square of each experimental factor effect by the MS of the error variance. This is the ratio of variation occurring between each of the experimental factors in comparison to the variation occurring within all the experimental factors being evaluated in the experiment. It is a form of signal-to-noise ratio in a statistical sense. The noise in this case is random experimental error, not variability due to the assignable cause noise factors in the Taguchi noise array.

factor effect The numerical measure (magnitude) of the contribution an experimental factor has on the variation of a functional response (CFR). In practical terms we use ε^2 to measure this, and in statistical terms we use the coefficient in the regression model.

feedback control system A method of compensating for the variability in a process or product by sampling output response and sending a feedback signal that changes a critical adjustment parameter to put the mean of the response back on its intended target.

feed forward control system A method of compensating for the variability in a process or product by sampling specific noise factor levels and sending a signal that changes an adjustment factor to keep the response on its intended target.

Fisher, Sir Ronald The inventor of much of what is known today as classical design of experiments. His major focus was on optimizing crop yield in the British agricultural industry.

fraction defective The portion of an experimental or production sample that is found to be outside of the specification limits. Counting and minimizing defects are not metrics that best reflect the fundamental, ideal function of a design and its performance. They are more commonly used in production and transactional Six Sigma applications for cost reduction projects.

Fractional Factorial design A family of two- and three-level orthogonal arrays that greatly aid in experimental efficiency. Depending on how a Fractional Factorial design is loaded with control factors, it can be used to study interactions or it can be manipulated to promote evaluation of additivity in a design or process. Fractional Factorial designs are a subset of all the possible treatment combinations possible in a Full Factorial array.

frequency distribution table A table of the frequency of occurrence of measured values within specific ranges that are used to generate a histogram.

Full Factorial design Two- and three-level orthogonal arrays that include every possible combination between the experimental factors. Full Factorial experimental designs use degrees of freedom to account for all the main effects and all interactions between factors included in the experimental array. Basically all interactions beyond two-way interactions are likely to be of negligible consequence; thus, there is little need to use large arrays to rigorously evaluate such rare and unlikely three-way interactions and above.

functional signal factor (M) A signal factor (CAP) used in the double dynamic Taguchi case where engineering knowledge is used to identify an input variable for a design that controls the major energy transformation in the design. This transformation controls a proportional relationship between the functional signal factor and measured response. A functional signal factor is used to tune the design's performance on to any desired target. Tunable technology is developed from a concept that possesses a functional signal factor. RSM is often used to do final optimization using these kinds of parameters.

fundamental The property of a critical functional response that expresses the basic or elemental physical activity that is ultimately responsible for delivering customer satisfaction. A response is fundamental if it does not mix mechanisms together and is uninfluenced by factors outside of the component, subassembly, subsystem, system design, or production process being optimized.

G

gate A short period of time during a product development process when the entire team reviews and reacts to the results from the previous phase and proactively plans for the smooth execution of the next phase. A gate review's time should be 20 percent reactive and 80 percent proactive. Gate reviews focus on the results from specific tools and best practices and manage the associated risks and problems. They also make sure the team has everything it needs to apply the tools and best practices for the next phase with discipline and rigor.

Gaussian distribution A distribution of data that tends to form a symmetrical, bell-shaped curve. See *normal distribution.*

goal post mentality A philosophy about quality that accepts anything within the tol-

erance band (USL-LSL) as equally good and anything that falls outside of the tolerance band as equally bad. See soccer, hockey, lacrosse, and football rule books.

grand total sum of squares The value obtained when squaring the response of each experimental run from a matrix experiment and then adding the squared terms.

H

half effect The change in a response value either above or below the overall experimental average obtained for an experimental factor level change. It is called the half effect because it represents just half of the factor effect. An example would be the average numerical response, obtained for a control factor that is evaluated exclusively at either its high set point or at its low set point, minus the overall experimental average.

histogram A graphical display of the frequency distribution of a set of data. Histograms display the shape, dispersion, and central tendency of the distribution of a data set.

House of Quality An input–output relationship matrix used in the process of quality function deployment.

hypothesis testing A statistical evaluation that checks the validity of a statement to a specified degree of certainty. These tests are done using well-known and quantified statistical distributions.

I

ideal/transfer function Fundamental functional relationships between various engineering control factors and a measured critical functional response variable. The math model of $Y = f(x)$ represents the customer-focused response that would be measured if there were no noise or only random noise acting on the design or process.

independent effect The nature of an experimental factor's effect on the measured response when it is acting independently with respect to any other experimental factor. When all control factors are producing independent effects, the design is said to be exhibiting an additive response.

inference Drawing some form of conclusion about a measurable functional response based on representative or sample experimental data. Sample size, uncertainty, and the laws of probability play a major role in making inferences.

inner array An orthogonal matrix that is used for the control factors in a designed experiment and that is crossed with some form of outer noise array during robust design.

inspection The process of examining a component, subassembly, subsystem, or product for off-target performance, variability, and defects either during product development or manufacturing. The focus is typically on whether the item under inspection is within the allowable tolerances or not. Like all processes, inspection itself is subject to variability, and out-of-spec parts or functions may pass inspection inadvertently.

interaction The dependence of one experimental factor on the level set point of another experimental factor for its contribution to the measured response. Interaction may be synergistic (mild to moderate and useful in its effect) or antisynergistic (strong and disruptive in its effect).

interaction graph A plot of the interactive relationship between two experimental factors as they affect a measured response. The ordinate (vertical axis) represents the response being measured and the abscissa

(horizontal axis) represents one of the two factors being evaluated. The average response value for the various combinations of the two experimental factors are plotted. The points representing the second factor's low level are connected by a line. Similarly, the points representing the second factor's next higher level are connected by a line.

K

k factor The economic constant for Taguchi's Quality Loss Function.

L

larger-the-better (LTB) A case where a larger response value represents a higher level of quality and a lower amount of loss.

Latin square design A type of Full or Fractional Factorial orthogonal array that has been modified for use in designed experiments. There are two kinds of order in designed experiments: Yates order and Latin square order. Latin squares are most commonly associated with Taguchi experiments.

level The set point at which a control factor, signal factor (CAP), or noise factor is placed during a designed experiment.

level average analysis See *ANOM (analysis of means)*.

life cycle cost The costs associated with making, supporting, and servicing a product or process over its intended life.

linear combination This term has a general mathematical definition and a specific mathematical definition associated with the dynamic robustness case. In general, a linear combination is the simple summation of terms. In the dynamic case it is the specific summation of the product of the signal level and its corresponding response $(M_i y i_{i,j})$.

linear graph A graphical aid used to assign experimental factors to specific columns when evaluating or avoiding specific interactions.

linearity The relationship between a dependent variable (the response) and an independent variable (e.g., the signal or control factor) that is graphically expressed as a straight line. Linearity is typically a topic within the dynamic cases of the robustness process and in linear regression analysis.

local maximum A point in the experimental space being studied, within the constraints of the matrix experiment, where the response to a given factor is at a maximum value (either in S/N or engineering units). Response surface methods can be used to hunt for global optimum points by first locating these local maximum points.

local minimum A point in the experimental space being studied, within the constraints of the matrix experiment, where the numerical response to a given factor is at a minimum value (either in S/N or engineering units). Response surface methods can be used to hunt for global optimum points by first locating these local minimum points.

loss to society The economic loss that society incurs when a product's functional performance deviates from its targeted value. The loss is often due to economic action taken by the consumer reacting to poor product performance but can also be due to the effects that spread out through society when products fail to perform as expected. For example, a new car breaks down in a busy intersection due to a transmission defect, and 14 people are 15 minutes late for work (cascading loss to many points in society).

lower specification limit (LSL) The lowest functional performance set point that a design or component can attain before functional performance is considered unacceptable.

M

main effect The contribution an experimental factor makes to the measured response independent of experimental error and interactive effects. The sum of the half effects for a factor is equal to the main effect.

manufacturing process capability (Cp$_m$) The ratio of the manufacturing tolerances to the measured performance of the manufacturing process.

matrix An array of experimental set points that is derived mathematically. The matrix is composed of rows (containing experimental runs) and columns (containing experimental factors).

matrix experiment A series of evaluations conducted under the constraint of a matrix.

mean The average value of a sample of data that is typically gathered in a matrix experiment.

mean square deviation (MSD) A mathematical calculation that quantifies the average variation a response has with respect to a target value.

mean square error A mathematical calculation that quantifies the variance within a set of data.

measured response The quality characteristic that is a direct measure of functional performance.

measurement error The variability in a data set that is due to poorly calibrated meters and transducers, human error in reading and recording data, and normal, random effects that exist in any measurement system used to quantify data.

meter A measurement device usually connected to some sort of transducer. The meter supplies a numerical value to quantify functional performance.

monotonicity A functional relationship in which the dependent parameter (a response) changes unidirectionally with changes in the independent parameter (an experimental factor). In robust design, this concept is extended to include consistency in the direction of change of the response even in the presence of potential interactions. Monotonicity is a necessary property for additivity.

multi-colinearity The measure of a response that is correlated to two or more input variables. When data is gathered outside the context of an orthogonal matrix experiment, multi-colinearity becomes a potential problem when one is trying to quantify precisely which input variables are controlling the response.

multidisciplined team A group of people possessing a wide variety of technical and experiential background and skills working together in the product commercialization process. They help form the basis of concurrent engineering.

multifunctional team A group of people possessing a wide variety of business and commercialization responsibilities working together in the product commercialization process. They help form the basis of concurrent engineering.

N

noise Any source of variability. Typically noise is either external to the product (such as environmental effects) or a function of unit-to-unit variability due to manufacturing, or it may be associated with the effects of deterioration. In this context, noise is an assignable, nonrandom cause of variation.

noise directionality A distinct upward or downward trend in the measured response depending on the level at which the noises are set. Noise factor set points can be compounded depending upon the directional effect on the response.

noise experiment An experiment designed to evaluate the strength and directionality of noise factors on a product or process response.

noise factor Any factor that promotes variability in a product or process.

nominal-the-best (NTB) A case in which a product or process has a specific nominal or targeted value.

normal distribution The symmetric distribution of data about an average point. The normal distribution takes on the form of a bell-shaped curve. It is a graphic illustration of how randomly selected data points from a product or process response will mostly fall close to the average response with fewer and fewer data points falling farther and farther away from the mean. The normal distribution can also be expressed as a mathematical function and is often called a *Gaussian distribution.*

O

off-line quality control The processes included in preproduction commercialization activities. The processes of concept design, parameter design, and tolerance design make up the elements of off-line quality control. It is often viewed as the area where quality is designed into the product or process.

on-line quality control The processes included in the production phase of commercialization. The processes of statistical process control (loss function based and traditional), inspection, and evolutionary operation (EVOP) are examples of on-line quality control.

one factor at a time experiment An experimental technique that examines one factor at a time, determines the best operational set point, locks in on that factor level, and then moves on to repeat the process for the remaining factors. This technique is widely practiced in scientific circles but lacks the circumspection and discipline provided by Full and Fractional Factorial designed experimentation. Sometimes one factor at a time experiments are used to build knowledge prior to the design of a formal factorial experiment.

optimize Finding and setting control factor levels at the point where their mean, standard deviation, or S/N ratios are at the desired or maximum value. Optimized performance means the control factors are set such that the design is least sensitive to the effects of noise and the mean is adjusted to be right on the desired target.

orthogonal array A balanced matrix that is used to lay out an experimental plan for the purpose of designing functional performance quality into a product or process early in the commercialization process.

orthogonality The property of an array or matrix that gives it balance and the capability of producing data that allow for the independent quantification of independent or interactive factor effects.

outer array The orthogonal array used in dynamic robust design that contains the noise factors and signal factors. Each treatment combination of the control factors specified in the inner array is repeated using each of the treatment combinations specified by the outer array.

P

p-diagram A schematic representation of the relationship between signal factors, control factors, noise factors, and the measured response. The parameter (P) diagram was introduced to the robustness process by Dr. Madhav Phadke in his book, *Quality Engineering Using Robust Design* (1989, Prentice Hall).

parameter A factor used in the design, optimization, and verification of capability processes. Experimental parameters are CFRs, CFPs, CTF specs, and noise factors.

parameter design The process employed to optimize the levels of control factors against the effect of noise factors. Signal factors (dynamic cases) or tuning factors (NTB cases) are used in the two-step optimization process to adjust the performance onto a specific target during parameter (robust) design.

parameter optimization experiment The main experiment in parameter design that is used to find the optimum level for control factors. Usually this experiment is done using some form of dynamic crossed array design.

phase A period of time in a product development process where specific tools and best practices are used to deliver specific results.

Phase/Gate product development process A series of time periods that are rationally divided into phases for the development of new products and processes. Gates are checkpoints at the end of each phase to review progress, assess risks, and plan for efficiency in future phase performance.

pooled error Assigning a certain few experimental factors, which have low or no statistical significance in the main effect values, to help represent or estimate the experimental error taking place in the experiment.

population parameter or statistic A statistic such as the mean or standard deviation that is calculated with all the possible values that make up the entire population of data in an experiment. Samples are just a portion of a population.

probability The likelihood or chance that an event or response will occur out of some number (n) of possible opportunities.

process signal factor (M*) A signal factor (CAP) that is a known adjustment factor in a design or process. Typically this factor modifies or adjusts the energy transformation but does not cause the function itself. The process signal factor is used in the double dynamic robustness case to modify the linear relationship of the functional signal factor to the response by changing the slope.

product commercialization The act of gathering customer needs, defining requirements, conceiving product concepts, selecting the best concept, designing, optimizing, and verifying the capability of the superior product for production, delivery, sales, and service.

product development The continuum of tasks, from inbound marketing to technology development to certified technology being transferred into product design to the final step of the certified product design being transferred into production.

project management The methods of planning, designing, managing, and completing projects. Project management designs and controls the micro timing of tasks and actions (underwritten by tools and best practices) within each of the phases of a product development process.

proximity noise The effect of one or more component, subassembly, or subsystem within a system design on a neighboring component, subassembly, or subsystem.

Pugh process A structured concept selection process used by multidisciplinary teams to converge on superior concepts. The process uses a matrix consisting of criteria based on the voice of the customer and its relationship to specific, candidate design concepts. The evaluations are made by comparing the new concepts to a benchmark called the *datum*. The process uses the classification metrics of "same as the datum," "better than the datum," or "worse than the datum." Several iterations are employed wherever increasing superiority is developed by combining the best features of highly ranked concepts until a superior concept emerges and becomes the new benchmark.

Q

quadratic loss function The parabolic relationship between the dollars lost by a customer due to off-target product performance and the measured deviation of the product from its intended performance.

quality The degree or grade of excellence, according to the American Heritage Dictionary (2nd college edition, 1985). In a product development context, it applies to a product with superior features that performs on target with low variability throughout its intended life. In an economic context it refers to the absence or minimization of costs associated with the purchase and use of a product or process.

quality characteristic A measured response that relates to a general or specific requirement that can be an attribute or a continuous variable. The quantifiable measure of performance that directly affects the customer's satisfaction. Often in DFSS, quality characteristics are converted to an engineering scalar or vector.

quality engineering Most often referred to as *Taguchi's approach to off-line quality control* (concept, parameter, and tolerance design) *and on-line quality control.*

quality function deployment (QFD) A disciplined process for obtaining, translating, and deploying the voice of the customer into the various phases of technology development and the ensuing commercialization of products or processes during product design.

quality loss cost The costs associated with the loss to customers and society when a product or process performs off the targeted response.

quality loss function The relationship between the dollars lost by a customer due to off-target product performance and the measured deviation of the product from its intended performance. Usually described by the quadratic loss function.

quality metrics Defects, time-to-failure, yield, go–no go; see *quality characteristics.*

quality plan The document that is used to communicate specifications to the production supply chain. Often the component House of Quality is converted into a quality plan.

R

random error The nonsystematic variability present in experimental data due to random effects occurring outside the control factor main effects. The residual variation in a data set that was induced by unsup-

pressed noise factors and error due to human or meter error.

randomization The technique employed to remove or suppress the effect of systematic or biased order (a form of assignable cause variation) in running designed experiments. Randomizing is especially important when applying classical DOE in the construction of math models. It helps ensure the data is as random as possible.

reference point proportional A case in the dynamic family of robustness experiments that uses the same approach as the zero point proportional case except that the data does not pass through the origin of the coordinate system. This case is built around the equation: $y = mx + b$, where b represents the non-zero y intercept.

relational database The set of requirements and fulfilling data that is developed and documented during Critical Parameter Management. The database links many-to-many relationships up and down throughout the hierarchy of the system being developed.

reliability The measure of robustness over time. The length of time a product or process performs as intended.

repeat measurement The taking of data points where the multiple measured responses are taken without changing any of the experimental set points. Repeat measurements provide an estimate of measurement error only.

replicate The taking of data in which the design or process set points have all been changed since the previous readings were taken. Often a replicate is taken for the first experimental run, then again at the middle and end of an experiment (for a total of three replicates of the first experimental run). Replicate measurements provide an estimate of total experimental error.

reproducibility The ability of a design to perform as targeted throughout the entire development, design, and production phases of commercialization. Verification tests provide the data on reproducibility in light of the noise imposed on the design.

residual error A statistical term that means the same as the mean square error or error variance. This is the measure of random and nonrandom effects due to running experiments in a specific environment independent of the main effects of the control factors.

response The measured value taken during an experimental run. Also called the *quality characteristic*. In DFSS, we prefer to focus on critical functional responses (CFRs).

Resolution III design A designed experiment that is structured such that some or all of the experimental factor main effects are confounded with the two-way and higher interactions of the experimental factors. A saturated array is a Resolution III design. This type of experiment uses the least degrees of freedom to study the main effects of the experimental factors. Using a Resolution III design in parameter optimization experiments requires the heavy use of engineering principles to promote additivity between the control factors. These designs are only recommended during robust design experiments and only when you already understand the interactions that exist in your ideal function. These are screening experiments, not modeling experiments!

Resolution IV design A designed experiment that is structured such that all the experimental factor main effects are free from being confounded with the two-way interactions of the experimental factors. In a Resolution IV design, there is confounding between all the two-way interactions of the experimental factors. Using a Resolution IV design in parameter optimization experiments requires many more degrees of freedom (treatment combinations) to study the same number of control factors as compared to a Resolution III design. These are also considered screening experiments.

Resolution V design A designed experiment that is structured such that all the experimental factor main effects are free from being confounded with all the two-way interactions of the experimental factors. In a Resolution V design, there is also no confounding between the two-way interactions. There will be confounding between the two-way interactions and the three-way interactions, which is typically viewed as being inconsequential. Using a Resolution V design in parameter optimization experiments requires many more degrees of freedom to study the same number of control factors as compared to a Resolution IV design. These designs are excellent for modeling.

robust design A process within the domain of quality engineering for making a product or process insensitive to the effects of variability without actually removing the sources of variability. Synonymous with *parameter design*.

robustness test fixture A device used in the robust design processes to stress a representation of an element of the production design. They are heavily instrumented to facilitate the direct physical measurement of function performance as noise is intentionally imposed during designed experiments.

S

sample A select, usually random set of data points that are taken out of a greater population of data.

sample size The measure of how many samples have been taken from a larger population. Sample size has a notable effect on the validity of making a statistical inference.

sample statistic A statistic such as the mean or standard deviation that is calculated using a sample from the values that make up the entire population of data.

saturated experiment The complete loading of an orthogonal array with experimental factors. There is the possibility of confounding main effects with potential interactions between the experimental factors within a saturated experiment.

scalar A continuous engineering variable that is measured by its magnitude alone (no directional component exists for a scalar).

scaling factor A critical adjustment parameter known to have a strong effect on the mean response and a weak effect on the standard deviation. The scaling factor often has the additional property of possessing a proportional relationship to both the standard deviation and the mean.

scaling penalty The inflation of the standard deviation in a response as the mean is adjusted using a CAP.

screening experiment Typically a small, limited experiment that is used to determine which factors are important to the response of a product or process. Screening experiments are used to build knowledge prior to the main modeling experiments in sequential DOE methodology.

sensitivity The change in a CFR based upon unit changes in a CFP, a CAP, or a CTF

spec. Also a measure of the magnitude (steepness) of the slope between the measured response and the signal factor (CAP) levels in a dynamic robustness experiment.

sigma (σ) The standard deviation; technically a measure of the population standard deviation.

signal factor A critical adjustment parameter known to be capable of adjusting the average output response of the design in a linear manner. Signal factors are used in dynamic robustness cases as well as in response surface methods.

signal-to-noise ratio A ratio or value formed by transforming the response data from a robust design experiment using a logarithm to help make the data more additive. Classically, signal-to-noise is an expression relating the useful part of the response to the nonuseful variation in the response.

Six Sigma (6s) A disciplined approach to enterprise-wide quality improvement and variation reduction. Technically, it is the denominator of the Cp index.

sliding level technique A method for calibrating the levels of an experimental factor that is known to be interacting with another related experimental factor. Sliding levels are engineered by analysis of the physical relationship between the interacting control factors. This approach is used to improve the additivity of the effects of the control factors.

slope Array of data that quantifies the linear relationship between the measured response and the signal factor (CAP); see *beta*.

smaller-the-better (STB) A static case in which the smaller the measured response is, the better the quality of the product or process.

specification   A specific quantifiable set point that typically has a nominal or target value and a tolerance of acceptable values associated with it. They are what results when a team tries to use tools and best practices to fulfill a requirement. Requirements are hopefully completely fulfilled by a final design specification, but many times they are not.

standard deviation   A measure of the variability in a set of data. It is calculated by taking the square root of the variance. Standard deviations are not additive while variances are.

static robustness case   One of the two major types of experimental robustness cases to study a product or process response as related to specific design parameters. The static case has no predesigned signal factor associated with the response. Thus, the response target is considered fixed or static. Control factors and noise factors are used to find local optimum set points for static robustness performance.

statistical process control (SPC)   A method for measuring attribute or continuous variables to determine if they are in a state of statistical control.

subassembly   Any two or more components that can be assembled into a functioning assembly.

subassembly requirements document   A document that contains the requirement, both critical and all others, for a subassembly.

subsystem   A group of individual components and subassemblies that perform a specific function within the total product system. Systems consist of two or more subsystems.

subsystem requirements document   A document that contains the requirement, both critical and all others, for a subsystem.

sum of squares   A calculation technique used in the ANOVA process to help quantify the effects of the experimental factors and the mean square error (if replicates have been taken).

sum of squares due to the mean   The calculation of the sum of squares to quantify the overall mean effect due to the experimental factors being examined.

supply chain   The network of suppliers that provide raw materials, components, subassemblies, subsystems, software, or complete systems to your company.

surrogate noise factor   Time, position, and location are not actual noise factors but stand in nicely as surrogate "sources of noise" in experiments that do not have clearly defined physical noises. These are typically used in process robustness optimization cases.

synergistic interaction   A mild-to-moderate form of interactivity between control factors. Synergistic interactions display monotonic but nonparallel relationships between two control factors when their levels are changed. They are typically not disruptive to robust design experiments.

system   An integrated group of subsystems, subassemblies, and components that make up a functioning unit that harmonizes the mass, energy, and information flows and transformations of the elements to provide an overall product output that fulfills the customer-based requirements.

system integration   The construction and evaluation of the system from its subsystems, subassemblies, and components.

system requirements document   A document that contains the requirements, both critical and all others, for a system.

T

Taguchi, Genichi The originator of the well-known system of Quality Engineering. Dr. Taguchi is an engineer, former university professor, author, and global quality consultant.

target The ideal point of performance which is known to provide the ultimate in customer satisfaction. Often called the nominal set point or the ideal performance specification.

technology development The building of new or leveraged technology (R&D) in preparation for transfer of certified technology into Product Design programs.

technology transfer The handing over of certified (robust and tunable) technology and data acquisition systems to the design organization.

tolerance design The use of analytical and experimental methods to develop, design, and optimize production tolerances. This process occurs long before geometric dimensioning and tolerancing methods (tolerance communication).

total sum of squares The part of the ANOVA process that calculates the sum of squares due to the combined experimental factor effects and the experimental error. The total sum of squares is decomposed into the sum of squares due to individual experimental factor effects and the sum of squares due to experimental error.

transducer A device that measures the direct physical action that produces functional performance in a design or process. Transducers are typically connected to a meter or data acquisition device to provide a readout and/or storage of a measured response.

transfer function Fundamental functional relationships between various engineering control factors and a measured critical functional response variable. The math model of $Y = f(x)$ represents the customer-focused response that would be measured if there were no noise or only random noise acting on the design or process. Sometimes these relationships are called *transfer functions* because they help model how energy, mass, and logic and control signals are transferred across system, subsystem, and subassembly boundaries.

treatment combination A single experimental run from an orthogonal array.

two-step optimization process The process of first finding the optimum control factor set points to minimize sensitivity to noise and then adjusting the mean response onto the customer-focused target.

U

unit manufacturing cost (UMC) The cost associated with making a product or process.

unit-to-unit variability Variation in a product or process due to noises in the production and assembly process.

upper specification limit (USL) The largest functional performance set point that a design or component is allowed before functional performance is considered unacceptable.

V

variance The mean squared deviation of the measured response values from their average value.

variation Changes in parameter values due to systematic or random effects. Variation is the root cause of poor quality and the monetary losses associated with it.

vector An engineering measure that has both magnitude and directionality associated with it. They are highly valued metrics for Critical Parameter Management.

verification The process of validating the results from a model or a designed experiment.

voice of the customer (VOC) The wants and needs of the customer in their own words. The VOC is used throughout the product commercialization process to keep the engineering requirements and the designs that fulfill them focused on the needs of the customer.

voice of the process The measured functional output of a manufacturing process.

voice of the product (VOP) The measured functional output of a design element at any level within the engineered system.

Y

Yates, Frank British mathematician who had a significant impact on the early development of the science and practice of classical DOE.

Yates order The standard format for laying out a designed experiment. Most experimental layouts in Minitab are structured in Yates order (prior to randomization).

yield The percentage of the total number of units produced that are acceptable; percent good.

Z

Zappa, Frank World's greatest practitioner of experimental music (not to be confused with Zombie, Rob).

zero point proportional A specific and foundational case in the dynamic approach to technology and product robustness development. This case focuses on measuring the linearity, sensitivity, and variability associated with a design that contains a signal factor, various control factors, noise factors, and a response that is linearly related to the signal factor. The model that defines this case is: $y = \beta x$. The y intercept is zero.

Index

About the Authors

CLYDE "SKIP" CREVELING is President of Product Development Systems & Solutions, a full service product development consulting firm. Previously, he was an independent consultant, DFSS Product Manager, and DFSS Project Manager with Sigma Breakthrough Technologies, Inc. He served as the DFSS Project Manager for 3M, Samsung SDI, Sequa Corp., and Universal Instruments and worked at Eastman Kodak for 17 years as a product development engineer within the Office Imaging Division. Mr. Creveling co-authored the text *Engineering Methods for Robust Product Design* and authored the world's first comprehensive text on analytical and experimental methods for developing tolerances for products and manufacturing processes, *Tolerance Design*.

JEFF SLUTSKY is CEO of Product Development Systems & Solutions Inc., a full service product development training and consulting firm. He was a Master Consultant at Sigma Breakthrough Technology, Inc., where he taught, consulted, and managed the deployment of DFSS programs at companies across the world. Mr. Slutsky has 20 years of experience designing and developing complex medical and imaging products for the Eastman Kodak company. He is an adjunct instructor at the Rochester Institute of Technology, he is an IEEE ABET program evaluator, and he holds an MSEE from RIT. His specialties include DFSS, statistical engineering, robust design, and discrete event simulation.

DAVID ANTIS, JR., CEO and President of Uniworld Consulting, Inc. and former Vice President of Operations for Sigma Breakthrough Technology, Inc., has deployed DFSS for over a dozen clients, drawing on best practices from Motorola, Kodak, GE, Black & Decker, and other leading firms. He formerly served as European Director of Operational Excellence and Total Quality for the Engineered Materials Sector of AlliedSignal, and as New Products Launch Manager for Motorola AIEG.